Bioimaging

Current Concepts in Light and Electron Microscopy

Douglas E. Chandler, PhD, and Robert W. Roberson, PhD
School of Life Sciences
Arizona State University

JONES AND BARTLETT PUBLISHERS
Sudbury, Massachusetts
BOSTON   TORONTO   LONDON   SINGAPORE

World Headquarters
Jones and Bartlett Publishers
40 Tall Pine Drive
Sudbury, MA 01776
978-443-5000
info@jbpub.com
www.jbpub.com

Jones and Bartlett Publishers
Canada
6339 Ormindale Way
Mississauga, Ontario L5V 1J2
CANADA

Jones and Bartlett Publishers
International
Barb House, Barb Mews
London W6 7PA
UK

Jones and Bartlett's books and products are available through most bookstores and online booksellers. To contact Jones and Bartlett Publishers directly, call 1-800-832-0034, fax 978-443-8000, or visit our website www.jbpub.com.

Substantial discounts on bulk quantities of Jones and Bartlett's publications are available to corporations, professional associations, and other qualified organizations. For details and specific discount information, contact the special sales department at Jones and Bartlett via the above contact information or send an e-mail to specialsales@jbpub.com.

Production Credits
Chief Executive Officer: Clayton Jones
Chief Operating Officer: Don W. Jones, Jr.
President, Higher Education Professional Publishing: Robert W. Holland, Jr.
V.P., Design and Production: Anne Spencer
V.P., Sales and Marketing: William Kane
V.P., Manufacturing and Inventory Control: Therese Connell
Publisher, Higher Education: Cathleen Sether
Acquisitions Editor, Science: Shoshanna Goldberg
Managing Editor, Science: Dean W. DeChambeau
Associate Editor, Science: Molly Steinbach
Editorial Assistant, Science: Caroline Perry
Production Manager: Louis C. Bruno, Jr.
Production Assistant: Leah Corrigan
Senior Marketing Manager: Andrea DeFronzo
Text and Cover Design: Anne Spencer
Illustrations: Elizabeth Morales
Composition: International Composition Corporation
Printing and Binding: Malloy, Inc.
Cover Printing: Malloy, Inc.
Cover Illustration and Image Design: Jacob Sahertian

Library of Congress Cataloging-in-Publication Data
Douglas E. Chandler
  Bioimaging : current concepts in light and electron microscopy / Douglas E. Chandler and Robert W. Roberson. — 1st ed.
    p. ; cm.
  Includes bibliographical references.
  ISBN 978-0-7637-3874-7 (alk. paper)
  1. Microscopy—Technique. 2. Electron microscopy. I. Chandler, Douglas E. II. Title.
  [DNLM: 1. Microscopy—methods. 2. Microscopy, Electron—methods. QH 207 R638b 2008]
  QH207.R535 2008
  570.28'2—dc22   2007025865

6048

Printed in the United States of America
12 11 10 09 08   10 9 8 7 6 5 4 3 2 1

We dedicate this book to people we love

To my wife Anne and my children Martin, Diane, and Marissa (Doug)
To my children Patricia, Erin, Morgan, and Lucien (Robby)

Brief Contents

Contents

Preface

THIS BOOK GREW out of our desire to present at the graduate level a course that would provide a broad understanding of bioimaging and would integrate both light and electron microscopy. After 10 years of developing such a course, it was natural to consider writing a book that would be useful to students of bioimaging and to scientists working in molecular and cellular biology. It was Stephen Weaver, however, then an editor at Jones and Bartlett, who first suggested the idea and encouraged us in this endeavor. The rewards have been great, and we thank Steve for his confidence in our work and in our initial proposal.

Our goal in writing has been to provide the conceptual framework for imaging techniques in a reasonably concise and understandable format using examples that have contributed important information to the study of cellular structure and function. While we aim to provide a realistic understanding of the technical requirements and the experimental context of the methods we discuss, our goal is not to provide detailed protocols because these are available in numerous books on methodology, some of which are referenced at the end of each chapter. Indeed, we would prefer to stimulate you to look into the details of what you find interesting rather than to shower you with a blizzard of detail right from the start.

This cuts a little bit against the grain for most imaging experts because, without a doubt, success in bioimaging can often be found by paying attention to experimental or instrumental detail. But we strive for enthusiasm first! With that said, our firm opinion is that actual laboratory work in microscopy is best done with the guidance of an experienced professional who can provide hands-on training aimed at achieving high-quality results and avoiding pitfalls that endanger instruments or waste time. Indeed, as codirectors of a community-use bioimaging facility, we expect our students to take laboratory courses in light and electron microscopy in addition to digesting the conceptual knowledge found in this book.

In a sense, the writing of this text is an experiment. As cell biologists, we have integrated both light and electron microscopy into the variety of approaches we use to answer research problems. We expect the reader to be doing the same, and, for this reason, we present many techniques as being complimentary to each other, not as unrelated approaches that are better or worse than one another. Finally, we wish that our book will encourage you to DO bioimaging and take the steps to see what no one else has seen before!

Acknowledgments

Putting together a book like this is a complex process, and there are many people to thank. First and foremost we thank our family and friends for their encouragement at every step of the way and their willingness to let us drop many other projects to see this book to fruition. We could not have done it without you!

Second, the publishing team at Jones and Bartlett gave us confidence and support all along the way and we thank specifically Stephen Weaver, Cathleen Sether, Shoshanna Goldberg, Dean DeChambeau, and Louis Bruno for their generous help. Lou was particularly pleasant in getting us through production—the last 20% of the work and the culmination of everyone's efforts. Another vital member of the team was our illustrator

Elizabeth Morales who time after time produced clear, clean renditions that make this book stand out.

Our many sources of scientific images are listed in the back of the book, and without their generosity this book would be a lesser volume. We thank these publishers, companies, and individuals for their courtesy in allowing us reproduce their work for our readers' benefit.

Third, we thank several colleagues who have contributed to our efforts in many ways. Charles Kazilek and William Sharp helped with photography and technical input, while Jacob Sahertian helped with graphic design. To acknowledge these last minute tasks alone would be to ignore the many years of professional advice (and on occasion criticism), mixed with friendship and humor, that they have given us.

In addition, we do not forget the many students in our bioimaging courses who provided the inspiration for writing this book. Most notable are Dr. Debra Page Baluch and Mr. David Lowry who rose to become managers of our community-use bioimaging facility and who stepped in to assume many responsibilities neglected by us during the writing of this book.

Finally, we would be remiss not to acknowledge the generosity of our employer (Arizona State University) and the supporters of our research programs (the National Science Foundation, Arizona State University, Science Foundation Arizona, the W.M. Keck Foundation and others) that allowed us the many years of research experience brought to this book.

At the same time, we make clear that the opinions and representations provided in this book are solely our own and are not that of any of our supporting individuals or organizations. Blame the errors on us! The writing of a technical book always brings the possibility that some errors of fact or concept have crept in, and we take full responsibility with the realization on your part that none of these were intentional. We appreciate the support of all our readers and welcome any corrections or suggestions for improvement.

Douglas E. Chandler
Robert W. Roberson

Introduction

THIS BOOK FOCUSES on essential details of modern light and electron bioimaging in what we hope is a presentation that is both informative and straightforward. Chapters 1 and 2 provide a concise yet comprehensive look at the key historical developments that have led to the sophisticated microscopes and image reproduction methods we use today.

Advancements made in science are linked to the instrumentation available to researchers as well as the methods used for sample preparation and interpretation. In Chapter 3 an overview of conventional protocols used in sample preparation for light and electron microscopy are discussed. Also in this chapter are descriptions of how the chemicals used interact with cellular components and the precautions that must be taken by the microscopist when using them. No method of tissue preservation is flawless, and Chapter 3 discusses the common types of artifacts encountered and how to recognize them to avoid misinterpretation.

An important aspect of bioimaging is being able to identify what you are viewing under the microscope; this is particularly true when using the transmission electron microscope. As we tell students in our bioimaging classes, "unlike what is seen in text books, cellular structures under the microscope do not come pre-labeled." Thus, in Chapter 4 we include a brief overview of the subcellular structures of eukaryotic and prokaryotic cells. This is not an attempt to provide an atlas of cellular structure because such compendiums are available in other works but rather to make available some basic information to be used in discussing microscopic techniques.

Because light microscopy has become such a powerful tool for the cell biologist, we have devoted Chapters 5, 6, and 7 to those characteristics of electromagnetic radiation that are important in determining the quality of the images we see. Time is also devoted to explaining how light interacts with the specimen and the components of the microscope (e.g., condensers, lenses, and filters) important for generating both detail and contrast.

The development and use of the electron microscope for biological investigation opened a window into the world of cell structure that is analogous to what the Hubble telescope has done for the study of the heavens. In Chapters 8 and 9 we provide an overview of how transmission and scanning electron microscopes work, how they generate images, and a practical guide for getting started.

In Chapter 3 we outline the conventional protocols for preparing samples for light and electron microscopy; however, are there alternative approaches that provide a higher quality of preservation, a preservation that is more true to life? In Chapter 10 we introduce cryogenic methods for preparing cells and tissues for ultrastructural analysis. Indeed, when successfully executed, these methods result in superior preservation of cellular details over conventional chemical methods. This is particularly true for labile components such as endomembranes and elements of the cytoskeleton. We do point out, however, that this approach is not without its own pitfalls and provide examples of common artifacts that occur as a result of suboptimal conditions.

Not only have there been significant improvements in the microscopes and specimen preparation protocols we use today, but also electronic imaging has revolutionized the ways we capture, process, and store our data. Chapter 11 identifies these improvements that range from the use of video cameras to capture cell and cytoplasmic movements to digital still cameras that provide a remarkable ease of use and important technical advances in spatial and temporal resolution and light sensitivity.

Chapters 12, 13, and 14 outline how the characteristic fluorescence identification of specific molecules is used today for determining subcellular localization and dynamics of important organelles, macromolecules, and signaling compounds such as proteins, nucleic acids, membrane lipids, and ions. Indeed, it is this ability that has allowed modern light microscopy to revolutionize molecular biology.

The shape of macromolecules plays a fundamental role in their function and interaction with other cellular components. Chapter 15 is devoted to advanced methods and how they are used to determine the structure of macromolecules and supermolecular complexes. In this chapter we discuss rotary platinum shadowing, negative staining, single particule analysis using cryoelectron microscopy, electron tomography, x-ray and electron diffraction, scanning probe microscopy, and near field microscopy.

The ultimate goal of bioimaging is to relate cellular structure to cellular dynamics. Today, the use of electronic imaging, computer analysis, and sophisticated software is mandatory. Chapter 16 provides possibilities for processing your data both quantitatively and qualitatively. Finally, we end this chapter and our book with suggestions of how to present your beautiful images in the formats used today.

In summary, this book provides a broad and relatively comprehensive look at both traditional and state-of-the-art techniques in bioimaging. Our goal has been to provide you with enough information to achieve an understanding of how bioimaging methods work and how they are used but not so much information that you fail to appreciate the excitement of this field and the truly remarkable advances that are possible.

About the Authors

Drs. Chandler (right) and Roberson (left) are codirectors of the W.M. Keck Bioimaging Laboratory and the Life Sciences Electron Microscopy Laboratory at Arizona State University. Dr. Chandler received his undergraduate education in chemistry at the University of Rochester and his graduate education in biochemistry and physiology at the Johns Hopkins School of Medicine and the University of California at San Francisco. After postdoctoral fellowships in San Francisco and at University College London School of Medicine, he assumed his current position at Arizona State University in 1980. His research specializations include the roles of egg extracellular matrices in fertilization, the structure and activity of sperm activating proteins and membrane structural changes during exocytosis. He has published over 70 research articles in the professional literature, has been the recipient of NSF, NIH, and private foundation research grants, and has served on the editorial board of Microscopy Research and Technique.

Dr. Roberson received his undergraduate education at Stephen F. Austin State University, his graduate education at the University of Georgia, and accepted his current position at Arizona State University in 1989. His research specializations include the biology of filamentous fungi, structural dynamics in the cytoskeleton, the mechanisms of vesicle transport, and membrane biosynthesis in cyanobacteria. He has published over 45 research articles, has received funding from the NSF and a number of private corporations, and has displayed his micrographs in art exhibitions throughout the Phoenix area. Drs. Chandler and Roberson have jointly taught bioimaging for over 10 years and have served as instructors in graduate and undergraduate courses in molecular and cellular biology.

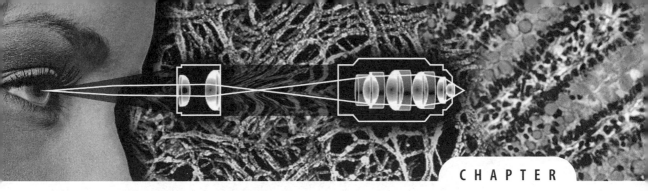

The History of Microscopy

There have likely been many unrecorded prehistoric incidents in which humans found that objects could be seen through a piece of glass or amber. Indeed, the making of glass and ceramics dates from as early as 3000 BC, as museum collections attest. The Egyptians made the first convex lenses from crystals in about 2600 BC, and the Greeks and Romans continued to use such lenses through the end of the Roman Empire. The Arabian scientist Alhazen first described the use of lenses in his writing "Optics Thesaurus Alhazeni Arabius Basil" in about 1000 AD. Spectacles (eye lenses) were thought to have been invented in Florence, Italy in 1285 by Salvano D'Aramento degli Amati, and the telescope is considered to have been built first by Jacob Adriaanzoon at Pisa in 1608 (Galileo followed with his a year later).

■ Development of the Light Microscope

The late 16th century can be credited with evolution of the crude hand lens into a somewhat sophisticated instrument that included a base and a device on which to fasten a specimen. A major advance occurred in 1590 when Zacharias Janssen developed a compound lens system having the ability to magnify 10 times. Robert Hooke put this invention to good use 70 years later to observe such structures as cork and such vermin as lice using a microscope built by Christopher Cock (**Figure 1.1**). Hooke's studies on the louse were well publicized (e.g., in "Micrographia," published in 1665) and led to an increased awareness of personnel hygiene in the population as a whole in England in the 1600s.

More remarkable was the ability of his Dutch contemporary Antoine van Leuwenhoek to make a single lens microscope that could magnify over 250 times and had a resolution of about 1 micrometer. This microscope revealed for the first time numerous "animalcules" and cells that we take for granted today. His reports to the Royal Society in London between 1673 and 1720 detail the structures and lives of bacteria, protozoa, sperm, and numerous neural tissues. His exquisite lens-making ability was guarded closely, and thus, few equaled his success for the next 100 years.

Microscopes in the 17th century were frequently made with wood, ivory, leather, and pasteboard, with only small amounts of metals—generally brass and iron—incorporated. Both single-lens and compound-lens models were made throughout the century, most

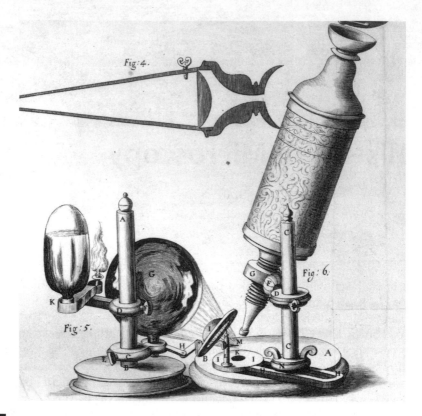

Figure 1.1
Hooke's compound microscope used for the research reported in his work *Micrographia* published in 1665.

notably by craftsmen in England, Holland, France, and Italy. Their bases were typically tripods or small wooden cabinets for storing specimens.

Microscopy became of immense interest to the curious, the learned, and the wealthy in 18th century England, and numerous microscopes from the simple to the extravagant were manufactured to meet the demand. Some of the variety available ranged from a field microscope for children to the extravagantly decorated baroque model made for King George III by George Adams in the 1760s. Famous makers in England included Culpeper, Adams, and Yarwell, whereas in France and Italy, such names as Depovilly, Bonanni, and Divinni were associated with the finest instruments. Design became increasingly sophisticated, incorporating such features as rack-and-pinion stage movement (Bonanni, 1690), concave mirrors for gathering light (Adams, 1740s), multilens eyepieces (Italian design, 1690s), and the lens turret (Adams, 1746). In addition, brass, and other metals became the construction material of choice, most models incorporating a sturdy brass base.

By the early 19th century, microscopes were adequate to identify many microscopic organisms and to discover a number of new cell organelles such as the nucleus, mitochondrion, and chromosomes. They were also adequate to catapult light microscopy into common use in medicine and natural history. Major advances included the development of the achromatic objective lens, the inverted microscope, and the polarizing microscope with its use of polarizing and analyzing prisms. These microscopes did not, however, reach the pinnacle of light microscopy resolution until a glass maker in Germany,

Carl Zeiss, teamed up with an expert optical scientist, Abbe, to form the Zeiss Microscopy works. Abbe's advances in lens design were put in place by Zeiss and resulted in a resolution of 0.25 μm. By the end of the 19th century, the three largest microscope manufactures were Zeiss and Leitz in Germany and Bausch and Lomb in the United States. All three manufacturers were producing microscopes that were designed for medical students and scientific laboratories.

The early 20th century saw the wedding of two great technologies—microscopy and photography—and recording of microscopic information entered a new age. In 1910, Ernst Leitz was the first to market a large-format camera using glass plates with a compound laboratory microscope. Soon afterward (1935), Zernike developed phase-contrast optics, a new technique that manipulated diffracted and nondiffracted light separately, which provided contrast in images of biological specimens without the use of stains. The Zeiss firm capitalized on this discovery by incorporating the new design into their microscopes within a few years. Then in 1955, George Nomarski developed an improved Wollaston prism that served as the basis for differential interference contrast optics, a technique that proved to have an even larger impact on the imaging of live cells. This method not only provided contrast but also a very thin focal plane, for the first time allowing "optical sections" to be used to reconstruct a three-dimensional specimen.

With the advent of television in the 1940s, the new field of video imaging arose, and by the 1970s, pioneers such as Shinya Inoue and Robert Allan were hard at work interfacing video imaging with microscopy of live and moving cells. High-speed cinematographic cameras were also used to capture motion under the microscope on film, although at this point, video and other electronic imaging methods have outstripped the use of film for recording both moving and static images from the light microscope. Film still offers greater spatial resolution than electronic image capture; however, digital cameras are gaining in this department, and soon there will be no reason to use film in any microscopic application.

In the 1970s and 1980s, the refinement of immunocytochemistry and the development of in situ hybridization techniques that allow localization of specific proteins and nucleic acids in a cell prepared the way for light microscopy to become one of the workhorse technologies of molecular and cellular biology. Absolutely necessary for these techniques was the ability to make polyclonal and monoclonal antibodies and to synthesize labeled nucleic acids of defined sequence using in vitro methods for replication of DNA. These developments were coupled with the synthesis of a new generation of fluorescence dyes with improved optical properties, for example, resistance to photobleaching and greater brightness. Synthetic dyes sensitive to calcium and hydrogen ions inside cells became important tools to decipher intracellular signaling events found to occur throughout the cellular world.

The 1990s saw the rapid development of laser scanning confocal microscopy, a technique that couples fluorescence microscopy with the acquisition of optical sections that can be used to reconstruct a three-dimensional fluorescence map of the specimen with submicron resolution. This technology required advances in laser technology and computer hardware and software capabilities and the automation of microscopic controls to come to fruition. This, combined with new digital cameras of high sensitivity, now allow dynamic analysis of single, specific macromolecules and ions within living cells—capabilities that have paved the way for a continued renaissance in the use of light microscopy for cell and molecular biology. Today, at least four major companies—Leica, Nikon, Olympus, and Zeiss—manufacture a full range of automated and computer-controlled light microscopic equipment.

■ Development of the Electron Microscope

In 1873, when Abbe demonstrated that the ultimate resolution achieved in an image is related to the wavelength of radiation used, it opened a search for new short-wavelength technologies. Twenty-four years later, Thompson discovered the electron, although he considered it to be a particle, not a form of radiation. The 1920s brought two important advances. De Broglie demonstrated that electrons do have wavelike properties like those of light, and Busch in 1926 discovered that electron trajectories could be bent by electromagnetic "lenses," much in the same way that glass lenses bend light; however, it remained for Max Knoll and Ernst Ruska, two German physicists, to solve serious lens aberration problems associated with electromagnets and produce usable magnetic lenses for the first electron microscope in 1933. By 1934, Martin in Belgium had published the first image of a biological specimen using an electron microscope, and a small number of electron microscopes were being built at university laboratories such as the one at the University of Toronto (**Figure 1.2A**).

The Metropolitan Vickers Company in England constructed a commercial prototype of an electron microscope in 1937, but it was Ruska now at the Siemens Company in Germany and RCA in the United States that actually marketed the first electron microscopes in 1939 and 1940. These microscopes achieved a resolution of 5 to 10 nm, and the value improved with each new generation of microscope. World War II saw advances in electron microscopy ostensibly for purposes of the war effort. For example, Ladd built an improved model that was used to study the structure of rubber so that the United States could make better tires (Figure 1.2B). After the war, commercial electron microscopes were again produced by Siemens and RCA and later by Philips. In addition, during the war, Carl Zeiss teamed up with AEG, a company that had worked as early as 1931 on commercializing an electron microscope based on Ruska and Knoll's original design. Notably, Model 1A from Siemens, Model EM B from RCA, Model EM 75 from Philips, and Model EM7 from AEG-Zeiss saw wide use in biological research laboratories. Japan Electron Optical Laboratories (JEOL) and Hitachi soon entered the market with their own models.

Improvements in the transmission electron microscope during the 1950s included automated photography with glass-plate emulsions, an increase in resolution from 1 to 0.4 nm, allowing higher magnifications, and improved pumping systems to achieve higher vacuums. This era was dominated by the Siemens Elmiskop Model 1A, which brought electron microscopy to many pioneering scientists in the field. During the 1960s and 1970s, JEOL's model 100B and Philips' model EM300 became the new standards (Figure 1.2C). By 1970, fierce competition had driven RCA out of the market, and later, even Siemens bowed out. Today, the major manufacturers of electron microscopes include JEOL, Hitachi, Carl Zeiss, and FEI, a company that bought out Philips in the late 1990s.

Concurrently, developments that led to the scanning electron microscope took place. Max Knoll in 1935 built a scanning electron beam to image the surface of cathode ray tube targets and demonstrated that a scanning electron beam could produce a secondary electron map of a specimen surface. A year later, Manfred von Ardenne in Berlin began a 2-year project with Siemens and within this short time was able to produce a submicron electron probe that was used in a combined scanning-transmission electron microscope (STEM) prototype. Likewise, Zworykin, director of research at RCA, picked up the quest for a high-resolution scanning microscope and made many important advances from 1936 to 1942. Ironically, both of these pioneers stopped their work on scanning electron microscope (SEM) technology because advances in the transmission electron microscope (TEM) area, especially in resolution, were coming so fast that the SEM work seemed doomed to make little impact.

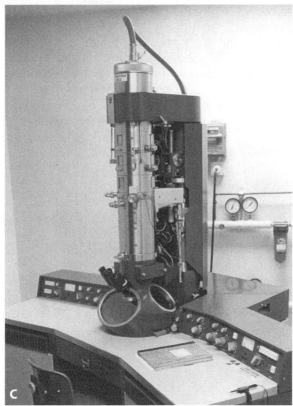

Figure 1.2
The earliest electron microscopes. (A) Custom-built electron microscope at the University of Toronto in 1939.
(B) Microscope built by Ladd for research during World War II. (C) The Philips model EM300, an electron microscope
used extensively in the 1970s and 1980s.

As a result, SEM development lay relatively dormant until 1950, when Oatley at Cambridge University did not buy the bum rap that SEM had gotten and started to design prototypes using improved electronics developed during World War II. With the help of some pioneering students (McMillan, Smith, Wells, Everly, and Thornton), he built increasingly better instruments, some of which were actually sold to industry. By 1965, Cambridge Instruments (with Oatley's oversight) marketed the first commercial SEM—the StereoScan I. JEOL followed soon with their first commercial SEM, and the era of widespread SEM use had begun. Subsequently, other companies such as Amway in the United States entered the SEM manufacturing business (and later bowed out), and Cambridge Instruments joined Wild Leitz to form Leica (1990). Today, four major manufacturers of scanning electron microscopes exist—Zeiss (which partnered and then bought out Leica's SEM business), JEOL, Hitachi, and FEI.

The 1970s brought many new advances for both TEM and SEM, including new vacuum pumps, magnetic lens designs, and electron beam guns, as well as x-ray spectrometers for elemental analysis of specimens. Ion getter pumps, based on adsorption of gas molecules to charged surfaces, and turbo-molecular pumps that avoid the use of contaminating oils now achieved clean vacuums up to 1000 times better than that provided by oil diffusion pumps. This in turn allowed production of the electron beam by use of lanthanum hexaboride crystals and by cold cathode field emission, methods that are many times more efficient than the tungsten filaments previously used. This resulted in much brighter images that facilitated work at higher magnifications. By 1980, high-resolution sheet film had replaced the emulsion-coated glass plates for photography, thereby increasing the number of exposures from 12 to 50 between-"plate" changes.

The 1980s brought computer control to both light and electron microscopes, and today all commercial electron microscopes employ computer control of beam generation, specimen position, magnification, focusing, and photographic exposure among other features. The ability to look at frozen or heated specimens was instituted by temperature-controlled stages, and the incorporation of differential vacuum chambers allowed specimens to be viewed at moderate vacuums in "environmental" SEMs. Finally, in the 1990s, charge-coupled-device (CCD) cameras had reached sufficiently high pixel counts (e.g., 2048 $\times$ 2048) to make electronic image storage a reality even for semi–high-resolution work. Film still outcompetes CCD capture for the highest resolution work, but this may soon change.

Finally, the development of high-voltage electron microscopes in the 1990s paved the way for three-dimensional imaging of specimens at high resolution. Most modern electron microscopes accelerate electrons through a potential ranging from 60 to 120 kilovolts, enough energy to allow electron penetration of a specimen about 100 nm thick. In contrast, "intermediate"-voltage microscopes use more highly energetic electrons and accelerating voltages between 200 and 400 kilovolts, whereas the largest "high"-voltage microscopes use as much as 7 megavolts. The high-energy electrons used for imaging by these microscopes can penetrate specimens as thick as 1000 nm and for this reason can readily image the three-dimensional structure in such specimens. These microscopes are extremely large and costly and are found only at nationally supported research centers in the United States and Europe.

■ Development of Other Imaging Technologies

In 1982, Rohrer and Binnig at IBM invented the scanning tunneling microscope. Within a decade, an entire family of scanning probe microscopies (SPMs) had emerged. The SPM technique employs an atomically sharp tip that is brought close to the surface of the specimen in a cyclic pattern that allows the data obtained to produce a topologic map of the specimen. This approach expanded rapidly in use from the material sciences to the biological sciences during the 1990s. Its versatility was increased by tips that could measure not only topological information but also chemical, electrical, magnetic, and intermolecular attraction data and map these data as a two- or three-dimensional array. The spectacular success of this technology is based on the fact that atomic resolution can be obtained even in specimens composed of light atoms such as carbon. In addition, this technology can be applied at atmospheric pressure (unlike electron microscopy) while the specimen is immersed in an aqueous buffer. Finally, the ability to measure minute forces between macromolecules has made this technology important to both biophysicists and microscopists. It represents a powerful adjunct to electron microscopy and in some ways (e.g., imaging molecules in a physiological environment) surpasses EM in its abilities.

Recently, x-ray microscopy has seen great strides in its ability to image cellular structure. X-rays have been used to image macroscopic body structures for over 100 years and, along with computer-assisted tomography, can provide detailed structural information in three dimensions. X-ray microscopy has great potential, as the wavelength of x-rays is shorter than that of light and, unlike electrons, passes easily through air, water and soft tissues, making possible imaging of cells in physiologic buffers. One hurdle that this technology faces is the development of high-quality "lenses" for x-rays. Magnetic fields and glass lenses are of no use here, and at present, the closest thing we have to an x-ray lens is a delicate array of metallic rings produced by nanofabrication methods that bend x-rays by diffraction and interference. Because of this, current imaging techniques using x-rays are limited to a resolution of 15 to 30 nm, a value that should improve as new designs for x-ray lenses are developed.

■ Development of Specimen Preparation Methods for Light and Electron Microscopy

In the days of Hooke and van Leuwenhoek, specimens received little or no preparation before being imaged. Specimens were glued to a stationary pointer and often remained with the microscope. Hooke did go so far as to cut sections of cork with a penknife, thereby allowing him to be the first to see cork "cells." In the 1700s, specimens were often compressed between sheets of mica and inserted into "sliders" that contained several specimens. Such sliders were prepared commercially for the amateur market. Also popular were "live" slides, having an actual compartment for enclosing small animals such as insects that could then be studied in vivo. In the early 19th century, glass slides with cover glasses secured by balsam finally came into use.

Tissue fixation was pioneered by Hannover, who in 1840 used chromic trioxide to harden tissues for cutting. Clark soon after developed an acetic acid and ethanol fixative that Carnoy further refined. Clark's fluid and Carnoy's fluid are still to be found in modern histotechnique manuals. In 1848, Quekett pioneered the use of tallow to infiltrate lung tissue, but it was not until 1869 that Klebs introduced paraffin as an embedding agent. The use of paraffin became routine thereafter, with Leukart developing embedding molds (1881), Fuhrmann developing a vacuum embedding oven for use with paraffin (1904), and Arendt inventing the first automatic tissue processor (1909).

The first known example of a microtome used to slice specimens was a device that Hill constructed in 1774. This machine was used to cut sections of timber (wood) as thin as 1/2000 of an inch. Later models even included a crude advance system that pushed specimens forward as they were cut! A sliding microtome was developed by Adams in 1798 and by Quekett in the 1840s. In use throughout the last half of the 19th century were handheld microtomes—simple, inexpensive, but imprecise. They required a handheld knife that earlier in the century was much like a scalpel. As early as 1871, Rutherford developed a freezing microtome for cutting frozen sections. The tissue was surrounded by a bath of ether, and the entire machine was cooled by ice and rock salt. After Minot developed the rotary microtome with automatic advance of the specimen, ribbons of paraffin-embedded sections could be turned out with little effort. Paraffin embedding was further perfected by the invention of the vacuum oven in 1904.

Staining of tissues for microscopic observation was carried out long before microtomy was perfected. As early as 1714, Muys reports using a stain in his study of muscle fibers; however, the use of stains in the modern sense of the word did not become popular until synthetic aniline dyes became available in the mid 19th century. Beneke first reported use of an aniline dye in 1862, but it was not until Erhlich's well-known work in the 1880s that these stains came into routine use. Staining with gold and silver, as developed by Golgi and Cajal, also constituted landmark achievements, and these histologists were awarded some of the earliest Nobel prizes.

With the advent of electron microscopy, new methods were sought that would be compatible with this new technology. In the 1950s, acrylic then epoxy resins were experimented with as embedding materials. The favored resins were Epon 812, produced by the Shell Oil Company, and Araldite 502, produced by Ciba. In 1969, Spurr introduced a new, low-viscosity epoxy resin that could infiltrate tissues much more easily, and today, it remains one of the most popular resins. Because most epoxy resins are polymerized by heat, the development of a new generation of ultraviolet-polymerizable resins such as LR Gold and Lowicryl, applicable to immunocytochemistry, was a significant step forward in the 1980s.

Necessary for the use of these hard resins was the development of the ultramicrotome. In the 1950s, Porter and Blum designed a highly sophisticated model that Sorvall marketed. This model and its later motorized version saw use in numerous electron microscopy laboratories in the 1960s and 1970s. LKB later designed ultramicrotomes that used thermal expansion of a metal rod to advance the specimen during cutting. Thermal advance, although competitive at the time, has proved to be less accurate than modern mechanical advances and is no longer used.

Preservation of tissue structure at the electron microscopy level required new fixatives. Noting the use of osmium tetroxide in light microscopy, Claude pioneered its use as an EM fixative. Later, Palade improved on Claude's efforts by using a buffering agent to control pH during fixation. An important breakthrough occurred in 1963 when David Sabatini pioneered the use of glutaraldehyde to crosslink proteins and provide excellent preservation at the ultrastructural level. Today, glutaraldehyde and formaldehyde are the most commonly used fixatives for both light and electron microscopy.

Equally noteworthy was the development of freezing techniques to rapidly halt cell processes. In the 1950s, Russell Steere developed the freeze fracture technique, whereby a tissue, when frozen, could be cracked open with a cold metal knife and the frozen surface replicated with metal for viewing in the electron microscope. Freezing the tissue required the use of cryoprotectants such as glycerol that prevented tissue damage due to the growth of large ice crystals. Subsequently, during the 1970s, van Hareveld and John Heuser independently experimented with rapid freezing of tissues without cryoprotectants using cryogen-cooled metal blocks. Heuser's rapid freezing device achieved tremendous success and was followed by other technologies for rapid freezing, such as the propane jet freezer in 1983 by Gilkey and Staehelin and the high-pressure freezer developed in 1990 by Hans Moor and the Balzers Company.

Conclusion

For quick reference, the major events in the development of light microscopy, electron microscopy and tissue preparation techniques are provided in **Tables 1.1, 1.2,** and **1.3,** respectively.

Table 1.1	Major Events in the History of Optical Microscopy
Year	**Event**
1590	Janssen builds the first compound microscope
1675	Hooke discovers cork "cells" with a compound microscope
1690–1710	van Leuwenhoek discovers bacteria and sperm with his own single-lens microscope
1690	Bonanni first to use rack-and-pinion focusing and dual-condenser lenses
1746	Adams builds all brass microscope featuring an objective "nosepiece"
1826	Lister builds the first achromatic objective lens—a breakthrough in image clarity
1833	Nachet builds the first polarizing microscope
1840	Donne publishes the first photomicrograph
1850s	Nachet builds a compound stereoscope
1867	Grunow manufactures an inverted compound binocular microscope
1870s	Zeiss produces lenses designed by Abbe that achieve a 250-nm resolution
1890s	Köhler prefects the condenser lens and describes how to achieve "Köhler" illumination
1910	Leitz combines a bellows camera with a microscope for photomicrography
1911–1913	Heimstaedt and Lehmann develop the fluorescence microscope
1929	Ellinger and Hirt are first to use epifluorescence illumination for fluorescence microscopy
1935	Zernicke invents phase-contrast microscopy
1955	Nomarski invents differential interference contrast microscopy
1960	Immunocytochemistry begins to be used widely
1970s	Inoue and Allan develop video microscopy for live cell imaging
1980s	Computer-controlled ratio imaging developed for detection of calcium signals in live cells
1985	Laser-scanning confocal microscope commercialized
1990s	Digital cameras become widely used for photomicrography
1990s	Computer control becomes a standard feature for research light microscopes
1995	Single-molecule fluorescence microscopy developed
2000	Multiphoton microscopy using infrared (IR) lasers commercialized

Table 1.2	Major Events in the Development of Electron Microscopy
Year	**Event**
1897	Thompson first reports the existence of electrons
1924	De Broglie demonstrates that electrons have wave properties
1926	Busch shows that electrons can be deflected by magnetic lenses
1931	Knoll and Ruska build the first electron microscope
1937	The company of Metropolitan Vickers develops a commercial prototype
1938	von Ardenne constructs the first scanning electron microscope
1939	The Siemens Corporation markets the first commercial transmission electron microscope
1940	RCA sells the first commercial transmission EM in the United States
1941–1963	Improvements in design allow a resolution of 0.2 nm to be achieved by transmission EM and 10 nm to be obtained by scanning EM
1954	Introduction of the Siemens "Elmscope IA," a popular and useful instrument for transmission EM of biological specimens
1958	Cambridge Instruments introduces the "StereoScan I," a scanning electron microscope for biological work
1960s	Elemental analysis by x-ray spectroscopy developed for EM
1977–1980	Large-format film replaces glass plates as the photography medium of choice for EM
1982	Development of the scanning tunneling microscope by Rohrer and Binnig
1985–1990	Computer-controlled electron microscopes come on the market
1990–2000	Large-format CCD cameras make digital imaging feasible for EM
2005	Advanced lens correction and energy filtering allow resolution of 0.08 nm

This brief history of microscopy serves to introduce the many facets of this science that are discussed in later chapters and to demonstrate that the history of microscopy is replete with new technologies that have been contributed by theoretical physicists and instrument designers. Indeed, if one is fortunate to develop a new imaging technology, one will have the privilege of seeing what no one else has seen before.

Table 1.3	Major Advances in Specimen Preparation for Microscopy
Year	Advance
1665	Hooke uses a pen knife to cut cork sections
1691	Bonanni develops the first "slider" for mounting specimens
1714	Muys uses a stain to study muscle fibers
1774–1775	Hill and Custance develop microtomes to cut timber sections
1827	Gould uses glass slides for mounting specimens
1840	Hannover uses chromium trioxide as a first chemical fixative
1848	Quekett uses tallow to embed lung tissue
1851	Clark uses acetic acid and ethanol as fixatives
1853	Curry develops hand microtome
1854	Hartig popularizes carmine as a stain
1862	Beneke is first to use an aniline dye to stain tissue
1864	Schultze uses osmium tetroxide as a fixative
1871	Rutherford develops first "freezing microtome" for producing frozen sections
1873	Golgi develops the silver staining method
1875	Ranvier uses picric acid as a stain; Bouin later incorporates it into a fixative
1876	Wissowsky is first to use the double stain hematoxylin and eosin
1879	Erhlich popularizes many synthetic dyes, especially hematoxylin
1881	Leukart develops embedding molds for paraffin
1886	Golgi and Mann develop perfusion fixation methods
1887	Minot develops rotary microtome
1894	Galeottl uses neutral red as a vital stain
1895	Cajal develops gold stain for neurons
1897	Coplin develops staining jar
1904	Fuhrmann develops the vacuum embedding oven
1909	Arendt invents the automatic tissue processor
1934	Martin publishes the first electron micrograph of biological tissue
1947	Claude introduces fixation with osmium tetroxide and ultramicrotomy
1949	Methacrylate is developed as an embedding medium
1950	Latta and Hartmann use glass knives for thin sectioning
1952	Palade introduces buffering systems for fixation
1953	Porter and Blum build the first widely used ultramicrotome, the Sorvall MT-1
1953	Fernandez-Moran is the first to use diamond knives for thin sectioning
1956	Glauert and later Luft use epoxy resins as embedding media
1957–1963	Steere introduces freeze fracturing; Moor designs the first commercial freeze fracture unit for Balzers AG
1958	Watson uses lead and uranium as heavy-metal stains for EM
1959	Horne introduces negative staining for EM
1961	Epon introduced as an epoxy embedding medium
1963	Sabatini uses glutaraldehyde as a primary fixative
1978	Heuser develops an ultrarapid freezing device
1984	Gilkey and Staehelin develop propane jet freezer marketed by RMC
1990	Moor and others develop a high-pressure freezing unit marketed by Bal-Tec

References and Suggested Reading

Bracegirdle B. *A History of Microtechnique.* Ithaca, NY: Cornell University Press, 1978.

Bradbury S. *The Evolution of the Microscope.* New York: Pergamon Press, 1967.

Hartley WG. *The Light Microscope: Its Use and Development.* Oxford, UK: Senecio, 1993.

Hawkes P, ed. *Beginnings of Electron Microscopy.* New York: Academic Press, 1985:633.

Hawkes PW, Mulvey T, Kazan B. *The Growth of Electron Microscopy, Volume 96 (Advances in Imaging and Electron Physics).* New York: Academic Press, 1996:885.

Jacker C. *Window on the Unknown: A History of the Microscope.* New York: Charles Scribner & Sons, 1966.

Molecular Expressions Web site: Museum of Microscopy. An excellent history of light microscope models and their designers over three centuries. http://micro.magnet.fsu.edu/primer/museum.

Rasmussen N. *Picture Control: The Electron Microscope and the Transformation of Biology in America, 1940–1960.* Palo Alto, CA: Stanford University Press, 1999:436.

The Paper Project Web site: A brief history of microscopy and some stunning confocal scanning images of paper. http://paperproject.org/microscopehistory/index.html.

Turner G L'E. *The Great Age of the Microscope: The Collection of the Royal Microscopy Society Through 150 Years.* Bristol, UK: Taylor and Francis, 1989:270.

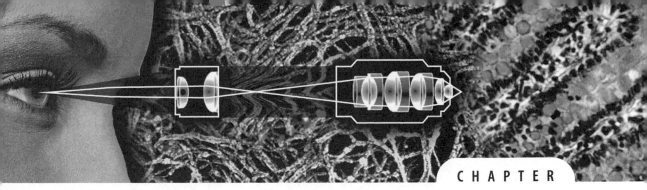

History of Microscopic Image Reproduction

Undoubtedly, early observations made with the use of magnifying devices have been lost to history because the images were not recorded. In some cases, we know from written descriptions that such microscopic observations were fascinating and lent themselves to imaginative descriptions that in some cases proved insightful and in other cases fanciful, although incorrect. Antoni van Leeuwenhoek, in observing the movement of sperm cells, provided the following thoughts: "The Human Creature is enveloped in an Animalcule from the Male sperm, but it is incredible to me that human intelligence will penetrate this great mystery so deeply that by accident or upon dissection of an Animalcule from the Male sperm we shall come to see a whole Human being" (van Leeuwenhoek, letter dated June 9, 1699). He further portrayed his ideas of sperm providing substance to the egg that could generate a new human being as a minute fetus in the head of the sperm. Nicolas Hartsoeker (a contemporary of Leeuwenhoek) gave life to this concept in a famous drawing (**Figure 2.1**).

We would not, however, mistake these as actual microscopic images. Nevertheless, the first images attempting to actually reproduce the structures seen by microscopy were drawn freehand. A historic example is Hook's drawing of dead cork cells from his treatise *Micrographia* published in 1665 (**Figure 2.2**). Such drawings are truly artistic portraits and attempted to record observational fact in an age before the invention of photography. The value of these drawings was recognized to be related to their degree of accuracy in order to convey image data independent of the conclusions derived from such data. Because accuracy was the *sine quo non* for data to remain reproducible and constant (as opposed to conclusions and interpretations that might differ from one year to the next), increasing the accuracy of such drawings was of interest. One invention that helped to ensure accuracy was the camera lucida, a diagram of which is provided in **Figure 2.3**. This optical device consists of a mirror that reflects the image back onto a half-silvered mirror. The half-silvered mirror in turn reflects the microscopic image onto the retinas of the observer, who at the same time can see the drawing paper through the half-silvered mirror. The microscopist could then use a pen to trace an outline of the superimposed image and its features onto the paper, providing crosshatching for differences in image density.

The essence of the pen and ink drawing could then be captured in an engraving to be used for the production of prints contained in a monograph or journal. The engraving was prepared using a sharp burin to cut grooves into a soft copper plate. Each print would be produced by inking the plate to fill the grooves with ink and then pressing the

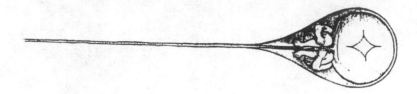

Figure 2.1
Conceptual diagram of a sperm.

plate onto paper using a manual press. Because these prints were produced on a different press than the text, illustrations often were segregated from the text in the final product, usually coming at the end. Based on this method, each engraved figure was referred to as a *plate*, and the use of this term continues even today. An illustration produced by an engraved plate (as well as those produced by other printing methods discussed later) is essentially black and white. As shown in **Figures 2.4A** and **2.4B**, shades of gray ("gray levels") in the print are actually an illusion created by black lines on the white paper background, the darkness of the gray being determined by the line density or crosshatching pattern used by the engraver in each area of the plate.

An alternative style in scientific illustration, quite popular between 1800 and 1970, was the use of dots for shading—a process referred to as "stippling." This technique uses various densities of tiny dots to create a three-dimensional sculptured appearance (Figures 2.4C and 2.4D). Dots were efficiently created by pen and ink and later by specialized reciprocating pens made by the Rapidograph Company. Like line drawings, stippled drawings allowed black and white representation of many "gray" levels and were easily reproduced by either engraving or lithography (discussed below). This style of illustration was taught, learned, and used by many microscopists during its "heyday."

Figure 2.2
Engraving of the structure of cork from *Micrographia* by Robert Hooke, published in 1665.

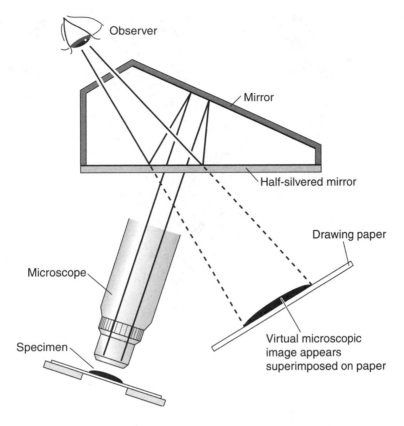

Figure 2.3
Design of the camera lucida, a device used to project microscopic images onto paper to aid accuracy in drawing.

Copper plates for printed illustrations could also be prepared by etching. In this case, the plate was coated with a black acid-resistant ground containing asphalt, wax, and rosin, and the image was drawn into the ground with a needle. After dipping the plate into nitric acid, these lines would admit acid, thereby etching the drawing into the plate in the form of a network of gullies while the remainder of the surface was protected by the ground. During printing, ink would fill the gullies and be transferred onto paper to reproduce the drawing. A variety of methods for applying the ground as fine particles followed by etching also allowed the printing of patterns and gray tones as shown in **Figures 2.5A** and **2.5B**; however, scientific and microscopic illustrations much more commonly used engraving techniques to produce crisp lines than etching techniques to produce patterns, mainly because engraving is a faster technique.

Another technique for illustration was wood engraving. Wood had actually served as the original surface for printing in the 16th and 17th centuries (preceding copper plate engraving). In preparing a wood block to be printed, the carver would remove wood from areas that were to remain white (i.e., uninked), while the printing surfaces remained uncut and in relief. These raised surfaces would then be coated with ink, and the ink was transferred to paper in the printing press. Even early microscopic illustrations came along a little too late to enjoy the status of being reproduced by wood block techniques; however, wood engraving enjoyed a golden age in the second half of the 19th century and was extensively used for scientific illustration during that period. In this technique, a block of boxwood, very hard and extremely fine grained, was polished to a smooth, flat surface. The wood was then engraved with a fine knife to produce grooves that held ink

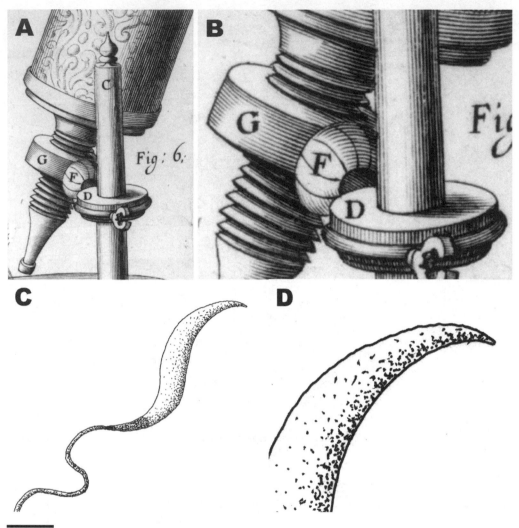

Figure 2.4
The use of cross-hatching and stippling for shading. (A) Detail of an engraving of Hooke's microscope. (B) At higher magnification one can see that cross hatching is used for shading. (C) Stippled drawing of a *Xenopus laevis* sperm. (D) Detail of (C) showing the use of dots for shading.

much like those in a copper plate engraving. In Figure 2.5C, the delicate detail of a butterfly wing is rendered by wood engraving. The grace and warmth of the fine lines achieved represent a hallmark of wood engraving. The softness of the wood surface compared with that of either copper or steel allowed a suppler rendition on the part of the engraver.

Even greater ease in producing artistic if not scientific drawings came with the invention of lithography in 1798. This process largely replaced engraving as the chief method for reproduction of illustrations by the middle of the 19th century and was the main competitor to wood engraving later in the century. This technique required the illustrator to prepare a flat limestone surface to which water but not oil-based inks would adhere. If the drawing was reproduced using a grease crayon on the stone and the grease drawing enhanced with asphalt and baked into the stone, the stone could then be inked, and the ink would adhere only to the greasy drawing on the stone. Paper could then be pressed against the stone using a manual lithographic press, resulting in transfer of the ink to the paper. In contrast, those areas of the stone representing the background,

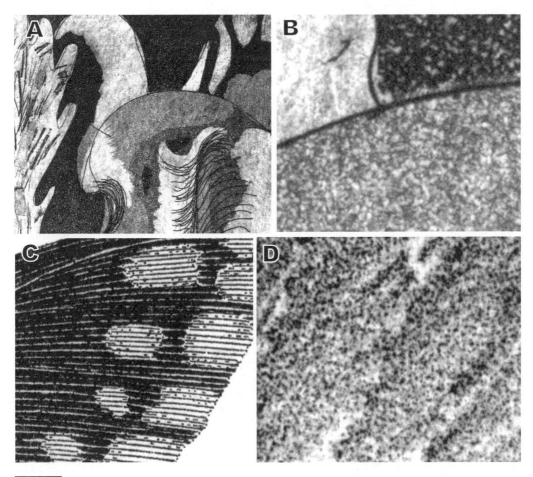

Figure 2.5
Etching, wood engraving, and lithography as techniques for 18th- and 19th-century scientific illustration.
(A) A copperplate etching of leaves by Linda Chandler. (B) The detail in (A) reveals the complex printed patterns
achieved by "aquatint" to represent gray tones. (C) Wood engraving showing detail of a butterfly wing. At this
magnification, gray tones are seen to consist of black and white regions of variable thickness, the black regions
being printed from ink filled crevices cut into the wood surface. (D) A lithograph at high magnification exhibits
crayon-like textures created by hydrophobic islands on the stone bearing ink.

having been treated with acid to roughen the surface to hold water and resist the adhesion
of the greasy ink, thereby remaining white in the print. As shown in Figure 2.5D, an
enlarged view of a lithographic print consists of tiny islands of ink. Their density, as
determined by the grain of the stone and distribution of grease on the stone, controls the
shade of gray obtained at the macroscopic viewing level. Lithographs are easily distinguished from engraved prints because the later, when enlarged, can be seen to consist of
distinct lines rather than islands (Figure 2.4A).

The techniques described to this point are monotone (i.e., black and white). Any
color information would have to be added to the black and white engraving, etching, or
lithograph by hand, typically using watercolor pigments and a brush. This type of "hand-
colored" print is simulated in **Figure 2.6A** (see Plate 1 in the Color Addendum) using a
copper plate engraving illustrated previously. Because "hand-colored" prints were labor
intensive to produce, the editions of such scientific monographs were relatively small,
usually no more than 500 to 1000 copies.

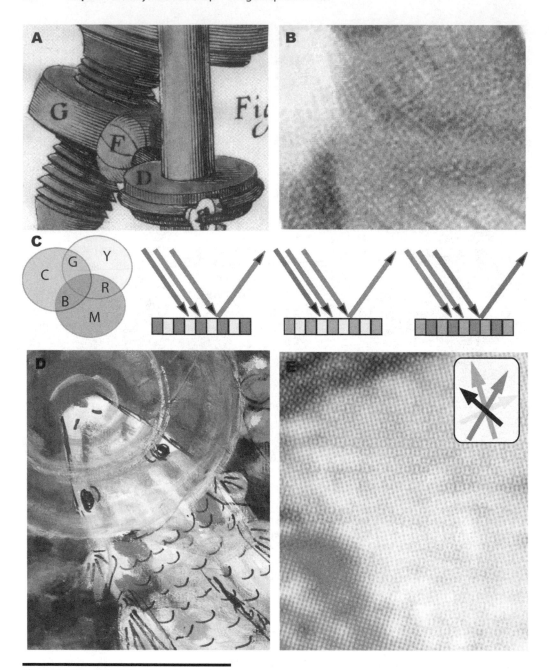

Figure 2.6 (See Plate 1 for the color version.)
Techniques for color illustration. (A) Hand coloring of an engraved illustration. (B) A chromolithograph at high magnification shows evidence of colored islands of ink. (C) A diagram showing which colors are reflected and seen in images printed with cyan, magenta, yellow, and black inks. (D) A four-color reproduction of a carp by a high-speed offset lithographic press. (E) At high magnification, the four-color print in (D) exhibits multicolored hexagonal patterns of dots composed of a screen pattern for each primary color. Inset: the orientation of each screen in E.

Black and white lithography, used extensively from the early 19th century onward, however, led naturally to color or "chromolithography." Chromolithography uses the same basic principles as black and white lithography, except that a separate stone must be prepared for each color. The drawings on each stone must be identical, except for their inking pattern, and they must be printed in sequence and in register on the same sheet of paper requiring multiple passes through a manual press. Chromolithography results in small islands of ink, each color being interspersed with islands of other colors (Figure 2.6B). The resulting prints are often beautiful from an aesthetic point of view. From a scientific aspect, lithography required both an accurate hand drawing on the part of the microscopist and reproduction by the lithographer on stone or multiple stones. Because this left quite a bit of room for subjective judgment if not error, photography was embraced soon after its invention as a technique providing more detailed images and less subjectivity.

■ Photographic Illustration

Photography was pioneered by Joseph-Nicephore Niepce and artist Louis-Jacques-Mande Daguerre in the 1830s using silver-coated copper plates that when sensitized by iodine or bromide vapor were exposed to light and then developed by fumes of heated mercury to produce a visible image that was then made permanent by exposure to sodium hyposulfite. This process was used extensively from the 1830s to the 1870s for portrait photography, and many so-called daguerreotypes still survive and are avidly sought after by collectors, antique shops, and museums. They offered several disadvantages for recording images, including long exposure times (minutes), the fact that only one copy could be produced, and that significant labor was required to produce each photographic plate. Later, this process was modified by using a tin backing rather than a copper backing. This led to the term *tintype* that was used for portrait, landscape, and historical photography even into the early 20th century. The tin surface was coated with black lacquer before applying the light-sensitive emulsion, the background serving to turn the negative image formed by silver grains of the emulsion into a positive image on the black background. Although production of tintypes became streamlined compared with that for daguerreotypes, tintypes were still "one of a kind" and not capable of duplication. The metal sheet on which they were produced is opaque and therefore not capable of supplying an image for reproduction by either photographic or mechanical printing. Despite this, large-scale production of tintypes, accomplished by multilens cameras, was occasionally undertaken, a notable example being that 300,000 tintype buttons of Abraham Lincoln were distributed during his 1860 presidential campaign, making him one of the earliest "media-savvy" presidents.

An alternative was the production of photographic images on an emulsion-coated glass. This was pioneered by Herschel in the 1840s and was considerably improved by Frederick Archer, who used a collodion emulsion that adhered firmly to the glass surface. Like the tintype, a coating of lacquer provided the black background that served to turn the negative image into a positive image. The dark background, however, could also be provided not by a coating but by using a darkly colored glass such as "coral" glass (deep violet) or "ruby" glass (deep red). These earliest methods on glass were collectively referred to as "ambrotypes," coined from the Greek word *ambryo* meaning "immortal."

Ambrotypes, tintypes, and daguerreotypes, however, were all "one of a kind" photographic processes and were rarely used to record microscopic images because photographers and microscopists seldom teamed up with one another until the late 19th century. The use of light-sensitive emulsions on a glass surface, however, did provide the technical breakthrough for photographic reproduction through the use of intermediate negative images, so-called negatives, which are still used today. If the negative image on glass is projected onto a second light-sensitive substrate, typically emulsion-coated glass or paper, a positive image is formed. Providing that the silver grains of this positive image can be "developed" (treated chemically) so as to produce a blackened silver deposit, the positive image will be visualized correctly. Such emulsion-coated glass plates have been routinely used for recording telescopic, macroscopic, and microscopic scientific images for over a century (1870s to 1970s). The negative images are typically printed as positive images on emulsion-coated paper using an "enlarger" to illuminate the negative and focus its image onto the photographic paper. The enlarger itself works much like a large-scale microscope, with the negative plate being the "specimen" and a real image being projected onto the paper at a magnification determined by adjusting the projection lenses.

The technology of printing photographs on emulsion-coated paper was developed in the 1850s and has been used to this day. In contrast, various transparent plastics have replaced glass as the substrate for making "negatives." The earliest plastic was cellulose nitrate, extensively used for cinematographic photography in which a flexible negative strip (the "film") was an absolute requirement for the projection of moving images. Cellulose nitrate, developed about 1900, was later replaced by other plastics because of its flammability and its tendency to harden and yellow with age. A plastic base for negatives was used in most consumer-grade still photography as early as the 1920s, thus permanently installing into our vocabulary *film*, which can be exposed and developed to produce negatives. By the 1950s, light microscopists also converted to using plastic-based films, especially the 35-mm format. On the other hand, applications requiring the highest standard of accuracy such as electron microscopy and astronomy continued to use emulsion-coated glass plates up until the 1980s; however, with the advent of greatly improved plastic bases for film, these sciences reluctantly gave up their beloved heavy, thick glass plates for the ease of film. Today, digital technologies are eliminating the need for film altogether; the spatial resolution of film, however, has not yet been matched by that of digital technology, and film is still a requirement for high-resolution microscopy.

■ Photographic Darkroom Procedures

Because film and printing onto paper remain requirements for the most demanding applications, we provide a short description of the processes and skills involved. Required is a dark room illuminated by orange or red light that is equipped with an enlarger and a sink with trays containing chemicals for developing glass plates or paper prints.

Loading and Exposing the Film Cassette

Film cameras for light microscopy generally use the 35-mm format, and for that reason, manufactured cassettes of rolled film can be loaded into the camera at full room light. In contrast, the sheet film used for electron microscopy, having standard dimensions of 3.25 × 4 inches, is loaded into a light-tight metal cassette under red-light conditions in a darkroom, and the cassette is then closed and loaded into the microscope under ambient light. Depending on the manufacturer, cassettes hold between 15 and 60 sheets of film. Exposure during microscopy is usually automatic, with beam intensity and magnification being determined by the operator. The exposed film is deposited in a receiving cassette and transferred into the darkroom for development.

Processing the Film

Developing and stabilizing the photographic emulsion is carried out chemically by the use of a "developer" and a "fixer." The developer, typically containing metol (monomethyl-p-aminophenol hemisulfate), hydroquinone, and a small amount of potassium bromide, serves to convert the "latent" image, consisting of a pattern of silver nitrate grains that have been photosensitized by light absorption, to reduced metallic silver grains. The density of these grains, determined by the amount of light impinging on each region of the emulsion, will determine the "gray level" in each region of the negative. The film is loaded into a rack, immersed in the developer and agitated to promote continuous chemical contact, and removed either after a set amount of time (four minutes is typical) or when the desired gray level caused by reduced silver grains is achieved. The gray of the film is best compared with a reference sheet to determine when to stop. The film rack is then transferred to a weakly acidic stop bath that immediately neutralizes the developer and stops its action. The film is then transferred to a fixative, usually sodium hyposulfite, to dissolve all silver salts that have not been reduced to metallic silver. During this process, the film turns from a milky gray to clear; in this manner, all silver nitrate crystals that did not interact with light are removed, and the film is no longer light sensitive. Regions of the negative that have been exposed to little or no light are now transparent. The fixative also hardens the protein in the emulsion so as to turn it into a clear but toughened matrix in which the black silver grains are embedded. The film is then washed thoroughly with water and dried with warm air. The film now has embedded in it a negative image when viewed against a light background (**Figure 2.7A**).

Enlarging and Printing the Negative

The negative is then placed in the film carrier of an enlarger, and a strong incandescent bulb and condenser lens are used to illuminate the negative in a uniform manner. An objective and projector lens is then used to enlarge and focus the negative image onto the photographic paper to be printed. The focus of this image on the paper holder is checked by a magnifying viewer that allows the individual silver grains of the negative to be seen; when in focus, they appear as sharply delineated black strings that trace the path of single electrons or photons through the emulsion (inset, Figure 2.7A). The paper is then placed in the carrier; the light source is turned on and off automatically by a timer, and the exposed paper is then removed from the holder and developed and fixed in the same manner as previously described for film. The print is then washed thoroughly and dried on a warm, rotating cloth drum. Photographic prints are now made on a plastic base rather than paper, allowing for much more rapid processing and drying.

The printed image is a positive image maintaining a detailed pattern of grays similar to that found in the specimen photographed (Figure 2.7B). The darkest parts of the specimen or landscape have the least amount of light reflected or scattered from them, and thus, these regions in the negative have the fewest reduced silver grains and are relatively transparent. When the negative is printed, these regions allow the greatest amount of light to pass and strike the photographic paper during exposure. Development of the paper then leads to the highest density of silver grains in these regions, producing dark grays to black in the printed image.

Production of a good photographic print is an art and requires practice. The intensity of the light source in the enlarger and the exposure time determines the overall density (darkness) of the print. Often a "test strip" of photographic paper exposed for a series of different times is used to get in the right ballpark. Second, the overall contrast of the print (the extent to which the dark and light features of the print are different in density) is determined by the how long the print remains in developer. A longer time in the developer means that a higher contrast is achieved. Thus, if one wanted a print of low contrast, one would increase the exposure time and decrease the amount of time needed in the developer to arrive at a reasonable density. If one wanted a print of high contrast, exposure time of the paper to light would be reduced, but the time needed in the developer would be increased to arrive at the same overall density but at a higher contrast.

Photographers also can subjectively modify application of the image to the photographic paper by exposing some regions of paper to less light ("dodging") and

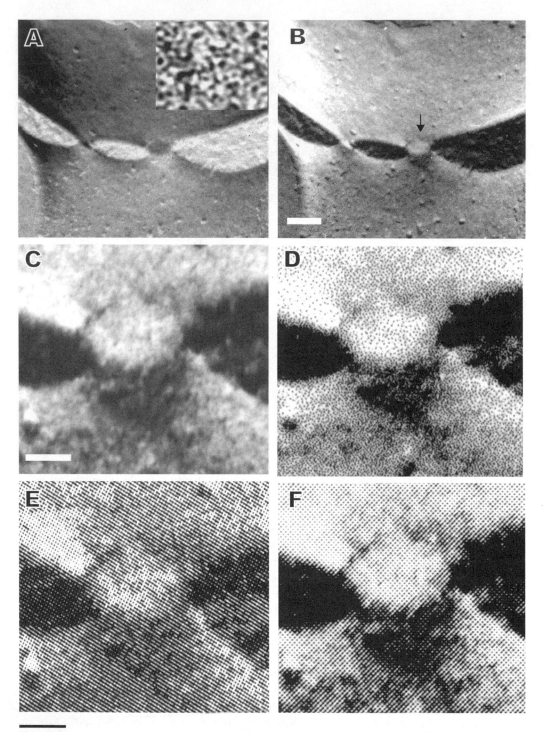

Figure 2.7
Modern printing methods for scientific illustration. Shown here are details of a freeze-fracture electron micrograph of fusing secretory granule membranes in a rat mast cell. (A) The image captured on a 3.25 × 4-inch film negative. The inset shows the silver grains of the negative at high magnification. (B) Positive image printed on photographic paper. The arrow points to the detail shown in (C). Bar = 50 nm. (C) Detail of (B) showing the "continuous tone" of the photograph. Bar = 20 nm. (D) An ink jet reproduction of (C) consists of small dots at high magnification. (E) A laser jet reproduction of (C) showing the dot-like screen pattern used. (F) A photolithography reproduction of the image in (C) exhibits a grid of dots created by the screening process.

other regions of the paper to more light ("burning"). These manual processes are time consuming, require practice and experimentation, and use ad hoc tools (usually pieces of cardboard) that block light from certain areas of the paper.

■ Digital Photography

Although film and printing are still highly valuable for both artistic and high-resolution imaging work, the ease and reduced expense of digital processes have made them the current technology of choice for most imaging, consumer or scientific. Light detection by a photosensitive solid-state chip camera, conversion of the electrical signal into a specific image-related format, and the use of such electronic files to both print and project images now require little material expense beyond instrumentation. A darkroom with sink and chemicals is not needed—nor are specialized photographic films or papers.

Digital files consist of numeric data that indicate the density and color of each position within the image. By necessity, the image must be divided into picture elements (pixels) arranged as a grid in columns and rows. The detail of an electronic image file is determined by its number of pixels, with consumer- and scientific-grade digital cameras having about 2 to 8 million pixels per image. In contrast, a high-resolution 35-mm film negative has the equivalent detail of 3 million pixels, whereas a 3.25×4-inch electron microscopy negative has the equivalent detail of 25 million pixels. A photographic image printed from a large-format, fine-grain film negative exhibits continuous gray tones even when enlarged up to 10 times (Figure 2.7C). Clearly, large-format film achieves much greater image detail than does digital imaging at this point, but that is likely to change in the future as new solid-state chips are developed. Digital files, however, can be stored conveniently (thousands on a single disk), rather than requiring file cabinets for film or (worse yet) heavy glass plates. Manipulating these images in contrast or density or even by dodging and burning can be carried out by software such as Adobe Photoshop on a laptop computer rather than in a darkroom. Furthermore, automated image analyses by quantitative methods that would be impossible using photography are easily carried out on digital files with appropriate software (see Chapter 16). Furthermore, digital files are easily sent worldwide by the Internet, and figures, images, and text for publication are now submitted electronically to most journals; therefore, regardless of whether image files are generated by drawing, film photography, or digital photography, they will all have to be converted to an electronic file before publication.

An important concern in capturing, communicating, and printing electronic files is maintaining adequate spatial resolution (i.e., maintaining an adequate number of pixels). The total number of pixels in an image is related to the image format rather than the physical size of the image. Because the appearance of an image to the human eye is dependent on image size, the usual gauge of detail in a printed image is linear resolution, given in "pixels (dots) per inch" or "pixels (dots) per centimeter." Publishers usually request a linear resolution of 600 dots per inch for images and 1200 dots per inch for line drawings or text within images. Thus, if one is working from a digital file having 1200×1800 pixels, one could expect it to appear sharp when printed at a size no larger than 2×3 inches. If printed at a considerably larger size (e.g., 8×12 inches), the individual pixels may become just barely visible, imparting a fuzziness to the image. This indicates that one must either capture a larger electronic file in the first place (use a 6- or 8-megapixel camera instead of a 2-megapixel camera) or increase the number of pixels in the image using software such as Photoshop that can create pixels whose density and color are interpolations from nearby pixels.

■ Low-Volume Printing of Electronic Image Files

Three types of printers are commonly available: ink jet, laser jet, and dye sublimation. Ink jet printers use microfluidic circuits that allow electronic signals to regulate the flow of tiny ink droplets onto the paper surface. These dots vary in spacing (density) and color and can be observed using a 10× magnifying loupe (Figure 2.7D). The spacing of dots can be controlled to produce resolutions as high as 2400 dots per inch in high-resolution ("photographic") ink jet printers. This resolution comes within range of matching that achieved by photographic processes. Laser jet printers operate by rastering a laser across the surface of a drum that picks up carbon based on the pattern of laser illumination. The drum then applies the carbon to the paper and the carbon baked into the paper to achieve a strong, waterproof bond. Using complex patterns of carbon fill, resolutions as high as 2400 dots per inch combined with high rates of printing can be achieved. Laser jet printers, however, use a "screening" process that divides the image into cubic units, and these may be seen as a grid pattern if the print is magnified to any great extent (Figure 2.7E). Dye sublimation printers heat and vaporize waxy dyes and apply a stream of dye vapor to the paper. On contact, the dye mixture condenses and bonds tightly to the paper as a thin, solid waxy layer. Dye sublimation printers achieve precise, high-density pigmentation of the paper and are considered to produce the highest quality reproduction of the three types of printers. All of these printers achieve excellent results but are relatively costly and achieve production rates only as high as 10 to 20 copies per minute. In order to print much larger editions of books or journals, one must use high-speed commercial printing processes.

■ High-Speed Commercial Printing

Commercial printing since the late 19th century has used either photo etching or photolithography processes—technologies that combine elements of photography with either etching or lithography. During the development of lithography, it was noticed that metal plates made of zinc could substitute for limestone as an inking surface and offered the advantage of being lower in cost and much lighter and thinner. Today, zinc or aluminum plates are more common than limestone for lithography. Copper plates have been traditionally used for photo etching. These technologies were developed in the last half of the 19th century and now form the basis of modern, high-volume printing.

In photolithography, a photographic negative on glass or film is placed in direct contact with a zinc or aluminum plate that has been coated with a light-sensitive emulsion. Exposure to light produces a "contact" print of the image on the metal plate emulsion, and development of the plate results in a hardened emulsion in the areas exposed to light, whereas the emulsion not exposed to light is washed away, exposing the metal surface of the plate. The hardened emulsion is water repelling, whereas the zinc or aluminum surface is water adhering and can be made more so by a slight roughening through etching with acid. When ink is applied to the plate with a roller, the greasy ink adheres to only the emulsion and is repelled by the water-dampened metallic surfaces of the plate. On contact with paper, the plate transfers the ink to the paper, producing a positive image.

This process can be repeated numerous times at high speed if the plate is attached to the surface of a cylindrical drum such that with each turn of the drum the plate is inked by a roller and then the ink on the plate is transferred to the surface of a rubber drum, which in turn transfers the pattern of ink to paper as it moves past the drum (**Figure 2.8A**). As the inked image is transferred from plate to rubber drum, it is flipped left to right and when transferred to paper is flipped again back to the same orientation seen on the plate.

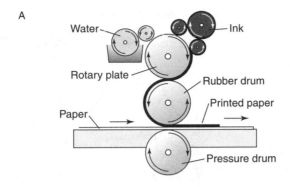

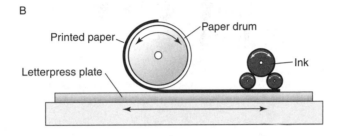

Figure 2.8

Two common printing presses. (A) Diagram of a high-speed offset printing press using a rotary plate prepared by photo lithography. (B) Diagram of a high-speed reciprocal printing press using a plate prepared by photoetching.

Similar high-speed printing can be achieved by photo etching. In this case, a copper or zinc plate coated with a light-sensitive emulsion is contact printed using a film negative and developed so as to leave a hardened emulsion in the exposed areas but bare metal in the areas of the image not receiving light. Next, the plate is coated with chemicals that adhere only to the emulsion-coated areas and further harden it into a layer that is resistant to being etched by acid. The plate is then dipped in acid, which then attacks and dissolves the bare metal surfaces while leaving the resistant surfaces untouched. Etching is sufficiently strong to actually lower the surface of the metal below that of the unetched regions. As a result, when the plate is inked with a roller, only the higher unetched surfaces will pick up ink from the roller, and only these areas of the image will transfer ink to the rubber drum and paper (Figure 2.8B).

Both processes work best if the entire image is printed in pure black and pure white corresponding to the inked and uninked surfaces of the plate. Thus, these processes can easily reproduce text in pure black or line art. However, what about images like photographs that have numerous levels of gray in them that are neither black nor white? The solution is to turn these areas into microscopic patterns of black and white dots that when viewed by our eye are integrated and perceived as gray levels. This can be done using a "screening" process. The screen is a semitransparent optical filter that consists of a microscopic grid of squares, each square having a dome-like pattern of optical density from one side to the other. This pattern is repeated throughout the screen in both its vertical and horizontal dimensions. This optical device, when used to print the image onto lithographic film, results in each square in the screen producing a single black dot on the developed lithographic film whose size is related to the average intensity of light emerging from the image negative at that point.

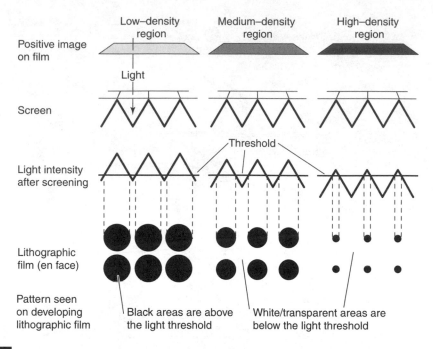

Figure 2.9
Production of half-tone illustrations by screening and photolithography. Results are shown for three regions of a film positive exhibiting low, medium, and high silver grain densities. Light emerging from the positive passes through the screen to expose the photolithography film below. The screen modifies the image by superimposing a grid of squares, each square having an optical density that is high at its center but that decreases to zero at its edges (shown as a sawtooth pattern). When developed, the photolithography film exhibits a negative image with no gray levels, just black or transparent regions. As a result, each pixel in the screen produces a black dot on the lithography film whose size is related to the light intensity from the orignal positive. This pattern is now transferred to an emulsion-coated zinc plate by photography producing a positive image whose light-exposed areas will be coated with a ground that will hold the ink to be transferred to paper. Alternatively, this pattern can be etched into a copper plate to produce a pattern in relief that will be coated with ink.

This principle is illustrated by a positive image on film having three zones of different density (**Figure 2.9**). The low-density region of the positive allows a lot of light to pass through the screen, resulting in a pattern of large black dots on the lithography film. The medium-density region of the positive allows through an intermediate level of light, producing a pattern of medium-sized black dots. High-density regions of the positive allow through only a low intensity of light, which when modified by the screen produces very small, black dots on the film. Thus, the screen serves to divide the image into pixels represented by a pattern of dots of various sizes on the lithographic film after it is developed. The properties of the lithographic film are critical to the formation of this pattern. When developed, this film produces only black regions and clear regions, with no intermediate gray levels. Above a critical threshold of light, the film will develop black, and below this threshold, the developed film will remain completely transparent. This purely black and clear array of dots on the lithography film, when printed onto a sensitized metal plate, will result in the printed picture being composed of pixels each represented by differently sized black and white dots. If one magnifies an image from such a printed page as little as 10 times, one can clearly see this pattern (Figure 2.7F).

The detail in the picture depends on the number of pixels per unit length, with typical screens ranging from 60 to 300 per inch in both vertical and horizontal dimensions. Course screens (50 pixels per inch) are used for newspapers. Medium screens (150 per inch) are used for reproduction in scientific journals and magazines, and fine screens

(250 to 300 per inch) are used in fine-art reproductions. This screening process is also referred to as "half-tone" reproduction, referring to the fact that all gray levels are produced by a microscopic pattern having only two tones—black and white. In contrast, normal photographic images, whether on film or paper, are not subdivided into pixels and gray values, but are related to silver grain densities that can change smoothly from one region to the next. For this reason, these are referred to as "continuous tone" images. Because the screening process required for high-speed reproduction limits the resolution of the printed image, the prints or electronic files submitted for publication need to have a resolution of no greater than twice that of the screen to be used, with 600 dots per inch being more than adequate in all cases.

■ Color Reproduction

In the early days of lithography, when each printed color required a different stone, much experimentation was carried out to determine the minimum number of colors (stones) needed to get high-quality color reproduction. As pointed out in Chapter 5, the human eye has photoreceptors that are sensitive to three colors—red, green, and blue. Indeed, even before these physiologic facts were known, lithographers had determined that three-color printing was the minimum required to get good reproduction of virtually any shade of color that our brain can perceive! This principle, discovered very early, continues to be used up to this day! Color reproduction techniques, whether carried out by high-speed presses, laser jet printers, ink jet printers, or photography, usually rely on the use of only three colors—cyan (light blue), yellow, and magenta.

Why these three colors? After all, televisions and computer monitors add patterns of red, green, and blue light together to stimulate our red, green, and blue light-sensitive photoreceptors to form what our brain interprets as any color possible. This is referred to as "additive" color production. In contrast, the pigment on a printed page absorbs light, and the reflected light that your eye sees in the image is just the leftover wavelengths that were not absorbed by the pigment. This is referred to as "subtractive" color production because what we see is what is "leftover." Thus, if red is to be perceived, the pigments must subtract/absorb both green and blue light. One can do this with a combination of magenta and yellow ink. In an analogous manner, green light will be reflected from the print by using a combination of yellow and cyan inks, and blue light will be reflected from the print by using a combination of cyan and magenta inks (Figure 2.6C). Computer programs (such as Adobe PhotoShop) designed to handle electronic-image files can "separate" color images into their red, blue, and green components for use in additive color reproduction (termed RGB color systems) or into the magenta, yellow, and cyan ink patterns used to create reflected color images by a subtractive process (termed CMY color systems or CMYK color systems if black is added as a fourth pigment).

In this manner, all reproduction processes use patterns of three different inks to produce a printed color image. Although application of equal amounts of all three inks produces black, high-speed printing processes often add a pattern of black ink that produces shading, lines, and edges without the highly precise registration of colored inks that would otherwise be required. This means that four different plates printing in sequence on a paper must be used produce color illustrations. Preparation of these plates requires four different screened images. This accounts for the fact that color illustration is much more expensive than black and white illustration, and in the case of scientific journals, this cost is often passed onto the scientific author.

Color illustration using screened images purposely tries to position the array of dots so that the dots of one color are not superimposed on the dots of another color. If superimposition of dots occurred, the color of the dot printed first would be hidden by the dot printed second! To avoid this, the grid of the screen used for each color is oriented at a different angle relative to the vertical and horizontal dimensions of the image (inset, Figure 2.6E). The result achieved can be of very high quality as demonstrated by the illustration in Figure 2.6D. If one magnifies the image to visualize the screen pattern (Figure 2.6E) one can verify that the screen orientation for each ink color is as indicated in the inset, and that appropriate offset of each screen produces a hexagonal pattern thus ensuring each primary ink makes its intended contribution to the image color.

References and Suggested Reading

Astrua M. *Manual of Colour Reproduction.* Kings Langley: Fountain Press, 1973.

Brunner F. A *Handbook of Graphic Reproduction Processes,* 6th ed. Teufen, Switzerland: Arthur Niggli, 1984.

Ford BJ. *Single Lens, The Story of the Simple Microscope.* London: William Heinemann, 1985.

Ford BJ. *The Leeuwenhoek Legacy.* London: Biopress and Bristol and Farrand Press, 1991.

Fournier M. *The Fabric of Life: Microscopy in the Seventeenth Century.* Baltimore: The Johns Hopkins University Press, 1996.

Hook R. *Micrographia or Some Physiological Descriptions of Minute Bodies Made by Magnifying Glasses.* London: John Martin and James Allestry for the Royal Society, 1665.

Irvins WM. *How Prints Look* (revised by Cohn MB). Boston: Beacon Press, 1987.

Lee JB, Mandelbaum M. *Seeing Is Believing: 700 Years of Scientific and Medical Illustration.* New York: The New York Public Library, 1999.

Mace OH. *Collector's Guide to Early Photographs.* Iola, WI: Krause Publications, 1990.

Reilly JM. *Care and Identification of Nineteenth Century Photographic Prints.* Rochester: Kodak Books/Silver Pixel Press, 1986.

Ross J, Ramano C, Ross T. *The Complete Printmaker.* New York: The Free Press/Roundtable Press, 1990.

van Leeuwenhook A. Concerning the animalcula in semine humano. *Phil Trans Roy Soc* 1699; 2:301–308.

Yount L. *Antoni van Leeuwenhoek: First to See Microscopic Life.* Berkeley Heights, NJ: Enslow Publishers, 1996.

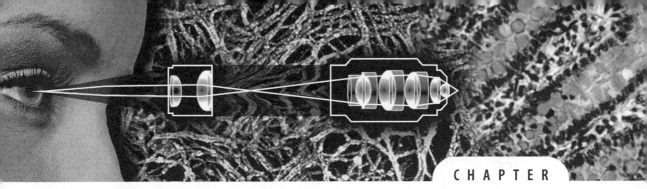

Preparation of Specimens for Light and Electron Microscopy

At the dawn of microscopy there was no specimen preparation. One simply placed the critter on the microscope stage and took a look—this is still done today. The technical word for this approach is using a "whole-mount" preparation. Microscopists soon learned that most specimens were too thick to transmit light (and certainly not electrons) and that they needed to be dissected so that the parts could be examined. Also, some specimens that would either move too fast or deteriorate in time needed to be killed and preserved for future observation. These needs eventually led to the relatively standard approach for tissue preparation employing chemical fixation (to kill and prevent deterioration), embedding (to stabilize the tissue structurally), sectioning (to provide a thin specimen through which light or electrons could pass) and staining (to make tissues more visible and to distinguish between their structures). There are two basic methods used for fixing tissues: physical (such as cryofixation) and chemical. The standard procedures for chemical fixation of tissue are described in this chapter.

■ Fixation

Fixation means to stabilize, to kill rapidly, and to harden and preserve *in toto* for microscopic study. Compounds that perform all requirements to render a tissue "fixed" do not exist, and thus, the protocols of sample preparation employ multiple chemicals that act independently toward the common goal of fixation. The goal of the microscopist is to study the structure of cells and tissues in as "life-like" state as possible—that is, normal in structure and function and free from as many autolytic processes as possible that come with death and are not present in living cells. In the strictest sense, this is an unattainable goal, and it is incumbent on the microscopist to recognize artifacts of death or fixation and to minimize them when present. Either fixation protocols must be modified to eliminate them, or if unavoidable, the microscopist must live with them and interpret the data in light of them.

To avoid artifacts, the specimens must be kept under optimal conditions appropriate for the goals of the study. If the tissue is harvested directly from life, the organism must be kept under optimal physiological conditions before use. If cells or tissues are grown in culture or incubated *in vitro*, they must be maintained in a medium that has the right nutrient requirements, salt composition, and pH and that is oxygenated (for aerobic organisms) and held at an appropriate temperature. The parameters that are optimal can vary from specimen to specimen, and it is important to examine the literature for conditions that others may have already found suitable. If this information is not available, choosing conditions that are optimal for a similar organism or tissue would be a good place to start. The buffer conditions that are used for live cells and tissues are also likely those that are optimal for chemical fixation, although there are exceptions to this. For specimens to be collected outside the laboratory (e.g., in the natural habitat, at the site of infestation or infection, or at the site of medical or agricultural procurement), the appropriate fixation materials must be taken to the specimen and the same precautions to minimize artifacts used.

Thus, the main objectives of fixation are to kill the tissue quickly, stabilize its structure, and protect its constituents from subsequent sample processing and observation procedures. In the case of pathogenic organisms, a further goal may be to render them harmless during microscopic observation. For these reasons, chemical fixatives tend to be highly reactive toward proteins, lipids, and nucleic acids and to have a low molecular weight so that they can diffuse into tissues quickly. Because proteins are the most abundant constituent of cells, it is not surprising that the most widely used chemical fixatives are formaldehyde and glutaraldehyde, which react with the amino groups of proteins. Before launching into a full-scale comparison, we summarize the properties of common fixatives in **Table 3.1.**

Aldehydes

Glutaraldehyde ($C_5H_8O_2$), a colorless liquid with a pungent odor that smells like bleach with a hint of apples, has two aldehyde groups linked by a three carbon chain. Because each aldehyde group can react with an amine group to form a very stable, essentially irreversible covalent bond, this compound can cross-link amine groups within a single protein or link neighboring proteins, turning them into a large, stable, polymeric complex (**Figure 3.1A**). Thus, glutaraldehyde is an excellent compound for initial fixation of cells and tissues for light and electron microscopy. (Interestingly, it is also used in the leather tanning process, in hospitals as a cold sterilizing agent to disinfect heat-sensitive medical and dental equipment, and in industry as a chemical preservative.) A major disadvantage of glutaraldehyde is that because of its relatively large size (molecular weight = 100.1) it penetrates tissues slower than other aldehydes, and, thus, physiologic and structural activities of cells are stabilized more slowly than is optimal, resulting in moderate to severe artifacts in some cases. In addition, changes to cells from autolytic activity can occur during glutaraldehyde fixation because membranes become leaky and lytic enzymes that are not rendered inactive by glutaraldehyde can damage cellular structure. Fixation at reduced temperatures (4°C) is recommended in some cases to allay this potential. Like all aldehydes, glutaraldehyde does not contain a heavy metal and does not impart amplitude contrast, or so-called electron density, to tissues. Glutaraldehyde is toxic and can cause severe eye, nose, throat, and lung irritation, along with headaches, drowsiness, and dizziness if not used safely.

Formaldehyde (CH_2O), a one-carbon monoaldehyde, is the smallest and simplest of the aldehydes (molecular weight = 30.03). It is able to cross-link only weakly proteins by connecting a primary amine group with a nearby nitrogen in the protein or DNA through a $–CH_2–$ linkage called a Schiff base. Formaldehyde is less desirable as a primary fixative for electron microscopy because of its weak cross-linking ability. Thus, proteinaceous structures are not as well preserved at the ultrastructural level; however, its small size allows it to penetrate tissues rapidly and the proteins fixed to remain closer to

	Table 3.1 Common Fixatives			
Fixative	Structure or Composition	Properties	Uses	Comments
Formaldehyde	H_2CO, 2% to 4% in aqueous solution	Reacts with amines on proteins; fast penetration	Primary fixative for both light and electron microscopy; fixative of choice for immuno-cytochemistry	
Glutaraldehyde	$CHO(CH_2)_3CHO$, 1% to 3% in aqueous solution	Cross-links amine groups on proteins	Primary fixative for both light and electron microscopy	Most common single fixative for EM
Karnovsky's fixative	Formaldehyde 4%; glutaraldehyde 5%	Fixes proteins	Frequently used at full or half strength for electron microscopy	Combines fast penetration with cross-linking
Osmium tetroxide	OsO_4, 1% in aqueous solution	Fixes and stains lipids and proteins	Secondary fixative for electron microscopy; occasionally used as a primary fixative	Adds heavy metal contrast
Uranyl acetate	1% in aqueous solution	Fixes nucleic acids	More commonly used as a stain than a fixative	Mild radiation hazard
Acrolein	$CH_2 = CHCHO$	Fixes proteins and lipids; active at low temp; penetrates rapidly	Used occasionally for freeze substitution	Highly volatile lachrymator
Cold methanol	CH_3OH, 100%	Rapid precipitation of proteins and nucleic acids	Used as primary fixative for light microscopy; can be used for immunocyto-chemistry	Disrupts membranes and causes shrinkage artifacts at EM level
Bouin's fluid	Picric acid Formalin Glacial acetic acid	Fixes proteins and precipitates macromolecules	Versatile primary fixative for light microscopy	Used for many years in medical histology
Carnoy's fluid	Ethanol Chloroform Glacial acetic acid	Rapid penetration; precipitation of proteins and nucleic acids	Emergency biopsy specimens, nucleic acid staining	
Formalin	Formaldehyde Methanol	Rapid penetration; useful for large specimens	Common fixative for light microscopy and specimen storage	

their native structure. The later quality has proved to be an advantage for cytochemical and immunocytochemical procedures.

Karnovsky (1963) showed that a combination of glutaraldehyde (e.g., 2.5%) and formaldehyde (e.g., 2%) in a hypertonic buffer can provide excellent results in dense,

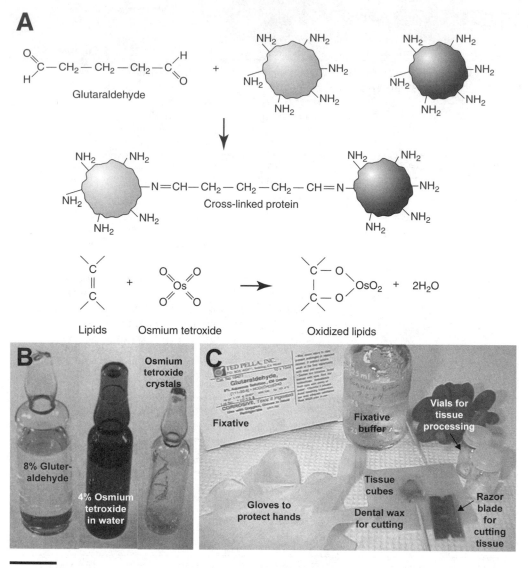

Figure 3.1
Chemical fixatives. (A) Glutaraldehyde reacts with amino groups to crosslink proteins. Osmium tetroxide reacts with unsaturated lipids. (B) Fixatives are best kept in glass vials under nitrogen. (C) Supplies needed for fixing tissues.

hard to penetrate tissues and cells. In this situation, the formaldehyde enters the tissues quickly to fix cytoplasmic materials, followed by a second wave of more slowing diffusing glutaraldehyde that cross-links and stabilizes the material. Like glutaraldehyde, formaldehyde does not contain a heavy metal and does not impart contrast to the sample. Because both formaldehyde and glutaraldehyde are volatile and reactive and have the ability to form homopolymers, they are often purchased from microscopy supply houses as "EM grade" aqueous solutions in glass vials sealed under nitrogen (Figure 3.1B). Stored in vials at 4°C, these fixatives can remain stable for a number of years. A range of concentrations are available, although higher percentages of glutaraldehyde (>25%) are prone to spontaneous polymerization if stored after the vial is opened. For routine work, small volumes of 4% to 8% glutaraldehyde can be kept in a sealed glass bottle at 4°C for about 6 months. For formaldehyde, an alternative is to purchase it as paraformaldehyde, a solid trimer that can be heated in water at alkaline pH to form the monomer. When making

Table 3.2	Common Buffers for Fixation	
Buffer	**pK/pH Range**	**Comments**
Phosphate (Na_2HPO_4/NaH_2PO_4)	5.7 to 8.0	Mixture of Na_2HPO_4 and NaH_2PO_4 as determined by pH desired
Sodium cacodylate ($Na(CH_3)_2 AsO_2 \cdot 3H_2O$)	5.0 to 7.4	Poisonous; contains arsenic; still commonly used but HEPES and PIPES are safer and are replacing cacodylate
S-Collidine (2,4,6-trimethylpyridine)	6.0 to 8.0	Toxic; mild smell is disliked by some; used infrequently
HEPES (N-2-hydroxyethylpiperazine-N′-2-ethanesulfonic acid)	7.35	Reacts slowly with glutaraldehyde; can be used between 5 and 50 mM
PIPES (Piperazine-1,4-bis-2-ethanesulfonic acid)	6.8	Reacts slowly with glutaraldehyde; can be used between 5 and 50 mM
Tris-Maleate (Tris (hydroxymethyl) aminomethane maleate)	5.2 to 8.6	Reacts slowly with glutaraldehyde; can be used between 5 and 50 mM

fresh formaldehyde in this way, understand that the monomer will start to repolymerize spontaneously and has a very short shelf life. This is also the case for formaldehyde purchased in glass vials; therefore, it is common to use formaldehyde that has been prepared fresh the same day. The common buffers used to maintain the pH of fixatives are listed in **Table 3.2.** Phosphate, cacodylate, PIPES, and HEPES are the most frequently used, despite that cacodylate is toxic because of its arsenic content and HEPES contains amino groups that react slowly with the aldehyde fixatives. Bicarbonate/CO_2 buffering is not commonly used because of the need for gassing the medium during fixation and processing.

Acrolein is a three-carbon monoaldehyde (C_3H_4O; molecular weight $= 56$) that has a piercing, acrid smell. It was first used as a primary fixative in 1959 because it can penetrate tissues rapidly and reacts quickly with the amino groups of proteins even at low temperature. Indeed, this compound is the fastest acting of all of the fixatives used. On the surface, acrolein appears to be an ideal fixative; however, its use is avoided because it is so dangerous. This compound is highly flammable and a severe pulmonary irritant and lacrimating agent, and it can polymerize violently in the presence of air, light, acids, and bases. In fact, it was used as a chemical weapon during World War I as a component of tear gas.

Aldehydes are commonly used as the first or "primary" fixative. For the fixative to penetrate easily, the tissue to be fixed must be cut into small cubes approximately 0.5 to 1 mm on a side. With any larger dimension, the specimen is likely to be poorly fixed in the interior. Cutting of the tissue before fixation is carried out with a razor blade on a sheet of dental wax in a small puddle of fixative, as illustrated in Figure 3.1C. Isolated cells, tissue culture cells, and thin tissues can be fixed in suspension without cutting. Fixation is often carried out for 1 to 6 hours at room temperature or, depending on the sample, at 4°C. The tissue is then washed thoroughly in buffer without fixative before proceeding.

Fixatives are not just poisonous; they will actually embalm you, sometimes without your knowledge until it is too late. One of the authors (D.C.) has a poor sense of smell after years of using glutaraldehyde. All who work with fixatives, heavy metals, and embedding resins should use protective hand and eye ware and perform the work in a functioning fume hood. Laboratory coats and closed-toed shoes are also recommended.

Fortunately, the many hazardous chemicals that the microscopist works with are mostly used in low volumes and concentrations. Additional data on chemicals used for processing tissues can be obtained from the Material Safety Data Sheets (MSDS) available from suppliers of the materials. For appropriate storage and disposal of these materials one should contact the hazardous materials office of their institution.

Osmium Tetroxide

Osmium tetroxide (OsO_4), also known as osmium tetraoxide, osmium (VIIi), or osmic acid, is an oxide of the element osmium. In its pure form, osmium tetroxide crystals are colorless, but it is usually contaminated by a small amount of yellow-brown osmium dioxide (OsO_2), giving it a slight yellow coloration. It smells like ozone, and in fact, the name osmium is derived from *osme,* Greek for odor. Osmium tetroxide is a strong oxidizing agent that reacts with and cross-links proteins and lipids, particularly phospholipids containing double bonds (Figure 3.1A). This property makes the fixative important for stabilizing membrane structure. Although osmium tetroxide was originally used as a fixative for light microscopy, it is now used almost exclusively for specimens to be prepared for electron microscopy. In this case, the fixative not only stabilizes tissue structure, but the reduced osmium produced as a result of fixation acts as a heavy metal stain to provide contrast in the image (see below). Tissues generally turn brown to black, as they are "osmicated" during fixation.

Osmium tetroxide can be purchased either as crystals (the most economical approach) that are soluble in warm water (dissolution is aided by sonication) or as an aqueous solution sealed in vials under nitrogen gas. Cleanliness is very important when working with osmium, as it is such a strong oxidant that it reacts with many buffers and biological materials. It is also light sensitive and should be stored at 4°C. Because it is highly volatile, stock solutions should be double sealed in two glass containers. If not stored properly, the walls of the refrigerator will eventually be stained black because of the action of osmium vapors, even at 4°C. Indeed, some fixation protocols use the crystalline or aqueous forms of osmium tetroxide to produce vapors that are directly absorbed by the tissue leading to their fixation. In fact, workers not using a hood or eye protection can suffer from impaired vision because of the osmification and darkening of the corneal epithelium. This could clearly be a serious problem, although fortunately the condition eventually resolves as fixed corneal epithelial cells slough off and new cells form to reestablish clear vision. All of the work with osmium tetroxide must be performed in a fume hood with appropriate protective clothing. If the odors of osmium are detected, one is too close, and the problem should be dealt with immediately.

Commonly, tissues are first fixed with aldehydes, washed thoroughly, and then treated with osmium tetroxide as a "secondary" fixative. Secondary fixation is often carried out with 1% OsO_4 in either buffer (typically) or distilled water for 1 to 2 hours at room temperature or 4°C. With a molecular weight of 254.23, osmium is the slowest penetrating fixative used to prepare tissue for electron microscopy. Washing out the primary fixative is important because osmium tetroxide reacts with aldehydes and will form precipitates both in solution and within the tissue. Despite this, protocols for fixing isolated cells occasionally use a mixture of glutaraldehyde and osmium tetroxide prepared on ice so that the reaction between the two fixatives is slow. Fixation time is short (less than 15 minutes), and the fixative is quickly washed out.

Cold Methanol/Acetone/Ethanol

Some specimens to be viewed at the light microscope level need only be fixed by methanol, acetone or ethanol cooled to −20°C. The cold solvent rapidly precipitates all proteins and nuclei acids, thereby stabilizing their structure. These solvents are called "coagulating" fixatives to differentiate them from "chemically reactive" fixatives such as glutaraldehyde. They also disrupt cell membranes, thereby permeabilizing cells, a feature that is useful if one is performing immunocytochemistry (see Chapter 13). Another advantage of these solvents is that dehydration and fixation are accomplished simultaneously, thus, shortening the time needed for processing. Rapid penetration and permeabilization of cells by these solvents also make them useful in combination with other fixatives such as formaldehyde. Formalin, a combination of methanol and formaldehyde, allows faster penetration and fixation than formaldehyde alone. Membrane damage, shrinkage artifacts, and lipid extraction make cold solvent fixation unsuitable for electron microscopy.

Bouin's Fluid

Picric acid is the active constituent in this fixative, reacting and binding to all tissue proteins to cause denaturation. This fixative also contains ethanol for rapid penetration and glacial acetic acid to help precipitate proteins. This fixative is not used for electron microscopy but has been used extensively in light microscopy for over a century and sees routine use in medical histology. Picric acid is not volatile, and this fixative can be used outside of a fume hood as long as care is taken to avoid skin contact. One caution is that picric acid is explosive when allowed to dry, and preparations more than a few years old should be discarded. The bright yellow color of picric acid automatically makes this fixative "color coded."

■ En Bloc Staining

This step is optional and is carried out only on specimens for electron microscopy. After fixation and washing, the tissue is exposed to 0.5% to 1% uranyl acetate in water for 1 or 2 hours in the dark at room temperature. Precautions against light are due to light-triggered precipitation of uranyl acetate. The positively charged uranyl ions bind tightly to negatively charged membranes, nucleic acids, and proteins to provide high-contrast staining with a heavy metal that will easily deflect electrons during imaging. This salt is also thought to stabilize chromatin and nucleic acids and is considered a specialized fixative for these structures. In addition, phospholipids are less likely to be extracted during acetone dehydration and resin infiltration when uranyl acetate is used. Staining with uranyl acetate has little effect on specimens viewed by light microscopy.

■ Dehydration

After the tissue specimen has been chemically fixed and thoroughly washed, it must be dehydrated with a nonaqueous solvent such as ethanol or acetone. The reason for this is that most specimens will need to be embedded in a solid matrix so that they can be sectioned. The matrix components used (paraffin for light microscopy and epoxy resin for electron microscopy) are not soluble in water, and their ability to harden is dramatically reduced in the presence of any water.

Dehydration of specimens is usually carried out slowly in graded stages to avoid differential shrinkage of cell components that can lead to structural tears in the tissue even within individual cells. Even though the tissues are well fixed at this point of preparation, they are still susceptible to shrinkage and extraction artifacts during dehydration. Ethanol is the dehydration agent of choice, as it is relatively gentle and is low in toxicity. The specimen is usually passed through a series of aqueous-based ethanol solutions such as 10%, 30%, 50%, 70%, and 90%, followed by two changes of 100% ethanol, spending 10 to 15 minutes in each solution at room temperature. As a rule, the shorter the time spent in dehydration solutions the better. To facilitate the dehydration process, the samples can be placed on a rotating rack that continually mixes the solutions. Some laboratories perform dehydrations at cold temperatures such as 4°C, stating that the gentle nature of water removal at these temperatures reduces artificial undulations of membranes, shrinkage of tissues, and extraction of materials.

■ Infiltration and Embedding

The purpose of embedding is the same in both light and electron microscopy. The tissue must be completely infiltrated with liquid components, usually with the aid of a solvent, and then the components must be hardened or polymerized to provide a solid block that can be cut into slices called "sections." If this were not done, the tissue would be damaged during slicing much like a loaf of bread being squashed by a dull knife. Numerous types of embedding media are in use (**Table 3.3**); the most common are paraffin for light microscopy and epoxy resin for electron and light microscopy.

Table 3.3	Common Embedding Media		
Embedding Medium	**Type/Solubility**	**Components**	**Uses**
Paraffin	Wax; soluble only in nonpolar solvents such as xylene	Long-chain hydrocarbons	Standard for light microscopy
Paraplast	Paraffin marketed for embedding	Long-chain hydrocarbons	Standard for light microscopy
Eponate 12	Epoxy resin; soluble in nonpolar solvents such as propylene oxide or acetone	Eponate 12 Resin DDSA (hardener) NMA (hardener) BDMA (accelerator)	Light or electron microscopy
Araldite 502	Epoxy resin; soluble in nonpolar solvents such as propylene oxide or acetone	502 Resin DDSA (hardener) DBP (hardener) BDMA (accelerator)	Light or electron microscopy
Spurr's	Epoxy resin; soluble in nonpolar solvents such as propylene oxide or acetone	VCD/ERL 4221 Resin DER 786 (flexibilizer) NSA (hardener) DMAE (accelerator)	Low viscosity; light or electron microscopy; excellent for plant and fungal tissue
Methacrylate	Acrylic resin; some water solubility	One-component resin	Earliest resin used
LR White	Acrylic resin; compatible with small amounts of water	One-component resin; catalyst added when used at 4°C	Heat polymerizable; low viscosity; oxygen inhibits polymerization; used for EM immunocytochemistry
LR Gold	Acrylic resin; compatible with small amounts of water	One-component resin; catalyst added when used at 4°C	Ultraviolet polymerizable; oxygen inhibits polymerization; used for EM immunocytochemistry
Lowicryl	Acrylic resin; compatible with small amounts of water	May be purchased as three components or just one "mono" mixture	Ultraviolet polymerizable; oxygen inhibits polymerization; used for EM immunocytochemistry

Paraffin

Paraffin is relatively soft and has a melting point between 40°C and 70°C (depending on the type chosen by the microscopist), and its infiltration into the tissue can be aided by mild heat and vacuum. A typical embedding routine with paraffin is provided in **Figure 3.2A**. Because paraffin is not soluble in ethanol or acetone, the tissue must be first passed through a transition solvent such as xylene. These solvents are sometimes referred to as "clearing agents" because they often extract pigments resulting in a transparent specimen. After passage through two steps of 100% xylene, the tissue is placed in xylene-paraffin 2:1, 1:1, and 1:2 for 6 hours each, then in a bath of liquid paraffin kept a few degrees above its melting point to allow for infiltration of the tissue with 100% paraffin. Each specimen is then placed in a disposable plastic mold (Figure 3.2B) and an identifying label is inserted before filling the mold with molten paraffin. The final step with 100% paraffin is best carried out in a vacuum oven at low pressure so that any microscopic gas bubbles in the tissue are removed. Some laboratories enjoy the use of automated tissue processors for dehydration, infiltration, and embedding like that shown in Figure 3.2C. The tissue, after fully infiltrated with paraffin, is then cooled to room temperature over a period of several hours to harden the paraffin.

Epoxy Resin

Epoxy resins were introduced in 1956 and remain very popular despite certain drawbacks such as high viscosity, intolerance to water, and their toxicity when not in a solid polymerized form. They are versatile resins producing a hard, highly cross-linked matrix that is usable for either light or electron microscopy. The most common resins are mixtures of four liquid ingredients—a monomer, a flexibilizer, a hardener, and an accelerator. The liquid mixture, once fully infiltrated into the specimen is polymerized by heat. The liquid components are volatile and allergenic, and many are considered carcinogenic; when working with these compounds, gloves are mandatory to prevent skin contact. The workstations must be located in a fume hood. After the resin is polymerized, it is no longer hazardous and is stable for years. Resin waste is often polymerized before being disposed of to eliminate its hazard. All liquid or solid resin waste must be disposed of through hazardous materials protocols. Epoxy resins commonly used for microscopy, along with their properties, are listed in Table 3.3.

A typical infiltration and embedding protocol for epoxy resins is provided in **Figure 3.3A**. In the past, propylene oxide has often been used for epoxide resin infiltration as a transitional solvent that is miscible with both alcohol and epoxy resins. One advantage of propylene oxide is that it is not miscible with water, and water is a detriment to resin polymerization. Resin polymerized in the presence of even trace amounts of water will be soft, rubbery, and impossible to section, resulting in embedded specimens that are not retrievable; however, most laboratories opt for acetone that performs almost as well and is much less toxic than propylene oxide. As a precaution, the acetone used is stored in the presence of molecular sieves that ensure the removal of all water from the solvent (Figure 3.3B). For infiltration, fixed and dehydrated specimens are typically incubated in resin:acetone mixtures at ratios of 1:2, 1:1, 2:1, and finally 100% resin. The incubation times used for the infiltrations steps vary from very brief (0.5 to 2 hours) to very long (up to 12 hours), depending on the resin used and whether the sample possesses barriers to infiltration such as cell walls or cuticles. Because epoxy resin components are highly viscous, specimens are placed in vials in a rotating rack for slow, continuous mixing (Figure 3.3C). During these steps, the accelerator is sometimes omitted to avoid premature polymerization, although this is dependent on the type of resin used and is done at the risk of uneven polymerization and blocks that are difficult to section. Spurr's resin has proved to be an excellent and widely used epoxy resin because it penetrates many

A. Light Microscopy Preparation Protocol

1. Chemically fix tissue for 2 to 10 hours depending on specimen size and fixative.

2. Wash with buffer three times, 15 min per wash.

3. Dehydrate in graded series of ethanols, 10 min each step until 100% ethanol is reached.

4. Transfer to clearing agent such as xylene in three steps, each step 30 min until 100% xylene is reached.

5. Infiltrate with paraffin in steps of 1:2, 1:1, 1:2, and 100%; use vacuum oven at 60°C.

6. Cool paraffin over a 12-hour period. Mount block on stub and trim.

7. Section with microtome, warm sections on water to relax, and pick up on glass slides.

8. Remove section paraffin using xylene.

9. Rehydrate in steps of ethanol/water.

10. Stain with aqueous dyes and wash thoroughly.

11. Dehydrate in graded steps of ethanol/water.

12. Add mounting medium and apply cover glass.

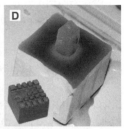
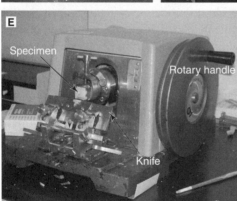
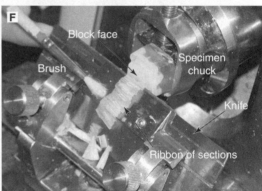
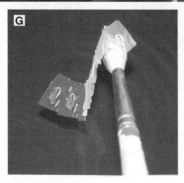

Figure 3.2

Preparation of tissue for light microscopy. (A) Preparation summary. (B) Disposable molds for embedding tissues in paraffin. (C) Commercial instrument for paraffin embedding. (D) Trimmed paraffin-embedded specimen mounted on a metal block (inset) and ready for sectioning. (E) Standard rotary microtome for sectioning paraffin-embedded tissues. (F) Microtome detail showing paraffin block, the metal knife, and a ribbon of sections that have just been cut. (G) A dry brush is used to transfer the ribbon of sections to the surface of a warm water bath. (H) Picking up the "relaxed" sections onto a glass slide.

A. Electron Microscopy Preparation Protocol

1. Chemically fix 1 mm tissue cubes for 1 to 6 hours in buffered glutaraldehyde, wash three times in buffer, then fix in osmium tetroxide.

2. Wash with buffer three times, 15 min per wash.

3. En bloc stain with uranyl acetate for 1 hour in the dark (optional).

4. Dehydrate in a graded series of ethanols or acetones, 15 min per step, until 100% EtOH/acetone reached.

5. Transfer into propylene oxide or acetone, two steps 15 min each.

6. Infiltrate with epoxy resin in four steps, 1:2, 1:1, 2:1, and 100%. Specimen is rotated slowly during infiltration.

7. Specimen placed in mold with fresh resin and cured in an oven at 60°C.

8. Cured blocks removed from mold and trimmed and faced with a glass knife.

9. Thin sections cut on an ultramicrotome and picked up on a copper grid.

10. Grid stained with uranyl acetate and lead citrate.

11. Grid observed in TEM.

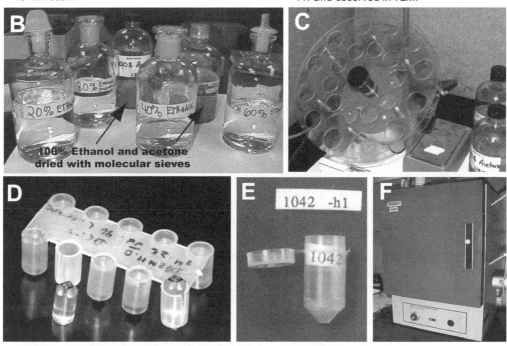

Figure 3.3
Preparation of tissue for electron microscopy. (A) Preparation summary. (B) Ethanols and acetone used for tissue dehydration. (C) Rotary mixing device for epoxy resin infiltration of tissue. (D) Cylindrical molds for epoxy resin blocks. (E) Individual mold labeled with specimen number. (F) Oven for curing epoxy resin blocks.

tissues fairly rapidly because of its relatively low viscosity; therefore, this is the resin of choice when working with tissues that are difficult to infiltrate or when one is trying to shorten infiltration time. This resin also has a long shelf life (up to 4 months when stored at −20°C) and has good sectioning characteristics. On the other hand, Spurr's resin exhibits some shrinkage during polymerization and does not post stain well with aqueous-based uranyl acetate.

Several changes of 100% resin with an accelerator over a period of 8 hours or more are followed by embedment: placing the specimens in molds that will determine the shape of the final resin block. There are two common types of molds—those for flat embedding that provide for blocks with flat sides and those for producing cylindrically shaped blocks (Figures 3.3D and 3.3E). The final selection of the type of mold is dependent on the specimen and the orientation of the specimen during final sectioning. At this point, a small paper label written in pencil can be inserted into each mold to provide definitive identification of the specimen (Figure 3.3E). The mold may be placed in a desiccator under a vacuum for 6 hours to remove microbubbles and then is polymerized in an oven at the temperature and time specified for that particular resin (Figure 3.3F). After curing, the molds are cooled, and the blocks are removed for storage.

Water-Compatible Resins

These acrylic-based resins are used instead of epoxy resins when immunohistochemistry or cytochemistry is the primary goal of the research; however, acrylic resins have a number of disadvantages compared with epoxy resins. They can exhibit uneven polymerization characteristics and tend to extract lipids during the infiltration process. In addition, acrylic resins must be polymerized in the absence of oxygen and are not as hard as epoxy resins. As a result, these resins can be more difficult to section and are more likely to be unstable under the electron beam, thus requiring that sections be supported on the grid by an electron transparent film such as Formvar. Their advantage is that some (e.g., Lowicryl K4M and LR White/Gold) can tolerate a small amount of water that is bound to macromolecules within the specimen. This allowance for water is thought to help preserve specimen proteins in their normal conformation, thereby allowing antibodies to bind to them for immunocytochemistry at the EM level (Table 3.3). Flexibility in polymerization conditions provides an additional advantage that aids in the retention of antigenicity. LR White/Gold and Lowacryls K4M, HM20, K11M, and HM23 are formulated so that they can be polymerized by ultraviolet radiation (360 nm) at temperatures ranging from $-85°C$ to $-10°C$. Researchers also have the choice of using either hydrophilic varieties of Lowacryl (K4M and K11M) or hydrophobic varieties (Lowacryl HM20 and HM23). In addition to ultraviolet-induced polymerization at low temperatures, LR White can be polymerized at 4°C with the addition of an accelerator, at room temperature with ultraviolet radiation, or at 50°C without ultraviolet light. Furthermore, acrylic resins have very low viscosity and are much less toxic than epoxide resins. Fixation protocols that promote the preservation of antigenicity (e.g., cryofixation) and are followed by embedding in acrylic resin at low temperature are considered to be optimal for immuno-EM (Figure 13.8).

■ Microtomy—Cutting Sections

Because light microscopy is limited in the resolution and because light can pass through relatively thick specimens (compared with electrons), the sections used range from 0.5 to 20 μm in thickness and are referred to as "thick" or "semi-thick" sections. In contrast, electron microscopy (to allow electrons to pass) requires sections ranging from 60 to 80 nm (0.06 to 0.08 μm) that are referred to as "thin" or "ultrathin" sections. The technology needed to produce thick sections and ultrathin sections is different, and for thick sections, the instrument used is a "microtome" (Figure 3.2E), whereas for thin sections, the instrument used is an "ultramicrotome" (**Figure 3.4A**).

Paraffin blocks are removed from their mold and mounted on a metal stud in a pool of molten paraffin, cooled, and the excess paraffin trimmed away (Figure 3.2D). They are cut using a permanent metal knife that is periodically resharpened using a lubricated stone or an automated knife sharpener. During cutting, the knife is held stationary while the paraffin block is held in a carriage that moves past the knife edge, advancing 5 to 20 μm with each cycle (Figure 3.2F). As a result, each time a section is shaved off the block it is pushed onto the knife blade. After a series of such cycles, ribbons of sections are transferred by brush to the surface of a warm water bath and allowed to "relax" for a few minutes (Figure 3.2G). The flattened sections are then picked up on a glass slide (Figure 3.2H).

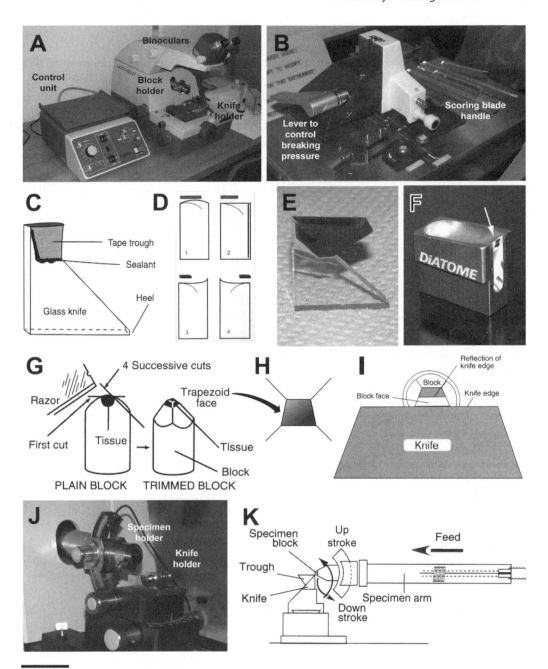

Figure 3.4
Cutting ultrathin sections. (A) A typical modern ultramicrotome. (B) Glass knife maker with glass strips. Glass squares are scored with a diamond and broken by pressure. (C and E) A knife boat can be formed either by tape or by molded plastic and sealed to the knife with wax or fingernail polish. (D) Glass knife edges vary in shape and quality. Convex edges have good cutting ability in most regions (bar). Straight and concave knife edges that are more common are sharp in only specific regions (bars). (F) Diamond knives have an edge (arrow) that is sharpened at the factory. (G) Epoxy blocks are trimmed to a pyramid shape with a razor blade. (H) The face of the block should be trapezoidal in shape. (I) Using the reflection method to line up the knife edge with the block face. (J) Detail of ultramicrotome showing specimen holder and knife holder. (K) The cutting and retraction cycle of the ultramicrotome.

Epoxy resin blocks, being much harder, are cut using either a glass knife or a diamond knife. Glass knives are made by a glass knife maker (Figure 3.4B) onsite no more than a few days before use. Strips of glass are scored, broken into squares, and the squares broken on a 45-degree diagonal to produce a sharp cutting edge at one end and a blunt "heel" at the other (Figure 3.4C). Not all glass knives are usable, but some indication is given by the shape of the knife edge (Figure 3.4D). Knife edges that are convex are the sharpest, whereas those that have concave edges are sharp in only certain regions (bars, Figure 3.4D). Even the best glass knife will remain sharp only for the cutting of about 20 to 30 sections and then must be discarded. For each knife, a small trough (the "boat") is made behind the knife that will be filled with water to provide a surface for cut sections to float onto during sectioning. To form the trough, either tape or preformed plastic collars are sealed to the knife body using melted wax or finger nail polish (Figures 3.4C and 3.4E).

Diamond knives are far more desirable than glass knives for ultramicrotomy; indeed, a diamond is a microtomist's best friend. They are, however, fragile and expensive, usually costing $1000 per millimeter of cutting edge and should be used only by trained personnel. The diamond is polished to a sharp edge and mounted permanently with epoxy into a soft metal trough (Figure 3.4F). Diamond knives are extremely hard and will remain sharp through thousands of cuttings if they have proper care. The knife edge must never be used to cut thick sections and must be cleaned after each use. Cleaning should be performed very carefully and according to the manufacturer's recommendations; overall cleanliness of tools, supplies, and water used in sectioning is necessary if one is to prevent debris and oil from spoiling the sections. During cleaning, one must never apply pressure on the knife edge in a manner that would flex the edge. Rather, all cutting and cleaning must place only vertical, not horizontal, pressure on the knife edge. Nevertheless, even a knife that is cared for under the best conditions will eventually dull with use. Today, most diamond knives can be resharpened at approximately half their original cost. A diamond knife is required for producing the highest quality and thinnest sections for electron microscopy. Sapphire knifes are sometimes used as a compromise between cost and durability.

Before an epoxy resin block is cut, it must be carefully trimmed by hand usually using a single-edge razor blade under a dissecting microscope (Figure 3.4G). The razor must be held firmly to avoid cuts, which can be severe. Also, the razor's edge must be cleaned with 95% ethanol and a Kimwipe to remove the fine layer of oil that coats its surface. The experienced microtomist knows the importance of a well-trimmed block in obtaining high-quality sections with minimal problem. Experience shows that the optimal trimming strategy is to cut away excess resin by roughly trimming the block to expose the tissue and then making cuts that produce a four-sided, truncated pyramid. The top "face" of the pyramid should be shaped like a trapezoid whose apex and base are parallel (Figure 3.4H).

The trimmed block is then mounted in the ultramicrotome and correctly oriented, and a glass or diamond knife is fastened in place. The block should be mounted so that the knife will enter the resin at the broad base and cut toward the narrower apex. The apex and base must be parallel to one another because this allows multiple cuts to produce a linear "ribbon" of connected sections that float off together onto the trough (**Figure 3.5C**). A pair of binoculars mounted just above the knife allows the operator to observe the process at high magnification (Figure 3.4A). The first objective is to "face" the block, that is, to cut partial sections off the block until the knife is cutting an entire section each cycle. Because this can be a delicate process, one usually uses a different part of the glass or diamond knife edge to face a block than one will use to produce thin sections. Many microtomists use a glass knife to face a block and cut thick sections before using a diamond knife to cut thin sections. Cutting thick sections with a diamond knife can quickly dull that part of the edge so that it cannot be used for cutting thin sections.

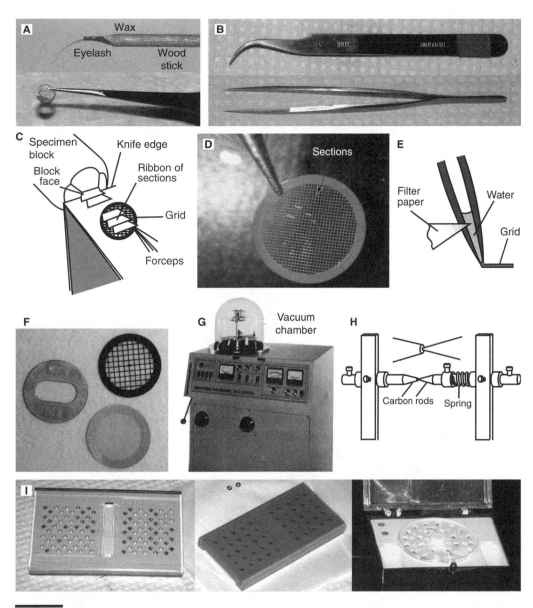

Figure 3.5
Picking up ultrathin sections on grids. (A and B) Tools used in picking up sections. (C–E) Cutting, picking up, and draining ribbons of sections. (F) Types of grids. From 9 o'clock clockwise: a slot grid, a 100 mesh grid, and a 400 mesh grid. (G) A carbon evaporator for coating grids. (H) Carbon rod electrodes used for carbon evaporation. (Inset) One pointed electrode contacts the flat surface of the other electrode. (I) Boxes for storage of EM grids.

After facing the block, one wants to cut thick sections that can be stained and observed by light microscopy to determine whether the expected tissue structure is being cut. This becomes extremely important if a particular feature is sought (e.g., an islet of Langerhans in the pancreas or a neuromuscular junction in muscle) or if a certain orientation of the tissue is desired. Sections from 0.5 to 1.0 μm thick are cut and transferred to a slide in a loop of water; the slide is heated briefly to adhere the sections to the slide and then stained with 1% toluidine blue or methylene blue. A drop of the dye is added to the slide on a hot plate and allowed to stain for about 20 seconds or until the edges dry, and then the stain is washed off with water. As shown in **Figure 3.6A** (see Plate 2 in the Color Addendum),

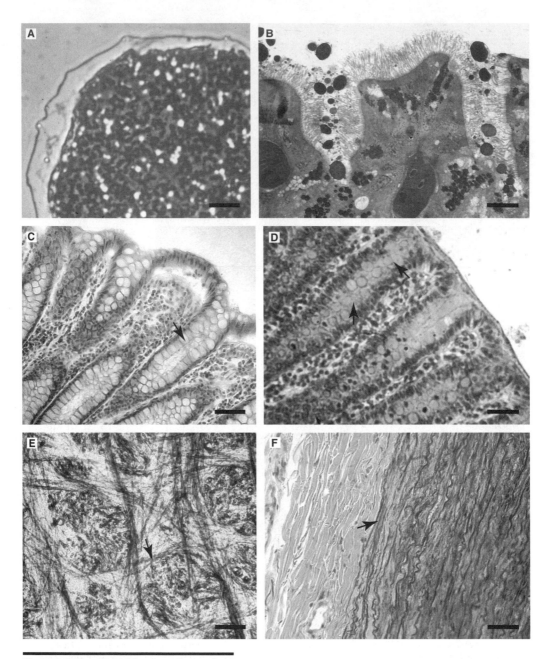

Figure 3.6 (See Plate 2 for the color version.)
Examples of stains for light microscopy. (A) Epoxy section of a fertilized sea urchin egg stained with toluidine blue. The fertilization envelope seen in cross-section protects the early embryo. (B) Epoxy section of the ciliated frog oviduct epithelium stained with a one-step staining mixture of eosin and toluidine blue. (C) Paraffin section of the large intestine stained with hematoxylin and eosin. The nuclei are stained dark purple with hematoxylin, while other proteinaceous structures are stained pink with eosin. The mucous-containing goblet cells in the epithelium (arrow) are clear because of lack of staining and extraction of the mucous. (D) Paraffin section of the large intestine stained by the periodic acid-Schiff (PAS) method. This technique stains the polysaccharides in mucous a brilliant pink to magenta (arrows). (E) Section of the hind brain stained with silver to reveal the numerous nerve axons (arrow) coursing through the tissue. (F) Transverse section through the wall of the aorta stained by van Gieson's method for elastin. The endothelium (left) contains a thick bed of collagen fibers that stain bright pink with eosin. The muscular layers (right) exhibit numerous concentric bands of elastin (arrow) that are stained dark violet to brown by this method. (A, B) Bars = 10 μm. (C–F) Bar = 20 μm.

toluidine blue stains most organelles and provides a good general layout of the tissue structure. As an alternative, one can use a commercial mixture of stains to provide an aesthetically pleasing record of specimen structure at the light microscopy level (Figure 3.6B). Based on what is seen, one might decide whether to trim the block further or go straight to thin sectioning. In either case, the final block face used for producing thin sections should be no larger than 0.5 mm on the top and 1.0 mm on the bottom.

To produce thin sections, the microtomist chooses an appropriate portion of the knife edge (particularly important with glass or diamond knives that may have damaged areas) and carefully advances the knife toward the block face. The lighting must be adjusted to produce a reflection off the smooth block face so that the shadow of the approaching knife edge can be clearly seen. Fortunately, modern microtomes have excellent systems for illuminating the specimen and knife from both the front and overhead. Proper manual adjustment of the knife edge to make it parallel to its shadow and to make the shadow as thin as possible without actually contacting the block is a skill that the beginning microtomist must practice (Figure 3.4I). After approach of the knife edge, the block is passed through successive cutting cycles, each cycle bringing the block closer to the knife edge (Figures 3.4J and 3.4K). Cycles can be carried out manually using a rotary handle on the side of the ultramicrotome, or they can be carried out automatically with a motor. Although motor speed can be controlled, some microtomists prefer to "feel" the knife cutting through the block by manual control.

Each section as it is cut slides off onto the surface of the water in the trough. The angle of the water reaching the knife edge can be critical to success. Filling of the trough with double-distilled water by syringe and then removing just the right amount before sectioning are other skills that must be learned. The water angle is assessed by reflection. With too much water, the block face may become wet during cutting; this necessitates drying the block and often a reapproach with the knife. With too little water, the sections will not float onto the trough and may get crumpled at the knife edge. Some practitioners find it easier to achieve the right water angle if surface tension is lowered with a small drop of Tween 20 detergent (one drop per 100 ml). This can be particularly helpful when the knife's edge is difficult to wet. The cleanliness of the water is of further importance. Indeed, cleanliness throughout all electron microscopy procedures is of utmost importance if one is to have a specimen that is free from dirt or grease from your fingers that shows up as boiling oil when hit with the electron beam during microscopy. In addition, the characteristics of the room and work area in which an ultramicrotome is used are very important. Generally, the machine should be isolated from vibration and located in a draft-free area as much as possible. Filter paper, a source for clean double-distilled water, high-quality, fine-tip forceps, Kimwipes, and an eyelash brush for positioning thin sections that are floating on the surface of the water should be close at hand.

The thickness of sections can be selected on the microtome and verified by their interference colors because only the wavelength of light that will reflect off of the top and bottom surfaces of the section and undergo positive interference will predominate. Silver sections are just the right thickness for electron microscopy (about 60 to 80 nm thick), whereas gold sections (about 80 to 100 nm) are still usable. Violet, green, and red sections (120 to 200 nm) are too thick for electron microscopy but can be used for light microscopy. Sections that are thinner than 60 nm grow increasingly less silver and become gray to colorless. Silver-gray sections are about 40 nm in thickness and can offer higher resolution but are difficult to cut and are prone to being damaged by the electron beam.

Sections floating on the trough water generally suffer from compression. They can be "expanded" or "relaxed" back to their original dimensions by wafting vapors of carbon tetrachloride over the trough. Individual or ribbons of sections can then be

Table 3.4 Types of Grids for Electron Microscopy			
Type	**Material**	**Use**	**Comments**
Bare, 200 mesh	Copper	Standard	Good compromise between support and viewing area; copper can be poisonous to live cells
Formvar-coated, 100 mesh	Copper; Formvar coating	Fragile or small sections or replicas needing support	Excellent support and viewing area; some loss of contrast
Hexagonal mesh	Any	Standard	More support for same amount of viewing area compared to square mesh
Bare or Formvar coated, 200 mesh	Gold	For tissue culture cells; nonpoisonous	Cells can be grown directly on grid
Bare or Formvar coated, 200 mesh	Nickel	For tissue culture cells; less poisonous	More costly than copper but less costly than gold
Slot grid; Formvar coated	Copper	For serial sectioning	Ultimate in viewing area (100%), less support
Numbered grid squares	Any	For relocation of specific specimen features	Expensive; grid location monitor on microscope can provide similar information as an alternative
Bare	Molybdenum	Elemental analysis by x-ray spectroscopy	Expensive, but x-ray spectrum conflicts less with metals in specimens

picked up on copper, gold, or nickel grids for observation. Sections on the water surface can be maneuvered into position using an eyelash brush (Figure 3.5A). Then the grid, held by forceps (Figures 3.5A and 3.5B), is brought face down on the sections and lifted away, or it can be dipped into the water, brought up under the sections, and lifted out of the trough (Figures 3.5C and 3.5D). Filter paper is used to wick away the excess water, and the grid is placed in a storage box until staining (Figure 3.5E). Alternatively, sections can be picked up in a loop of water and placed on a grid while water is wicked away by a filter paper support. Grids come in many patterns and materials, but all grids are 3 mm in diameter and stamped out of a thin metallic foil. Variations on grid geometry and coverings are listed in **Table 3.4** and are illustrated in Figure 3.5F. The size of grid openings, usually square or hexagonal, are designated by mesh size—the number of openings per inch—and usually range from 100 to 400. Larger openings (100 mesh) give more viewing area, but smaller openings (400 mesh) give more support for the sections. Because copper is poisonous to cells and some enzymes, work involving cytochemistry, immunocytochemistry, or culturing of cells on grids is usually done with gold or nickel grids. Regardless of the type of grid, oxidation or oils from their surface must be cleaned before using them to retrieve sections. If this is not done, sections will likely not stick very well. Grids are usually cleaned by sonication in acetone for 5 minutes before each use, allowed to dry on clean filter paper, and stored covered in a Petri dish.

Increased support for sections can also be obtained by coating grids with thin plastic films through which electrons can pass. The most common coating is Formvar, prepared as a stock solution in ethylene dichloride. A dry, clean glass slide is dipped into the Formvar solvent, drained of excess, and allowed to air dry with the plastic hardening into a thin sheet. The Formvar sheet is then floated off onto the surface of a trough of clean water. Grids are individually placed on the plastic sheet using forceps, and the sheet is rolled up on a clean plastic centrifuge tube and allowed to dry. Grids with Formvar coating can then be broken out of the sheet and used immediately or coated with evaporated carbon on either their front or back side for further strength. Most carbon evaporation units found in electron microscopy laboratories consist of a vacuum chamber in which a carbon electrode is placed (Figure 3.5G). One electrode is flat, and the other electrode is sharpened to a point, the point pressing against the flat end of the other electrode in a spring-loaded holder (Figure 3.5H). When high voltage is applied to pass current, the pointed electrode heats up to over 5000°C, and atomized carbon is showered onto the specimen about 20 cm away.

An alternative route to embedding and sectioning is the freezing of tissue and its sectioning at low temperature without thawing. The standard procedure (**Figure 3.7A**) is to fix the tissue with formaldehyde, wash it free of fixative, infiltrate with sucrose, which acts as a cryoprotectant, and then place it on a metal stub in a bed of OCT freezing medium, a viscous hydrophilic gel containing glycerol and resin polymers that after freezing forms a block that does not break easily during sectioning. The stub with OCT-covered specimen is then frozen in liquid nitrogen or in a fluorocarbon cryogen and mounted within the chamber of a refrigerated microtome held at −40°C (Figures 3.7B and C). The block is sectioned with a cold metal knife, and the dry frozen sections are transferred from the knife to a slide using a chilled artist's brush. After thawing, the section is further processed, usually by immunocytochemistry. Specimen proteins, not having undergone dehydration, better retain their antigenicity and label more efficiently with antibodies by this procedure. Ultrathin frozen sections can also be cut on an ultramicrotome modified to chill the knife and specimen holder with liquid nitrogen (Figure 3.7D). The frozen block is mounted and cut with a glass knife, and then sections are transferred to a grid for further processing. An additional consideration of frozen sectioning for EM immunocytochemistry is provided in Chapter 13.

■ Staining

Light and electron microscopy use inherently different stains. In the former, stains that either absorb light (they are colored) or fluoresce (they give off light) are used. These stains would be of no use in electron microscopy because most are organic compounds that contain carbon, oxygen, and nitrogen—atoms that have no ability to scatter or stop electrons. Instead, electron microscopists must use compounds that contain heavy metals such as uranium, lead, osmium, and bismuth that can easily deflect electrons because of their massive nuclei; however, an important feature of both LM and EM stains is that they be somewhat selective in what they stain. Some tissue structures should be stained strongly and others weakly because of the ability of stains to bind to some chemical constituents but not others. Ionic bonding between stains and the structures they interact with can play an important role in this selectivity.

Staining of Paraffin Sections for Light Microscopy

Tissues embedded in paraffin are inaccessible to most stains, and therefore, the paraffin must be removed beforehand, usually by a harsh, nonpolar solvent such as xylene. After a 30-minute wash in xylene, the tissue sections must then be reintroduced to an aqueous solvent in a stepwise manner usually starting with 100% ethanol and working downward until 100% water is reached. Now in water, the tissue sections can be exposed to one or more stains usually in solutions buffered at a specific pH. Stains for light microscopy are numerous, and only the most commonly used are listed in **Table 3.5**. Particularly useful

A. Protocol for Producing Frozen Sections for Immunocytochemistry

1. Chemically fix tissue with formaldehyde for 30 to 60 min.

2. Wash and infiltrate with 2.3 M sucrose. Place on stub with cork pad and cover with OCT.

3. Freeze in liquid nitrogen or in fluorocarbon.

4. Mount in cryotome prechilled with dry ice or ultramicrotome cooled with liquid nitrogen.

5. Cut frozen sections and transfer with artist's brush to a glass slide or grid.

6. Carry out immunocytochemistry procedure washing carefully between each step. Fix with 0.1% glutaraldehyde (optional).

7. Dehydrate in graded ethanols and mount using mounting medium, or for electron microscopy infiltrate with resin.

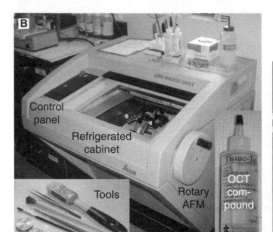

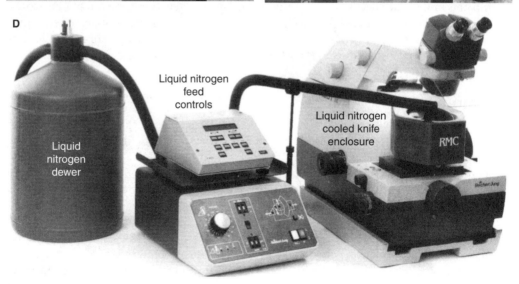

Figure 3.7
Cutting frozen sections. (A) Procedure summary. (B) Cryomicrotome for cutting frozen thick sections. Left inset shows brushes, forceps, and razor blade for trimming the frozen block and transferring the sections. Right inset shows bottle of OCT compound used as a "glue" and tissue embedding matrix that when frozen becomes hard and of high tensile strength. (C) Internal detail of the cryomicrotome within the freezer compartment. The specimen is frozen to the stub using OCT and moved forward across the edge of a cold metal knife (not shown) to cut a section with every turn of the rotary arm. (D) Liquid nitrogen cooling of an ultramicrotome used for cutting frozen thin sections.

Table 3.5 Noted Stains for Light Microscopy

Stain	Composition	Use	Comments
Eosin Y	0.5% in 90% ethanol	Acid stain for standard histology	Very popular as counter stain for hematoxylin
Orange G		Acid stain for standard histology	
Hematoxylin-alum (Ehrlich's, 1886)	Hematoxylin Potassium alum	Basic stain for standard histology	Very popular as part of hematoxylin and eosin staining
Methylene blue	1% in aqueous solution	Basic stain for standard histology	
Toluidine blue	1% in aqueous solution	Basic stain for epoxy thick sections	
Periodic acid-Schiff	Periodic acid-Schiff reagent (potassium metabisulfite and basic fuchsin)	Stain for neutral carbohydrates	Bright pink staining of mucus
Alcian blue	1% in aqueous buffer, pH 2.5	Stain for acid carbohydrates	Copper containing dye; excellent for bone and cartilage
Feulgen's stain	HCl hydrolysis Schiff's reagent	DNA stain	Mitosis
Methyl green	Methyl green Pyronin	Basic stain for chromatin	
Sudan black		Neutral lipids	Lipid droplets; must use frozen sections
Osmium tetroxide	1% in aqueous buffer	Unsaturated lipids	Lipid droplets; must use frozen sections
Mallory's azan	Azocarmine G Aniline blue Orange G	Trichrome stain for standard histology	Stains extracellular matrix blue
Masson's trichrome	Celestine blue Hematoxylin Phosphomolybdic acid Methyl blue	Trichrome stain for standard histology	Stains extracellular matrix green
Flemming's trichrome	Safranin O Crystal violet Orange G	Botanical specimens	
Wright/Giemsa	Eosin Y Methylene blue Methylene azures	Parasites and blood cells	Used for differential leukocyte counts in clinical hematology
Gram stain	Crystal violet Lugol's Iodine (I/KI) Carbol fuchsin/ Neutral red	Bacteria	Key stain for identification of bacteria
Golgi-Cox stain	Potassium dichromate/ Osmium tetroxide Mercuric chloride/ Silver nitrate	Neurons and glia	A variation is used to "silver stain" polyacrylamide electrophoresis gels
Cajal stain	Mercuric/Gold chloride	Astrocytes	

are staining protocols that use counterstaining—staining with two or more dyes each of which binds to different tissue structures. The standard example for such counterstaining is the use of hematoxylin and eosin in medical histology; we use this example to gain a better understanding of the process.

Like many dyes, eosin and hematoxylin bind to tissue structures by charge–charge interactions. Because negative and positive charges attract each other, a negatively charged dye will bind to a positively charged organelle or structure. Organic dyes that carry negative charge at a neutral pH are acids typically having carboxyl or sulfate groups and for this reason are called *acid dyes*. One such dye is eosin. These dyes bind preferentially to positively charged organelles such as secretory granules and mitochondria that contain a lot of positively charged proteins. As a group, these organelles and structures that bind acid dyes are referred to as *acidophilic*, meaning "acid loving." On the other hand, organic dyes that carry a positive charge at neutral pH are bases, generally having amino groups, and therefore are called *basic dyes*. One such dye is hematoxylin. These dyes bind preferentially to negatively charged organelles such as nuclei and endoplasmic reticulum because they contain DNA and RNA with many negatively charged phosphate groups. Those organelles and structures that bind basic dyes are referred to as *basophilic*, meaning "base loving."

If a tissue, therefore, is stained with both hematoxylin and eosin (referred to as H&E), basophilic organelles will be stained purple with hematoxylin, and acidophilic organelles will be stained pink to red with eosin. An image shown in Figure 3.6C of the large intestinal epithelium stained with hematoxylin and eosin demonstrates the ease with which tissue organization is distinguished in such preparations. Other common stains such as Mallory's azure and Masson's trichrome use three different dyes and achieve quite colorful results, staining cells in red and purple and extracellular materials blue or green.

Other stains do not rely on charge–charge interactions but on other chemical affinities or reactions. For example, Sudan black, a nonpolar dye, stains lipid droplets because of its affinity for hydrophobic environments. A common stain for detecting glycoproteins and sugar polymers is the periodic acid-Schiff (PAS) stain. This procedure specifically identifies sugars with vicinal hydroxyl groups. Periodic acid treatment of the tissue oxidizes any sugars with vicinal hydroxyl groups to aldehydes and breaks the sugar ring. Application of Schiff reagent links these aldehyde groups to the dye basic fuchsin. Tissues and cells containing large amounts of mucous, such as goblet cells in the large intestine, stain bright pink with this procedure (Figure 3.6D). Tissues of the nervous system are often stained with silver or gold whose salts bind tightly to proteins and often are observed to stain selectively neurons and nerve axons (Figure 3.6E). These stains, pioneered by Golgi and Cajal have been so important that these histologists were awarded Nobel prizes for their efforts at the turn of the 19th century. Other stains may react optimally with particular proteins such as van Gieson's stain, which binds to bands of elastin found in walls of arteries (Figure 3.6F).

Stains for light microscopy are numerous, and the protocols used for each can be quite specific. Before staining, one is best served by looking for exact methods in a microtechnique handbook, several of which are referenced at the end of the chapter. These are much like cookbooks, but are essential for producing the best results. Here we provide a general method for staining with hematoxylin and eosin to provide a feeling for the type of procedure involved:

1. Place sections in acidified alum hematoxylin for 20 minutes to achieve overstaining. The alum hematoxylin consists of 0.5% hematoxylin, 0.3% aluminum ammonium sulfate, and 0.6% mercuric oxide in 50% methanol that has been aged ("ripened") appropriately.

2. Wash with water to remove excess stain. Pass through 15%, 30%, and 50% ethanol, and then destain in 70% ethanol containing 0.5% vol/vol hydrochloric acid. Destaining time is typically 5 to 10 minutes, but this must be determined empirically.

3. Transfer into alkaline 70% ethanol (very dilute ammonium hydroxide added) to "blue" the stain, and then wash in 70% ethanol.

4. Place sections in 90% ethanol containing 0.5% wt/vol of eosin Y (pH carefully adjusted to 5.5) for 60 seconds or until overstained.

5. Rinse in 95% ethanol to destain (time determined empirically) and then in 100% ethanol.

Some staining procedures include dehydration (as in the one above for hematoxylin and eosin), whereas others take place entirely in an aqueous medium. If the latter, the slides are again dehydrated by a series of ethanol:water mixtures until 100% ethanol is reached. The stained sections are then passed through a clearing medium that acts as a solvent for the resinous mounting medium to follow. The mounting medium, a natural product such as Canadian balsam or a synthetic resin such as DMX, is applied, and a cover glass is laid in place, with care taken to avoid the trapping of air bubbles. The mounting medium gradually hardens to a clear transparent encasement, and the slide can be observed for many years without biological degradation. The mounting medium serves not only to preserve the specimen but also to bring the refractive index of the specimen layer to 1.45 to 1.55, about the same as the cover glass, thereby avoiding any refraction at the specimen/coverglass interface.

Staining of Epoxy Resin Sections with Heavy Metals for Electron Microscopy

Epoxy sections on grid are stained by floating the grid (section side down) on a drop of stain and then rinsing the grid thoroughly with water from a squirt bottle (**Figures 3.8A** and **3.8B**). A common means of staining resins that are moderately cross-linked (e.g., epon/araldite epoxy and acrylic-based resins) is with 2% uranyl acetate (wt/vol) in water

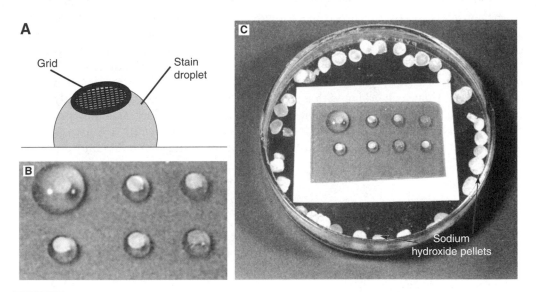

Figure 3.8
Staining EM grids. (A and B) Diagram and photograph of EM grids on drops of stain. (C) Petri dish with sodium hydroxide pellets for staining with lead citrate.

Table 3.6	Common Stains and Stain Enhancers for Electron Microscopy	
Stain	**Use**	**Comments**
Uranyl acetate	En bloc and/or on grid stain	Excellent stain for membranes and nucleic acids
Lead citrate	On grid stain	Excellent for protein containing organelles
Bismuth subnitrate	On grid stain	Excellent overall stain
Ruthenium red	Basic stain for extracellular matrix	Usually added during aldehyde fixation
Tannic acid	Mordant that increases binding of osmium tetroxide	Usually added during aldehyde fixation

for 15 minutes. For more highly cross-linked resins such as Spurr's epoxy resin, aqueous uranyl acetate is often not adequate because of the lack of penetration into the section. In this case, uranyl acetate salts are dissolved in ethanol or methanol. Alcohol-based stains are typically used at a 2% (wt/vol) saturated solution dissolved in 50% ethanol, and staining times range from 5 to 15 minutes. After washing, the grids are then stained with lead citrate in aqueous medium for 5 minutes and washed again. One common problem in staining is the deposition of stain precipitate; this is avoided by microfiltration of staining solutions by syringe immediately before use. Another precaution is to avoid adsorption of CO_2 from the air by the lead citrate stain that can result in production of lead carbonate precipitates. Staining with lead is usually done inside a Petri dish with a few sodium hydroxide pellets added to adsorb the CO_2 (Figure 3.8C). Common stains and stain enhancers for EM are provided in **Table 3.6.**

■ Methods That Facilitate the Overall Tissue Processing

If one considers the overall procedure for preparing tissues for light and electron microscopy, it is apparent that there are numerous passages through solutions and that solution changes need to be convenient and without loss of small specimens. Typically, the specimen is placed in small, tightly capped vials where they stay put as solutions are changed. Glass Pasteur pipettes are used both to remove the old solution and to add the new solution. If one is working with isolated cells, these can be centrifuged gently to the bottom (or allowed to settle by gravity) before changing the fluid. If this is not easily done, then fixed cells may be pelleted; the pellet is mixed with warm 1% agar, and the agar is allowed to cool. The agar, with cells, is cut up into 1-mm cubes and processed much like a cube of tissue.

■ Nonstandard Methods for Tissue Processing

Recently, microwave heating has been used to greatly shorten the time for tissue processing. Fixation times can be reduced by 75%, and infiltration and curing of epoxy resins can be shortened by a similar amount. This can reduce processing time from 3 or 4 days to a single day.

Another useful development is automated tissue processing units that can fix, dehydrate, infiltrate, and embed tissues using a preset time sequence. Likewise, they can also be used to automate multistep staining procedures in histology. Automated units are often available in clinical laboratories with a high throughput of specimens but are not always present in smaller settings.

■ Artifacts of Tissue Processing

Common artifacts that are created by tissue processing and are not a part of the specimen structure are illustrated in **Figure 3.9.** First, one must learn to recognize artifacts produced by tissue fixation and dehydration. Aldehyde fixatives do not stabilize membrane structures very well, and at the electron microscopic level, this can be readily seen as membrane wrinkles and blebs. Likewise, cold alcohols used as fixatives can disrupt and remove membranes. Neither of these effects is apparent at the LM level but can be seen at the electron microscopy level (see Chapter 10 for examples). Second, materials that are poorly fixed can be extracted from cells during processing. Carbohydrates such as mucous are often extracted by aqueous fixatives, whereas fats and lipids are extracted by nonpolar solvents during dehydration and embedding. An example of this can be seen in Figure 3.9A (arrows) in which secretory granule contents have been dissolved leaving vacant spaces within the cells.

Dehydration, if carried out too quickly, can result in shrinkage of tissue components, producing tears and empty spaces usually at the border between tissue layers (LM) or between organelles (EM) (arrows, Figure 3.9B). Inadequate dehydration can also result in faulty resin polymerization and a soft block. Such blocks do not section properly, and chunks of the section can even be knocked out, as shown in Figure 3.9C. Problems can also accompany sectioning of well-polymerized blocks. Vibration during sectioning can lead to "chatter"—alternating bands of section thickness resulting in a wave-like pattern, as shown in Figure 3.9D. Different problems arise if the knife edge has microscopic burrs, has dull regions, or simply has some dirt here and there. This leads to "knife marks" that are oriented parallel to the direction of the cut. Generally, these marks are linear and of almost any width and may be so deep that they actually cut the section into pieces (arrows, Figure 3.9E).

A frequent occurrence is the folding of sections as they are picked up on a grid or adhered to a glass slide (Figure 3.9F). Folds are usually easy to spot and cannot be mistaken for biological structure; they hide only information that is desired. On the other hand, cleanliness of the trough water is of utmost importance; otherwise, grease spots may show up in all areas of the section and occasionally be mistaken for specimen information (arrows, Figure 3.9G). A similar problem is the occurrence of dirt or stain precipitate because these can also involve extensive regions of every section (e.g., Figure 3.9H). All stains must be filtered through paper for light microscopy and through 0.22-μm pore filters for electron microscopy to avoid precipitates. As pointed out previously, precautions should be taken to avoid CO_2 in the atmosphere during lead staining to prevent lead carbonate precipitate.

■ Interpretation of Three-Dimensional Tissue Structure Using Two-Dimensional Sections

One challenge for the microscopist is to gain a three-dimensional understanding of a specimen's structure from two-dimensional sections. A rigorous way to do this is to cut a series of sections completely through the structure of interest and to look at every section. This, referred to as "serial sectioning," is very time-consuming and technically demanding. Each section must be picked up from the trough in the same order as cut, and no section must be faulty. If each section is photographed and the images are properly

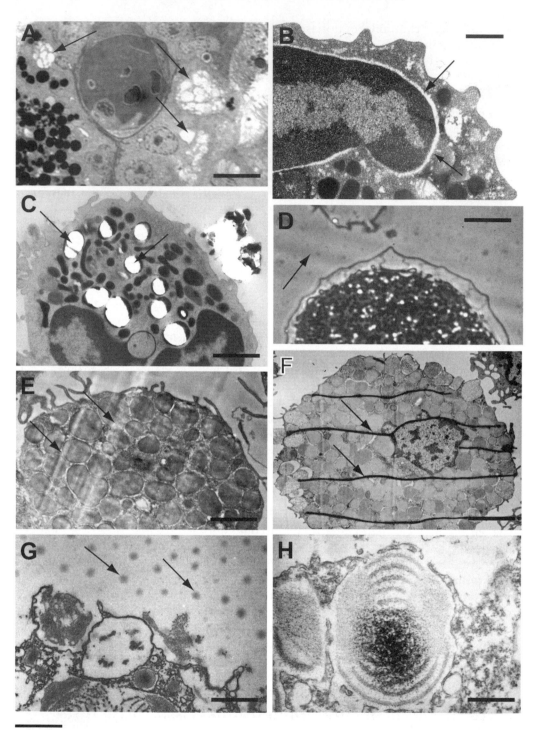

Figure 3.9
Common artifacts encountered in preparing sections for microscopy. (A) Clear areas (arrows) where materials have been extracted by solvents. (B) Tissue tear (arrows) caused by shrinkage during dehydration. (C) Inadequate embedding or hard tissue, components result in removal of material during sectioning (arrows). (D) "Chatter" marks (arrow) caused by vibrations during sectioning. (E) Knife marks (arrows) due to an imperfect knife edge. (F) Folds in the section (arrows) created during adhesion of section to coated grid. (G) Grease from fingers deposited on sections (arrows). (H) Stain precipitate caused by failure to filter stain before use or by stain precipitation during staining. (A–C and E and F) Bars = 2 μm. (D) Bar = 20 μm. (G) Bar = 1 μm. (H) Bar = 0.2 μm.

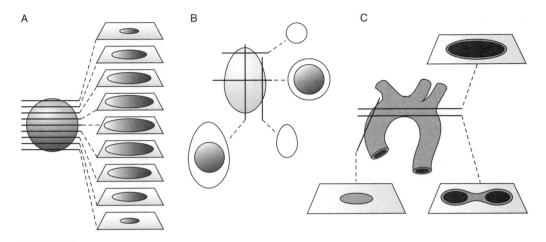

Figure 3.10
Three-dimensional interpretation of two-dimensional sections. (A) A stack of serial sections representing a three-dimensional structure. (B) The appearance of an egg in various sections. (C) The appearance of the aortic arch in sections at several depths and orientations.

registered in the x,y plane, a vertical stack of positional information can be constructed that will reproduce the structure of the specimen (**Figure 3.10A**). To facilitate this process, the microtomist must cut ribbons of sections, and each ribbon positioned within a slot grid so that all sections are visible. The order in which ribbons are collected and the thickness of the sections must be recorded. Production of a three-dimensional model from section images is usually carried out by dedicated computer software.

Because serial sectioning is an arduous process, most three-dimensional models of tissue structure are deduced informally by the microscopist. First, an important source of information is the combined interpretation of both light and electron microscopy images. Light microscopic sections that extend over millimeters, even centimeters, provide a low-resolution view of the "forest," much like a mapmaker would want. Electron microscopic images at high resolution provide the detail at the cellular and subcellular level, which allows the individual "trees" in the forest to be studied. Complex tissues composed of multiple cell types are often understood better if the microscopist is familiar with the same cell types in other tissues (i.e., an appropriate knowledge of histology is important).

Second, the microscopist must assemble two-dimensional information obtained from multiple angles and depths into one three-dimensional model. For example, as shown in Figure 3.10B, one could section an egg in many ways. In a transverse section, both the yolk and the surrounding white would appear circular, whereas in vertical and horizontal planes, the yolk would appear circular and the white would appear like a tapered ellipse (egg-like). Taken together, an experienced microscopist might envision the three-dimensional structure of this object without ever having seen an egg. Another example in Figure 3.10C shows that a bent tube (such as the aortic arch) might appear as a solid piece of tissue in a glancing section, an elliptical tube in a superficial section, and two tubes joined together in a deeper section.

Three-dimensional conceptualization is simpler if the planes of section are orthogonal to one another and are along natural axes of tissue structure. For this reason, tissues are often oriented before sectioning rather than being sectioned randomly. Orientation can be achieved at three different stages. First, the tissue may be grown, dissected or cut in an oriented manner. Second, the tissue can be oriented during embedding. To maintain or produce orientation, a flat mold may be advantageous for larger specimens, or embedding on the surface of a glass slide may prove useful for a thin specimen. Third, the embedded specimen can be oriented or reoriented before sectioning. This can be done by cutting out an epoxy-embedded tissue with a jig saw and gluing the specimen with correct orientation onto the end of a new epoxy block or by using a warm knife to trim and reorient a paraffin block before mounting on a stub.

Nevertheless, much three-dimensional thinking comes after the fact using tissues sectioned at random or oblique orientations. Here, the microscopist, through practice and experience, can construct mental models that are based on many different transects of the tissue structure. These "many views" are often incorporated into a single specimen. For example, if one is imaging a suspension of cells, each cell will be randomly oriented and sectioned differently. In viewing a single slide of solid tissue, one has at hand the raw material for envisioning a three-dimensional model because similar units of structure are sectioned at different depths and different orientations. With practice, you too will soon start thinking three dimensionally whenever looking at two-dimensional sections. Remember, however, that the model you envision must be confirmed or revised by sections at multiple angles.

The importance of three-dimensional information is often great enough to warrant a systematic approach to collection of such data rather than relying on the microscopist's intuition. In recent years, laser-scanning confocal microscopy at the light level and high-resolution tomography at the electron microscopy level, coupled with advances in computer reconstruction, have resulted in a revolution in the area of biological three-dimensional modeling. Electron tomography is analogous to medical computerized axial tomography (CAT-scan imaging) except that the work is on a far finer scale. It uses the transmission electron microscope to assemble multiple projections of an object that when combined produce a three-dimensional representation of the specimen. For organic and biological specimens, tomography enables the highest three-dimensional resolution (5-nm spatial resolution) of internal structures in relatively thick slices of material (i.e., 0.5 μm) without requiring the collection and alignment of large numbers of precisely oriented, serial thin sections. Tomography, coupled with cryoelectron microscopy of specimens embedded in vitreous ice can provide high-fidelity information about frozen hydrated samples such as macromolecular complexes, isolated organelles, small bacterial cells (approximately 0.5 μm) and cryosectioned materials of larger eukaryotic cells and tissues (see Chapters 15 and 16 for further information).

■ Storage and Documentation of Microscopic Specimens

Weeks, if not years, of work can be in vain if specimens prepared for microscopy are not stored and documented properly. The first step in documentation is to keep a laboratory notebook (preferably bound) describing how each specimen was treated experimentally (or the source of the specimen) and the procedures used for processing that specimen. Experiments should be identified by both consecutive numbering and date. Individual specimens within experimental groups should be clearly identified by number followed by a description of the parameters by which they differ from other specimens. In the computer age, an electronic notebook can be a convenient tool, but it in no way substitutes for a hardcopy record. It is far too easy to erase, alter, corrupt, or otherwise create problems in retrieving accurate information not the least of which is future outdating of the electronic document format or medium on which it is stored.

Second, embedded tissue blocks must be stored in a safe, organized, and labeled manner. Here labeling each block with experiment and specimen number is essential.

Third, all sections cut and picked up on grids or glass slides must be stored in boxes made for the purpose. Slide boxes typically contain slots for 100 slides with a numbered index; preferably, each slide and its index should be identified with experiment and specimen number. EM grid boxes contain diamond-shaped slots that are also numbered and/or lettered (Figure 3.5I). An index book for all grid and slide boxes should be kept, and for each experiment and specimen number, the box number and slot numbers of all grids or slides for that specimen should be listed. An additional notation should indicate those grids that have been stained. For example, the location entry for grids of specimen 3 in experiment 21 might read: #21/3 IIIA2UL,A3UL,A4,A5. In this entry, the Roman numeral III would stand for the grid box number, and A2, A3, A4, and A5 would indicate the slots having grids for that specimen. The designation of UL, U, or L at the end would indicate whether the grid had been stained with uranyl acetate or lead citrate.

Finally, the images taken of each grid (whether in film or electronic form) must be noted. Each exposure by a film or digital camera on a microscope is associated with a serial exposure number. Thus, in the same grid or slide log, the range of exposure numbers for that grid or slide should be recorded. Working prints made from each exposure should display the exposure number so that one can identify the specimen and grid/slide it came from. Electronic data files from many electron microscopes and computer-controlled light microscopes contain a number of operational parameters as well as the image itself. These are almost always in a proprietary format; thus, it is important to keep the essentials in hardcopy form as well.

References and Suggested Reading

General
Hammersen F. *Histology: Color Atlas of Microscopic Anatomy,* 3rd ed. Munich: Urban and Schwarzenberg, 1985:260.

Spector DL, Goldman RD. *Basic Methods in Microscopy: Protocols and Concepts from Cells, a Laboratory Manual.* Woodbury, NY: Cold Spring Harbor Laboratory Press, 2005:375.

Spector DL, Goldman RD, Leinwand LA. *Cells: A Laboratory Manual.* Woodbury, NY: Cold Spring Harbor Laboratory Press, 1997:2136.

Young B, Lowe JS, Stevens A, Heath JW, Deakin DJ. *Wheater's Functional Histology: A Text and Color Atlas,* 5th ed. Edinburgh: Churchill Livingston, 2005:448.

Light Microscopy Technique
Bancroft JD, Stevens A. *Theory and Practice of Histological Techniques,* 2nd ed. Edinburgh: Churchill-Livingston, 1982:662.

Galigher AE. *Essentials of Practical Microtechnique,* 2nd ed. Philadelphia: Lea and Febiger, 1971:531.

Horobin RW. *Conn's Biological Stains.* London: BIOS Scientific Publishers, 2002:502.

Locquin MV, Langeron M. *Handbook of Microscopy.* (Translated by Hillman H from the French Manuel de Microscopie, Masson, 1978). London: Butterworth, 1983:322.

Sanderson JB. *Biological Microtechnique (Microscopy Handbooks, No. 28).* London: BIOS Scientific Publishers, 1994:244.

Electron Microscopy Technique
Bozzola JJ, Russell LD. *Electron Microscopy: Principles and Techniques for Biologists,* 2nd ed. Sudbury, MA: Jones and Bartlett, 1999:670.

Glauert AM, Lewis PR. *Biological Specimen Preparation for Transmission Electron Microscopy.* Princeton, NJ: Princeton University Press, 1998:316.

Hayat MA. *Principles and Techniques of Electron Microscopy: Biological Applications,* 4th ed. Cambridge, England: Cambridge University Press, 2000:543.

Hunter EE. *Practical Electron Microscopy: A Beginner's Illustrated Guide,* 2nd ed. Cambridge: Cambridge University Press, 1993:185.

Kuo J. *Electron Microscopy: Methods and Protocols,* 2nd ed. (*Methods in Molecular Biology,* volume 369). Philadelphia: Humana Press, 2006:669.

Nasser Hajibagheri MA. *Electron Microscopy Methods and Protocols.* Philadelphia: Humana Press, 1999:283.

Reid N, Beesley JE, Glauert AM. *Sectioning and Cryosectioning for Electron Microscopy (Practical Methods in Electron Microscopy,* Vol. 13). Amsterdam, Netherlands, Elsevier, 1991:322.

Ruzin S. *Plant Microtechnique and Microscopy.* New York: Oxford University Press, 1999:336.

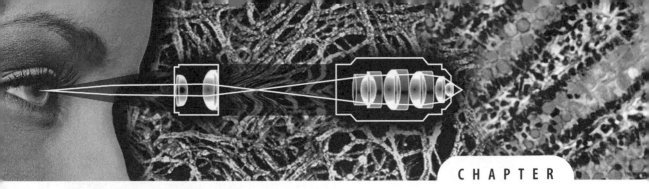

A Brief Introduction to Cell Structure

Atlases of tissue histology and cell structure and textbooks of cell biology are present in nearly all scientific libraries, and it is not our goal to reproduce these here. A few of the many available are referenced at the end of the chapter. Rather, this chapter is meant to serve as a short pictorial guide to the classic cell organelles that are likely to be encountered while being introduced to the field of bioimaging.

■ The Animal Cell

A typical animal cell is illustrated in **Figure 4.1** with many of its major organelles labeled. Organelles common to most animal, plant, and fungal cells are discussed in this section.

■ The Cell Surface and Cytoplasm

The Plasma Membrane

This unit membrane serves as a boundary between the cytoplasm inside the cell and the extracellular space outside. It controls the cytoplasmic ion and metabolite composition through its permeability properties and its numerous ion channel and transport proteins. The plasma membrane is the site of signal transduction mechanisms, including receptors for hormones and neurotransmitters. It is also involved in bulk movement of materials into and out of the cell by exocytosis and endocytosis. The membrane itself is composed of two monolayers of phospholipids back to back, creating a hydrophobic interior and two polar surfaces, one surface in contact with the extracellular space and the other in contact with the cytoplasm, as illustrated in **Figure 4.2A**. When treated with osmium tetroxide and uranyl acetate and observed in thin-section with transmission electron microscopy, the membrane takes on a three-layered ("railroad track") appearance because of staining of unsaturated phospholipids in each monolayer that are separated by a un-stained central layer representing the hydrophobic interior. Freeze fracture replicas provide a panoramic view of membrane architecture. In Figure 4.2B, the contours of a neutrophil plasma membrane are clearly defined in this platinum replica. At high magnification (Figure 4.2C), intramembrane particles (arrowheads) represent proteins that traverse the membrane bilayer.

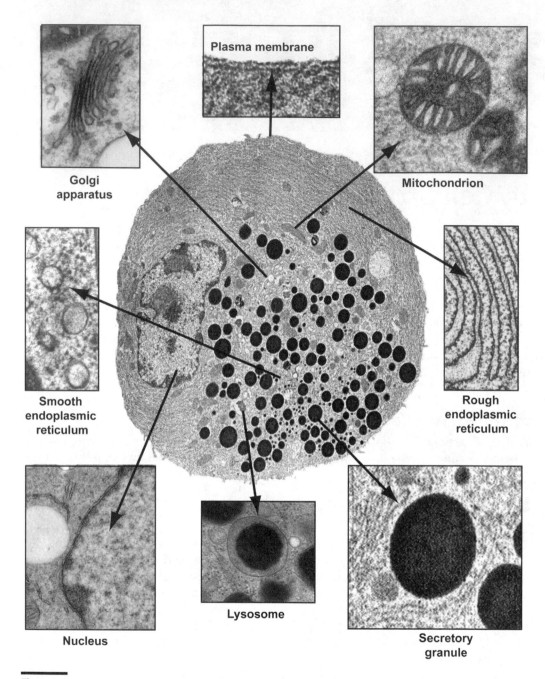

Figure 4.1
Organelles of the typical animal cell. An isolated pancreatic acinar cell at the center exhibits a number of the organelles found in all animal, plant, and fungal cells.

The Extracellular Matrix

Although some cells have relatively bare plasma membranes, others have extracellular matrices anchored to their surface. Classic examples include the cell walls of plant and fungal cells made of cellulose and chitin, respectively. These long-chain sugar polymers, along with associated proteins, usually appear as a network of fibers in thin sections

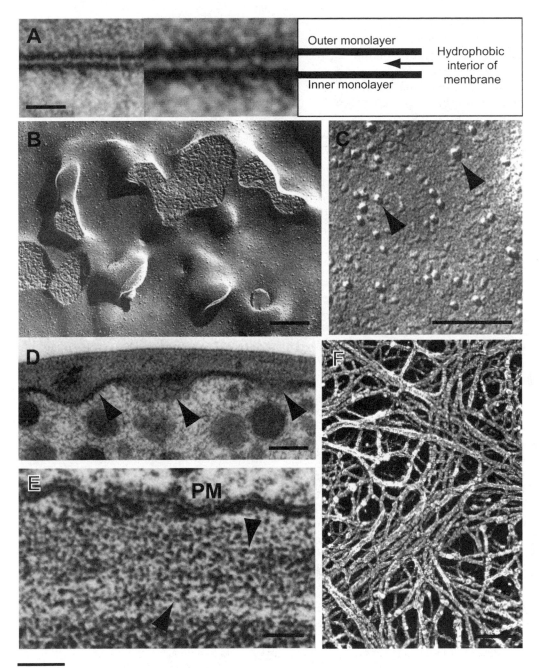

Figure 4.2
The plasma membrane and cell surface. (A) The plasma membrane is a typical bilayer membrane that stains in a trilaminar ("railroad track") pattern. Two osmophilic regions, one in each monolayer, are separated by the unstained hydrophobic interior. (B) Aerial view of a neutrophil plasma membrane as seen in a freeze fracture replica. Undulations of the membrane are easily observed. (C) The replica in (B) viewed at high magnification reveals transmembrane proteins seen as "intramembrane particles" (arrowheads). (D) Transmission electron micrograph of an ultrathin section of a fungal cell plasma membrane and cell wall. New membrane and cell wall materials are delivered by spherical exocytotic vesicles to the cell's surface where they undergo fusion with the existing cell membrane. Arrowheads point out regions along the plasma membrane where exocytotic events have occurred. (E) The cell wall of a plant cell is composed of cellulose fibrils (arrowhead) and lies in contact with the plasma membrane. (F) Fibrous extracellular matrix of a frog egg as visualized by quick freezing, deep etching, and rotary platinum shadowing. Bars = 20 nm in (A), 200 nm in (B) and (C), 100 nm in (D) and (F), and 40 nm in (E).

(Figures 4.2D and 4.2E). Cell wall synthetic enzymes and polymers are delivered to the plasma membrane by transport vesicle fusion (arrowheads, Figure 4.2D) or are synthesized at the plasma membrane surface. Individual cellulose fibers of the plant cell wall can be seen either in thin sections (arrowheads, Figure 4.2E) or by scanning electron microscopy. In animal cells, and some fungi, the extracellular matrix is composed of a mixture of fibrous proteins such as collagen and high molecular weight proteoglycans that contain complex polysaccharide chains in addition to protein. These networks are synthesized by the cells themselves and function to adhere cells to substrates, one another, organize tissues, bind signaling molecules, and act as a support on which immune cells can move and fight infections. Not merely static, this matrix is continually remodeled by the cells it serves. As shown in Figure 4.2F, the extracellular matrix of a frog egg that serves to bind sperm before fertilization consists of a network of fibers with several diameters. The tissue was prepared by freeze fracture, deep etching, and rotary platinum shadowing of a frog oocyte coat (for methods refer to Chapter 10).

Membrane Junctions

Plasma membranes of neighboring cells are often joined at junctions. These structures provide functional links between neighboring cells, particularly those making up epithelia (**Figure 4.3A**). Tight junctions isolate the apical surface of epithelial cells from their basal and lateral surfaces, thereby preventing movement of molecules between cells. In freeze-fracture replicas, tight junctions are shown to consist of protein ridges composed of integral membrane proteins (arrowhead, Figure 4.3B). Ridges can range in number from a few to over a dozen, the number of ridges determining the "tightness" of the junction and its ability to retard movement of solute molecules between the neighboring cells. Zona adherens junctions, composed of cadherins (calcium-sensitive adhesion proteins), link the actin-containing microfilament network of one cell with that of neighboring cells, thereby creating a multicellular web that can contract or expand to produce shape changes in the epithelium. Desmosomes consist of adhesion proteins that link the intermediate filament networks in each cell. These junctions, like the intermediate filaments themselves, are mechanically strong, in fact, stronger than any other components in the cell. Desmosomes are found in profusion in tissues that must resist mechanical stress such as the epidermis and cardiac muscle. In freeze-fracture replicas, desmosomes are revealed by dense plaques of large intramembrane particles that represent desmogleins, the linking proteins. Tight junctions, zona adherens and desmosomes are often found together near the apical surface of epithelial cells and in such combination are termed "junctional complexes" (Figure 4.3A). The neighboring cells of most solid tissues are linked by a fourth type of junction, termed a gap junction. Gap junctions consist of an array of connexon proteins forming tunnels that traverse two plasma membranes to provide ionic continuity between the cytoplasms of adjacent cells. In freeze-fracture replicas, connexons are represented by intramembrane particles arranged in dense plaques (arrowhead, Figure 4.3C). Gap junctions are used for passage of metabolites and ionic signals between cells.

Figure 4.3D summarizes the differences between these junctions diagrammatically. Tight junctions form ridges between closely apposed membranes. Zona adherens proteins link microfilament networks in adjacent cells. Desmosomes link intermediate filament networks in adjacent cells, and gap junctions provide tunnels connecting cell cytoplasms.

The Cytoskeleton

The cytoplasm of all cells is filled with a rich network of filaments. These filaments are of three types that serve different purposes. The microfilament, about 6 nm in diameter and composed of a double helix of actin, is involved in contractile interactions with myosin and with formation of bundles and three-dimensional networks that

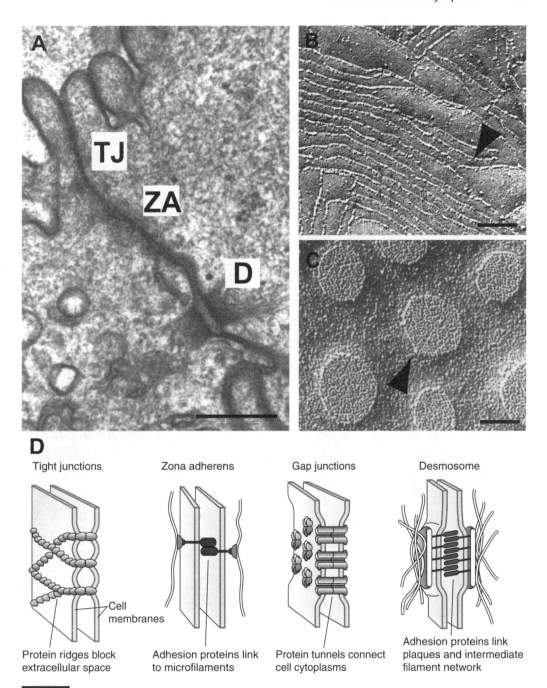

Figure 4.3
Membrane junctions between epithelial cells. (A) Apical and lateral surfaces of adjacent epithelial cells are segregated by a series of membrane junctions that link together the two cell plasma membranes. Tight junctions (TJ) block the passage of solutes between the cells. Zona adherens (ZA) link the microfilament cytoskeleton of the two cells. Desmosomes (D) link the intermediate filament networks of the two cells. (B) Freeze-fracture replica revealing the multiple protein ridges (arrowhead) that make up the tight junction. These ridges are interconnected as shown in (D), left panel. (C) Gap junctions appear as high-density plaques of intramembrane particles in freeze fracture replicas. (D) Diagrammatic representations of cell junctions. Bars = 0.3 μm in (A), 150 nm in (B) and (C).

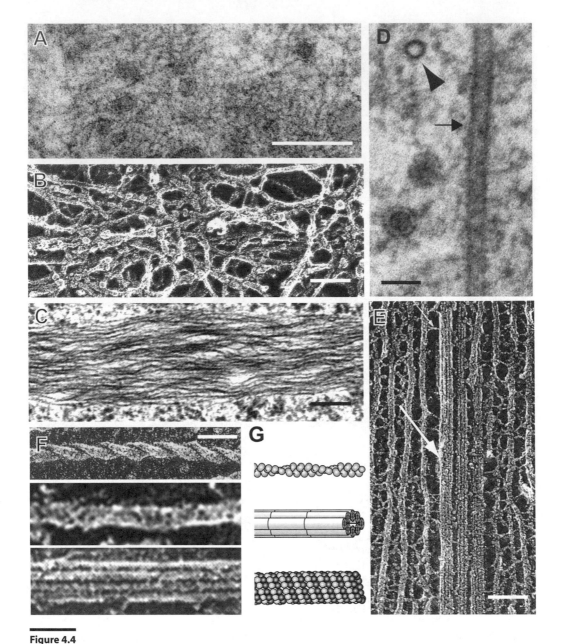

Figure 4.4
The cell cytoskeleton. (A) Thin-section micrograph illustrating a network of microfilaments in the cytoplasm of a fungal cell. (B) Networks of microfilaments in the cytoplasm of a neutrophil as revealed by deep etching and rotary platinum shadowing. (C) Intermediate filament bundles in the cytoplasm of a fungal cell. (D) A single microtubule (arrow) as seen in thin section of a fungal cell cytoplasm. Note the microtubule in cross-section (arrowhead). (E) The highly polarized cytoskeleton of a nerve axon contains microtubules (arrow) and intermediate filaments. (F and G) Deep-etched and rotary-shadowed cytoskeletal elements demonstrating their relative diameters and substructure. A myosin head-decorated microfilament (upper panel), an intermediate filament (middle), and microtubule (lower panel) are shown both microscopically in (F) and diagrammatically in (G). Myosin head decoration of the microfilament reveals its double helical nature. Bars = 0.3 μm in (A), 60 nm in (B) and (D), 150 nm in (C), 80 nm in (E), and 25 nm in (F).

shape the cell in a dynamic manner (**Figures 4.4A** and **4.4B**). Intermediate filaments (Figure 4.4C), about 10 nm in diameter, form strong, static networks that can withstand large forces. Thus, intermediate-filament networks help to join epithelial cells together in tissues exposed to deformation and mechanical stress. Although not dynamic in nature (as are microfilament and microtubule networks), they also serve to position organelles such as the nucleus and to link epithelial cells to the underlying basal lamina and connective tissue. The microtubule, about 24 nm in diameter, consists of a cylinder of α-tubulin and β-tubulin (arrow, Figure 4.4D). Microtubules are rigid and serve as tracks on which organelles and vesicles can move. They also form temporary structures such as asters and the mitotic spindle, which can transmit force to move cellular components and cells themselves. Microtubules and microfilaments can be assembled and disassembled as needed.

Platinum replicas of rotary-shadowed and deep-etched cells have been used extensively to visualize cytoskeletal networks and the individual components. Figure 4.4E reveals the parallel array of microtubules (arrow) and intermediate filaments that course down the core of a nerve axon. Figure 4.4F demonstrates the double helix of myosin-decorated actin in a single microfilament (top panel), a single intermediate filament (middle panel), and the individual protofilament chains of tubulin in a single microtubule (bottom panel). Scale diagrams of each filament architecture are shown in Figure 4.4G for comparison. At the light-microscopy level immunocytochemistry employing anti-actin and anti-tubulin antibodies is used extensively to map cytoskeletal organization (see Chapter 13).

Cell Shape and Motility

The cytoskeleton is also a prime determinant of cell shape. Microvilli, finger-like extensions of the cell surface (**Figure 4.5A**), contain a core of bundled microfilaments. These microfilaments run parallel to the axis of the microvillus (arrow, Figures 4.5C and 4.5D) and are linked by protein crossbridges. Cross-fractured microvilli in deeply etched and rotary platinum-shadowed specimens reveal an array of microfilament stubs representing these bundles (Figure 4.5B). Networks of these filaments are also seen in sheet-like extensions of the cell surface (Figure 4.5C). Cell motility is also served by cytoskeletal elements. Cells that move by amoeboid crawling such as the neutrophil extend pseudopods at their leading edge (**Figure 4.6A**) filled with dynamic microfilament networks as demonstrated in replicas of the sheared open cell cortex (Figure 4.6B). Likewise, microtubule doublets form the axoneme—the dynamic superstructure of cilia and flagella that propels rapidly swimming cells such as protozoa and sperm. At the light microscopy level, flagella and cilia appear phase dense, as demonstrated by the single flagellum of a mouse sperm in Figure 4.6C. At the EM level, longitudinal thin sections demonstrate darkly stained microtubule doublets coursing from tip to base that make up the axoneme (Figure 4.6D). Viewed diagramatically in cross-section (Figure 4.6E), the axoneme typically exhibits nine outer microtubule doublets and a central pair of microtubules. Adjacent doublets, by virtue of the forces generated by dynein motor proteins, slide against one another to produce rhythmic bending patterns that propel the cell. Also shown is that the outer doublets are linked to the central microtubule pair by radial spoke proteins, thereby converting the sliding motion of adjacent doublets into a bending motion of the flagellum.

The Cytoplasm

In addition to membrane bound organelles, the cytoplasm harbors a number of structures, including (but not limited to) the centrosome, ribosomes, proteosomes, glycogen, and fat droplets.

Centrosomes   These organelles act as microtubule organizing centers from which arrays of microtubules radiate. At their core is a pair of barrel-shaped centrioles consisting of an array of microtubule triplets surrounded by a cloud of proteins required for anchoring

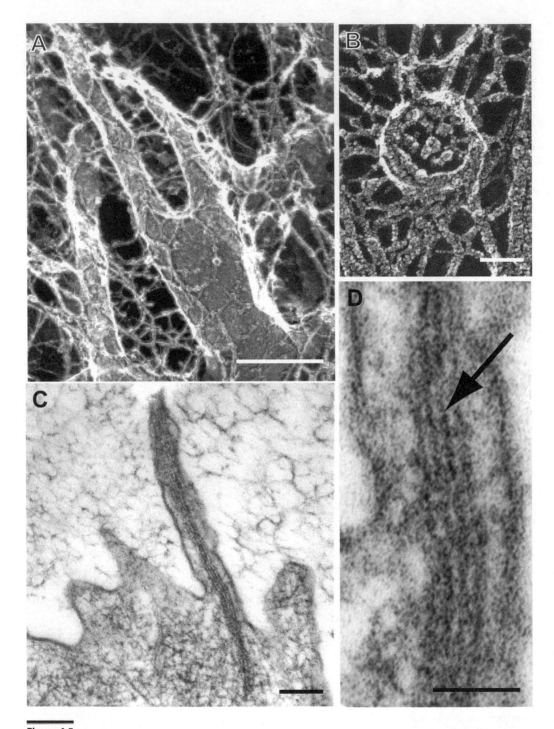

Figure 4.5
The shapes of microvilli are determined by the cytoskeleton. (A) Microvilli on the surface of a fertilized sea urchin egg as seen in a rotary-shadowed platinum replica. (B) A cross-fractured microvillus on the surface of an unfertilized frog egg reveals its cytoplasmic core of cytoskeletal filaments. (C) An internal bundle of microfilaments runs the length of a sea urchin egg microvillus. (D) Detail of (C) showing the microfilament bundle (arrow). Bars = 1.0 μm in (A), 0.2 μm in (B) and (C), and 80 nm in (D).

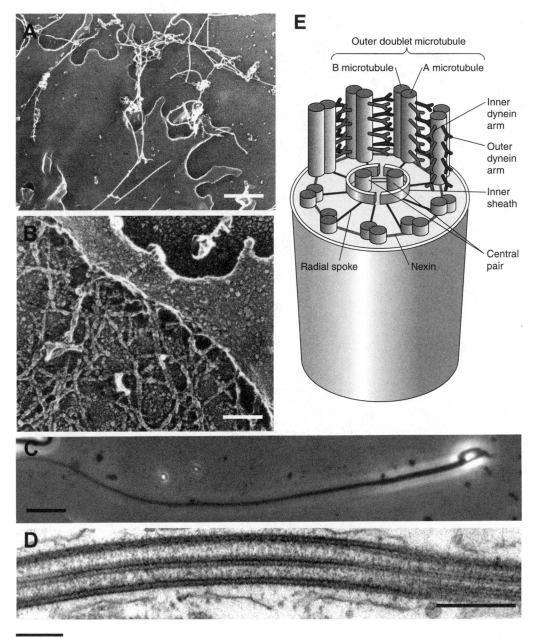

Figure 4.6
Pseudopods, cilia, and flagella—motile structures formed by the cytoskeleton. (A) The leading edge of a locomoting neutrophil that has been freeze dried and rotary platinum shadowed. (B) A network of microfilaments is seen at the leading edge of a neutrophil that has been sheared open. (C) A mouse sperm in phase contrast exhibits a long flagellum. (D) Flagellum of a fungal zoospore exhibiting longitudinal microtubules doublets that make up the axoneme. (E) A three-dimensional diagram of axoneme structure. Bars = 0.3 μm in (A), 60 nm in (B), 5 μm in (C), and 0.4 μm in (D).

of microtubules (**Figure 4.7A**). During interphase, the centrioles act as an organizing center from which numerous microtubules radiate (large arrows, Figure 4.7B). These microtubule networks serve as tracks on which vesicles and organelles are moved (arrows, Figure 4.7C), thereby providing means for transport of proteins and membrane within the cell. At high magnification (arrow, Figure 4.7D), one can see a putative motor protein linking the vesicle to the microtubule, a mechanoprotein that provides motive force through ATP hydrolysis.

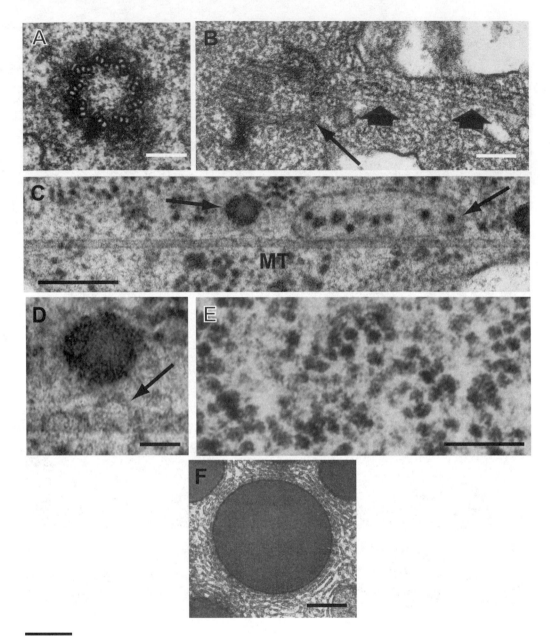

Figure 4.7
Cellular components in the cytoplasm. (A) Cross-section of a centriole showing an array of microtubule triplets and the surrounding cloud of proteins that make up a centrosome, a nucleation site for microtubules. (B) Microtubules (large arrows) radiating out from the perinuclear centrosome (thin arrow) of a neutrophil. (C) Microtubules act as tracks for the movement of numerous transport vesicles (arrows) found in the cytoplasm of this fungal cell. (D) Detail of (C) showing a transport vesicle attached to the microtubule via a putative motor protein (arrow). (E) Thin section showing numerous osmophilic ribosomes, the site of protein synthesis. (F) A fat droplet surrounded by smooth endoplasmic reticulum in a steroid secreting cell. Fat droplets are filled with triglycerides, surrounded by a phospholipid monolayer and serve as a storehouse of these calorie-rich substrates. Bars = 100 nm in (A), 150 nm in (B), 0.2 μm in (C), 50 nm in (D), 0.4 μm in (E), and 0.3 μm in (F).

Ribosomes Too small to be seen at the light microscope level, these enzymatic workbenches for protein synthesis appear as electron-dense dots in electron micrographs (Figure 4.7E). Ribosomes are found attached to the endoplasmic reticulum where they manufacture membrane and secretory proteins and free in the cytoplasm where they synthesize nuclear, peroxisomal, mitochondrial, chloroplast, and cytosolic proteins. Proteosomes (not shown) are small cylindrical structures that specifically degrade proteins that have been marked by ubiquination. Entry into the cylinder where proteolysis takes place is limited to ubiquinated proteins, thereby protecting other cytosolic proteins from degradation. In many cells, the cytoplasm is the site of nutrient provisions in the form of glycogen (a glucose polymer), triglycerides, or lipoproteins. Glycogen, stored in liver, muscle or white blood cells forms osmophilic clusters, whereas triglycerides are stored as fat droplets surrounded by a monolayer of phospholipids (Figure 4.7F). Similarly, lipoproteins, stored as yolk platelets in eggs, serve to nourish the new embryo after the egg is fertilized.

■ Cell Organelles

The Nucleus

The largest organelle, typically 3 to 5 μm in diameter, houses chromosomes consisting of the DNA that encodes the genome arrayed on a protein scaffold. In stained sections at the light microscope level, this organelle is usually round and found at the cell center or toward the base of epithelial cells (arrowhead, **Figure 4.8A**). Differential interference optics usually do not reveal substructure in the nucleus, although the nucleolus (a region in the nucleus specialized for making ribosomal subunits) can be quite prominent (arrowhead, Figure 4.8B). On the other hand, phase-contrast optics often reveals both the pattern of heterochromatin and the nucleolus (Figure 4.8C). DNA-binding stains such as 4′,6-diamidino-2-phenylindole (DAPI) and Hoescht 33342 light up nuclei (Figure 4.8D) and are often used in fluorescence microscopy to reveal the location of nuclei in tissues stained for specific proteins by immunocytochemistry. In transmission electron microscopy, the nucleus is surrounded by a nuclear envelope composed of two unit membranes (Figure 4.8E). Within the nucleus, darkly stained areas of condensed chromatin (heterochromatin) line the envelope, whereas lightly stained regions of euchromatin are found throughout the interior. The heterochromatin is condensed and transcriptionally inactive, whereas euchromatin is decondensed and actively undergoing transcription. Also present is the nucleolus, which represents a site of rRNA transcription and ribosomal subunit assembly (arrowhead). Nucleoli are particularly prominent in metabolically active cells and can be seen in both light and electron micrographs. The nuclear envelope itself contains numerous pores that provide a passageway for protein and RNA transport between the cytoplasm and the interior of the nucleus. "En face" views of these pores in thin sections show that they are circular, occur at high densities throughout the envelope, and join the inner and outer unit membranes of the envelope (Figure 4.8F). Pores appear to have material within (arrows, Figure 4.8G), and in "en face" views (Figure 4.8F), this superstructure consists of an octagonal protein complex that controls the size and identity of substances passing through.

The Mitochondrion

Known as the site of ATP production and fatty acid oxidation, these thread-like organelles are 1 to 5 μm in length and have an inner and outer membrane (**Figure 4.9A**). The inner membrane contains numerous infoldings, termed cristae, that represent the site of electron transport, a process that acquires electrons from substrates and passes them to oxygen to produce water in a series of energy yielding steps. The energy is temporarily stored as a proton and voltage gradient across the inner membrane and is then used to synthesize ATP from ADP. Alternatively, the voltage gradient can be used to transport calcium into the mitochondrial matrix, where it is often deposited as granules of calcium phosphate. Mitochondria come in many shapes and sizes depending on cell type.

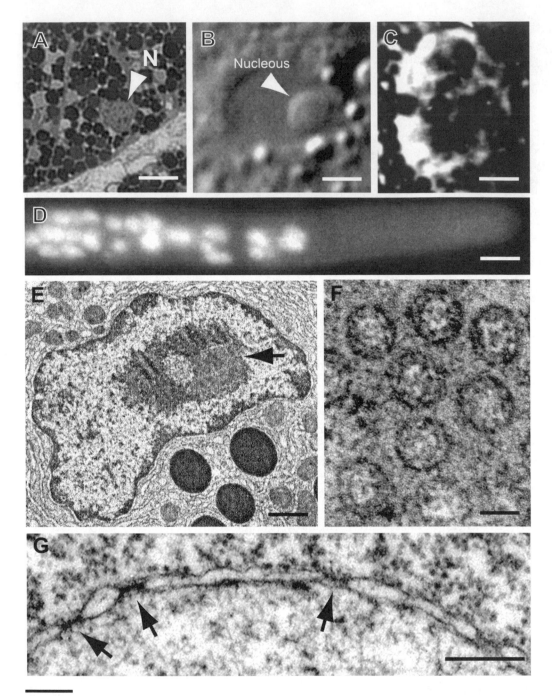

Figure 4.8
The nucleus. (A) The nucleus of a secretory cell in the frog oviduct as seen in light microscopic section (arrowhead). (B) The nucleus of a fungal cell as seen by differential interference contrast optics. Note the prominent nucleolus (arrowhead). (C) Phase-contrast optics reveal chromatin substructure in the nucleus of a pancreatic acinar cell. (D) Nuclei of a fungal hyphal tip fluorescently stained with the DNA binding dye DAPI. (E) Electron micrograph of the nucleus in a pancreatic acinar cell. Note the prominent nucleolus (arrow) having both fibrillar and nonfibrillar regions. (F) Pores of the nuclear envelope revealed in an "en face" thin section. Note the octagonal superstructure within the pore. (G) Thin section of the nuclear envelope showing that it consists of two unit membranes that are joined at the nuclear pores. Each nuclear pore (arrows) is filled with a large protein complex that controls entry and exit of substances from the nucleus. Bars = 5 μm in (A) and (D), 2 μm in (B) and (C), 1 μm in (E), 100 nm in (G), and 0.4 μm in (G).

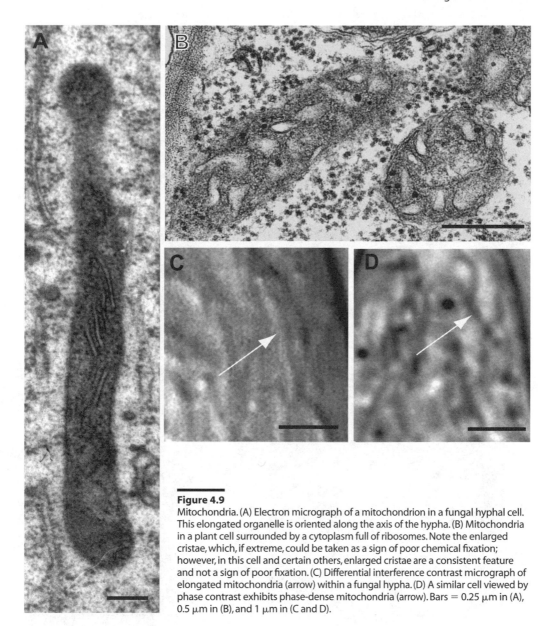

Figure 4.9
Mitochondria. (A) Electron micrograph of a mitochondrion in a fungal hyphal cell. This elongated organelle is oriented along the axis of the hypha. (B) Mitochondria in a plant cell surrounded by a cytoplasm full of ribosomes. Note the enlarged cristae, which, if extreme, could be taken as a sign of poor chemical fixation; however, in this cell and certain others, enlarged cristae are a consistent feature and not a sign of poor fixation. (C) Differential interference contrast micrograph of elongated mitochondria (arrow) within a fungal hypha. (D) A similar cell viewed by phase contrast exhibits phase-dense mitochondria (arrow). Bars = 0.25 μm in (A), 0.5 μm in (B), and 1 μm in (C and D).

In Figure 4.9B, mitochondria in a plant cell contain tube-like cristae that are fewer in number and have larger lumens. Enlarged mitochondrial lumens can be a sign of poor fixation, but in cells such as these, it is simply a morphologic variation. In differential interference (arrow, Figure 4.9C) or phase-contrast (arrow, Figure 4.9D) microscopy, mitochondria appear thread-like, and occasionally exhibit branching. Mitochondria are easily visualized by specific fluorescent dyes because of the large voltage potential across their inner membrane (see Chapter 13). These organelles have their own genome that codes for a portion of their proteins, whereas the remaining proteins are coded for by the nuclear genome.

Rough Endoplasmic Reticulum

This convoluted network of membrane-enclosed sacs often piled one on top of another (**Figure 4.10A**) is a major site of protein synthesis. Its cytoplasmic surface is decorated with ribosomes, and the proteins synthesized here are threaded into the endoplasmic reticulum lumen, folded properly and delivered by transport vesicles to the Golgi apparatus for additional modifications and packaging. The complex network of membranes that make up the endoplasmic reticulum can be nicely visualized by fluorescence microscopy using lipophilic dyes that diffuse throughout the organellar membrane much as a phospholipid would (see Chapter 13).

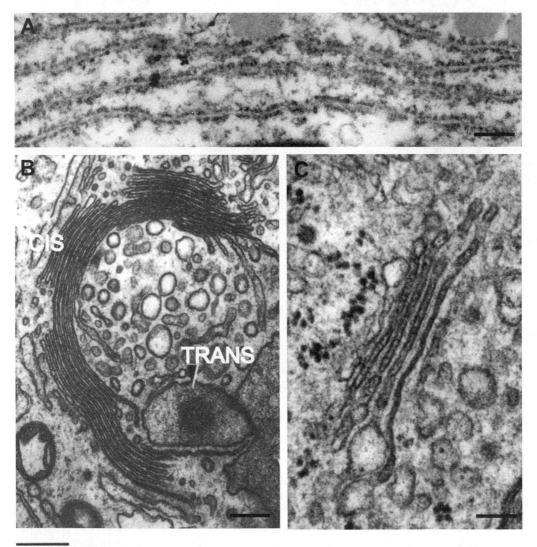

Figure 4.10
Rough endoplasmic reticulum and Golgi apparatus. (A) Pancake-like stacks of endoplasmic reticulum studded with ribosomes in a plant cell. (B) The Golgi apparatus in an animal cell consists of multiple polarized cisternae. The cis side of the Golgi apparatus receives transport vesicles from the rough endoplasmic reticulum, whereas the trans side of the Golgi produces vesicles that carry glycoprotein products to appropriate organelles. (C) Golgi apparatuses in plant cells are often smaller and contain fewer cisternae compared to those in animal cells. Bars = 0.7 μm in (A), 0.3 μm in (B), and 100 nm in (C).

The Golgi Apparatus

This organelle consists of a pile of flattened membrane-enclosed sacks that act like an assembly line for the glycosylation (sugar addition) of proteins (Figure 4.10B). The proteins arrive at the cis side of the Golgi apparatus from the rough endoplasmic reticulum and after glycosylation leave via transport vesicles from the trans Golgi side. Evidence of glycosylation is nicely shown by cytochemical visualization of glycosylated lysosomal enzymes on the trans side but not the cis side. Golgi membranes can also be visualized at the light microscopy level by using ceramide-linked dyes that specifically intercalate into the membranes of this organelle. Golgi "bodies" in plant cells (Figure 4.10C) have fewer stacks and are smaller in size than their counterparts in animal cells but are much greater in number, while "Golgi equivalents" in fungi have a unique morphology appearing as spherically inflated cisternae.

Smooth Endoplasmic Reticulum

Smooth endoplasmic reticulum can be visualized at the light microscope level using fluorescent dyes that diffuse throughout the membrane network in live cells to reveal the lace-like structure of their organelle (**Figure 4.11A**). Predictably, this network can be

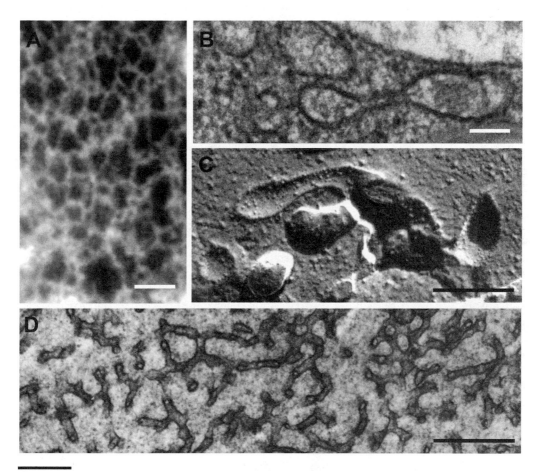

Figure 4.11
Smooth endoplasmic reticulum. (A) A fluorescence micrograph of the smooth endoplasmic reticulum network in a live cell. (B) Smooth endoplasmic reticulum as seen in thin sections of the cortex of a sea urchin egg. (C) A freeze-fracture replica of the reticulum shown in (B). This reticulum sequesters then releases calcium ions at fertilization to trigger secretion of cortical granules. (D) The extensive smooth endoplasmic reticulum (SER) of a steroid-secreting cell. Enzymes for steroid synthesis are found embedded in the SER in these cells, whereas enzymes for detoxifying drugs are found in the SER of other cells such as hepatocytes. Bars = 0.5 μm in (A), 50 nm in (B), 150 nm in (C), 1 μm in (D).

difficult to identify in thin sections because the tubules are usually seen in circular or elliptical cross-section (Figure 4.11B). The branching network of membrane-enclosed sacs is distinguished from the rough endoplasmic reticulum by its lack of ribosomes and serves different purposes in different cells. In some cells, such as the sea urchin egg cortex (Figures 4.11B and 4.11C), this organelle is a repository for calcium that is released to initiate contraction, secretion or other cell responses through calcium signals. In other cells, such as hepatocytes, this network is studded with oxidizing enzymes that serve to detoxify drugs. In steroid secreting cells, such as the Ledig cell shown in Figure 4.11D, smooth endoplasmic reticulum is the site of enzymes critical to the synthesis of steroids from cholesterol. The classic example of smooth endoplasmic reticulum of the calcium-sequestering type is the sarcoplasmic reticulum of skeletal muscle that releases its calcium to initiate contraction.

Secretory Granules and Exocytosis

Endocrine, exocrine, and neural tissues that synthesize proteins or neurotransmitters for export generally store these secretions in membrane bound granules or vesicles. These range in size from 0.1 to 3 μm, depending on the cell but are of a characteristic size for each cell type. They may be either translucent as in the case of synaptic vesicles, or darkly stained because of their densely packed secretory materials. Secretion is accomplished by exocytosis—the fusion of secretory granules with the plasma membrane. This process is often studied by freeze-fracture electron microscopy in cells that exhibit a rapid burst of activity in a short period. One such cell, the mast cell (**Figure 4.12A**), is packed full of histamine-containing granules that undergo exocytosis both singly and in groups. At higher magnification (Figure 4.12B), many granules are seen to lie just under the plasma membrane perfectly positioned for plasma–granule membrane fusion. Exocytosis begins with formation of a small, cylindrical pore connecting the granule interior with the extracellular space (Figure 4.12C). Moments later, granules in the interior of the cell fuse to more superficial granules to release their histamine as well, a process referred to as "compound" exocytosis (Figure 4.12D).

Endocytosis and the Lysosomal System

All cells retrieve materials from their surface and the extracellular space by a process of endocytosis. As can be seen from the diagram of receptor-mediated endocytosis in **Figure 4.13A,** materials taken in are frequently moved through endosomal compartments to lysosomes. These organelles are 1 to 2 μm in diameter and contain acid hydrolases, enzymes that degrade unwanted macromolecules to amino acids, sugars, and nucleotides. Lysosomes serve as recycling centers for materials from the cell surface, for unwanted/nonfunctional organelles, and for digestion of organisms consumed by professional phagocytic cells. Endocytosis at the cell surface is accomplished by formation of a clathrin-coated pit (arrow, Figure 4.13B) that serves to internalize membrane patches having specific proteins, often ligand-bound receptors. The vesicle formed delivers its cargo to the endosomal network containing multiple membrane compartments that sort and deliver cargo to the lysosome. Membrane-bound lysosomes (Figure 4.13C) enzymatically digest the material. The acid environment of the lysosome interior speeds digestion and makes these organelles easily detected by weak acids that are fluorescent (Figure 4.13D). Lysosomes that are in the process of degrading cellular components are also termed "autophagocytic" vacuoles or "multivesicular bodies" based on appearance (Figure 4.13E). Any material that is nondigestible is retained in cellular "garbage cans," termed lipofuscin granules (Figure 4.13F). Professional phagocytes such as neutrophils

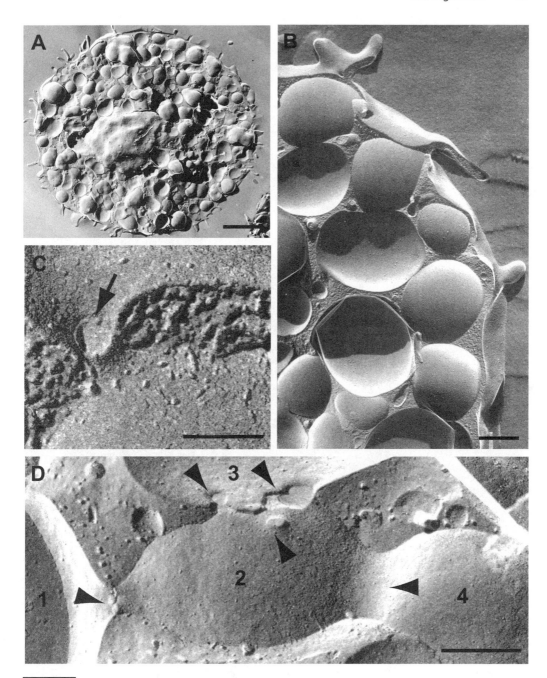

Figure 4.12

Secretion and exocytosis in the mast cell. (A) Freeze-fracture replica of a mast cell viewed by transmission electron microscopy. The cytoplasm is full of histamine-containing secretory granules and the nucleus is centrally located. (B) Higher magnification view showing granules positioned just under the plasma membrane. (C) Exocytosis of each granule begins with formation of a localized pore seen here as a thin tube connecting the extracellular space with the granule interior. (D) The mast cell exhibits compound exocytosis wherein multiple interior granules fuse to peripheral granules to discharge their contents. Four granules in this figure (numbered) have joined their membranes via small pores (arrowheads). Bars = 2 μm in (A), 0.5 μm in (B), 150 nm in (C), and 1 μm in (D).

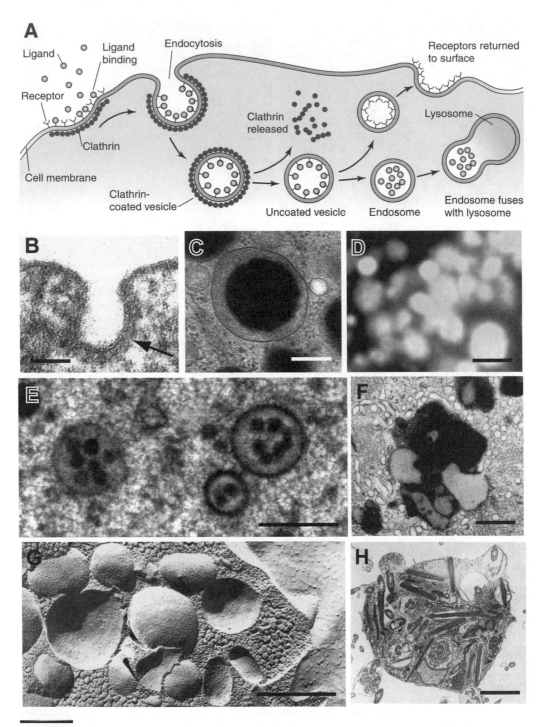

Figure 4.13

The endocytic pathway. (A) Diagram illustrating receptor-mediated endocytosis whereby molecular ligands are brought into the cell and degraded in lysosomes while the receptors are recycled to the cell surface. (B) A clathrin-coated endocytic pit that will pinch off the plasma membrane to form a coated vesicle. (C) After passing through a series of endosomes, materials will be degraded in lysosomes, the acidic membrane bound organelle shown here. (D) Fluorescence micrograph of lysosomes stained with a dye that accumulates in acidic organelles. (E) The multivesicular body, a lysosome-like site for degradation of transport vesicles. (F) Lipofuscin body containing undigested waste after lysosomal degradation. (G) Lysosome-like granules in a neutrophil that provide digestive enzymes used during phagocytosis. (H) Phagocytosis of sperm by macrophages in the female reproductive tract. Bars = 80 nm in (B), 500 nm in (C), 1000 nm in (D) through (G), and 5 μm in (H).

(Figure 4.13G) and macrophages (Figure 4.13H) contain large numbers of lysosome-like granules that fuse with and deliver digestive enzymes to the phagocytic vacuole. Phagocytosed sperm are easily visualized within the macrophage of Figure 4.13H that was isolated from the female reproductive tract after copulation.

■ Plant Cells

The typical plant cell (**Figure 4.14**) contains all of the organelles described in the previous section but, in addition, a number of unique organelles and structure not found in animal cells. These include the cell wall, chloroplasts, vacuoles, storage organelles, and plasmodesmata.

The Cell Wall

Composed of extracellular sugar polymers and associated proteins, this network of fibers coats the plasma membrane of the plant cell. This layer is synthesized by assembly of glucose polymers, notably cellulose, at the plasma membrane and serves to prevent these turgid, water-rich cells from bursting. The filamentous substructure of the plant and fungal cell walls is best seen in Figures 4.2D and 4.2E.

Chloroplasts

These organelles are at the heart of plant physiology because they are the site of photosynthesis. Although chloroplasts, like mitochondria, have their own genome and undergo division, the internal photosynthetic membranes are produced and assembled only in response to light. Plants kept in the dark exhibit etioplasts—precursor organelles that do not have mature functional internal membranes but rather have a highly developed architecture for storing membrane components (**Figure 4.15A**). At higher magnification, the etioplast appears to contain a core of membrane structures laid out in a triangular lattice (Figure 4.15B). After the plant is exposed to light, there is a rapid remodeling of these membranes both from a molecular and architectural point of view. The chloroplasts take up positions at the cell periphery squeezed between the plasma membrane and the large central vacuole (see center, Figure 4.14). Within each chloroplast, stacks of pancake-shaped membrane compartments (grana) form and position themselves throughout the chloroplast matrix (arrowheads, Figure 4.15C). As shown in Figure 4.15D, each membrane compartment has a distinct interior space, and each grana stack (single arrow) is joined to other grana by membrane connections (double arrow). At the molecular level, these thylakoid membranes house the photosystem proteins that harvest light energy, generate high-energy electrons, and use these to power ATP synthesis, production of NADH used in sugar synthesis, and the generation of oxygen. In freeze-fracture replicas, the photosystem proteins of thylakoid membranes can be seen as very large intramembrane particles. Light induces the assembly of these photosystem proteins and synthesis of photopigments such as chlorophyll that harvest the light energy and deliver it to the photosystem proteins. Chlorophyll makes these organelles easily identifiable by light microscopy both because of its green color (because of selective absorption of blue and red light) and because of its fluorescence.

Vacuoles

Vacuoles typically serve to hold water, metabolic products or wastes and less frequently gas. Plant cells often exhibit one or more large central vacuoles filled with water that weigh in as the largest organelles and make some plant cells appear "hollow" (center, Figure 4.14). Vacuoles in plant and fungal cells, however, can be smaller, more numerous and distributed

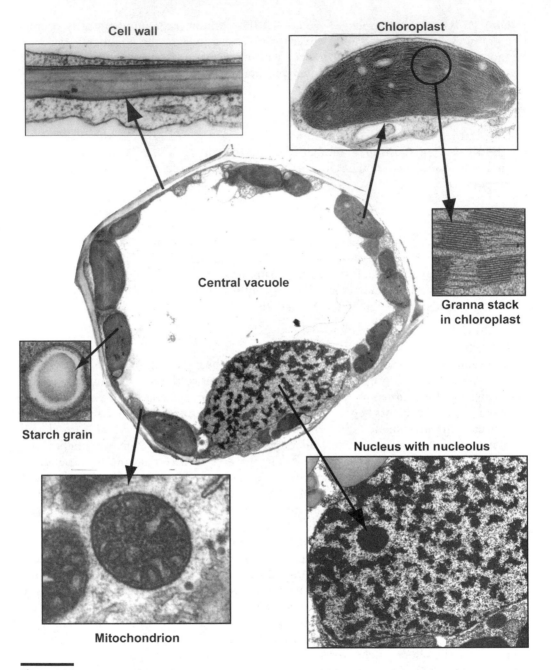

Cell wall

Chloroplast

Central vacuole

Granna stack in chloroplast

Starch grain

Nucleus with nucleolus

Mitochondrion

Figure 4.14
Organelles and cellular components of plant cells. A plant cell at the center exhibits specialized organelles found only in plant cells as well as those found in both plant and animal cells. Organelles are typically found in the layer of cytoplasm sandwiched between the cell wall and the central vacuole, which is filled with fluid.

throughout the cell. These storage organelles are usually smooth, circular or elliptical in cross-section and are surrounded by a single bilayer membrane.

Storage Organelles
Storage of metabolites in plant and fungal cells is, in many ways, comparable to that in animal cells. Starches stored in granules are present in chloroplasts of most plant cells (**Figure 4.16A**). In some cases, like the potato, tubers filled with such granules represent

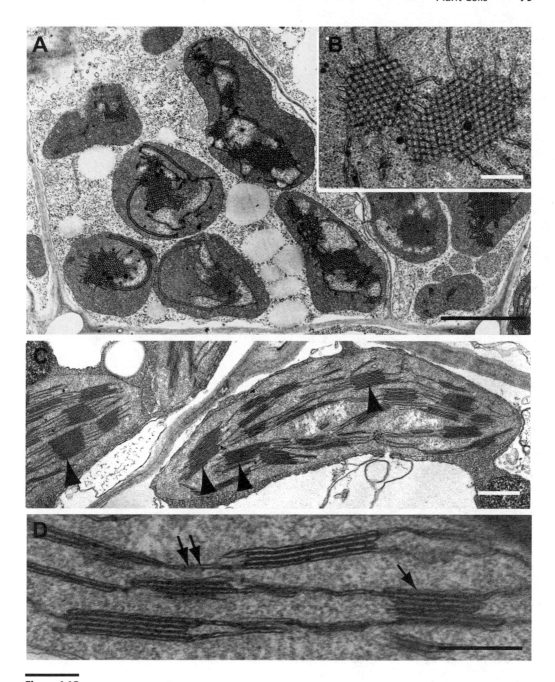

Figure 4.15

The chloroplast. (A) Cells in a leaf of a normal plant placed in complete darkness for several days. These cells possess modified plastids called etioplasts, which are capable of differentiating into mature chloroplasts following long-term exposure to light. (B) Enlarged view within an etioplast of a complex structure called a pro-lamellar body, which is a geometric grid of membranes that can transform upon exposure to light into grana stacks of membrane containing photosynthetic pigments and proteins. (C) View of differentiating chloroplasts forming grana (arrowheads) in a leaf cell of a plant placed in the light for 1 day. (D) Enlarged view of grana (single arrow), which are composed of stacked, flattened membrane compartments called thylakoids. Light energy harvested by the photosystem pigments (chlorophylls and carotenoids) embedded in the grana membranes is used to pump hydrogen ions out of the lumen of each thylakoid to generate hydrogen ion (proton) gradients used to synthesize ATP in the light. Stroma thylakoid membranes (double arrows) join grana stacks. Bars = 3 μm in (A), 0.3 μm in (B), 1 μm in (C), and 0.5 μm in (D).

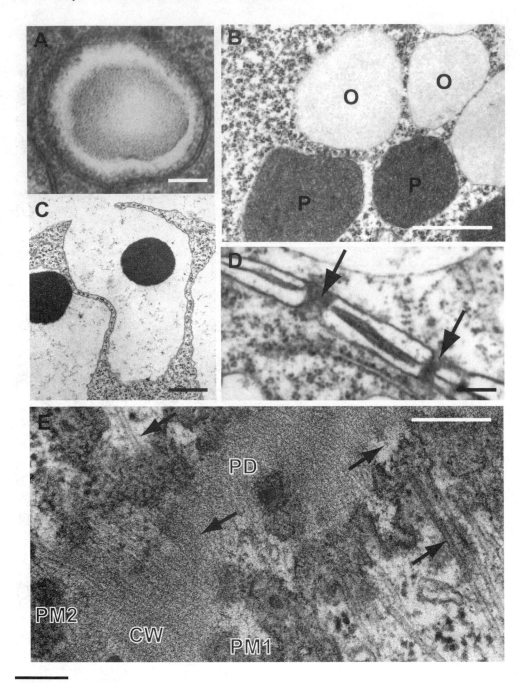

Figure 4.16
Other typical organelles and structural components found in plant cells. (A) A single starch grain within the stroma of a chloroplast. Each of these organellar inclusions is composed of thousands of glucose-containing starch molecules serving collectively as an enormous carbohydrate energy reserve. One or more starch grains also occur commonly in nongreen plastids such as within the stroma of amyloplasts in potatoes and rice grains. (B) Two peroxisomes (P) adjacent to two oil droplets (O) in a cotyledon cell of a cottonseed seedling. Peroxisomes, specifically referred to as glyoxysomes in oilseed seedlings, contain oxidative and other metabolic enzymes that convert the stored oils into carbohydrates for seedling growth into mature autotrophic plants. (C) Two protein bodies in a cucumber seedling cell. These protein storage organelles provide a vast source of amino acids for protein synthesis in growing plants. (D) Plasmademata (arrows) are tubular membranous passageways through the cell wall that separates adjacent plant cells. At these junctions, the cytoplasms of the two cells actually are in confluent contact allowing transport of metabolites between the cells. (E) "En face" thin section view of a plant cell wall (CW) showing a cross section of a plasmodesmatum (PD). The cellulose microfibrils that compose the cell wall line up with microtubules (compare arrows) just under the plasma membranes (PM1, PM2) of the neighboring cells. Bars = 150 nm in (A), 1 μm in (B) and (C), and 0.3 μm in (D) and (E).

specialized storage tissues. A high molecular weight sugar polymer, starch, serves similar metabolic storage purposes as glycogen does in animal cells. Lipid storage in plants, particularly in seeds, is accomplished by oil droplets that are filled with a variety of unsaturated hydrocarbons and surrounded by a single monolayer of phospholipids ("O," Figure 4.16B). Often in contact with these droplets are peroxisomes, organelles specialized for metabolizing lipids through the use of oxidizing enzymes such as catalase ("P," Figure 4.16B). Because plant peroxisomes metabolize oils by the glyoxylate cycle of reactions to provide ATP, they are also referred to as "glyoxysomes." The close proximity of these ATP-generating factories to the oils that they use is an example of cellular efficiency. Finally, seeds also contain stored protein in the form of "protein bodies" (Figure 4.16C) that is used as a source of amino acids during germination.

Plasmodesmata

Surrounded by cell walls, plant and fungal cells need some means to communicate by direct contact. This purpose is served by plasmodesmata, regions where the cell wall is interrupted by a pore (arrows, Figure 4.16D). At these pores, the cytoplasms of neighboring cells are continuous, allowing for the diffusion and transport of small metabolites and signaling molecules. The grazing section of a plant cell in Figure 4.16E provides a unique look at the spatial relationships between the plasmodesmata (PD), the fibers of the cell wall, and microtubules that lie just underneath the plasma membranes of cell one (PM1) and that of its neighbor, cell two (PM2). The plasmodesmata are seen as an electron dense hole in the cell wall through which membranes run. Fibers making up the cell wall have an orientation that is identical to that of the microtubules on the opposite (cytoplasmic) side of the plasma membrane (arrow pairs, Figure 4.16E). Indeed, microtubules are thought to anchor the cell wall synthesizing enzymes (e.g., cellulose synthetase) that construct these fibers, and, thus, this similarity in orientation is functional, not coincidental.

■ Fungal Cells

Filamentous Fungi

Fungal cells are unique in operating much like animal cells but yet manufacture cell walls, like plants, using fibrils of chitin, a polymer of N-acetyl glucosamine. The growing fungal hyphal tip in **Figure 4.17** (center) exhibits some of the specialized organelles found in these cells. The hypha is a rapidly elongating structure, analogous to a growing nerve axon, that is served by an array of parallel microtubules coursing through its length that serve as tracks for rapid vesicle transport and delivery of such cargo as cell wall synthetic enzymes and precursors. Powering transport is ATP produced by the numerous mitochondria than line up along the microtubules. The sources of the cell wall precursors and transport vesicles are the so-called Golgi equivalents that are analogous to the Golgi apparatus in other eukaryotic cells. Receiving and sorting of these vesicles at the distal tip are carried out by the Spitzenkörper (apical body). This organelle is thought to organize the reception and deployment of the raw materials required for plasma membrane and cell wall growth at the hyphal apex. Thus, the tip is the site of nascent chitin deposition whereas a more mature cell wall covers the remainder of the hypha. Fungal cells are also unique in exhibiting fluid-filled vacuoles much like plant cells and in having spindle pole bodies that take the place of centrosomes as sites of cytoplasmic microtubule nucleation and organization.

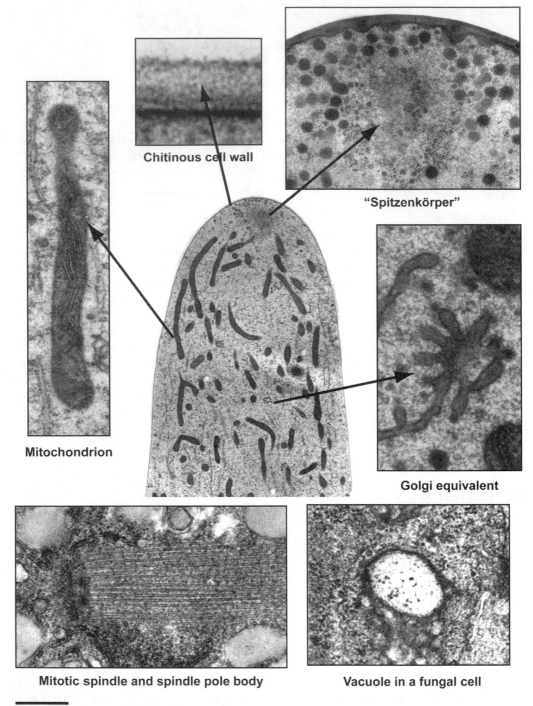

Chitinous cell wall

"Spitzenkörper"

Mitochondrion

Golgi equivalent

Mitotic spindle and spindle pole body

Vacuole in a fungal cell

Figure 4.17

Cellular structures and organelles found in most fungal hyphae. Hyphae are filamentous cells that grow by apical expansion. The Spitzenkörper is composed of secretory vesicles, a core of actin filaments, and ribosomes. This complex structure is directly involved in hyphal growth through its regulation of vesicle flow to and ultimate exocytosis at the apical plasma membrane. Like other eukaryotic cells, Golgi bodies are abundant in the cytoplasm of fungal cells. Interestingly, the morphology of fungal Golgi bodies is different than in other eukaryotes and are referred to as Golgi equivalents. Mitochondria are often elongate and are positioned parallel to the growing axis of the cell. They contain plate-like cristae. A chitinous cell wall is always found at the exterior of the cell during all portions of the fungal life cycle.

■ Prokaryotic Cells

The prokaryotic world is diverse as the eukaryotic world and here we highlight only a few examples.

Bacteria

Escherichia coli are not only the best known of the bacteria in regard to genetics and physiology but also in ultrastructure. Typically, about 1 to 3 μm in length, this cell is surrounded by a double membrane (**Figure 4.18A**). The inner membrane that separates

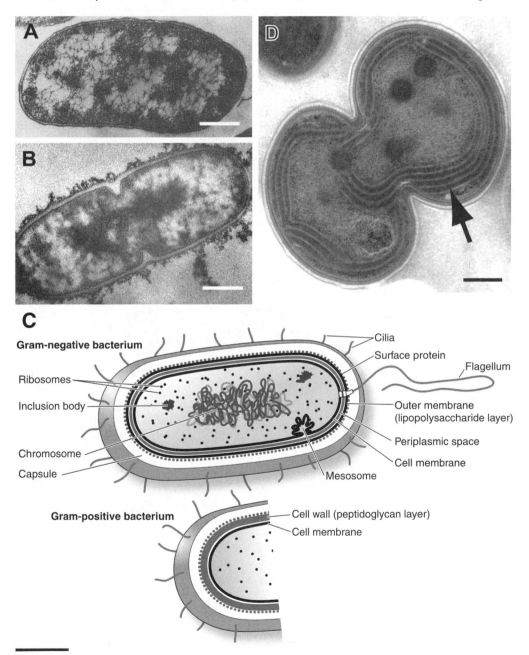

Figure 4.18

Bacteria. (A) Thin section micrograph of a gram-negative *E. coli* cell. (B) A dividing gram-positive *Streptococcus mutans* cell. (C) Diagram comparing the structural features of gram-negative and gram-positive bacteria. (D) Thin section of a photosynthetic bacterium undergoing division. The photosynthetic thylakoid membranes are layered (arrow). Bars = 0.3 μm in (A), (B), and 0.5 μm in (D).

cytoplasm from periplasm is a classic unit membrane composed of two monolayers with a hydrophobic interior. This membrane serves as a platform for electron transport, hydrogen ion gradient formation, and ATP synthesis, much as the inner membrane of a mitochondrion does. The outer membrane includes a layer of lipopolysaccharides in which are embedded porins—proteins that form channels for the passage of nutrients and other small molecules. These lipopolysaccharides are characteristic of "gram-negative" bacteria that do not stain with Gram stain. In contrast, gram-positive bacteria such as *Streptococcus mutans* (Figure 4.18B) lack this lipopolysaccharide layer, thereby allowing a thick peptidylglycan layer to stain heavily with the crystal violet dye in Gram stain. The structural differences between gram-positive and gram-negative bacteria are summarized in Figure 4.18C. Although bacteria do not have membrane-bound intracellular compartments, the cytoplasm is highly organized, with set locations for different regions of the bacterial genome. Many bacteria are replete with flagella seen as long filaments in shadowed or negatively stained transmission electron microscopy preparations. Bacterial flagella are self-assembled helical chains of proteins, each attached to a rotary "motor" on the cell surface. Depending on the direction of motor rotation, these flagella act collectively to either propel the bacterium forward or cause it to tumble, resulting in a mechanism that can direct the *E. coli* to nutrients by chemotaxis. Of fundamental importance is the photosynthetic cyanobacterium *Synechocystis* sp. PC 6803, whose ancestors became symbionts of eukaryotic cells and as such were evolutionary precursors of chloroplasts. These bacteria exhibit concentric layers of thylakoid membranes containing photosystem proteins and pigments (arrow, Figure 4.18D).

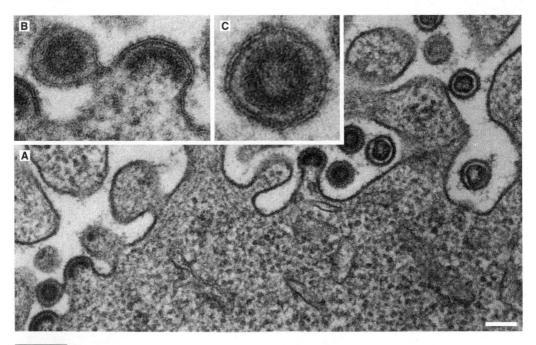

Figure 4.19
Retrovirus budding. (A) Thin section showing retroviruses (tumor causing RNA viruses) budding from a leukemia cell. Bar = 0.4 μm. (B) The viral nucleocapsid binds to and pinches off a portion of the host cell plasma membrane. (C) The mature viral particles released into the extracellular space consist of a core of nucleic acids (the viral genome) surrounded by nuclear capsid proteins. The capsid itself is linked to and enclosed within a protein-rich lipid bilayer.

Viruses

Not cells in their own right, these "particles" contain a DNA or RNA genome that can direct its own replication and the formation of new viral particles within a host cell. From a structural standpoint, the genome is surrounded by nuclear capsid proteins to form the core of the virus particle. This core is surrounded by the viral envelope consisting of a lipid bilayer containing a high density of viral proteins which in negatively stained particles can form geometric surfaces such as icosahedrons. In thin sections of retrovirus-infected cells, numerous viral particles can be seen budding off of the host cell (**Figure 4.19**). Budding is preceded by assembly and binding of the viral nuclear capsid to regions of the host cell plasma membrane at which viral membrane proteins have been inserted; the host cell membrane pinched off during budding becomes the viral particle membrane (Figure 4.19B). Thus, the mature viral particles released have a dense osmophilic core (the nuclear capsid) surrounded by a typical bilayer membrane (Figure 4.19C). Viruses attack prokaryotic cells as well as eukaryotic cells. Bacteria, for example, are infected by bacteriophage that consist essentially of a head containing the genome and a tail that serves as a device for injecting the genome through the cell wall and membrane of the bacterium.

References and Suggested Reading

Textbooks

Barton LL. *Structural and Functional Relationships in Prokaryotes*. New York: Springer, 2004:820.

Bozzola JJ, Russell LD. *Electron Microscopy: Principles and Techniques for Biologists,* 2nd ed. Sudbury, MA: Jones and Bartlett, 1999:670.

Evert RF, Eichhorn SE. *Esau's Plant Anatomy: Meristems, Cells, and Tissues of the Plant Body: Their Structure, Function, and Development,* 3rd ed. Wilmington, DE: Wiley-Liss, 2006:601.

Harrison M, Dashek WV. *Plant Cell Biology*. Gainesville, Florida: Scientific Publishers, 2006:494.

Hsiung GD, Fong CKY, Landry ML. *Hsiung's Diagnostic Virology: As Illustrated by Light and Electron Microscopy,* 4th ed. New Haven: Yale University Press, 1994:404.

Lewin B, Cassimeris L, Lingappa V, Plopper G. *Cells*. Sudbury, MA: Jones and Bartlett, 2007.

Young B, Lowe, JS, Stevens A, Heath JW, Deakin DJ. *Wheater's Functional Histology: A Text and Colour Atlas,* 5th ed. London: Churchill-Livingstone, 2005:448.

Atlases and Monographs

Fawcett DW. *The Cell*, 2nd ed. Philadelphia: W.B. Saunders, 1981:862.

Goldstein J, Newbury DE, Joy DC, et al. *Scanning Electron Microscopy and X-ray Microanalysis*. New York: Springer, 2003:689.

Hall JL, Hawes C, eds. *Electron Microscopy of Plant Cells*. New York: Academic Press, 1991:432.

Hoppert M, Holtzenburg A. *Electron Microscopy in Microbiology*. New York: Springer, 1998:112.

Kessel RG. *Tissues and Organs: A Text-Atlas of Scanning Electron Microscopy*. W.H. Freeman, 1979:317.

Kohen E, Hirschberg JG, Santus R, Ozkutuk N. *Atlas of Cell Organelle Fluorescence*. Boca Raton, FL: CRC Press, 2003:208.

Pavelka M, Roth J. *Functional Ultrastructure: An Atlas of Tissue Biology and Pathology*. New York: Springer Verlag, 2005:326.

Roy P, ed. *Virus Structure and Assembly (Advances in Virus Research),* Vol. 64. Academic Press, 2005:448.

Savidge T, Pothulakis C, eds. *Microbial Imaging,* vol. 34 (Methods in Microbiology). Academic Press, 2004:282.

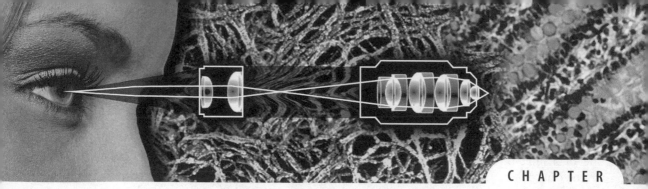

Electromagnetic Radiation and Its Interaction with Matter

In most applications, bioimaging involves use of electromagnetic radiation to image specimen structure. Electromagnetic radiation, a name coined by the famous English physicist James Maxwell in 1873, consists of propagating waves having both an electric and magnetic field oscillating in time as it propagates. An important feature of such radiation is that it can interact with atoms in a variety of manners, including scattering, refraction, and absorption. These interactions result in conversion of incident radiation coming to the specimen to processed radiation, whose magnitude, wavelength, and direction of propagation represent data that can be used to infer the structural features of the specimen with which it has interacted. Often, but not always, the radiation produced by these specimen interactions can be suitably formed into an image that maps the structural properties of the sample whether those properties are topology, chemical structure, or internal structure.

Electromagnetic radiation has energy associated with it that is related to wavelength and frequency by classic equations derived by Plank:

$$E = hf = hc/\lambda \tag{5.1}$$

where E is energy, h is Plank's constant, f is frequency, c is the speed of light, and λ is wavelength. This equation demonstrates that energy is proportional to frequency and is inversely proportional to wavelength. Thus, short wavelength radiation is of high energy, and long wavelength radiation is of low energy. The entire range or spectrum of electromagnetic radiation is extremely large, varying from high-energy gamma rays having a wavelength of 10^{-13} meters to low-energy radio waves with a wavelength of 10^4 meters. This spectrum is illustrated in **Table 5.1** and is sufficiently broad to require wavelength and frequency to be expressed exponentially in powers of 10. Although not a form of electromagnetic radiation, electrons have been included in this table for comparison.

Most regions of the spectrum are familiar for one reason or another: x-rays used in medical imaging, ultraviolet—so-called black light that makes minerals fluoresce—visible light that our eyes detect, infrared associated with heat, microwaves used in the telecommunications industry, and radio and television waves used in media broadcasting. Surprising, almost all wavelengths in this spectrum can supply information about a specimen with which it interacts. In addition to this spectrum, we must consider subatomic

Table 5.1 Electrons and the Electromagnetic Spectrum

	Electrons	Gamma Rays	X-Rays	Ultraviolet Light	Visible Light	Infrared Waves	Microwaves	TV and Radio Waves
Wavelength (nanometers)	0.001–0.01	0.00001–0.01	0.01–10	10–350	350–700	700–10^6	10^6–10^9	10^9–10^{14}
Frequency (Hz)	10^{23}–10^{20}	10^{23}–10^{20}	10^{20}–10^{17}	10^{17}–10^{15}	3×10^{15}	10^{14}–10^{11}	10^{11}–10^9	10^8–10^3
Interactions with matter	Subatomic particle; forms covalent bonds	Ionizing radiation; knocks out electrons	Emitted during electron scattering	Breaks covalent bonds	Absorbed or emitted during electronic transitions	Bond motions	Bond motions	None
Practical uses	Electricity	None	Medical imaging, CAT scans	Kills micro-organisms, sterilizing	Illumination	Heating	Telecommunications, cooking	Media broadcasting, local broadcasts
Bioimaging uses	Electron microscopy	None	X-ray microscopy, x-ray crystallography	Fluorescence excitation	Light microscopy	Heat-sensitive imaging	Rapid fixation	None

Table 5.2	The Visible Spectrum								
Color	UV	Near UV	Violet	Blue	Green	Yellow	Red	Near IR	Infrared
Wavelength (nanometers)	100–200	200–350	350–420	420–500	500–550	550–600	600–700	700–1500	1500–
Photon energy equivalent to	Breaking covalent bonds	Electronic transitions		Electron transitions in conjugated double bond systems				Stretching and bending of bonds; motion; heat	

particles such as the electron. Although subatomic particles have mass and electromagnetic radiations do not, they both have similar features such as kinetic energy, momentum, and wavelength. The imaging methods described in this book use electrons and ultraviolet, visible, and infrared light. Absorption or emission of electrons, x-rays, visible light, and infrared light from a specimen can all be used to determine or map the chemical composition of the material.

Of particular interest is the visible light spectrum that can be detected by our eyes. As shown in **Table 5.2,** this familiar rainbow of colors extends from the higher energy, shorter wavelength violet to the lower energy, longer wavelength red.

Although the visible spectrum is continuous in wavelength and energy (and thus, in color), our eyes do not detect wavelength in a continuous manner. Rather, patterns of light (images) impinge on the retina, a light-sensitive layer of tissue covering the internal surface at the back of the eye (**Figure 5.1A**). The function of this layer is carried out by three kinds of cells in retinal tissue, as shown in Figure 5.1B. Primary to the detection of light are the photoreceptors, cells that are elongated to contain stacks of membrane disks that act as an antenna for capturing light. Detection of light by the photoreceptors is transduced into changes in voltage across the plasma membrane of the photoreceptors, which in turn leads to diminished secretion of a neurotransmitter. The changes in neurotransmitter concentration lead to electrical potentials that are processed by two sets of neurons within the retina, the bipolar cells and the ganglion cells, before the processed signal in the form of action potentials is sent via the optic nerve to the lateral geniculate nuclei and visual cortex within the brain. The visual cortex of the brain carries out higher level image processing, leading to intensity, color, depth, and movement perception. In addition to photoreceptors and neurons, the retina also contains a layer of pigmented cells—the pigment epithelium—that absorb any stray light that the photoreceptors have not processed so as not to result in reflections that would otherwise confuse the image patterns detected.

At the core of how we detect light is the fact that the disc membranes of photoreceptors have embedded in them proteins that contain light-absorbing dyes (**Figure 5.2A**). The classic light detecting protein is rhodopsin, found in "rod" photoreceptors, which are most numerous in the retina and which detect low levels of light at most visible wavelengths. This broad wavelength detection pattern is optimal for sensitivity but not for color (wavelength) discrimination; therefore, rods are used for black and white vision, low light vision, and because of their great number, high-resolution mapping of image detail.

In contrast, color is detected by "cone" photoreceptors, which come in three types. These three classes of cone photoreceptors each detect a different wavelength of light. The reason for this is that each type contains a slightly modified rhodopsin having a different type of chromophore associated with it. All rhodopsins, whether in rods or cones, have sitting within their three-dimensional structure a small dye molecule (the chromophore) whose electronic structure is designed to absorb light over a certain range of wavelengths—the absorption spectrum of the dye. In the rhodopsin of rod photoreceptors, this light-absorbing dye is called 11 cis-retinal. Absorption of light by retinal leads to a change in shape of rhodopsin, the protein in which it is embedded. As shown in Figure 5.2B, this change in shape allows rhodopsin to rapidly interact with two other

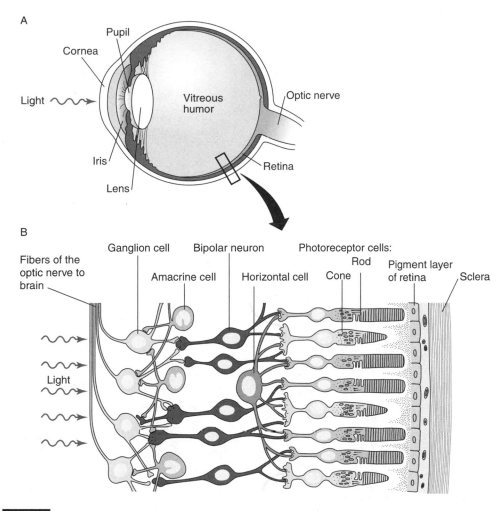

Figure 5.1
Anatomy of the human eye. (A) The lens focuses image rays onto the retina. (B) The retina consists of several layers of neuronal cells through which light rays have to travel to reach the photoreceptors. Nerve axons from the ganglion cells converge on and travel via the optic nerve to the brain.

proteins, transducin and phosphodiesterase, also located in the disc membrane. Transducin is a member of the GTP-binding protein family and, by interacting with rhodopsin, binds GTP available in the surrounding cytoplasm. In turn, GTP-bound transducin (specifically its α subunit) is now able to activate phosphodiesterase, an enzyme that breaks down cyclic GMP. The decrease in cyclic GMP then acts to trigger the closing of sodium ion channels, leading to changes in membrane potential of the photoreceptor. These membrane potential changes temporarily diminish the secretion of a neurotransmitter, which in turn leads to electrical changes in the bipolar and ganglion cells. Action potentials fired off by the ganglion cells are propagated to the brain, and it is the optic cortex of the brain that interprets the rate and pattern of action potentials as light and images.

Important to color vision is that electrical signaling to the brain by each of the three types of cone photoreceptors is absolutely dependent on the absorption spectrum of the chromophore within its rhodopsin. As shown in **Figure 5.3,** the absorption spectra of these three chromophores are separate and distinct, leading to the three classes of cone photoreceptors specialized for detecting blue, green, and red light, respectively. It is not a

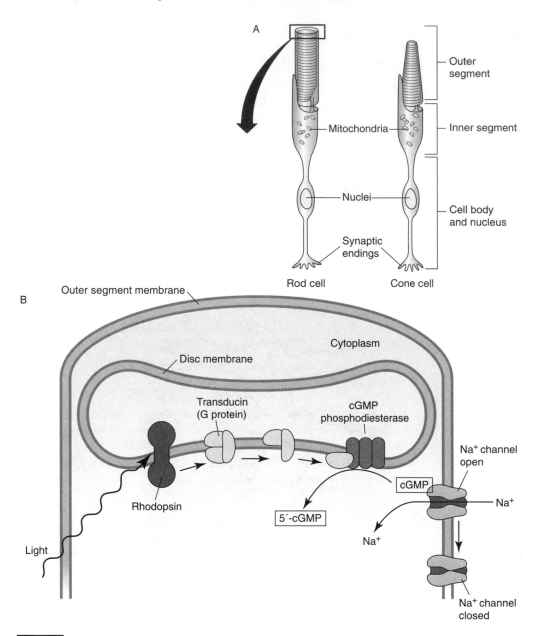

Figure 5.2
Photoreceptor cells and the mechanism of light detection. (A) Anatomy of rod and cone photoreceptors in the retina showing the outer segments where disc membranes reside. (B) Disc membranes contain the protein rhodopsin, which houses the light absorbing pigment retinol. Light absorption causes rhodopsin to activate the GTP-binding protein transducin which in turn activates the enzyme cyclic GMP phosphodiesterase. Breakdown of cyclic GMP by this enzyme switches off sodium channels located in the outer segment, resulting in hyperpolarization of the photoreceptor membrane. Hyperpolarization leads to a reduction of neurotransmitter release and formation of an action potential in the bipolar cell that is relayed to the brain.

coincidence that these three colors correspond to what are called the "primary colors" of the visible spectrum.

If there are only three primary colors and three classes of cones, why do we see the world in hundreds if not thousands of colors? The answer is that our brain analyzes all three of the primary color signals in regard to relative intensity from any given location

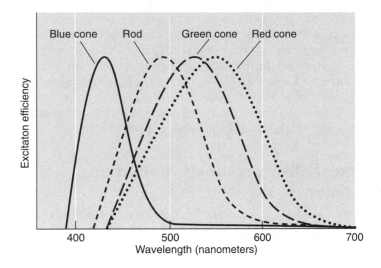

Figure 5.3
Excitation spectrum of human photoreceptor cells. The eye contains rod photoreceptors that respond to wavelengths in the middle of the visible spectrum and are responsible for low-light vision and detection of motion and edges. Also present are three types of cone photoreceptors, each type having a separate visual pigment that is most sensitive to either blue, green, or red light. They are responsible for color vision but require higher light intensities.

in an image. Based on a linear combination of these three colors, your brain assigns a certain color or shade to the overall signal. You might say that our perception of color is based on our brain assigning a reasonable color shade to every linear combination of primary colors. Equation 5.2 represents such a linear combination:

$$B \text{ (blue signal)} + G \text{ (green signal)} + R \text{ (red signal)}$$
$$\rightarrow \text{(color shade; hue) and (color intensity; saturation)} \tag{5.2}$$

The coefficients B, G, and R are used by the brain to determine both color shade and color intensity. If we arbitrarily restrict each coefficient to vary between 0 (no signal) and 1 (maximal signal), then it makes sense to say that the coefficients for pure blue are 1,0,0 for B, G, and R, respectively. In the same manner, the coefficients for red would be 0,0,1 and for yellow (a mixture of green and red) would be 0,1,1; therefore, we can see that color shade or hue is related to the *relative* values of the coefficients. The intensity of the color (referred to as saturation) would be dependent on the magnitude of the coefficients. For example, coefficients of 1,0,0 and 0.5,0,0 would both be blue, but the blue of the first set is more *saturated* or intense. Such coefficient information is described as CHROMINANCE information because it relates to color perception.

In contrast, the overall magnitude of all light detected at any given image point is termed LUMINANCE information and is related to how dark or light an image point is perceived. In other words, an image represented by a large range of "gray levels," typical of black-and-white photography, contains only luminance information. Surprisingly, luminance information is not entirely independent of chrominance information in visual processing by our brain. Empirical data from actual experiments on human vision have shown that green is perceived as being brightest while red is darker and blue is darkest. This accounts for the following relationship between the color intensities B, G, and R and the luminance (brightness) value Y of each image location:

$$Y = 0.30R + 0.59G + 0.11B \tag{5.3}$$

This interdependence suggests that color perception is not entirely independent of luminance. Indeed, blue could be perceived as almost black in some situations and almost white (a very light blue) in other situations. The fact that color perception is influenced by luminance gives rise to the fact that the same color can be found at different *saturations* that differ in luminance. Thus, both chrominance and luminance information can be carried in the relative values of B, G, and R. For this reason, virtually all color images under ideal conditions can be transmitted as a linear combination of three signals, whether in film or electronic or printed representations.

■ Interaction of Electromagnetic Radiation with Specimens

Experimental behavior of electromagnetic radiation can be modeled in two strikingly different manners. The first, and historically older model, is that of waves that have the implicit characteristics of magnitude, wavelength, and phase. This type of model was developed most notably by Christian Huygens, a Dutch physicist of the 17th century, and Sir Isaac Newton, an English mathematician most famous for his invention of calculus and equations of mechanics. Conventionally, this model is represented by a propagation vector representing direction and velocity, as well as a sinusoidal function that accounts for its electromagnetic and energy properties. As shown in **Figure 5.4**, a sine function has three properties. The amplitude of the sine wave (I) is related to the intensity of radiation, that is, luminance information. The period of the sine function is related to the wavelength (lambda) or energy of the radiation and, therefore, to chrominance information. Finally, the phase of the sine wave relates the positions of its maxima and minima to a fixed standard on the x-axis, x = 0. Phase cannot be detected directly by our eyes but, nevertheless, contains important information that can be used to predict the behavior of radiation as well as reveal the structure within a specimen that is interacting with radiation.

The wave model of electromagnetic radiation (and even subatomic particles) is well suited to understanding such phenomenon as refraction, diffraction, and interference as we shall see below. There are other phenomena, however, notably those occurring at

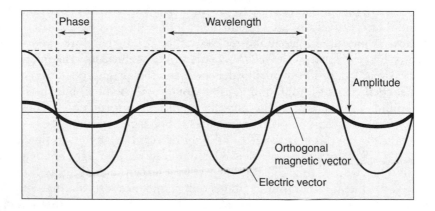

Figure 5.4

A light ray represented as a sine wave. The electric field vector (thin line) oscillates in a plane 90 degrees from the plane in which the magnetic field oscillates. Both can be represented as a sine wave, with one complete cycle (360 degrees) of the sine function representing the wavelength. The magnitude of the sine wave represents the amplitude (intensity) of the ray. The phase of the light ray is determined by the relationship of the wave function to the position of x = 0 and is useful for comparing whether the sine function of one light ray is offset from that of a second light ray. Phase differences between rays are expressed in degrees.

the atomic level, best explained by radiation acting as a particle. This is quite obvious in the case of the electron, which historically was described by a particle model that is still commonly used today. The particle model has served well in explaining phenomena where mass, velocity, momentum, collisions, and localization are involved. It is much easier to imagine that a particle-like electron can be scattered or bounced off heavy metal nuclei in an "elastic collision" event than it is to think of an electron wave interacting with a nuclear wave field so as to change the wavelength and direction of the electronic wave. We conceptually (and mathematically) attach a specific mass and momentum to a particle, and for the electron, these parameters are experimentally measurable.

The particle model can be applied to other forms of electromagnetic radiation, such as visible light, that traditionally are described by the wave model. These "particles" or packets of electromagnetic energy are more generally referred to as "quanta," while a quantum of light energy is specifically referred to as a photon. The photon represents a finite packet of light energy that has momentum but does not have mass. This may seem counterintuitive because usually one thinks of momentum as being mass multiplied by the velocity vector. This is certainly true for a traditional particle such as the electron, which has a mass and whose kinetic energy is related to its mass by way of momentum. Momentum, however, is actually a measurement of energy that does not require the existence of mass. The momentum of electromagnetic radiation (E), for example, is related to its frequency (f), wavelength (λ), and the speed of light (c) by the following equation:

$$E = hf/c = h/\lambda \tag{5.4}$$

Furthermore, the energy (and, therefore, momentum) comes in discrete units, just as they do for particles. Although kinetic energy levels for traditional particles (and, therefore, velocity) appear to be continuous, they actually are not. These units are just so small that there is no direct way to measure them—it can only be done indirectly. For example, if an electron moves from one orbit to another, it changes in energy by a defined, noncontinuous amount. This is represented by a change in kinetic and potential energy that is not random or continuous in magnitude but is predetermined. This exact amount of energy difference could be measured by the wavelength of light that is absorbed during the transition. Thus, the photon model quite nicely serves to explain such properties of light as absorption, fluorescence, and the photoelectric effect. For example, as shown in **Figure 5.5,** the photoelectric effect takes place when light "knocks" an electron out of its

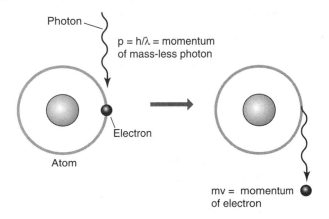

Figure 5.5
The photoelectric effect. Light rays can be represented as photons having no mass but having momentum. Photon momentum can be sufficient to knock electrons out of their orbit resulting in a "photoelectric" current.

usual location and sends it wandering to become what is a detectable current. This phenomenon can be understood intuitively by a transfer of momentum between "particles" involved in a "collision." In reality, we are discussing quantal energy transfer that is not directly related to mass and does not absolutely require that the participants be viewed as "particles." This convention is more a matter of conceptual convenience.

■ The Speed of Light

At the core of the particle and wave models of electromagnetic radiation is the fact that they describe the same phenomena. They differ in their emphasis, wave theory being most common for describing interactions of low-energy, no-mass radiation at the long wavelength end of the spectrum, whereas particle theory is common for describing high-energy, massive "particles" having short wavelengths.

A fundamental constant in nature is the speed of electromagnetic radiation in a vacuum, usually referred to as the speed of light. This speed has been measured very accurately and is now defined as 299,792,458 meters per second by international agreement. The value is rounded off to 300,000,000 meters per second in most calculations. How can radiation travel at such a high speed? Is this speed truly a constant?

Relativity theory, pioneered by Einstein in the early 20th century, not only provides for the concept that matter and energy are interconvertible by the now-famous $E = mc^2$ equation, but also predicts that as a particle is accelerated to the speed of light its "relativistic mass" increases. The term relativistic mass is a misnomer because it is actually a measure of energy and not of mass! Nevertheless, an ever-increasing amount of energy has to be fed into a "particle" as it is pushed toward the speed of light. This suggests that in order to be at or near the speed of light the entity must have little or no mass associated with it. Rather, it must be entirely a form of propagated energy that we describe as electromagnetic radiation.

This speed represents a limit that cannot be exceeded according to relativity theory because radiation that has no mass cannot be further accelerated regardless of the force applied; however, the speed of propagation for mass-less radiation is a constant only in a vacuum, that is, in the absence of matter. This speed is diminished by interactions with matter and is always less than that achieved in a vacuum for any real medium. For example, the speed of light in air, water, diamond, and lead sulfide is 299, 225, 124, and 77 million meters per second, respectively, suggesting that the relative interatomic distance and atomic number in the medium is related to this decrease in speed. The relative value of these speeds in different media relative to the speed of light in a vacuum is related to the refractive index of the medium in the following manner:

$$\text{refractive index} = \text{speed of light in vacuum/speed of light in the medium}$$

Thus, the refractive index of a vacuum is 1.00 and that for air, water, diamond, and lead sulfide are 1.01, 1.33, 2.42, and 3.91, respectively. This index is used to describe the phenomenon of refraction, hence its name.

■ Refraction

It is a long-standing observation that if light passes from one medium to another, for example, from air to water, there is a change in the direction of propagation. As shown in **Figure 5.6,** this can lead to an object appearing in the "wrong" position or appearing larger or smaller than it actually is. For example, the fisherman trying to spear a fish from above the water will observe the fish to be at position A, whereas the fish is actually at position B. After missing a few times, the fisherman may reasonably conclude that he needs to aim below where he sees the fish. The reason is that light rays passing from the fish to the fisherman eye pass through an air–water interface, at which point their direction is changed.

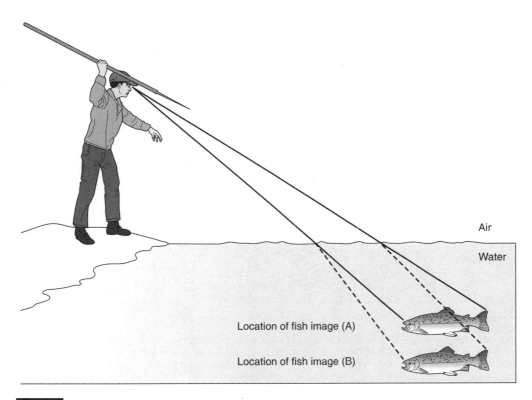

Figure 5.6
Refraction and the fisherman. Bending of light rays occurs at the interface of media having different indexes of refraction. This can result in the distortion or "mislocation" of object images as in this case of a fisherman who will miss his prey if he throws the spear directly at the fish he sees.

In the wave model, refraction is explained by Huygens's wavelet theory, which employs a wave front of sufficient size such that one part of the wavelet A enters the medium before another part B, as shown in **Figure 5.7A**. Providing that A is retarded because of the lower speed of light in the new medium sooner than B is, the wavelet must change direction if A and B are reunited as C and D to propagate together as a wavelet. Clearly, the extent to which the wavelet direction is changed depends on the incident angle and on the difference in refractive index between the two media, as given by the following geometric equation referred to as Snell's Law:

$$(n_1)(\sin \theta_1) = (n_2)(\sin \theta_2) \qquad \text{or} \qquad (\sin \theta_1)/(\sin \theta_2) = (n_1)/(n_2) = n_r \qquad (5.5)$$

where n_1 and n_2 are the indexes of refraction for medium 1 and medium 2, n_r is the relative index of refraction for the two media, and $\sin \theta_1$ and $\sin \theta_2$ are the angles of incidence and refraction, respectively (Figure 5.7A). From this equation, we can see that there will be no bending of light if the incident angle is 0 degrees (as in light passing straight through a pane of glass) or if there is no difference in refractive index between the two media.

Bending of light rays by refraction is of major importance to light microscopy. A magnifying glass is a simple biconvex lens that uses refraction to bend light to a focal point. We are all familiar with the fact that a magnifying glass can be used to burn a hole in a piece of paper (or fry ants if you are of that ilk). As diagrammed in Figure 5.7A, the parallel rays of the sun are being bent by this lens to a focal point of high intensity. In order to do this, the lens must be made of glass having a high refractive index compared with that of air and must have a curved surface so that the sunlight to be bent by refraction does not hit the lens with an incident angle of 90 degrees.

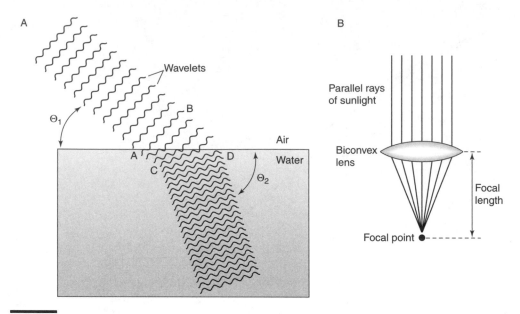

Figure 5.7
Refraction of light wavelets and the use of refraction by lenses. (A) Wavelets at an air–water interface are bent at an angle described by Raleigh's law. The angle is consistent with light in air traveling from B to D in the same amount of time that light in water travels from A to C. (B) A convex lens focuses parallel light rays to a point because the light rays are bent at the air-glass interface by refraction.

Refraction of light by the lens is also responsible for its ability to magnify. If one looks at a printed letter "A" with this lens, as illustrated in **Figure 5.8,** light coming from each point of the A will be bent by the lens toward the focal point. As a result, what our eyes see are light rays converging at a much sharper angle, similar to what would be the

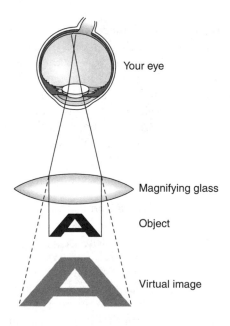

Figure 5.8
Formation of virtual images. A convex magnifying glass bends light rays (solid lines) that are focused as a "real image" onto the retina. In contrast, your brain interprets the image as if the lens were not there (dashed lines), and the object appears larger than it actually is. This larger image is a "virtual image" that is not composed of light rays.

case if the "A" were much larger than it actually is. Thus, what we see is a virtual image that is upright, floating in space behind the lens and is much larger than the object itself. Refraction resulting in the bending of light is therefore an essential phenomenon in the production of magnified images in a light microscope.

■ Reflection

Reflection is the ability of smooth surfaces to redirect radiation back into the medium from which it came. The classic example is a mirror consisting of a smooth polished metallic surface. In the modern era, extreme smoothness has been obtained by coating a glass surface with atomized or chemically deposited metal, usually silver. As described in **Figure 5.9A,** when radiation reaches a surface interface, part of the wave will continue into the object, and part of the wave will be reflected. The fraction reflected will depend on the smoothness of the surface, the angle of incidence, and to what extent the material can absorb or transmit light. Reflection is favored by high atomic number materials caused by the fact that light is elastically colliding with these atoms and being ricocheted off them like billiard balls. As predicted by Newtonian (particle) mechanics, the angle of incidence is equal to the angle of reflection. For those materials that are transparent to light, reflection rather than refraction will occur if the incident angle becomes greater than the so-called critical angle. This commonly occurs when light travels from a medium of high refractive index to one of low refractive index (Figure 5.9B). At the critical angle of incidence, the angle of refraction becomes 90°—parallel to the interface—and no light

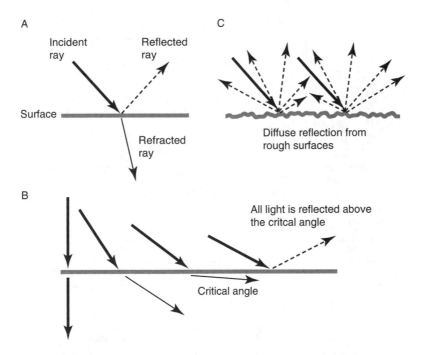

Figure 5.9
Reflection and the critical angle. (A) Light rays, arriving at an interface between media of differing refractive indexes, is split into two rays: a reflected ray that "bounces off" the interface at an identical angle to normal and a refracted ray that passes through the interface. (B) If the angle of incident light striking a medium interface is low enough, all light will be reflected. This "critical" angle is referred to as Brewster's angle. (C) Rough surfaces (even at the micrometer level) result in diffuse reflection at many angles. Such light is referred to as "scattered," although this term is also used to describe diffracted light as well (discussed later).

enters the second medium. At incident angles higher than the critical angle, all light is reflected back into the medium from which it came—there is total internal reflection.

The angle of reflection becomes more poorly defined as surface irregularities become larger. If surface irregularities are smaller than the wavelength of radiation (400 to 700 nm in the case of visible light), the efficiency of reflection is high at almost every incident angle. If the surface is bumpy or irregular, light is still reflected, but the direction of reflection is distributed through a wide range of angles. The light is said to be "scattered" because of interaction with the irregularities over a variety of incident angles on the microscopic level. Thus, the resulting reflection is termed diffuse, meaning that the surface can no longer be relied on to form images requiring precise reflection angles, as shown in Figure 5.9C.

Reflection is also commonly used in imaging through the use of mirrors that can either be flat, convex, or concave. Flat mirrors are used for redirecting light beams or images without loss of intensity or changes in dimensions. On the other hand, concave mirrors can be used to gather light or enlarge images while convex mirrors can be used to disperse light or miniaturize objects within panaramic views. Combinations of these types of surfaces can distort images in predictable but surprising manners, as can be attested to by any visitor to the hall of mirrors in a "fun house." Gathering of radiation by concave reflectors, as shown in **Figure 5.10A,** is routinely used in satellite dishes to concentrate radio and television signals coming in from a distance or to gather light from lamps, as in the parabolic reflectors of coastal light houses or lamp houses providing light for microscopes. Convex mirrors can be used to obtain wide angle images as in the modern rear view mirrors of autos, as in Figure 5.10B. A series of mirrors can be used

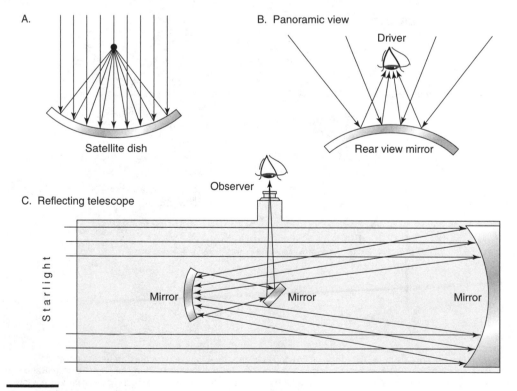

Figure 5.10
Mirrors and examples of their uses. (A) A concave mirror used to gather electromagnetic signals. (B) A convex mirror used to provide a wide angle view in an automobile. (C) A series of mirrors in a reflecting telescope used to concentrate image light.

to provide extremely large degrees of magnification in telescopes with very little loss of light as in the Palomar reflecting telescope diagramed in Figure 5.10C.

■ Diffraction

Electromagnetic radiation interacts not only with surfaces but also with edges of objects. This was demonstrated clearly by the classic experiments of Grimaldi, Fresnel, and Young in the late 18th century and early 19th centuries. These optical scientists demonstrated that when incident light passes directly through a slit cut in an opaque material and the light falls on a flat surface at some distance from the slit, the pattern observed is not that of a single band with sharp edges as one might expect. Providing that the slit is sufficiently narrow, there appears a diffraction pattern in which there is a central soft-edged band of maximal light intensity and parallel to it on each side, secondary and tertiary bands of lesser magnitude. If one plots the light intensity of this pattern along a transect 90 degrees to the slit (**Figure 5.11A**), one finds that the central maximum falls

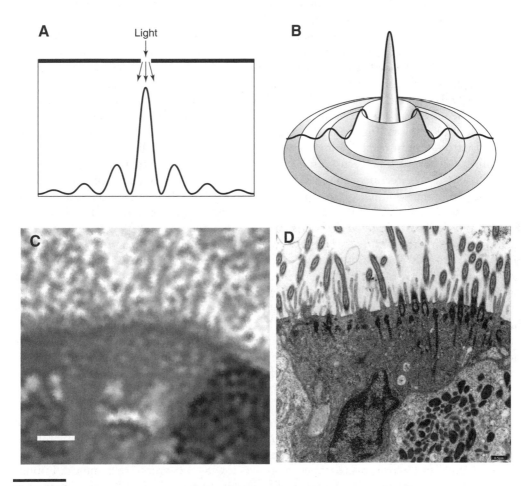

Figure 5.11
The one slit diffraction pattern and the airy disc. (A) Light passing through a narrow slit will produce a diffraction pattern consisting of a peak of high intensity surrounded by subsidiary peaks of decreasing intensity. (B) Light passing through a pinhole will produce an analogous two-dimensional diffraction pattern termed an "airy disc." The diameter of this pattern is dependent on the wavelength of light used. (C) A comparison of a light micrograph and electron micrograph of the oviductal epithelial cells of a frog. The light micrograph is blurry because the airy disc diffraction patterns formed at each image point are large given the long wavelength of visible light. Bar = 1 μm. (D) In contrast, the electron micrograph is sharp because the wavelength of the electrons used is more than 1000 times shorter and the airy disc diffraction patterns are too minute to be noticed.

smoothly to a valley of low intensity on either side and then rises to a secondary peak, falls to a valley, and rises to a further series of peaks, all separated by valleys and all of descending intensity. The envelope of intensity falls quickly enough that only the central, secondary, and tertiary peaks need be considered in most applications. This type of pattern is referred to as a diffraction pattern. It was determined that the diffraction angles at which maxima occurred could be calculated by the following equation:

$$\sin \theta = n\lambda/d \tag{5.6}$$

where θ is the angle of diffraction, n is an integer, λ is the wavelength, and d is the width of the slit. This pattern can be accounted for by positive and negative interference of light waves emanating from each edge of the slit (see the next section).

An experiment carried out by Young in 1803 provided further insight to the difficulties that diffraction could pose for use of light (or any radiation) in imaging. His "two-slit" diffraction experiment showed that two rays (actually formed by slicing a single beam of light into two halves) produced a typical diffraction pattern caused by interference. Further experimentation showed that these diffraction patterns limited one's ability to distinguish one beam from the other. If light from two slits that are far apart is observed, it is clearly seen that there are two central maxima, each surrounded by secondary and tertiary peaks providing definitive evidence that the pattern came from two different slits (**Figure 5.12A**); however, if the slits are brought closer to one another, the resulting overlapping patterns become difficult to distinguish from one another (Figure 5.12B). In fact, if the central maximum of one pattern is positioned at the first valley of the other pattern, it becomes impossible to tell whether there is one large or two smaller central maxima (Figure 5.12C). It also becomes impossible to determine whether the diffraction pattern results from one wide slit or two narrow slits.

Similar diffraction patterns can be set up by radiation interacting with any edge, regardless of shape. Pertinent to microscopy and imaging is the passage of light through a circular opening, whether it is an aperture or a lens. The edges of such a circular opening give rise to a circular diffraction pattern termed an Airy pattern, named after an English physicist who studied diffraction in the 19th century. As expected, the pattern has radial symmetry with a central peak of highest intensity and surrounding circular patterns of intensity commonly known as diffraction rings (Figure 5.11B). The lenses of all microscopes have a finite light gathering ability. In light microscopes, this is based on the fact that glass lenses can gather light only over a restricted angle, termed the acceptance angle. This angle is limited for three reasons. First, the lens has a practical limit on how large a diameter it can be and how close it can get to the specimen. Second, at large angles, the light being gathered can hit the lens surface at an angle too low for diffraction to bend the light adequately for capture. Third, light coming from the specimen may undergo internal reflection at the surface of the coverglass and, therefore, be deflected away from the lens. In an electron microscope, the limit on acceptance angle from which electrons can be harvested is based on the extent to which an electromagnetic lens can bend the path of a charged electron. This in turn is dependent on the strength of magnetic field that can be achieved having a perfect radial distribution. Objective lenses in light microscopes have an acceptance angle that is no higher than 67 degrees, while in electron microscopes, angles as low as 3 degrees are common. In both cases, a finite acceptance angle makes diffraction patterns inevitable in the images formed by real lenses.

Such diffraction patterns limit the ultimate detail that can be generated in the images formed. The crux of the matter is that each image point at best is represented by an Airy pattern rather than an infinitely small point. If image points are too close together, these patterns will merge, giving the appearance of only one point. How close

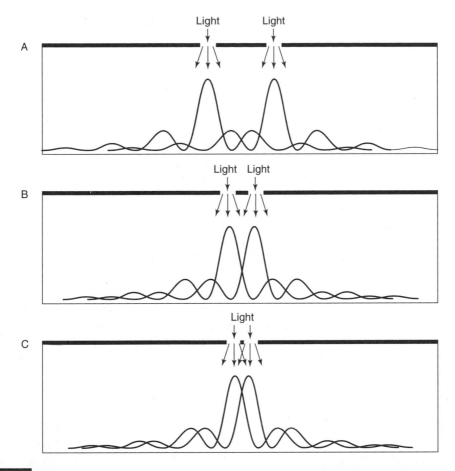

Figure 5.12
(A) The two-slit diffraction pattern and the limit of resolution. Diffraction patterns produced by two slits will be superimposed. If the two slits are far enough apart, the two peak intensities will be seen as separate. (B) If the slits are brought closer together, a point is reached at which the two major peaks are barely discernible as separate. (C) Slits closer together will produce patterns that appear to have only one central peak when summed. The distance between the two slits in (B) is referred to as the "diffraction-based" limit of resolution.

they can get without this happening is dependent on the geometry of the pattern itself, and this in turn is primarily dependent on the wavelength of the radiation, just as the geometry of our slit diffraction patterns discussed above was determined by wavelength. Distinguishing two central maxima apart is generally possible if the Raleigh criterion is met: The maximum of one pattern must fall at least at the first minimum of the other. Because the first trough of an Airy pattern is at a radius of $r = 1.22\lambda/2 \sin \theta$, the maxima of the two patterns must be at least a distance r from each other to meet this criterion.

This diffraction-based limit on detail in an image is referred to as resolution and is expressed in units of distance. If resolution is a small number, there will be much greater detail in the image, and this situation paradoxically is referred to as a "high resolution." If resolution is a larger number, there will be less detail in the image, and this (again paradoxically) is referred to as "low resolution." Clearly, it is dependent on the wavelength of the radiation and on the acceptance angle through which radiation coming from the specimen can be gathered. Without detailing the mathematic derivation and assumptions made, we present the equation usually used to calculate the ultimate limit of resolution achieved in an image:

$$d = 0.61\lambda/n \sin \alpha \qquad (5.7)$$

where d is resolution, λ is wavelength, n is the refractive index of medium between the specimen and the lens, and α is the acceptance angle.

As expected, resolution is proportional to wavelength, is inversely proportional to the acceptance angle α, and in a light microscope is inversely proportional to the refractive index of the medium positioned between the specimen and the front lens of the microscope. In order to make d as small as possible in the light microscope, we would want to maximize the acceptance angle and the refractive index of the immersion medium and use the shortest wavelength of radiation available. In order to estimate this limit of resolution for a light microscope, let us assume an acceptance angle of 90 degrees, a medium refractive index of 1.5 equivalent to that of immersion oil and a wavelength of 400 nm equivalent to blue light—the shortest wavelength available in the visible spectrum. By Equation 5.7, these parameters result in a resolution of about 0.25 μm, widely acknowledged as the limit in light microscopy.

One can radically change the resolution if one can use radiation of a much shorter wavelength such as electrons. This is indeed the basis for the 1000-fold increase in resolution routinely achieved in electron microscopes compared with light microscopes. This difference in resolution can be easily appreciated in Figure 5.11, in which ciliated epithelial cells imaged by light microscopy (C) are compared with the same cells imaged by electron microscopy (D).

■ Interference

The phenomenon of interference is best described by the wave model of radiation. If two wavelets are close enough to each other to interact, they will be summed to produce a new wavelet. Surprisingly, this summation can lead to either an increase or decrease in intensity. This is best described mathematically if the two wavelets are each represented by a sine function, as in **Figure 5.13.** Because sine functions oscillate between negative and

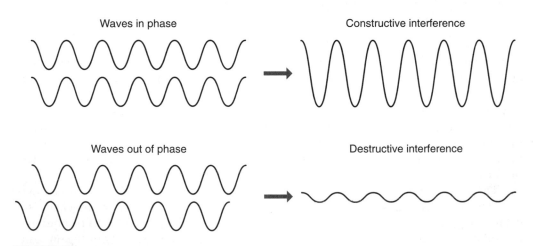

Figure 5.13
Constructive and destructive interference. Waves having the same phase add their intensities. Waves out of phase can produce subtractive intensities; such destructive interference can lead to a summed intensity of zero if the two waves are exactly 180 degrees out of phase.

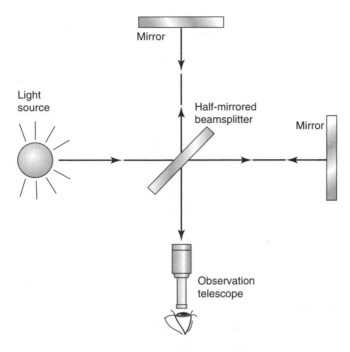

Figure 5.14
Michelson's interferometer. Destructive interference can result in easily detected changes in intensity if the waves being summed are slightly out of phase. In the case of Michelson's experiment, small changes in light velocity were being detected by their slight differences in phase at the detector.

positive values, how these functions sum depends very heavily on their relative phases. Both functions are said to be "in phase" or "coherent" with one another if their maxima and minima coincide with each other on the x-axis. In this case, summation results in a new wavelet having an intensity that is the sum of the two progenitor wavelets. Such an interaction is referred to as "constructive" or "positive interference" because the product wave has an intensity that is greater than that of either progenitor wave. On the other hand, if the two wavelets are "out of phase," that is, the maximum of one coincides with the minimum of the other, summation will result in a decrease of intensity in the product wave resulting in what is called "destructive" or "negative interference."

Interference can be observed by the experiment first performed by Michelson in the 1870s using what is referred to as an interferometer (**Figure 5.14**). If a coherent beam of light is split and each beamlet is required to follow a separate path before each can be recombined into a product beam for measurement, exceedingly small differences in path length for the two beamlets can result in negative interference. The path of one beamlet has to be only one-half wavelength greater than that of the other (about 200 nm for blue light). Figure 5.14 illustrates such an interferometer with the path of the two beamlets superimposed. Each beamlet follows a path involving reflection off of a mirror, and the position of each mirror is critical to the path length. Although it is almost technologically impossible to position each mirror with an absolute error of less than a micrometer, one can certainly measure relative changes in mirror position quite precisely. Let's suppose we move the mirror of one beamlet path ever so gradually to change path length in a predictable manner. As some point, the difference in path length between the two beamlets will become exactly an integral number of wavelengths, and as a result, the two beamlets will be exactly "in phase" when they recombine to form the more intense product beam. As we further move the one mirror, this difference in path length will change until the two beamlets are exactly "out of phase" when they recombine to form a product beam of lesser intensity due to negative interference. The overall result is that as we move the mirror the intensity of the product beam will go through a series of maxima and minima that are produced by interference between the two beamlets.

The interference phenomenon is used to great advantage in microscopy and physics. Note that Michelson's interferometer has two beamlet paths that differ in direction of travel with major portions being orthogonal to one another. The phase of each beamlet is just as dependent on the speed of light as on path length. Any difference in speed between the two beamlets could show up as a change in phase leading to negative interference. Using his interferometer, Michelson provided evidence that the speed of the two beamlets is identical, despite their difference in direction, and therefore, the speed of propagation is direction independent. In microscopy, interference is used to visualize organelles of cells because of their difference in refractive index. The methods of phase contrast and differential interference contrast are described in Chapter 7.

■ Polarization

The wavelength model of radiation propagation can be further modified to include the concept that the electromagnetic fields involved have x- and y-axis directionality if the z-axis is the direction of propagation. Normal, nonpolarized light is considered to be made of beamlets, each of which has an oscillating field oriented in a specific x/y direction; however, beamlet field orientation is random, and when observing a population of beamlets, there is no preferred orientation of the field for the group as a whole. This population overall will interact with matter as though the electromagnetic field is not oriented; however, one can select from this population only the beamlets that have a preferred field orientation by passing it through a polarizing filter. As illustrated in **Figure 5.15A**, the polarizing filter contains oriented chemical structures that allow only the correctly oriented radiation to pass through; the remainder is absorbed into the chemical structure of the filter, ultimately to be dissipated as heat. Polarizing filters are familiar from their use as sunglasses, which allow only a fraction of sunlight through in the preferred electromagnetic orientation. Because our eyes cannot detect the polarization of light, we see the result as simply a reduction in intensity.

After we have obtained so-called plane polarized light by use of a polarizing filter, we find from experimental observations that polarized light can reveal new structural features of a specimen that go unnoticed with nonpolarized light. If structural features of a specimen are lined up along a certain axis, as is the case in a variety of minerals and in many biological systems, such as the parallel alignment of myosin thick filaments in skeletal muscle cells, the specimen will exhibit what is called *birefringence*. Birefringence refers to the fact that polarized light may be either transmitted through the specimen or its transmission blocked because of the fact that its electromagnetic field alignment is inappropriate. The specimen is actually acting as its own polarizing filter! Also observed is the fact that some specimens, in letting light pass, can actually rotate the plane of polarization in a beam. This effect is not noticed if the incident light is nonpolarized, but if the light is polarized before it enters the specimen, this rotation can actually be measured (Figure 5.15A). To do so, we not only need a filter to polarize the incident light—the "polarizer"—but we also need a similar filter behind the specimen—the "analyzer" that monitors the polarity of light emerging from the specimen.

The orientation of the polarizing and analyzing filters must be known. The analyzer filter is then rotated until maximal light intensity is obtained; the orientation of the analyzer

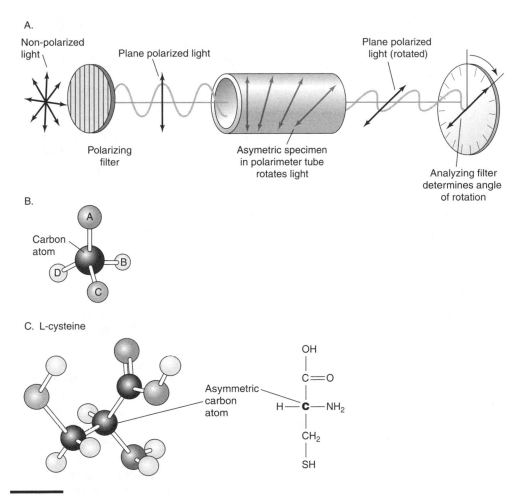

Figure 5.15
Asymmetric molecules and light polarization. (A) Optical design of a polarimeter that can measure the rotation of polarized light by an asymmetric specimen. (B) An asymmetric carbon atom having four optically different substituents. (C) Such an asymmetric carbon atom in L-cysteine (arrow).

is then noted and compared with the orientation of the polarizing filter. If the two filter orientations are different, then the plane polarized light entering the specimen must have had its field orientation rotated by the specimen. To perform this rotation, the structure of the specimen does not have to be highly oriented as does that of a polarizing filter; indeed, it can accept light polarized in any orientation and rotate it.

A simple example is the rotation of polarized light produced by a solution of sucrose. Sucrose, or table sugar, comes in two geometrically different forms, and table sugar is usually a mixture of both. Sucrose contains a carbon atom that is bonded to four different chemical groups and for that reason is referred to as an "asymmetric carbon" (Figures 5.15B and 5.15C). Three groups can be arranged either clockwise or counterclockwise around the fourth group, and these two structures are not superimpossible! This gives rise to two different geometrical forms of sucrose called enantiomers that have the same chemical formula but slightly different optical properties. If one isolates a pure enantiomer of sucrose and puts it in an aqueous solution, it is found that this solution will rotate polarized light and that the degree of rotation is dependent on the concentration of sucrose enantiomer. Despite the fact that sucrose molecules in solution are randomly oriented and that for any single molecule this orientation is changing rapidly with time, polarized light will still interact with these molecules in a non-random manner. Clearly, the absolute orientation of the molecule is not as important as the presence of an asymmetric carbon atom. This is emphasized by the fact that enantiomers are exact mirror images of each other, and they will both rotate polarized light but in opposite directions! Likewise, absence of an asymmetric carbon atom in any organic molecules will lead to it not being able to rotate polarized light.

An easy way to visualize the effects of such chemical asymmetry is to consider light to be propagating as two helices of opposite polarity, that is, a left-handed helix and a right-handed helix, but of equal magnitude. If these two helical components are in phase with each other, they will sum to produce what appears to be plane polarized light. If one helical component travels slightly slower than the other, however, they will become out of phase with one another. This phase difference will lead to a change in the perceived plane of polarization. This model of circular polarization of radiation is useful in describing differential interference contrast microscopy (see Chapter 7).

■ Absorption and Emission of Radiation

Using the particle model, we can describe radiation as consisting of discrete packets of energy that when applied to visible light are called photons. A photon is the minimal unit of light energy that can interact with matter. There is no experimental evidence for the existence of one half of a photon any more than there is evidence for the existence of one half of an electron. The amount of energy associated with a photon, however, is related to its wavelength (color), with greater energy associated with photons of shorter wavelength. Again, we must have integration of the particle model and wave model to describe electromagnetic radiation adequately.

The energy associated with photons and with the electrons in an atom or molecule is transferable! Electrons within an atom or molecule exist at a series of discrete energy levels referred to as electron orbitals or molecular orbitals. At the atomic level, electrons are not easily described as particles spinning around the nucleus, even though this analogy is frequently used in elementary chemistry texts. Indeed, it is the basis for the use of the word "orbital" in analogy to our solar system. At the molecular level, it is common to describe each electron in an orbital as a wave, albeit a localized wave. Consistent with a wave model, the electron cannot be localized to a particular position at a particular point in time like a planet can, but rather, is expressed as a probability distribution in space—some times referred to as an electron cloud. Electrons in orbitals associated with a specific nucleus (atom) exhibit a discrete series of energy levels referred to as "shells." These shells form a similar series in every atom, but the number of shells that are actually filled with electrons depends on the element and its net charge.

If a photon of light has the same amount of energy as the energy difference between two electronic shells and the quantum comes sufficiently close to the atom, its energy can be absorbed by an electron thereby "kicking" the electron up to the next energy level. As a result, individual atoms of noble gases and ions of many elements are able to absorb light only at very discrete series of wavelengths that are characteristic of the particular element. This principle is put to practical use in an atomic absorption spectrometer, whereby ions vaporized in a flame absorb light in proportion to the concentration of the element.

The same phenomenon can run backward such that loss of energy that accompanies an electron dropping from a shell of high energy to a shell of lower energy can be used to create a photon, a process known as emission. Light emission requires that the electrons of an atom or molecule first be "bumped up" to higher energy levels by input of chemical, electrical, heat, or light energy, with subsequent emission of light as the higher energy electrons fall back to lower energy orbitals. Again, the wavelengths of the photons emitted are precise and exactly equivalent to the discrete differences in energy between the electron orbitals involved. Practical examples of this include sodium vapor lamps used as street lights and fireworks. Sodium vapor whose electrons are continually excited by an electric discharge continually emit yellow light at a wavelength of 588 nm whose photon energy is exactly the energy difference between two electronic orbitals. Inorganic salts in fireworks, when heated to high temperature by gun powder explosions (thereby placing their electrons in high energy orbitals), emit light of specific wavelengths (colors) characteristic of the elements involved. Strontium salts produce bright red emissions. Barium chloride emits green light, whereas copper chloride provides blue emissions.

In organic molecules, electron orbitals are used to form covalent bonds. In this case, the orbital becomes delocalized to now include a relationship with two or more nuclei. The "sigma" orbitals that are involved in simple carbon–carbon and carbon–hydrogen bonds envelope only the two nuclei of the atoms they bond; however, formation of carbon–carbon, carbon–oxygen, or carbon–nitrogen double bonds involves the formation of "pi" orbitals. Pi orbitals associated with each C, O, or N atom have an axis that is orthogonal to the axis of the sigma orbitals, as shown in **Figure 5.16A**. The formation of double bonds involves the interaction of the pi orbitals associated with each nucleus to form extended pi orbitals that include both double-bonded nuclei. If double bonds are alternated ("conjugated" double bonds), the pi orbital electrons can become further delocalized to include interaction with all of the nuclei that make up the system of conjugated double bonds. A good example is the aromatic ring of benzene, as shown in Figure 5.16B. Electrons in the pi orbital are each associated with all six carbon nuclei of the ring, forming what is termed a "molecular orbital." The larger the system of conjugated double bonds, the more extended the molecular orbital and the more electrons involved. Also observed is the fact that the more extended an orbital is, the closer together are the separate energy levels within the orbital. As one might predict, the spacing between these energy levels determines what wavelengths of light can be absorbed; therefore, a molecule having a limited system of conjugated double bonds may absorb high-energy blue light because of the wide spacing of the orbital energies, while a molecule with an extensive conjugated double-bond system will absorb low-energy red light because of the narrow spacing of orbital energies.

As shown in Figure 5.16D, organic molecules in solution exhibit broad absorption peaks rather unlike noble gases and ionized elements that show absorption only at very specific wavelengths. Both types of data are referred to as an absorption spectrum. For gases or ions in the gaseous phase, these are referred to as "line spectra" because absorption is confined to specific wavelengths. In contrast, the spectrum from molecules in solution consists of broad peaks and is referred to as a "continuous spectrum."

Why are the absorption peaks so wide? Could they be due to very narrow spacing between orbital energies? The answer is no—broad peaks are not caused by narrow orbital spacing. Rather, it is due in part to the fact that the energy of any given orbital can fluctuate in time because of stretching and rotation of the bonds that these orbitals are involved in. The higher the temperature, the more frequent are these bond motions. Indeed, if we cool organic molecules to very low temperatures to stop all molecular motion, these absorption peaks become very narrow, reflecting that the molecular orbital energies are now relatively stable and no longer being bounced up and down by the twisting and bending of bonds. Even at low temperatures, however, the spectra of organic molecules will not become the line spectra seen in the gaseous state. This is because our organic molecules are embedded in solvent molecules, which if close enough will alter the orbital energies of our organic molecule. At higher temperatures, this influence of solvent molecules will be constantly changing with time because of their motion. At lower temperatures,

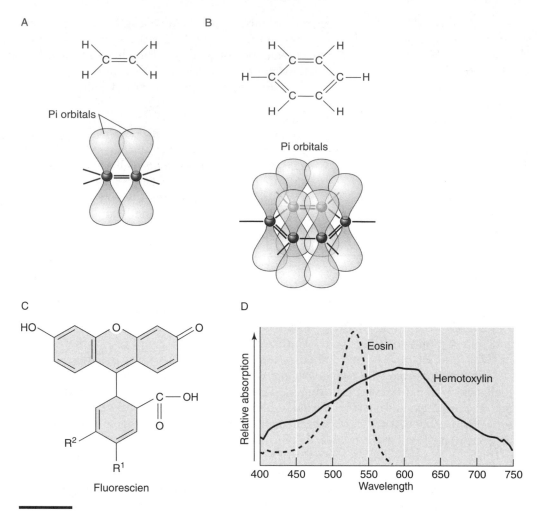

Figure 5.16
Pi orbitals in organic compounds. (A) The double bond in ethene involves two sets of overlapping pi orbitals. (B) Aromatic rings such as that of benzene have overlapping pi orbitals throughout resulting in delocalized electronic orbitals. (C) Dyes that absorb visible light have long chains of conjugated double bonds that represent highly delocalized molecular orbitals. In fluorescein, these molecular orbitals extend over four aromatic rings. (D) Absorption spectrum of eosin and hematoxylin. Dyes in solution have extremely broad absorption peaks which represent a probability distribution for absorption at each specific wavelength. The breadth is due to individual molecules having different kinetic and bond rotational energies as well as different interactions with multiple solvent molecules.

their influence is less but still present because of the fact that each individual organic molecule is "frozen" into a configuration of solvent molecules that is slightly different than the next.

Thus, our wide absorption peaks for organic molecules in solution are due to the fact that our spectral measurements are on a large population—millions—of molecules and each peak is a probability distribution for the numerous environmental differences within the population. If we could take a ride on a single photon as it passes though this population, we would find that the photon would be annihilated, that is absorbed, only when it came close to an individual organic molecule whose difference in orbital energy levels exactly matched that of the photon at that particular moment. Our photon may have had to pass thousands of other structurally identical molecules before it found the one with exactly the right amount of bond motion and solvent environment contribution that tuned the electronic transition energy to exactly that of the photon. If the wavelength of the photon was near the maximum of the absorption peak, it would have a much greater chance of finding that "right" molecule than if it had an energy at the shoulder of the peak; worse yet, if the energy of the photon was relatively far from the peak, it would have almost no chance of finding "Mr. Right."

The same arguments apply to why light emission from organic molecules in solution does not occur at a precise wavelength but rather is distributed over a broad range of wavelengths. Indeed, such emission spectra typically exhibit broad peaks rather than narrow lines of emission. It is not surprising then that colored dyes used in light microscopy are organic molecules containing systems of conjugated double bonds. As shown in Figure 5.16C, fluorescein, a yellow-green dye, has a conjugated double-bond system associated with 20 atomic nuclei. Such delocalized molecular orbitals are found in all organic dyes. Similar double-bond systems result in the absorption of blue light by eosin and the absorption of red light by hematoxylin (see absorption spectra in Figure 5.16D). Hematoxylin absorbs lower energy red light because it has a larger conjugated double-bond system than does eosin. Tissue components stained with hematoxylin appear blue because as white light (containing all wavelengths of light in the visible spectrum) passes through the stained area, hematoxylin absorbs the red, leaving the blue to pass on through to be detected by our eyes or camera. Conversely, structures stained with eosin absorb the blue portion of the incident white light and allow the red to pass through and be detected.

■ Qualities of Electromagnetic Radiation Summarized

Specific terminology is used to describe the qualities of electromagnetic waves and wave-like particles such as electrons (**Figure 5.17**). Wave populations that oscillate at the same frequency are said to be **monochromatic**. If the population oscillates at a variety of

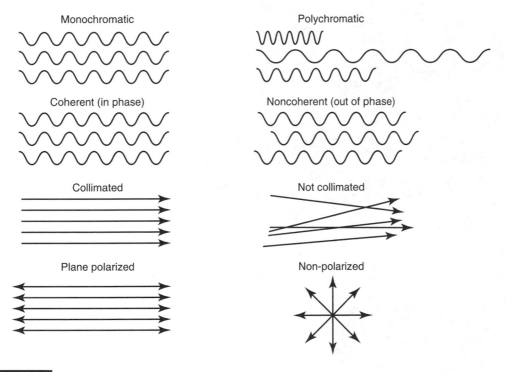

Figure 5.17
Terms used in describing light beams.

frequencies, they are referred to as **polychromatic.** A population of waves that are "in phase" with each other are said to be **coherent.** Waves that are not in phase with one another are considered to be **noncoherent.** Wave populations that are moving closer together are **convergent;** wave populations becoming further apart are **divergent.** Waves traveling exactly parallel with one another and neither converge or diverge are said to be **collimated.** Wave populations whose electric vectors oscillate in the same plane are said to be **polarized** or **plane polarized.** In contrast, wave populations whose electric field vectors oscillate at random orientations are said to be **nonpolarized.** These terms are used throughout this book to describe both wave-like radiation and particles.

References and Suggested Reading

Books

Bass M, Van Stryland EW, Williams D, Wolfe W, eds. *Handbook of Optics 1: Fundamentals, Techniques and Design,* 2nd ed. New York: McGraw-Hill, 1995.

Born M, Wolf E. *Principles of Optics,* 7th ed. New York: Cambridge University Press, 1999.

Falk DR, Brill DR, Stork DG. *Seeing the Light: Optics in Nature, Photography, Color, Vision and Holography.* New York: John Wiley and Sons, 1985.

Hecht E. *Optics,* 4th ed. San Francisco: Addison-Wesley, 2002.

Jenkins F, White H. *The Fundamentals of Optics.* New York: McGraw-Hill, 1976.

Pedrotti FL, Pedrotti LM, Pedrotti LS. *Introduction to Optics,* 3rd ed. New York: Prentice Hall, 2006:656.

United States Bureau of Naval Personnel. *Basic Optics and Optical Instruments.* Mineola, NY: Dover Publications, 1969.

Websites

Davidson MW. *Molecular Expressions Optical Microscopy Primer.* www.micro.magnet.fsu.edu/primer.

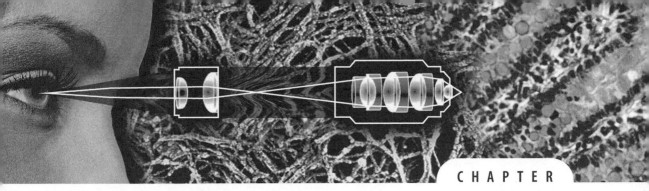

The Light Microscope and Image Formation

The first compound microscopes built in the 17th century represented a dramatic improvement over a single lens. The main advantage was that the magnifications achieved were much higher because they occurred in two stages: an objective lens and an ocular lens. A single lens, if properly shaped, can form a virtual image or a real image depending on the focal length of the lens and the distance of the lens from the object (**Figure 6.1**). A virtual image is one that is seen "through" the lens. It appears to be in front of the lens but cannot be projected onto film or used by another lens as an object. On the other hand, a real image is formed in back of the lens, and this image can be projected or used by another lens. The specimen, an arrow, is visualized by light that is either reflected off of the object or has passed through the object. From each point on the arrow, light rays emanate in all directions, and it is the job of the lens to capture as much of this light as possible and redirect the rays so that they can recombine to form an image point. The real image formed on the opposite side of the lens is inverted (both top to bottom and right to left), and it is formed at a specific distance from the lens.

The focal length of a lens, F, is the distance at which the lens can bend parallel rays to a point behind the lens, and this point is referred to as the focal point. The front focal point, F′, is a point an equal distance in front of the lens. Optical experiments have shown that the distance of an object from the lens, A, in relationship to the focal length of the lens determines what the lens can do with the light coming from the object. As shown in Figure 6.1A, if the object distance is greater than twice the focal length, the lens forms a real image, but the image is smaller than the object itself. As we bring the object closer to the lens so that A = 2F, the real image formed becomes exactly equal to the size of the object (Figure 6.1B). Bringing the object closer yet (2F > A > F) starts to magnify the real image formed, a situation that is desirable in microscopy (Figure 6.1C). As the object is brought closer and closer to the front focal plane of the lens, the real image gets larger and larger and is positioned farther and farther in back of the lens. Finally, when the object reaches the front focal plane (A = F), the image formed would be infinite in size and an infinite distance from the lens—that is, no image is formed (Figure 6.1D). Bringing the object any closer to the lens (A < F) results in no real image forming, but it does result in a virtual image that can be seen through the lens as it is in a magnifying glass (Figure 6.1E).

In a compound microscope, the objective lens gathers light from the specimen and diffracts these rays to form a real "intermediate" image at a plane within the tube of the

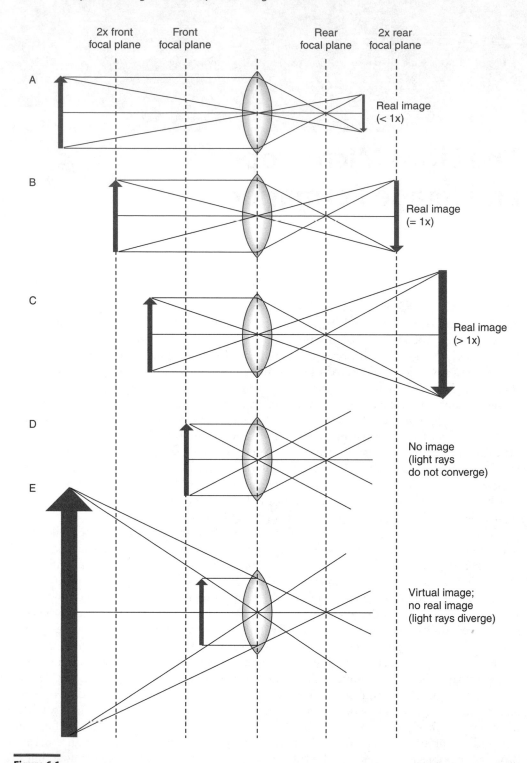

Figure 6.1

Image formation by a convex lens. Image formation depends on the focal length of the lens and the distance of the object from the lens. (A) If the object is farther from the lens than two times the focal length, a real, demagnified image is produced. (B) If the object is exactly two times the focal length from the lens, a real image of the same dimensions is formed. (C) If the object is between one and two focal lengths from the lens, a magnified real image is formed. This is the situation desired in a microscope. (D) If the object is exactly one focal length from the lens, no image is formed. (E) If the object is closer than one focal length to the lens only a virtual image is formed.

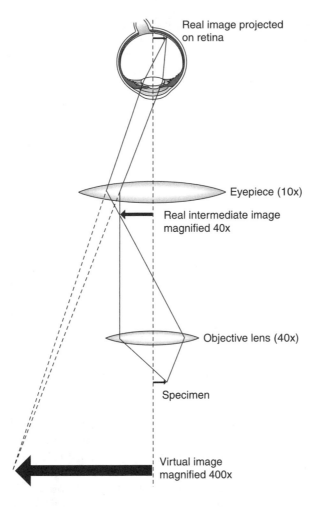

Real image projected on retina

Eyepiece (10x)

Real intermediate image magnified 40x

Objective lens (40x)

Specimen

Virtual image magnified 400x

Figure 6.2
Optics of a compound microscope. The objective lens produces a real intermediate image that is further magnified by the eyepiece and focused onto the retina as a real image. The magnification of the virtual image created by the brain can be calculated by multiplying the power of the objective lens times that of the eyepiece (ocular lens).

microscope (**Figure 6.2**). The real intermediate image is designed to occur at the front focal plane of the ocular lens, which provides the second stage of magnification. This lens gathers the rays, making up the intermediate image, and diffracts and projects these rays onto the retina of the observer, thereby forming a second real image—the image relayed to the brain. The acute convergence of these rays onto the lens and retina of the observer produces the perception of a virtual image that is greatly magnified compared with the specimen (dashed lines, Figure 6.2).

■ Image Formation Requires Diffraction and Interference

The theory of image formation used today was formulated in 1876 by Ernest Abbe. Abbe, a pioneer in both optical theory and design, worked initially at the University of Jena but later teamed up with entrepreneur Carl Zeiss to turn his new lens designs into production of the finest microscopic equipment available at that time. Remarkable is the fact that Abbe's designs for both condenser and objective lenses achieved an NA of 1.4 and allowed image resolution to reach the theoretical limit of 0.25 μm.

Abbe's basic assumption was that incident light either passed straight through the specimen or interacted with it by diffraction and deflected at an angle compatible with diffraction theory. Gathering of both 0th-order (nondiffracted) light and nth-order diffracted light from each specimen point by the lens enabled this light to be bent and redirected to converge at an image point. Summation of these light rays at the image point was accompanied by constructive and destructive interference phenomena that determined the overall light intensity of that image point. In order to understand image formation better, we must understand diffraction at the specimen, propagation of light waves through the lens, and summation of light rays by interference.

Diffraction at the specimen is the *sine quo non* of image formation. If no light is diffracted or absorbed by the specimen, there is no information in the light emerging from the specimen that the specimen actually exists. A good example is a flat piece of glass that is completely transparent. The surface of the glass is so smooth that any irregularities are much smaller than the wavelength of visible light. The internal molecular structure of the glass is much smaller in detail than the wavelength of visible light, and thus, from the standpoint of the light ray, it is uniform. The electronic orbital energies of glass are such that their differences do not correspond to the energy levels of visible light; therefore, light is not absorbed. (In contrast, glass does absorb energies in the ultraviolet and microwave regions of the spectrum.) This is fortunate because this means that the glass lenses of our light microscope will not absorb light that would otherwise be used to form a specimen image.

After we put a scratch in the piece of glass, however, the story is completely different! This is because we now have a feature that is as large or larger than the wavelength of light being used for imaging. A more subtle example is that of a specimen that contains numerous particles (it could even be a layer of dirt). If the particles are much smaller than the wavelength of light, then they are effectively invisible. As the size of the particles increase to near the wavelength of light, however, light rays are dispersed by the particle surfaces much like they would be by the edges of a slit. Rays interacting with the opposite edges of the particle can be recombined to form a diffraction pattern much like that formed by a slit (Chapter 5). The diffraction pattern is produced by constructive and destructive interference of the waves dispersed by opposite surfaces of the particle and provides evidence (from the standpoint of the light) that the particle exists. Of greater importance is that the diffraction pattern also provides information on the size and shape of the particle.

A good example of such a diffraction pattern can be seen if one shines a laser pointer through a copper electron microscope grid and the pattern is visualized on a screen placed in back of the grid. If the grid contains parallel slits, as shown in **Figure 6.3A,** it will give rise to the diffraction pattern shown in Figure 6.3D. Likewise, a grid containing an orthogonal pattern of opaque grid bars between which are square openings through which light can pass (Figure 6.3B) will give rise to the diffraction pattern shown in Figure 6.3E. In each case, edges of the grid bars disperse the light that forms the diffraction pattern shown. Diffraction spots show only when there is constructive interference between waves from two or more edges, and this occurs only at specific angles from the optical axis. The criterion for constructive interference, as shown in **Figure 6.4,** is that the angle should be such that the difference in pathlength between waves from different edges (slits in the case of Figure 6.4) should be an integral number of wavelengths so that the waves will remain in phase with each other. This criterion is fulfilled by angles for which the following equality holds where λ is the wavelength of light, d is the distance between diffraction sites in the specimen, and n is an integer:

$$\sin\theta = n\lambda/d \qquad (6.1)$$

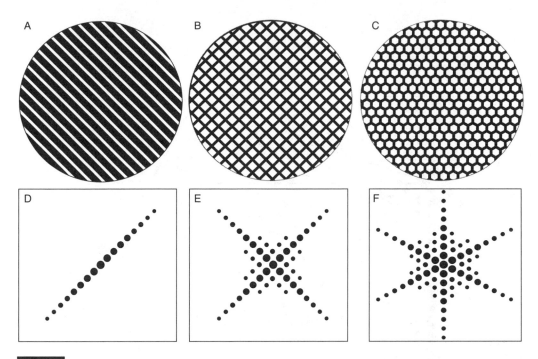

Figure 6.3
Diffraction patterns resulting from repeating specimen features. Simulated diffraction patterns, like those observed at the back focal plane of an objective lens are shown for an array of slits (B), a square array (D), and a hexagonal array (F).

This criterion and equation describes Rayleigh's Law named after the 19th-century physicist who discovered this relationship. By looking at this equation carefully, one can see that diffraction angles are related to wavelength and are inversely related to the distance between features in the specimen. Three important conclusions that can be drawn from this equation are easily verified by experiment. First, two or more specimen features, each diffracting light and each a distance d from each other, will produce not just one intensity but a series. The first intensity at the optical axis represents 0th-order light that passed right through the specimen without interacting with it. The first-, second-, and third-order diffraction intensities are seen at angles in which the path lengths of rays diffracted from adjacent features differ by 1, 2, and 3 wavelengths, thereby keeping them exactly in phase. The pattern of diffraction lines formed is orthogonal to the optical axis and to the vector d (i.e., a straight line passing through both of the diffracting features). For example, in Figure 6.3A, the slits in the grid are –45 degrees from vertical, whereas the diffraction pattern formed in Figure 6.3D is orthogonal at +45 degrees from vertical.

Second, the closer the features of the specimen are to each other (i.e., as d gets smaller), the larger are the diffraction angles. This suggests that the larger features of the specimen give rise to a diffraction pattern that is at a low angle, that is, close to the optical axis, whereas the smallest details of the specimen give rise to diffraction intensities that are at the periphery of the diffraction pattern. Mathematically, it can be shown that features in the specimen separated by a distance d give rise to a series of intensities in the diffraction pattern that are $(nf\lambda)(1/d)$ from the optical axis where f is the focal length of the lens. For this reason, the diffraction pattern is said to be an inverse transform of the image pattern. As a result of this, the ability of a lens to resolve fine detail of a specimen

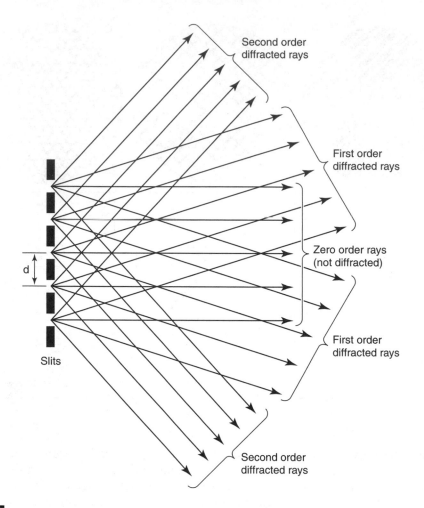

Figure 6.4
Angles of diffraction and the Raleigh criterion. Light diffracted from a series of slits is bent at angles to the optical axis determined by Raleigh's equation. Both diffracted (first and higher orders) and nondiffracted (zero order) light must be captured and refocused by the objective lens to a series of image points representing the slits.

is dependent on its ability to gather diffracted light at the periphery of the lens, and any geometrical factors that limit this (such as the acceptance angle) also limit resolution. In fact, for any detail that is so small that the lens cannot capture both its 0th-order rays and its first-order diffracted rays, an image cannot be formed because interference of these two groups of rays at the image plane will be impossible.

Applying these rules to the small particles discussed previously, the diameter of the particle, representing the distance between two diffracting surfaces, must be at least large enough such that its first-order diffraction waves, converging on the back focal plane of the objective at a distance of $(f\lambda)(1/d)$ from the optical axis is smaller than the radius of the back aperture, a radius that is usually not larger than that of the objective lens itself.

This represents the maximal resolution that the objective lens can obtain, a value that is achieved only when the acceptance angle of the condenser lens is at least as great as that of the objective lens. A lower acceptance angle in the condenser lowers the resolution of the objective lens in the same manner that an aperture of smaller radius at the back focal plane of the objective would.

Nevertheless, if the first-order diffraction waves from a small particle or specimen feature are excluded from the image, the feature may still be detectable because of its 0th-order waves. These zero-order waves, however, will form an Airy diffraction pattern in which the size is independent of the size of the feature. Thus, any particle or feature that is smaller than the ultimate resolution of the lens, if detectable, will appear artificially larger than it actually is, and no size or shape information will be available. An excellent example is the fact that microtubules, having a diameter of 0.025 μm, which is below the resolution limit of 0.25 μm, are detectable but will appear to have a diameter of 0.25 μm rather than 0.025 μm.

The ability of the lens to gather diffracted light is dependent not only on the physical size of the lens but also on its shape, index of refraction, and the index of refraction of the medium between the lens and the specimen. First, the index of refraction of the lens itself compared with that of air is ultimately responsible for making diffracted light reconverge to form image points. This will not happen if the index of refraction is too low, and in the 19th century, newly developed glass formulations having a higher index of refraction began to be used for making microscope lenses. An ultimate material for making lenses would be diamond, having a refractive index of 2.4, but its cost and hardness make this prohibitive. Second, the design of the lens is important in that light rays at the highest angles need to undergo internal reflection in order to be brought to a proper convergence angle. Third, the medium between the lens and specimen will determine the angles of refraction both at the coverglass–medium interface and at the medium-lens interface. Using a medium such as oil, in which the refractive index exactly matches that of the glass used, reduces these angles of refraction to zero, as is shown in **Figure 6.5**. This increases the maximum acceptance angle from that of a "dry" lens at about 40 degrees (NA = 0.95) to the acceptance angle of an oil immersion lens, about 66 degrees (NA = 1.4). The limitation on acceptance angle experienced by dry lenses actually occurs at the coverglass. Diffracted light rays at an angle of greater than 41 degrees from the optical axis undergo total internal reflection at the coverglass-air interface and never reach the lens.

Because of refraction, the lens will bend the gathered light, forcing it to converge and form a diffraction pattern at the back focal plane of the lens, as shown in **Figure 6.6**. The zero-order rays will converge on the optical axis, and the diffracted rays will converge

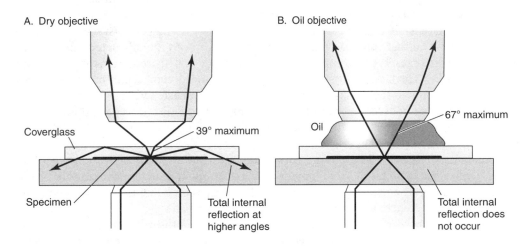

Figure 6.5
Acceptance angles of dry and oil objective lenses. (A) The acceptance angle of a "dry" objective lens is relatively low because of the sharp angles at which light from the specimen is refracted at the air-glass interfaces. (B) Oil between the objective and specimen eliminates refraction, thereby increasing the acceptance angle; however, oil can only be used with lenses that are designed for that purpose.

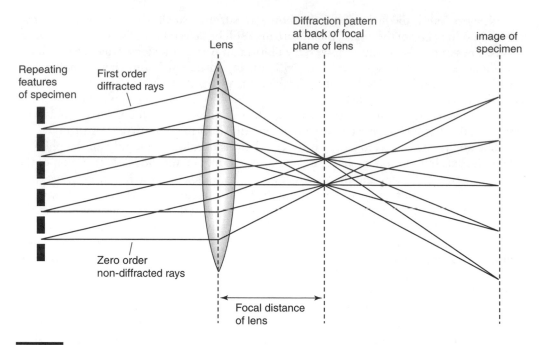

Figure 6.6
Image formation according to Abbe. Light striking a specimen is split into nondiffracted (zero order) and diffracted (first, second, and higher order) light rays that are captured by the lens. The lens bends these rays so as to converge and form an image at the image plane. The same light sampled at the back focal plane forms a diffraction pattern. For clarity, only one set of first-order diffracted rays is shown.

into a diffraction pattern off the optical axis. After passing through the back focal plane, both zero-order rays and diffracted rays will converge on each other (providing that they originated at the same specimen point) to form the image, typically in a plane that is 5 to 20 focal lengths from the lens so that the image will be magnified relative to the specimen.

Formation of the diffraction pattern at the back focal plane and formation of the image at the image plane is dependent on constructive and destructive interference of the converging rays so as to produce the variations in intensity that make up these patterns. Because interference is dependent on the relative phase differences between converging waves, it is critical that the relative phase differences of both the diffracted and nondiffracted waves be preserved throughout the entire process. This requirement starts with the light source. The laser pointer in our experiment meets this requirement extremely well—all light emitted from a laser is coherent, that is, in phase. Other light sources have varying degrees of coherence. Incandescent lamps and sunlight have partial coherence, whereas light emitted by uncoupled molecules such as fluorescence is completely incoherent and cannot be used as a light source for imaging. The most common light sources for microscopy are incandescent, and these are sometimes thought of as being incoherent. In reality, the fact that the hot filaments in these lamps contain large regions of atoms that are vibrating in synchrony results in their light emission being locally coherent, and this localized coherence of "raylets" is vital to image formation.

Second, the nth-order diffracted rays as well as the 0th-order nondiffracted rays must have their relative phase preserved as they are gathered and redirected by the lens. Maintenance of phase is not obvious because these rays, as shown in **Figure 6.7**, undergo refraction at lens surfaces, clearly follow geometrical paths of different lengths, and pass through points on the lens differing in thickness. Nevertheless, relative phase is, in fact, maintained for the following reason: The change in refractive index within the lens results

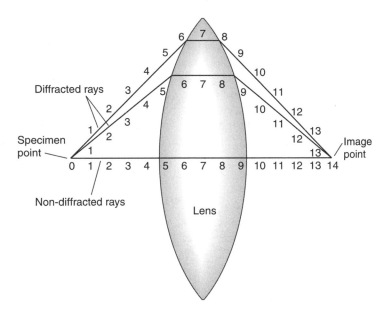

Figure 6.7
Paths of image-forming light rays. Light coming from a single specimen point (both diffracted and nondiffracted) is refocused by the lens to an image point. Regardless of the path, the number of wavelengths traveled is the same.

in a change in light velocity that exactly compensates for the difference in pathlength. Within the lens, light speed is reduced by a factor that is equivalent to the refractive index. Because the frequency of the electric field oscillations does not change, the wavelength of the light within the lens must become shorter. As a result, the nondiffracted 0th-order rays at the optical axis undergo many more oscillations within the lens due not only to the greater thickness of the lens in this region but also to this shortened wavelength. In contrast, the nth-order diffracted rays go through a much thinner region of the lens but a much longer geometrical distance in reaching the periphery. The surprising (but experimentally verified) finding is that all orders of rays, regardless of the point at which they passed through the lens, have undergone an equal number of oscillations at the time they recombine to form either a diffraction pattern at the back focal plane or a real image at the image plane. This concept is simulated in Figure 6.7 by the numbers along each optical path that in each case reach 14 at the image point.

Another interesting outcome of Abbe's theory is that wave interference, necessary for the array of light intensities we see in an image or in a diffraction pattern, in no way actually affects the light rays themselves! Because we sometime speak of "constructive" and "destructive" interference, we might be led to believe that light waves or photons are actually being created, annihilated, or combined during this process. This is clearly not the case. The light waves that are undergoing interference at the back focal plane to produce a diffraction pattern are exactly the same light waves that are propagating onward, unchanged, to form an image at the image plane! Likewise, light waves capable of forming an intermediate image within the microscope tube (with its implied interference patterns) are exactly the same as those being used by the eyepiece to project a real image onto the retina. Rather, interference is a spatially and temporally restricted process whereby the electric vectors of two waves transiently come close enough together to be registered as one summated wave. In fact, summated waves cannot be detected unless we put a barrier at the plane such as a piece of paper, film, or our retina and in doing so halt the propagation of the component waves.

■ Lens Aberrations

We have now described a perfect lens that performs in an ideal manner. Real lenses, however, are not perfect and have several types of technical shortcomings, called "aberrations." There are four common types of aberrations that microscopists have to deal with—chromatic, spherical, planar, and comac aberrations. The first, chromatic aberration, is due to the fact that real lenses often bend short-wavelength blue light more strongly than longer wavelength red light. As a result, the focal distance of a biconvex lens is shorter for blue light than it is for red light (**Figure 6.8A**). This means that an image

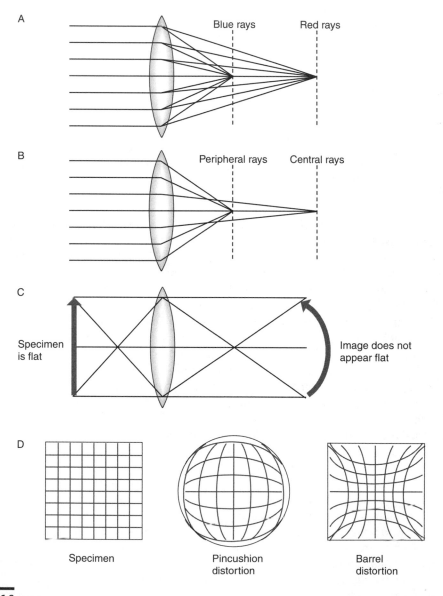

Figure 6.8

Common aberrations in lenses. (A) Chromatic aberration results in light of different wavelengths being focused at different distances. Usually short wavelength light (blue) is bent more strongly than long wavelength light (red). (B) Spherical aberration results in peripheral rays being bent more strongly than central rays. (C) Planar aberrations result in a flat specimen plane being visualized as a curved image plane. (D) Common planar aberrations include the pincushion and barrel distortions.

composed of multiple colors will never be completely in focus. There is no true focal plane. One way to minimize this problem is to use a blue or green filter to make the incident light monochromatic; another advantage of using a filter is that imaging with light of short wavelength provides better resolution if one is working at high magnifications. This solution is fine for black and white imaging but cannot be used if specimen color is important.

Fortunately, correction of chromatic aberration by appropriate lens design is not costly. Using two lenses made of differing types of glass, the chromatic aberration of one lens can be equal and opposite to the chromatic aberration of the other, thereby allowing these aberrations to cancel each other. Objective lenses corrected for chromatic aberration are referred to as **achromatic** lenses. It is actually difficult to buy an objective lens that has not been corrected in this manner unless you are shopping at a toy store.

The second, spherical aberration, is due to real lenses tending to bend light more strongly at their periphery than at their center. As a result, light rays gathered from the lens periphery converge at a shorter focal distance than do rays gathered from the center (Figure 6.8B). Because light interacting with each specimen point is processed by both central and peripheral parts of the lens in order to reconverge to a single image point, every single point in the image is affected. Peripheral regions of the image will be affected more drastically than central regions of the image because more light rays are being processed by peripheral parts of the lens. One way to minimize spherical aberration is to use only the central region of the lens. This is easy to do simply by using a circular aperture in front of the lens. This solution is at the cost of image detail because the angle of acceptance will be smaller and the resolution that can be achieved in the image reduced. In light microscopy, reduction of resolution is often unacceptable; instead, one often uses a convex lens and a concave lens in series to allow spherical aberrations to cancel. In electron microscopy, however, spherical aberration is severe and unavoidable and is one of the major reasons that the resolution obtainable in practice is approximately 3% of that theoretically possible (see Chapter 8).

Correction of objective lenses for spherical aberration comes at a cost. The price of corrected objective lenses is increased about threefold and ranges typically from hundreds of dollars to several thousand dollars. An objective lens that has been corrected for both spherical aberration and chromatic aberration is called an **apochromatic** lens. From a practical standpoint as well as a cost standpoint, there is no reason to make a lens that has spherical correction but not chromatic correction. A cutaway view of an apochromatic objective lens is shown in **Figure 6.9,** which illustrates the extra optical elements that increase cost.

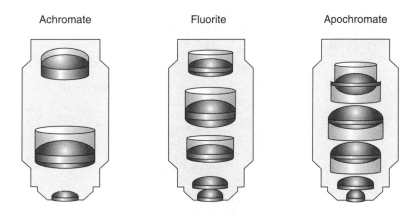

Achromate   Fluorite   Apochromate

Figure 6.9
Objective lens design. Achromate lenses are relatively simple, typically having three elements, including a front lens and two multicomponent elements. Fluorite and apochromatic lenses are more highly corrected and typically have five sets of elements each set having several components.

The third, planar aberration, is due to the fact that the image is systematically distorted compared with the specimen, even though it is clearly focused. This distortion may go unnoticed in many specimens, making the problem rather insidious; however, it is easily noticed if the specimen has repeating unit structure such as that of a diffraction grating or a miniature crossword puzzle (Figures 6.8C and 6.8D). Distances may be expanded either at the center of the field, giving rise to a "pincushion" effect, or at the periphery of the field, giving rise to "barrel" distortion. Thus, what is a flat plane in the specimen appears as convex or concave surface in the image; landmarks or edges in the specimen that should appear straight appear curved. Correction of these aberrations to produce a planar image is relatively costly because additional optical elements must be added to the objective lens. Lenses corrected for all three types of aberrations are costly indeed, typically fetching $5000 to $10,000. These lenses are referred to as planapochromats and are found only on high-end research microscopes.

The fourth, comac aberrations, results in formation of an image that is not centered on the optical axis. Coma aberrations, usually accompanied by planar aberrations, can result from a lens element that is slightly asymmetric or is not perfectly mounted in a plane orthogonal to the optical axis; however, another common source of comac aberration is the fact that lenses and other optical elements may focus light of differing wavelengths at slightly different positions in the x–y image plane. That is, light of one wavelength may be centered on the optical axis, whereas light of a different wavelength may be focused off-axis. The ability of an optical system to center and superimpose the image patterns formed by multiple colors of light is referred to as "accurate color registration." This property is very important to color imaging and to techniques such as immunocytochemistry that use widely differing wavelengths in the same image.

■ Objective Lenses

The objective lens is the single most critical optical component of the light microscope. It will be the largest determinant of resolution, magnification, image aberration, color fidelity, and light throughput. All modern objective lenses have multiple optical elements (lenses) that must be perfectly positioned within a metal barrel. Each optical element itself must be ground and polished to an exacting shape and mounted securely. The number and type of elements will depend on the magnification to be achieved, the aberrations corrected for, and the use of the objective lens. Several features are common to all objective lenses, as illustrated in **Figure 6.10.** First, every objective lens is designed for a specific

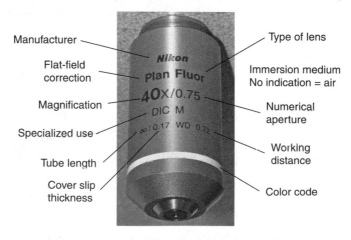

Figure 6.10
Markings on an objective lens.

Table 6.1	Characteristics of Objective Lenses				
Magnification	Type of Lens	Medium	Working Distance (mm)	Numerical Aperture	Resolution (µM)
5	Achromat	Air	9.9	0.12	2.80
10	Achromat	Air	4.4	0.25	1.34
20	Achromat	Air	0.53	0.45	0.75
25	Fluorite	Oil	0.21	0.8	0.42
40	Fluorite	Air	0.5	0.75	0.45
40	Fluorite	Oil	0.2	1.3	0.26
60	Apochromat	Air	0.15	0.95	0.35
60	Apochromat	Oil	0.09	1.4	0.24
100	Apochromat	Oil	0.09	1.4	0.24

magnification, working distance, numerical aperture, and immersion medium. Specifications for commonly used lenses are listed in **Table 6.1.** These specifications are engraved into the metal housing so that they are readily available to the microscopist at a glance. The magnification appears in large numbers (usually ranging from 4 to 100), the numerical aperture in smaller numbers (usually ranging from 0.15 to 1.4), and the immersion medium, if not air, written in capital letters. Frequently, the working distance is not noted unless it is larger than normal, in which case LWD or ELWD (for long working distance or extra-long working distance) will appear on the barrel above the magnification.

Special-use objective lenses have features that are generally not present in general use objectives. Some of the more common types for biological work include the following: immersion objectives, fluorescence objectives, phase-contrast objectives, differential interference contrast (DIC) objectives, correction-collar objectives, and plan apochromatic objectives.

Immersion Objectives

These lenses are used with a liquid medium placed between the specimen and the lens. Common immersion media include water, glycerol, and oil. Because each of these media has a different refractive index, the geometry of the lens design is specific to one medium only. The lens cannot be used effectively in air or any other medium. In addition, the mounting adhesive for the front lens, which contacts the immersion medium, is resistant to being degraded by that medium but not necessarily other media.

Fluorescence Objectives

Detection of weakly fluorescent signals requires optical components having an efficient light throughput. While objective lenses used for transmission microscopy often use only 60% to 80% of the incident light to form the image, objectives used in fluorescence microscopy have transmission efficiencies of 90% or higher. Second, the use of color in distinguishing separate signals in immunocytochemistry requires very low chromatic aberration. These features can be achieved by using lens materials that exhibit minimal color dispersion and by keeping the number of lens elements to a minimum. One strategy is to use lenses made of fluorite, calcium fluoride, or lanthanum fluoride. These materials naturally display high transparency over the entire visible spectrum and low color dispersion that is easily corrected. Low color dispersion minimizes both focal plane aberrations along the optical axis and registration problems in the x–y image plane. In order to keep the number of lens elements to a minimum, these objectives are usually not fully corrected for spherical and planar aberrations. A diagram of typical lens elements in a fluorite lens is shown in Figure 6.9.

Phase Contrast Objectives

As described in Chapter 7, these objective lenses are designed to change small differences in refractive index between features in the specimen into differences in light intensity that our eyes and cameras can detect. This is achieved by selectively retarding part of the light used to form the image using a phase plate positioned at the back focal plane of the objective. The plate has a circular well machined into it such that the nondiffracted image light that passes through the well is shifted 90 degrees in phase compared with the scattered image light. When the two components of image light are merged, they undergo positive and negative interference to produce contrast in the image, despite the fact that there was little contrast in the specimen. These objectives can also be used for brightfield microscopy.

DIC Objectives

DIC microscopy, also explained in Chapter 7, requires that polarized light be passed through the specimen. The optical components of objective lens must not depolarize this light in any manner, lest they be mistaken for a property of the specimen; therefore, the components of these objectives need to be strain free, and lens coatings must be formulated to avoid depolarization. DIC objectives can also be used for brightfield microscopy.

Correction-Collar Objectives

Most objective lenses are designed to be used with #1.5 coverglasses that have a thickness of 0.17 mm for optimal performance. Some specimens require viewing with coverglasses of different thicknesses. A correction collar, rotated to the proper position, allows a single objective lens to obtain optimal resolution with coverglasses thicker than the standard #1.5 coverglass.

Plan Apochromatic Objectives

Objective lenses that have been corrected for chromatic, spherical, and planar aberrations. These lenses avoid systematic distortion of the specimen in the image and are particularly important when accurate measurements of specimen distances, areas, and volumes are being made. The high degree of correction achieved requires a high number of lens elements, thus, increasing cost.

■ Oculars

Eyepieces or oculars produce the second stage of magnification in a compound microscope. Generally consisting of two or three lenses within a thin metal barrel, this component essentially acts like a magnifying lens. It visualizes not the specimen itself but instead the intermediate image formed by the objective lens at the front focal plane of the eyepiece. In this way, the magnifying power of the eyepiece (8×, 10×, 16×, and 20× are standard) multiplies the magnifying power of the objective lens producing total magnifications that range from 40 to 2000 times. Higher total magnification is possible but usually is not desirable. Two specimen points that are 0.25 μm apart can barely be distinguished as separate since this distance is at the limit of resolution for a light microscope. If these two points are magnified 2000 times, they will appear to be about 0.5 mm apart at a distance of 10 inches. This matches well with the visual acuity of a person that has 20/20 vision. If the image is magnified any further, our eye will begin to see unresolved specimen points for which there is no additional information in the image. This simply results in the absence of detail, and for this reason, further magnification is referred to as "empty magnification."

■ Light Sources and Lamp Houses

In order for an objective lens to properly form an image, the specimen must be illuminated with light, which may be transmitted through the specimen or may be oblique. Transmission or "brightfield" microscopy and DIC microscopy use light that penetrates the specimen, whereas phase-contrast, darkfield, and reflectance microscopy use oblique lighting. The job of the light source is to produce light having an appropriate spectrum (wavelength or color distribution), whereas the job of the condenser lens is to modify the distribution of light intensity and to direct the light at the specimen from the correct angle.

The earliest source of light for microscopy was sunlight gathered by a single concave mirror or a series of mirrors and focused onto the condenser lens. A number of bizarre devices were invented whereby light from candles or oil lamps could be employed, but these must have been far inferior sources. Electrical lamps, both filament and vapor, are the most common sources today and are discussed below. Lasers and light emitting diodes are more powerful and more expensive, but discussion of these is postponed to Chapter 12.

A regular desk lamp with incandescent bulb was used combined with a concave mirror throughout the 20th century in "recreational" and "student" microscopes. In the last half of the century, the bulbs were miniaturized and built into the base of the microscope but were still basically the same technology as the desk lamp and mirror. Research microscopes in the last 40 years have routinely used light sources enclosed in a separate compartment, the lamp house. The lamp house serves to isolate the light source from the rest of the microscope, important for the more powerful lamps that produce large amounts of heat. Heat, even in small amounts, can cause unwanted contraction and expansion of microscope parts, leading to a lack of stability in the optical system.

Before describing lamp house optics, we must first discuss the three most common types of electrically powered light sources used in light microscopy: tungsten-halogen, mercury vapor, and xenon vapor lamps. The tungsten lamp or incandescent light bulb, the least expensive and the earliest, was invented by Swan in England and perfected by Edison in the United States. The lamp consists of a tungsten wire (the filament) enclosed in a partially evacuated chamber (the bulb) that is filled with argon so as to remove all traces of oxygen and thereby prevent the lifetime of the tungsten wire being shortened by oxidation. Passage of an electrical current through the high-resistance wire heats the wire to very high temperatures (about 3000°K) and excites the electrons to very high energy states, and as they continually drop back to lower energy orbitals, they emit infrared and visible radiation. The emission spectrum is continuous and broad between 300 and 1600 nm and depends on the temperature of the filament (**Figure 6.11A**). Less current results in less heating and a spectrum that is shifted toward lower energies, that is, yellow and red and infrared. This accounts for the fact that lamps operating at low current produce images that have a severe yellow cast. Running the lamp at a higher current produces a spectrum that is shifted toward higher energies, including more green and blue, and the visible light produced now looks white to the eye. Because of this spectral dependence on temperature, spectra are often described by the temperature at which they are produced. For example, the low-energy spectrum produced by a "dimmed" bulb that appears yellow represents a 2800°K spectrum, whereas a tungsten bulb at full operating temperature may emit a 3200°K spectrum.

A. Halogen incandescent lamp

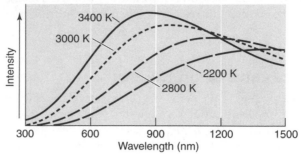

B. Mercury arc lamp

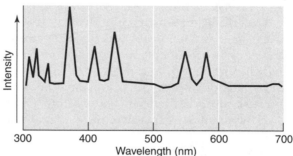

C. Xenon arc lamp

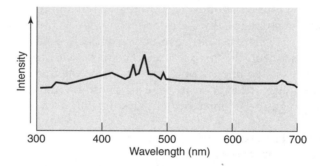

Figure 6.11
Spectra of common light sources. (A) Halogen-tungsten lamps have their strongest output in the infrared and have a skewed light output in the visible range, being yellow-red rich and blue-green poor. (B) Mercury arc lamps have a strong continuous spectrum throughout the visible and well into ultraviolet wavelengths. In addition, there are peaks of output 10 times higher in intensity that can provide powerful illumination for some fluorescent dyes. (C) Xenon arc lamps have a strong continuous spectrum without peaks and provide excellent service for fluorescence microscopy.

The tungsten-halogen lamp is of more recent development and represents a modification of the tungsten bulb. Like the latter, the tungsten-halogen lamp contains a tungsten filament that is heated by passage of electric current. The glass bulb, however, is made of glass that is doped with a halide, usually iodide, the halide preventing deposition of vaporized tungsten on the glass surface. This increases the lifetime of the tungsten filament, which otherwise would be gradually eroded by loss of its tungsten to the bulb. Typically, these lamps can be run very hot (3300°K), producing a whiter light (3000°K) that is consistent over the 2000-hour lifetime of the filament. Their high operating temperature requires a well-ventilated lamp house.

Mercury and xenon vapor lamps represent "arc" lamps that work on a principle different from that of a filament lamp. There is no filament, but instead, there are two electrodes with a gap between them. The gas between the two electrodes is at high pressure and is very concentrated. Firing of the lamp produces large voltage differences between the two electrodes, which in turn ionizes the high-pressure gas and conducts current between the two electrodes. The electrons of the ionized gas are bumped up to very high-energy states during ionization and subsequently drop back down to ground level, emitting light in the process. These lamps have very short lifetimes (200 to 500 hours) but produce very intense light at shorter wavelengths compared with filament-based lamps. The xenon lamp exhibits a continuous spectrum between 200 and 1000 nm in wavelength (Figure 6.11C). For this reason, its spectrum is balanced and not short wavelength poor, as is the spectrum of an incandescent lamp. The intensity of a 50-watt xenon lamp is almost 10 times that of a 50-watt tungsten-halogen bulb in the blue region of the spectrum and about two times greater in the yellow range. As a result, this light source is excellent for use in fluorescence microscopy because many fluorescent dyes are excited in the ultraviolet, blue, and green portions of the spectrum (see Chapter 12). The mercury vapor lamp also has a continuous spectrum from 200 to 800 nm similar in intensity to that of a xenon lamp (Figure 6.11B). Its chief difference is the presence of some extremely high-intensity peaks (a partial line spectrum) that represent very favorable electron transitions that have a high quantum yield. The major peaks at 313, 334, 365, 406, 435, 546, and 578 nm can be very useful for fluorescence excitation, provided that an appropriate dye is used.

Although arc lamps are routinely used for fluorescence microscopy, they are not useful for other forms of microscopy, including brightfield, darkfield, phase, DIC, or polarization microscopy. There are three reasons for this: (1) direct observation of the light emitted by these lamps can damage the retina because of their high ultraviolet content, (2) the nonuniform spectrum of a mercury arc lamp is not able to produce normal color representation of specimens, and (3) arc lamps produce relatively incoherent light that is problematic for image formation, as discussed previously in the chapter.

These light sources produce so much heat and (in the case of arc lamps) dangerous short-wavelength radiation that they need to be enclosed in a lamp house. As shown in **Figure 6.12A,** a typical lamp house is a metal housing covered with ventilation slits on the top and sides to let heat out. Each slit is covered with an angled lever to keep stray light from entering the room, thereby causing damage to the operator's eyes or unwanted reflections. Centrally located within the lamp housing is the lamp itself and at the back of the lamp is mounted a parabolic mirror that gathers light from the back of the lamp and redirects it toward the front (Figure 6.12B). The lamp house is connected to the microscope by an optical tube that feeds the lamp output to the microscope. In some designs (e.g., lamp houses for inverted microscopes or for epifluorescent microscopy), this tube must also be strong enough to support the lamp house in midair, thereby minimizing contact with the microscope and maximizing air cooling.

Contained within the optical feeder tube are collector/culminating lenses that help gather the light into parallel or convergent rays that propagate to the other end of the tube and are relayed to the condenser lenses. Also present is a heat filter, which absorbs infrared radiation that is not used for imaging and would otherwise heat the microscope. A shutter is required so that delivery of light can be interrupted without turning off the lamp. Also common within this feeder tube are positions for neutral density filters that transmit only a certain fraction of the available light without changing its spectrum. An entire series of neutral-density filters (e.g., reducing light intensity by 10%, 25%, 50%, and 75%) can be used singly or in combination to reduce light intensity to almost any level without changing the light output of the lamp. Changing the light output of a

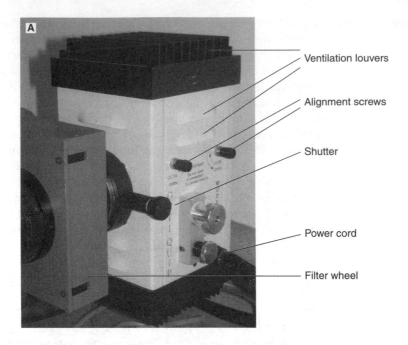

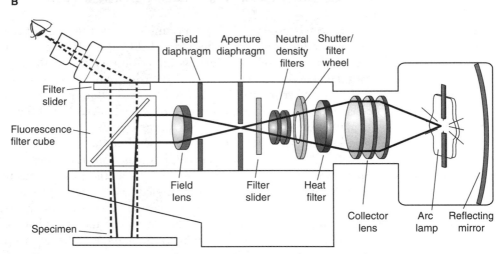

Figure 6.12
Lamp house design. (A) A typical xenon arc lamp house providing both safety from stray ultraviolet light and from possible lamp explosion resulting from high operating pressures. The housing must be well ventilated to prevent the lamp from overheating. (B) Lamp house optical path for epifluorescent illumination as seen from a side view. The main microscope tube is at the left.

filament lamp by controlling the current (i.e., dimming) changes the spectrum of the light produced, and changing the light output of a vapor phase lamp is technically difficult, hence the utility of these filters. In addition, there are usually positions for other filters that may be useful for specialized applications (e.g., a blue lens for doing monochromatic work at short wavelengths to get the highest possible resolution). Alternatively, the light may be processed by filters contained in a filter wheel. Such a wheel can hold 6 to 10 filters that can be rapidly placed in and out of the light path by computer control. From this point, the light beam generated is typically fed into the optical axis by a series of mirrors or prisms to the condenser lens that lies just under the specimen stage of an upright microscope.

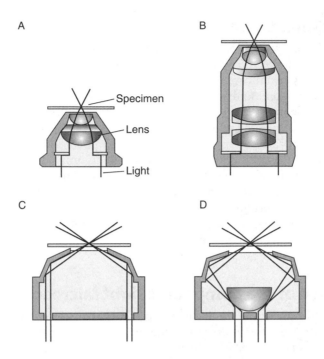

Figure 6.13
Condenser lens design. (A) Achromatic condensers having two elements can provide light for low-NA brightfield microscopy applications. (B) High-NA condensers contain larger numbers of elements but are required for high-resolution imaging. (C and D) Darkfield condensers provide light to the specimen at an oblique angle sufficient to prevent the objective lens from capturing zero-order light. Only scattered and diffracted light is imaged in darkfield microscopy.

The function of the condenser lens is to concentrate and condition the light further, forming it into a uniform, high-intensity beam that in brightfield and DIC microscopy is directed parallel to the optical axis and perpendicular to the specimen and the stage on which the specimen lies. As shown in **Figures 6.13A** and **6.13B**, the condenser can be a single lens or contain several elements much like an objective lens of simple design. Like objective lenses, condensers have a numerical aperture that in this case is a gauge of how large an acceptance angle is used for illuminating the specimen. For the best result, the numerical aperture of the condenser lens should be at least as high as that of the objective lens. Like objective lenses of the highest numerical aperture, high NA condensers are designed to be used with oil between the condenser and the specimen to increase the acceptance angle.

Some techniques, such as darkfield, reflectance, and phase-contrast microscopy, require that light strikes the specimen in an oblique manner. A darkfield condenser, for example, has a central pillar that blocks the passage of light straight through the specimen and prisms designed to funnel the light around this central pillar and impinge on the specimen from an oblique angle (Figures 6.13C and 6.13D). Other techniques such as phase contrast use a circular mask above the condenser lens to allow light that passes directly through the specimen without diffraction to be handled exclusively by the periphery of the objective lens, as discussed in Chapter 7. A series of masks having circular apertures of different sizes along with other useful filters and lenses are held in a condenser turret that can be rotated to put the desired mask, filter, or lens into the optical path.

■ The Specimen Stage

The typical specimen, a section of tissue or a suspension of cells, is placed on a glass slide that is 1 mm thick and 25 by 75 mm in width and length. The slide is either slid under one or two stationary metal clips (these are sometimes seen even on the most expensive microscopes) or more commonly are set into a spring-loaded caliper device. The entire stage itself is movable in the x and y directions by geared micrometers, or the calipered slide holder is moved across the stage surface by micrometers. Some stages contain plates that can be used to rotate the specimen around the optical axis. The stage is usually mounted on a rack-and-pinion gear set so as to be able to move the specimen into focus, that is, into the front focal plane of the objective lens. This gear set must be sturdy because the stage, more than any other piece of the microscope, must be free of vibration and play. In order to be certain of stability and to accommodate relatively heavy specimens (entire organs or small animals or plants), some stages are stationary with focus being achieved by movement of the objective lens.

■ Design Plan of the Compound Light Microscope

We have discussed now all of the basic components that are needed in a compound microscope. These are put together in two different architectures: the upright microscope illustrated in **Figure 6.14A** (see Plate 3 in the Color Addendum) and the inverted microscope illustrated in Figure 6.14B (see Plate 4 in the Color Addendum). Each of these microscopes is equipped with a tungsten-halogen lamp used for transmitted light microscopy and mercury-xenon lamp for fluorescence microscopy. Shown in colored arrows superimposed on each microscope are the optical paths that light takes for illumination of the specimen (the primary beam) and for image formation and recording (the secondary beam); further details on the color coding of the arrows are given in the figure legends. The upright microscope is the traditional and more common design. It features a simple, relatively straight optical path that lines up the condenser lens, specimen stage, objective lens, eyepiece, and either eye or camera that favors a maximum throughput of light. Its major limitation in some applications is the fact that the closeness of the objective lens to the specimen blocks easy access to the top of the specimen. This is not a concern if the specimen is a tissue section or dead cells that require no manipulation. On the other hand, if the specimen consists of live cells or tissues or cells in culture that need perfusion, microinjection, or electrical recording by microelectrodes, the space above the specimen is a necessity. It is for these purposes that the inverted microscope was introduced in the 1970s. An additional advantage is the large footprint of the inverted design, thereby minimizing vibration problems. This design plan inverts the optical axis and puts a number of 90-degree bends in the light path that must be handled by prisms or mirrors. The light source at the top rear is brought forward over the stage, bent 90 degrees downward, and passed through the condenser lens, the working space above the specimen, and the specimen itself. The secondary beam, consisting of light merging from the specimen, is gathered by the objective lens just under the specimen, the image forming rays bent through two 90-degree angles and a 45-degree angle and finally up into the eyepieces.

The advantage of the inverted design to certain applications does not come without some compromises. First, the large distance between the condenser and the specimen requires a person to use a long working distance condenser having a lower numerical aperture than would otherwise be the case. The resulting loss of resolution prevents high-quality imaging at the highest magnifications; however, the use of a long working distance objective lens in an upright microscope so as to gain space above the specimen would likely result in even a greater loss of resolution. Second, the specimen must be

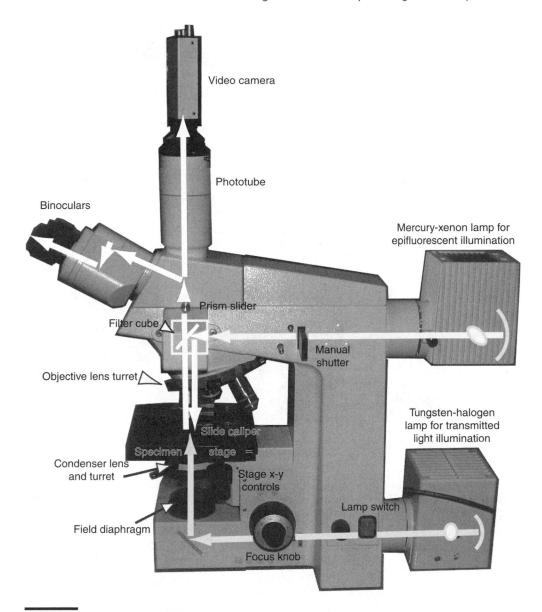

Figure 6.14   (See Plates 3 and 4 for the color versions.)
(A) The design of a typical upright microscope equipped for both fluorescence and brightfield microscopy. Optical paths in the microscope are traced by arrows superimposed on the microscope image. Light from the tungsten-halogen lamp (yellow arrows) is used for transmitted light microscopy. After passage through the specimen, image-forming light is gathered by the objective lens and transmitted through the binocular eyepieces to our retinas to form a final image (green arrows). Alternatively, images can be sent to a camera for recording (red arrow). For fluorescence microscopy, light is provided by a mercury-xenon lamp (white arrows) and the wavelengths desired for chromophore excitation (blue arrows) selected by the filter cube. As a consequence, emitted light from the chromophores in the specimen is transmitted either to the binocular eyepieces (green arrows) or alternatively to a camera (red arrow).

(Continued)

capable of being inserted upside down if one is to get the objective lens as close as possible to the coverglass and specimen plane to be imaged. More commonly, experiments needing an inverted microscope for access to the specimen require an upward-facing specimen. In this case, the image-forming rays gathered by the objective lens below must pass through the slide, dish, or transparent support on which the specimen rests. Typical

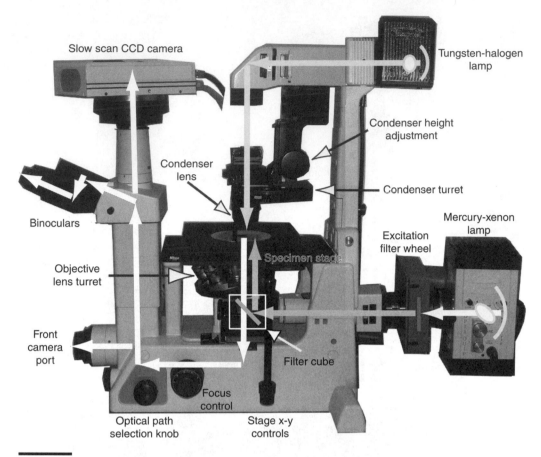

Figure 6.14 (B) The design of a typical inverted microscope equipped for both fluorescence and brightfield microscopy. Optical paths in the microscope are traced by arrows superimposed on the microscope image. Light from the tungsten-halogen lamp (yellow arrows) is used for transmitted light microscopy. After passage through the specimen, image-forming light is gathered by the objective lens and transmitted to our retinas to form a final image (green arrows). Alternatively, images can be sent to a camera for recording (red arrows). For fluorescence microscopy, light is provided by a mercury-xenon lamp (white arrows) and the wavelengths desired for chromophore excitation (blue arrows) selected by the filter wheel or filter cube. As a consequence, emitted light from the chromophores in the specimen is transmitted either to the binocular eyepieces (green arrows) or alternatively to a camera (red arrows).

slides and dishes that are 1 mm thick will prevent the highest quality imaging because of a larger working distance than is optimal and in some cases because of distortion. For this reason, experiments are often carried out on cells that have been cultured directly on coverglasses to avoid this problem. Other drawbacks include a slightly lower light throughput, increased expense because of the high number of prisms or mirrors needed to bend the light path, and a greater danger of damage to objective lenses because of collision with the underside of the stage and dripping of immersion oil.

■ Light Paths in the Compound Microscope

A better understanding of the processes important to image formation in a compound microscope can be obtained if one removes all of the 90- and 45-degree angles present in the light path of a real microscope to produce, in diagram, a straight path. As illustrated in **Figure 6.15A**, such a light path can be traced as rays gathered from the light source that are processed by the condenser lens and focused onto a single specimen point.

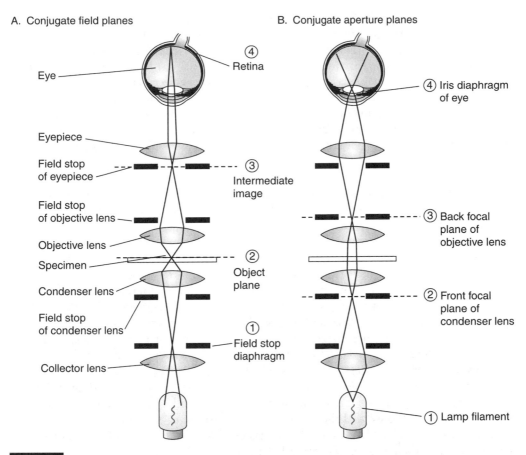

A. Conjugate field planes

B. Conjugate aperture planes

Figure 6.15
Conjugate planes in the optical path of the light microscope. (A) Conjugate planes that include the specimen plane are numbered. The image projected onto the retina can contain in-focus information from each of these planes, hence the need to make sure all components near these planes are dirt and dust free! The ray diagram shows that each specimen point is illuminated by light from the entire lamp and that all rays emanating from this point and captured by the objective lens are refocused to a point in the intermediate image and at the retina. (B) Conjugate aperture planes, including those of the condenser and objective lenses, and the iris of the eye are numbered. The ray diagram demonstrates the disposition of light coming from a single point on the lamp filament and illuminating the entire specimen.

This light interacts with the specimen within a minute restricted location and as a result is transmitted, absorbed, refracted, or diffracted. The transmitted light continues to follow the same path as the incident ray, whereas the diffracted light is spread through a variety of angles. The job of the objective lens is to gather as much of this light as possible, regardless of angle, and to force it to converge to form a single point in the intermediate image. This intermediate image is further magnified by the eyepiece. In order to do this properly, the front focal plane of the eyepiece must be properly positioned at a fixed distance relative to this image.

Light rays coming from a single point in the intermediate image again slowly diverge as they travel up the optical tube of the microscope. The eyepiece regathers this collection of rays and bends them dramatically so that they again converge very sharply. Before they are fully converged to focus by the eyepiece, the lens of the eye projects them onto the retina, at which point they fully converge to focus and form one single image point in our visual field. The fact that the eyepiece forces these rays to converge at a very steep angle before entering the eye accounts for their ability to form a magnified virtual image, much as does a magnifying glass (Figure 5.8). Unlike a magnifying glass, bending of the light rays severely requires eyepieces to have two or three lens elements rather than one. This scenario requires that the eyepiece be at a fixed distance from the objective lens. If microscope lenses (both objective and ocular) are to be interchangeable, they must be designed to use a standard distance between the objective lens mount and the eyepiece. This distance is referred to as "tube length" and for many years was set to 160 mm as a standard shared by microscopes produced by most major manufacturers. Because different objective lenses have different focal lengths and working distances, careful design is required for a specimen to be in focus for multiple lenses without changing the stage height. Objective lenses meeting this condition are said to be "parfocal," but only certain sets of objective lenses available from each manufacturer have been designed to offer this ease of use. A fixed standard tube length has several disadvantages, and, in the last decade, "infinity corrected" optics have been used to allow the tube length to be changed depending on application (see below).

Based on Figure 6.15A, one can see that the specimen is brought into focus at several positions on the optical axis, namely at the specimen stage (2), the intermediate image plane (3), and the retina (4). Also brought into focus at these planes is the field diaphragm (1) positioned between the light source and the front focal plane of the condenser lens. Because of this, the image formed at the retina will have "in-focus" objects positioned at any one of these planes, including dirt on the glass covering the field diaphragm, dirt on the coverglass, and dirt on the front focal surface of the eyepiece! For this reason, these planes are referred to as "conjugate field or image planes." They all contribute to the real image visualized whether part of the specimen or not.

This same optical process can be described by light that is emanating from a single point on the filament of the light source (1) or from a "point light source" (Figure 6.15B). Light from this single point spreads out in all directions and is gathered by the parabolic mirror in the lamp house and the culminating lens to focus on the front focal plane of the condenser lens (2). The condenser lens takes this light and spreads it to form incident rays that impinge on every part of the microscopic field and on every specimen point. These rays reconverge at the back focal plane of the objective lens (3), only to once again diverge and contribute light to every point of the intermediate image. Once again, as the rays progress up the optical tube, they are bent by the lenses of the eyepiece to converge at the iris diaphragm of our eye (4) or that of a camera lens. The light is then again spread by the lens of our eye or camera and distributed to every image point on the retina, film, or solid-state chip.

This illustration shows that each point of the light source is refocused to a point at a series of "conjugate aperture planes" along the optical axis. This implies that at each of these planes, one should be able to see an image of the light source superimposed on the diffraction pattern created by the specimen. In the best situations, the light source acts like a single point source, allowing the diffraction pattern of the specimen to look like a pattern of points rather than a pattern of little lamp filaments! It is at these planes where a finite aperture size can limit the extent of the diffraction pattern and thereby determine the resolution and contrast of the image. Comparing these two diagrams helps clarify two points: (1) If the light source is uneven in output from point to point, these differences will be shared by all specimen and image points equally and should not contribute to differences in brightness or contrast from one point in the image to another, and (2) the ability of the condenser lens to spread the light from a single point source evenly over the entire specimen is critical and dictates that the condenser must be of adequate numerical aperture and that it be properly positioned relative to the specimen. This last criterion, position of the condenser, was studied by August Köhler in 1893, who laid the theoretical ground for both design and proper positioning of the condenser to achieve what today is called "Köhler illumination."

◼ Köhler Illumination

A criterion for setting up Köhler illumination is to place the effective light source at the front focal plane of the condenser lens and the specimen to be viewed at the back focal plane of the condenser lens. The effective light source is usually the light coming through an iris diaphragm that acts as an aperture of variable diameter that is fixed in position. Because the specimen to be viewed is not at a predetermined, fixed position, the specimen must be brought into focus before the condenser can be adjusted. Because the condenser lens on most microscopes is adjustable in position along three axes, must one adjust its "height" along the z-axis (the optical axis) and "center" it in the x- and y-axis (the plane perpendicular to the optical axis and parallel to the specimen stage). A typical procedure for doing this is as follows (**Figure 6.16**):

1. Place the specimen securely on the stage, and rotate the 10×-objective lens into play. Bring the specimen into sharp focus by adjusting the stage height.
2. Close the iris diaphragm that controls the diameter of the light beam being fed to the condenser lens. This aperture is called the "field diaphragm."
3. Looking through the eyepieces again, if the beam of light coming through the field diaphragm is off center or not even visible, center the beam using the left and right (x and y) centering screws for the condenser.
4. Still looking through the eyepieces, adjust the condenser height until an image of the field diaphragm opening comes into focus. If you are using a white-light source, there will be a blue halo of diffracted light around the opening before focus is reached and a red halo after focus is passed. Proper condenser height is just at the point where the blue halo changes to red.

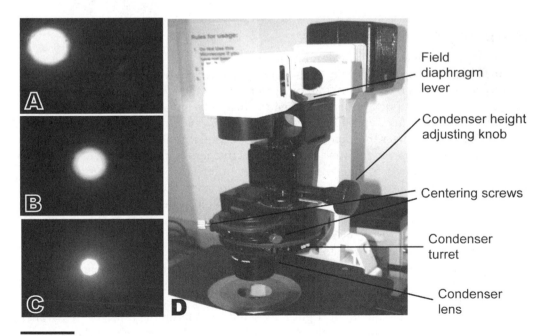

Figure 6.16
Köhler illumination. All adjustments must be made after bringing a test specimen into focus and closing down the field diaphragm. (A) Off-center illumination requires condenser centering via centering screws. (B) Next, an out-of-focus field diaphragm image requires adjusting the condenser height via the condenser height adjusting knob. (C) A centered, in-focus image of the field diaphragm aperture indicates that Köhler illumination has been achieved. (D) Condenser adjustment controls on a Nikon Ellipse 300 inverted microscope.

5. Open the field diaphragm enough so that light completely fills the field of view but open it no further than that.

6. If one wishes to use a different objective, this process should be repeated to fine tune the condenser lens height and centering at the new magnification; however, it is best to go through this process at low magnification first before repeating it at high magnification.

The importance of adjusting the field diaphragm (or condenser iris diaphragm) opening to spread the light is based on the following: If the opening is not large enough, the entire viewing field will not be filled. Filling a large field at low magnification requires a larger opening than does filling a small field at high magnification. If the opening is too large, that is, much greater than the diameter of the field of view, the image will be lower in contrast than it would be otherwise. The reason for this is that the diffraction patterns created by features in the specimen will have more information, leading to less dramatic extremes in positive and negative interference. Abbreviating the information in these patterns by using a smaller diameter light source leads to greater extremes in interference and more contrast. Usually, good contrast is important for adequate detection of specimen features by film, solid state chip, or your eyes—they are all based on detecting differences in light intensity; however, good contrast comes at the expense of resolution. By abbreviating the diffractions patterns set up by the specimen features, the information that would lead to the highest resolution is lost. Those seeking the highest resolution would be advised to (1) increase contrast by staining methods, (2) open the field diaphragm as wide as is consistent with feature detection, (3) use more sensitive light detection coupled with lower light levels, and (4) use both objective and condenser lenses that are of high numerical aperture.

■ Infinity-Corrected Optics

Previously here we learned that in many microscope designs the tube length (i.e., the distance between the back focal plane of the objective lens and the front focal plane of the eyepiece) was fixed at 160 mm by agreement. In the 1980s, it became apparent that there was an important disadvantage to this. New microscopic techniques and equipment, among them epifluorescence illumination and scanning confocal illumination, required a longer tube length to allow for adequate entry into the optical path. The solution to this problem was to design objective lenses and ocular lens such that the light rays produced by one and used by the other were focused at infinity (i.e., they were all parallel to the optical axis). In this manner, the relative position of all image-forming light rays would be maintained as they traveled through the tube. For that reason, they could be used for image formation at any distance from the objective lens. The handling of light rays from a single specimen point by such "infinity-corrected" lenses is diagrammed in **Figure 6.17B** and can be compared with the traditional fixed tube length situation diagrammed in Figure 6.17A. Light coming from the specimen is gathered by the infinity-corrected objective in the usual manner. After the light rays have passed through the back focal plane, they are bent by a concave lens to a course parallel to the optical axis and straight up the tube. These parallel rays are then focused by the tube lens to form the intermediate image that the eyepiece then further enlarges. In a sense, an infinity-corrected objective lens and the tube lens together do the same job as a standard objective used to do except for the addition of a region of parallel ray transmission that can be varied in length with no effect on image formation.

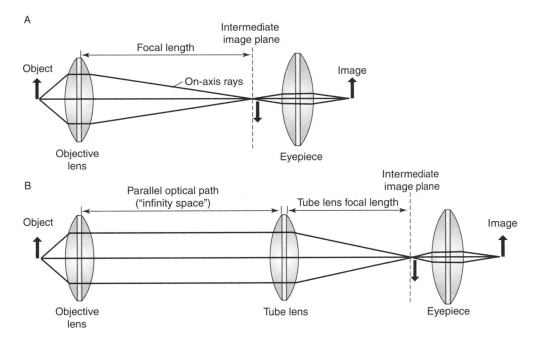

Figure 6.17
Design of an infinity-corrected microscope. (A) Image formation in a "fixed tube length" microscope. The focal length of the objective lens determines the position of the intermediate real image relative to the specimen. (B) Image formation in an "infinity-corrected" microscope. The "infinity-corrected" objective lens projects sets of parallel light rays through "infinity space" to reach a tube lens that focuses the rays to produce the intermediate image. Light redirected to other applications by beamsplitters and filters has little effect on intermediate image formation.

■ Prisms and Beamsplitters

There are many cases in which the geometry of the light path requires that the optical train be redirected at an angle (commonly 45 or 90 degrees) or be split into two components so that each component can be imaged or analyzed separately. For example, even in a binocular upright microscope of simple design, the image-forming rays are split into two beams, and each is bent through one or two 90-degree angles to form equivalent images on both left and right retinas. Angles and divisions within the light path are typically handled by prisms and beam splitters made of rather large pieces of glass whose surfaces are flat and meet at precise angles. The design of the prism or beamsplitter determines the angle at which the optical train is bent, the portions of the optical train allotted to each direction, and whether the image forming rays will be flipped or rotated.

For example, in **Figure 6.18A,** a right-angle prism is used to bend light at a 90-degree angle. A beam enters the prism normal to one of the right angle surfaces and undergoes total internal reflection off of the prism-air interface at the hypotenuse of the prism. Total internal reflection is achieved because the angle of incidence (45 degrees) is less than the critical angle for reflection. The critical angle is a function of the difference in refractive index between the prism and air and requires that the prism glass have a refractive index of greater than 1.41 according to Snell's and Brewster's laws. Lateral displacement of a light beam can be achieved by a rhomboid prism, as diagrammed in Figure 6.18B. The two 45-degree beveled surfaces at either end of the prism each act like a right angle prism, thus guiding the light beam through two 90-degree angles. Each time a light beam

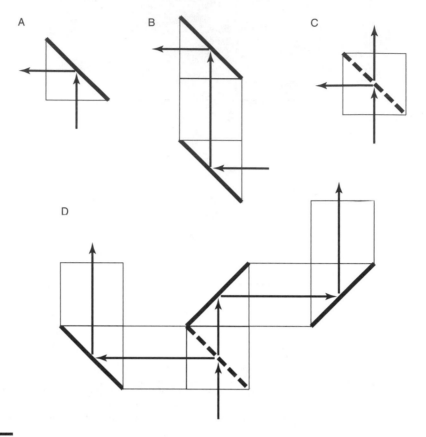

Figure 6.18
Prisms and beamspitters. (A) A 45-degree prism uses total internal reflection to redirect a light beam at 90 degrees. (B) A rhomboidal prism, exhibiting reflection at both ends, displaces a light beam laterally. (C) Two 45-degree prisms, half silvered at their common interface acts as a beam splitter, allowing part of the beam to be transmitted while another part is reflected at 90 degrees. (D) A combination of prisms and beamsplitters used in providing binocular vision.

is bent 90 degrees by these prisms, its symmetry is flipped left to right. Because this occurs twice in a rhomboid prism, the second flip canceling the first, the exit beam has exactly the same orientation as the entering beam. If one uses two right angle prisms, placed hypotenuse to hypotenuse, one has just designed a beam splitter (Figure 6.18C). The only minor modification required is that the hypotenuse of one right angle prism should be coated with a thin metallic coating that will reflect a certain portion of the light, for example, 50%, while letting the rest continue. If this metallic coating were not present, no light would be reflected at the prism–prism interface because there is no change in refractive index. With the metallic coating present, 50% will be reflected at 90 degrees, and the other 50% will pass straight through not having recognized the presence of an interface. The relative proportions of reflected versus nonreflected light is determined by the thickness of the metal coating. Even these three simple components can be combined to produce a binocular head, as diagrammed in Figure 6.18D. Mirrors can be used to carry out many of the functions of beam splitters but have a number of disadvantages. First, mirrors are harder to position and are more sensitive to vibration. Because the reflecting surface of a mirror is exposed to the air, dust and corrosion can be problems that are completely absent in prisms. For these reasons, high-quality optical instruments often employ prisms rather than mirrors to bend the light path.

References and Suggested Reading

Books

Bracegirdle B, Bradbury S. *Modern Photomicrography (Royal Society Microscopy Handbook).* Oxfordshire, England: BIO Scientific Publishers, 1995.

Bradbury S, Bracegirdle B. *Introduction to Light Microscopy (Royal Society Microscopy Handbook).* New York: BIOS Scientific Publishers, 1998.

Bradbury S, Evennett PJ, Haselmann H. *Dictionary of Light Microscopy (Royal Society Microscopy Handbook),* Oxford: Oxford University Press, 1989.

Goldstein DJ. *Understanding the Light Microscope: A Computer-Aided Introduction.* New York: Academic Press, 1999.

Lacey AJ, ed. *Light Microscopy in Biology,* 2nd ed. *The Practical Approach Series.* New York: Oxford University Press, 1999.

Murphy DB. *Fundamentals of Light Microscopy and Electronic Imaging.* New York: Wiley-Liss, 2001.

Pluta M. *Advanced Light Microscopy. Principles and Basic Optics* (vol. 1). Amsterdam: Elsevier, 1988.

Shotten D, ed. *Electronic Light Microscopy, Techniques in Modern Biomedical Microscopy.* New York: Wiley-Liss, 1993.

Spencer M. *Fundamentals of Light Microscopy.* Cambridge: Cambridge University Press, 1982.

Thomson DJ, Bradbury S. *An Introduction to Photomicrography (Royal Society Microscopy Handbook).* Oxford: Oxford University Press, 1987.

Websites

Abramowitz M. Microscope Basics and Beyond, revised. Olympus America, Inc., 2003. Available as a PDF file at www.micro.magnet.fsu.edu/primer. Also available for purchase in hardcopy at the Olympus website below.

Davidson MW. Molecular Expressions Optical Microscopy Primer. www.micro.magnet.fsu.edu/primer.

Nikon Microscopy U. Nikon USA. www.microscopyu.com/.

Olympus Microscopy Resource Center. Olympus America, Inc. www.olympusmicro.com/.

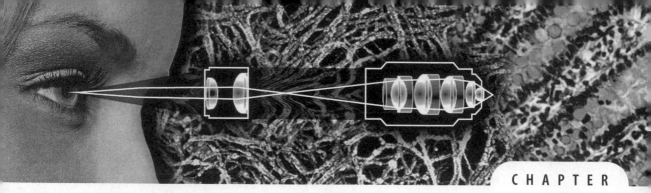

Phase, Interference, and Polarization Methods for Optical Contrast

Imaging live cells is the forte of light microscopy as much today as it was in Hooke and van Leeuwenhoek's era 300 years ago. Live cells, however, are typically transparent unless they contain photosynthetic pigments or high concentrations of colored metabolic products. This problem was originally solved by using vital dyes that bind to cells and organelles but that have little effect on the metabolic processes of the cell; however, during the last 60 years, imaging of live cells has been done chiefly by optical methods such as phase-contrast and differential interference contrast that rely on the fact that cellular structures differ markedly in their refractive index. These differences result in the light rays passing through an organelle becoming out of phase with light rays passing through the cytoplasm or surrounding extracellular medium. Our eyes (and cameras) do not perceive these phase differences, but if one uses optical technologies to change differences in phase into differences in light intensity, these organelles do become "visible" to both cameras and eyes. Thus, the light microscopist now has a variety of basic imaging techniques available; these are compared in **Table 7.1**.

■ Vital Dyes

The oldest method for enhancing the visibility of cells is to add a vital dye. The dye must either bind to the outside of the cell or be permeable to the plasma membrane and diffuse inside of the cell. The dye must not affect cell processes, at least not immediately. Some, but not all, of the early textile dyes were found to be tolerated by cells, and the use of these became popular in the 19th century before embedding and sectioning of tissue came into common use. **Table 7.2** provides a list of some of these vital dyes, the lowest concentrations at which they are used, and their toxicity as judged by how fast they cause cell injury or death. **Figure 7.1A** (see Plate 5 in the Color Addendum) illustrates vital staining of two species of protozoa, a *Didinium* and its prey, a *Paramecium*, using the dye neutral red. The transparency of biological materials in brightfield microscopy is illustrated well by the jelly layers surrounding the frog egg. In Figure 7.1B, methylene blue has been used as a vital dye to stain selectively J1, the innermost jelly layer. In contrast, this dye does not bind to J2 or J3 (the outer jelly layers), and they are essentially invisible. Other dyes such

Table 7.1	Comparison of Contrast Methods in Optical Microscopy			
Method	**Mechanism of Contrast**	**Advantages**	**Disadvantages**	**Uses**
Brightfield	Light absorption by pigments or dyes	Multiple dyes can be used to distinguish different structures; can be used for all tissues	Dye toxicity; fixation, embedding, and sectioning often necessary	Medical histology
Darkfield	Only scattered (diffracted) light is imaged	Live cell imaging possible; no added chemicals; high signal to noise ratio	Structures that do not scatter light will not be imaged; multiple cell layers cannot be imaged	Cell tracking; dynamics of light scattering organelles
Phase contrast	Conversion of optical path differences (usually due to refractive index differences) to intensity differences	Live cells can be imaged; no added chemicals; time-consuming tissue preparation not necessary	Large depth of focus; multiple cell layers can not be imaged	Tissue culture and other isolated cells
Polarization microscopy	Rotation of polarized light converted to intensity differences	Selective imaging of oriented cell structures; live cell imaging possible	Many cell organelles do not have oriented structure and cannot be imaged	Dynamics of the mitotic spindle, muscle sarcomeres, extracellular matrix fibers
Differential interference contrast	Rate in change in optical path length converted to intensity	Live cell imaging; no added chemicals; shadowing gives a quasi three-dimensional effect; thin focal plane allows "optical sectioning" and image clarity	Multiple cell layers cannot be imaged	Tissue culture and other isolated cells
Fluorescence microscopy (see Chapter 12)	Light emission from dye detected	Selective detection of specific macromolecules possible; high signal to noise ratio; high sensitivity	Low light output requires sensitive cameras; dye bleaching	Localization of specific proteins and structures by immunocyto-chemistry

as alcian blue and methyl red stain all three of the egg jelly layers (Figure 7.1C). Today, most vital dyes are localized by fluorescence emission (see Chapter 12), a much more sensitive technique than absorption of light. In Figure 7.1D, a fluorescent lectin from *Bonita* has been used to detect specific sugar polymers in all three jelly layers. For comparison, these same jelly layers can be visualized quite easily but with quite different results by optical contrasting methods such as phase-contrast microscopy (Figure 7.1E) and darkfield microscopy (Figure 7.1F).

Table 7.2	Examples of Vital Dyes	
Dye	**Minimum Useful Concentration (μg/ml)**	**Toxicity: Percentage of Cells Dead in 1 Hour**
Aniline yellow	182	0
Basic fuchsin	40	30
Bismark brown	7	0
Janus green B	6	40
Methyl violet	2	20
Methylene blue	10	5
Methylene green	27	5
Neutral red	7	3
Safranin	111	30
Toluidine blue	10	5

■ Phase-Contrast Microscopy

Light passing through each part of a specimen is subject to diffraction, that is, splitting of the beam into nondiffracted (zero order) rays and diffracted (first and higher order) rays. The diffracted rays differ from the nondiffracted rays in three respects. First, their intensity is highest at subcellular structures (organelles) having a high index of refraction compared to the surrounding cytoplasm. Second, their phase is shifted by an amount that is dependent on the refractive index of the organelle, typically about 90 degrees or $\lambda/4$. Third, the diffracted rays travel in a different direction, at an angle from the optical axis that depends on their wavelength (Figure 6.4). This third point means that it is physically possible to treat the diffracted rays in a manner that is optically different from the nondiffracted rays before they are recombined to form the image! This ability makes phase-contrast microscopy possible and led to the discovery of this technology by Zernicke in 1935.

In "brightfield" microscopy, the diffracted and nondiffracted rays recombine to form the image. Because they are out of phase, recombining these rays results in a certain amount of destructive interference giving rise to variations in light intensity from one point to the next that our eyes see as an image; however, this image is of very low contrast because the intensity of zero-order light that passed straight through the specimen is typically 10- to 20-fold greater than the intensity of the light diffracted by any organelle. This is shown in **Figure 7.2A** (top panel) in which the "surround" (nondiffracted, zero-order) rays are represented by the "S" wave and the diffracted (higher order) rays are represented by the "D" wave. The D wave is $\lambda/4$ (90 degrees) out of phase with the S wave, and it is of much lower intensity; therefore, when these two sets of waves recombine to form the product or "P" wave seen in the image, its amplitude is ever so slightly lower than that of the original S wave. Thus, zero-order light overwhelms the effects of interference, and the image produced has very low contrast—the "dark" spots representing organelles are almost as bright as the surrounding cytoplasm.

Zernicke thought that these effects of interference, usually only 5% to 10% of the image light, could be accentuated. He correctly reasoned that two methods would be necessary. First, one must decrease selectively the intensity of the zero-order rays so that their intensity is more comparable to that of the diffracted rays (Figure 7.2A, middle panel). Second, the phase difference between the diffracted and nondiffracted light rays must be increased to $\lambda/2$ (180 degrees) to maximize negative interference effects during

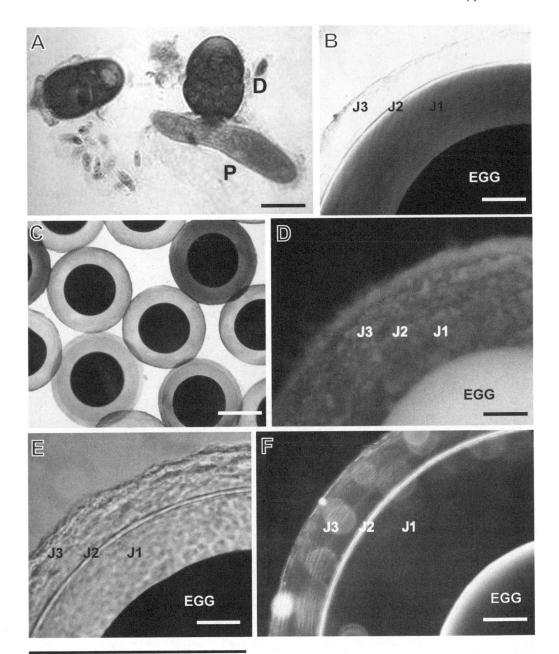

Figure 7.1   (See Plate 5 for the color version.)
Comparison of cells contrasted with vital dyes and optical contrast methods. (A) Three protozoa (a *Didinium* [D] is attacking a *Paramecium* [P] that it will soon eat while another *Didinium* stands by) are stained with neutral red. (B) The three jelly layers of a frog egg (J1, J2, and J3) are shown, as imaged by brightfield microscopy. The innermost layer, J1, has been selectively stained with methylene blue because of its high content of negatively charged sulfate groups. J3 and J2 are translucent, as are most biological structures in brightfield microscopy if they are unstained. Because of its size and pigment, the egg completely blocks transmitted light. (C) Alcian blue and methyl red staining of the frog egg jelly layers. (D) Fluorescence microscopy reveals staining by a dye-conjugated lectin that binds to sugars in all three jelly layers. The pigment of the egg fluoresces even brighter, an example of autofluorescence. (E) Phase-contrast microscopy reveals the boundaries and substructure of the three jelly layers without the use of dyes. (F) Darkfield microscopy (also without dyes) delineates the three layers because of the strong light scattering ability of J2. (A and B) Bars = 10 μm. (C–F) Bars = 100 μm.

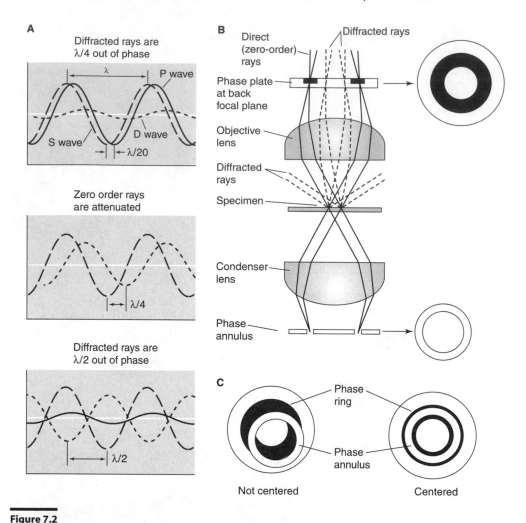

Figure 7.2

Phase-contrast microscopy. (A, top panel) The zero-order "S" wave coming from the specimen is about 20 times more intense and about 90 degrees out of phase with the diffracted light "D" wave. Their summation to form a "P" wave that is very little reduced in intensity accounts for the weak, low-contrast shadows in brightfield microscopy. (A, middle panel) The "D" and "S" waves have comparable intensities after the "S" wave is attenuated by the silver coating of the phase ring. (A, bottom panel) Further retardation of the "D" wave by the phase plate creates maximal destructive interference and formation of a very low-intensity "P" wave representing the dark contrast of organelles in a phase contrast image. (B) In phase-contrast microscopy, the specimen is illuminated with an oblique cone of light formed by a phase annulus in the front focal plane of the condenser lens. The ring of zero-order (nondiffracted) light (solid lines) is captured by the objective and focused into a well machined into a quartz disk at the back focal plane of the objective that is coated with silver. Diffracted light rays (dashed lines) are captured by the objective and pass through the thicker central area of the quartz disk. Phase retardation of the diffracted light at the specimen and at the quartz disk compared with the zero-order light produces optimal interference and contrast in the image. (C) Centering of the phase annulus (white ring) on the phase ring (shaded) is essential for good phase contrast.

image formation. If this is done, recombination of the S and D waves to form a P wave results in a P wave of very low intensity (Figure 7.2A, bottom panel) corresponding to the dark organelles of a phase-contrast image.

Technically, these two goals are achieved in the following manner. First, the specimen is illuminated with a hollow cone of light produced by a mask at the condenser lens that has a ring-shaped window (Figure 7.2B). This mask is referred to as the "phase annulus" and is located at the front focal plane of the final condenser lens. The oblique, zero-order rays that pass straight through the specimen will enter the periphery of the objective lens and then be selectively guided to pass through a circular well cut into the "phase plate," a quartz disk positioned at the back focal plane of the objective lens. This well is referred to as the "phase ring" of the lens. The bottom of the well is coated with a thin layer of silver that attenuates the zero-order light by 90% to make its intensity comparable to that of the diffracted light. In comparison, the diffracted light rays coming from the specimen are gathered by the center of the objective lens and do not pass through the well but rather through the entire thickness of the quartz disc at the back focal plane.

Because the quartz disc has a relatively high index of refraction (1.6), the refracted light rays, having to pass through a thicker region of the disc, get further retarded and out of phase compared with the zero-order light. The well in the quartz disc is machined just deep enough to produce an additional one-quarter wavelength shift in phase. This phase shift adds to any that is produced by an organelle (typically ranging from $+1/4$ to $-1/4\,\lambda$), thus, either enhancing destructive interference (organelle looks black) or eliminating destructive interference (organelle looks white). Clearly, how deep the well is machined determines the relative phase shift of the diffracted light and the type of contrast achieved. How dark the background of the image appears is determined by the degree to which the S wave is attenuated by the silver coating in the well, and phase objective lenses are often referred to as "dark phase" or "light phase" depending on the degree of attenuation.

None of this technology works, however, if the ring of light illuminating the specimen is not centered on the well or does not fall within the width of the well. For this reason, the circular mask that defines this ring of light is designed to be of a certain specified diameter and width. Because the dimensions needed differ for objective lenses of different magnifying power, most phase-contrast microscopes are equipped with three masks contained in a substage turret so that each can be rotated into place. Objective lenses are marked (e.g., Ph1, Ph2, Ph3) to indicate which mask they are to be used with. After the correct mask is in place, if not fixed in position, it must be centered on the well in the back focal plane of the objective. To do this, one uses a special lens (a "phase telescope" that takes the place of an eyepiece or a "Bertrand" lens that swings into place) to focus on the quartz disc at the back focal plane. As shown in Figure 7.2C, the mask is then adjusted in the x, y plane using thumbscrews to center it on the well.

Typical phase contrast images are shown in **Figure 7.3**. In Figure 7.3A, a growing hyphal tip of the fungus *Botrytis cinerea* exhibits a phase dense cell wall/plasma membrane with numerous spherical vesicles inside. In Figure 7.3B, detail of the tip shows the phase-dense "Spitzenkörper" (an organizing site for transport vesicles) and long phase-dense strands representing mitochondria. In Figures 7.3C and 7.3D, the flagellum of a frog spermatozoon (arrow) and the organelles of an isolated pancreatic acinar cell are visualized, including heterochromatin in the nucleus, spherical zymogen granules (arrows), and patches of endoplasmic reticulum. In Figure 7.3B, gelatin has been added to the extracellular medium to match its refractive index with that of the cytoplasm. This cuts down on the substantial phase effects at the plasma membrane that can obscure structural details (as seen in Figures 7.3E, F). Interestingly, in Figure 7.3G, neighboring cells in the hyphal tip can have different phase effects (one positive the other negative), suggesting that their cytoplasms are very different in refractive index.

■ Darkfield Microscopy

Darkfield microscopy, like phase contrast, uses oblique lighting of the specimen, as shown in **Figure 7.4A**. In this case, the oblique light is produced by a special darkfield condenser that does not require a mask to obtain a cone of light. In addition, the angle of incidence is more severe so that the vast majority of surrounding (zero-order) rays going

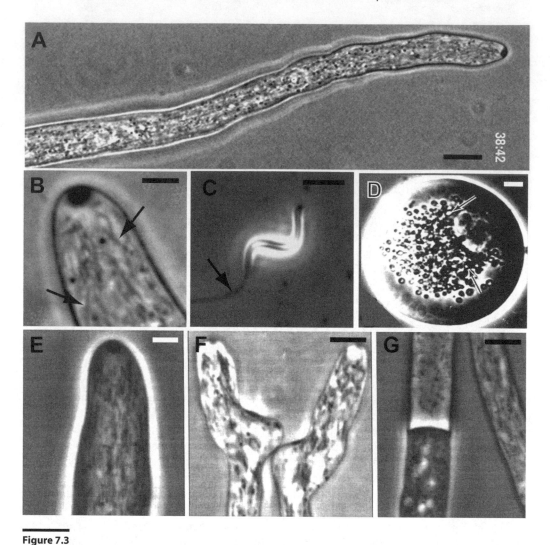

Figure 7.3
Phase-contrast micrographs. (A) A growing hyphal tip of the fungus *Botrytis cinerea*. (B) The distal end of the hyphal tip contains a phase-dense Spitzenkörper and numerous mitochondria (arrows). Note that the refractive index of the background (growth medium) and the cytoplasm of the cell are matched creating a high quality phase contrast image. (C) Sperm from the frog *Xenopus laevis* exhibiting a phase dense, corkscrew-shaped head and prominent flagellum (arrow). (D) Isolated pancreatic acinar cell exhibiting a large nucleus and numerous secretory granules (arrows). (E) Apex of fungal hypha exhibiting a large "phase halo" around the cell caused by the significant difference in refractive index between the cell and the extracellular medium: the cytoplasm has a higher index than the surrounding medium. (F) A reverse "phase halo" caused by the cytoplasm having a lower refractive index than the surrounding medium. The refractive index of the extracellular medium can be regulated by the addition of gelatin so as to equalize its refractive index with that of the cytoplasm. (G) Adjacent cells in the hyphal tip exhibit positive and negative phase halos caused by the difference in the refractive indexes of their cytoplasms. (A, C, E–G) Bars = 5 μm. (B and D) Bars = 2 μm.

straight through the specimen miss the objective lens entirely. What little zero-order light gets into the periphery of the lens is blocked by an iris diaphragm at the back focal plane of the lens. The rays processed by the objective lens to form an image are entirely of the diffracted ("scattered") variety; therefore, most regions of the specimen (e.g., the cytoplasm and extracellular space) will appear black because they diffract very little light. Organelles of high refractive index that scatter large amounts of light will be highlighted. An excellent example is the isolated pancreatic acinar cell of Figure 7.4B, in which the secretory granules are easily located (arrows).

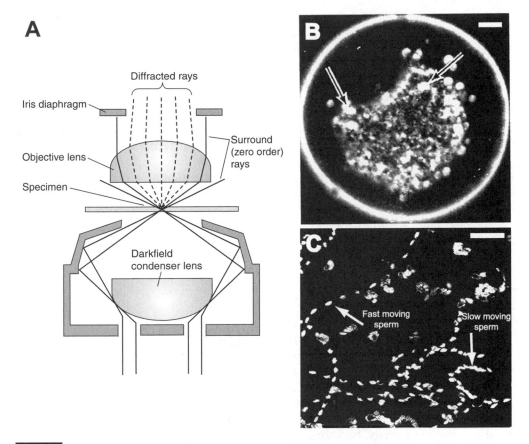

Figure 7.4

Darkfield microscopy. (A) Diagram of optics for darkfield imaging. Oblique illumination of the specimen allows only the diffracted rays to form the image. (B) A mouse pancreatic acinar cell exhibits an array of secretory granules that scatter light (arrows). Bar = 2 μm. (C) Mouse sperm, imaged at 0.5-second intervals as they swim, trace out their trajectories. This image represents a composite of 25 individual exposures. Bar = 50 μm.

An interesting use of darkfield microscopy is found in the tracking of highly refractive, motile cells such as sperm. The high signal-to-noise ratio experienced in darkfield microscopy is put to good use by illuminating the sperm with a stroboscopic light at high frequency (Figure 7.4C). Successive flashes of scattered light at set time intervals trace out the path and allow calculation of the velocity of each sperm.

■ Polarization Microscopy

As described in Chapter 5, light rays (photons) can have their electric vector oriented in any direction as long as it is perpendicular to the axis of travel. Typically, a population of rays will exhibit a randomly oriented set of electric fields and for this reason is referred to as nonpolarized light; however, after passage through a polarizing filter, light rays having a specific orientation (or polarization) will pass, whereas those having other orientations will not. In an inverse manner, polarized light can be used to query the organization of a specimen so as to reveal oriented or asymmetric structure. Usually this interaction consists of absorbing polarized light or rotating the axis of polarization. Rotation of polarization by the specimen can then be detected by analyzing the light passing through the specimen using a second polarizing filter as previously diagrammed in Figure 5.15.

The orientation of the electric vector for each light wave is of prime importance in determining how it is handled during absorption, reflection, or refraction by specimens that have repeating order in their structure that is oriented along a specific axis. We can illustrate these processes by using unpolarized light made up of waves having electric vector components that oscillate along both x- and y-axis—axes that are both perpendicular to the direction of propagation (represented by the z-axis). As shown in **Figure 7.5A**, if the light-absorbing chromophores (dye molecules) of a specimen are all lined up along the y-axis, the y component of the incoming light will be absorbed and will not pass through. In contrast, the specimen will be transparent to the x component, and it will pass right through.

The x and y components are also handled differently during reflection. As shown in Figure 7.5B, if the angle of reflection is low (less than so-called Brewster's angle) and the reflecting surface is smooth (compared with the wavelength of light), only the x component of the light oscillating parallel to the reflecting surface will be reflected, and, therefore, the reflected wave will be polarized. An example is the reflection of sunlight off the surface of a lake. These brilliant reflections can be completely blocked by sunglasses whose polarizing lenses are oriented vertically, that is, at a perpendicular to the horizontal surface of the lake. The lenses of the sunglasses contain polymer chains oriented in one axis something like a molecular version of a Venetian blind. Any light ray oscillating parallel to these blinds will get through; any light ray oscillating perpendicular to the blinds will be reflected and will not get through.

Asymmetric materials can be transparent to light but can have different indexes of refraction along different structural axes. This means that the x and y components of unpolarized light will travel with different speeds through the material and will be refracted at different angles upon entering the material. This phenomenon is known as double refraction, and the structural property of the material that produces double refraction is termed birefringence. Crystals such as calcite are well known for this property. If the crystal is oriented correctly, two separate images (the letter "e" in this case) can be seen through the crystal (Figure 7.5C). The x and y components of the image are refracted at two different angles; thus, the x component makes up one image, and the y component makes up the other. Both images consist of polarized light with their axes of polarization orthogonal to one another (Figure 7.5D, left panel). In this context, the x component is referred to as the "ordinary" or "O" wave, and the y component is referred to as the "extraordinary" or "E" wave.

In biological specimens, repeating structural features typically lead to a different index of refraction but not to differing angles of refraction, as shown in Figure 7.5D, right panel. This means that the O and E waves continue to have the same direction of propagation, but because of their different velocities within the specimen, they become out of phase with one another by the time they emerge. One might expect them to interfere with one another, but this is not the case. Electric vectors that are orthogonal to one another cannot interact or cancel one another by destructive interference and therefore cannot produce image contrast. They can only sum in their intensities.

Nevertheless, the fact that they are out of phase does result in differences in their polarization. O and E components that are in phase (0-degree or 180-degree phase difference) represent plane-polarized light. In contrast (Figure 7.5E, left panel), O and E components that are 90 degrees or 270 degrees out of phase, when summed, appear to have a circularly rotating electric vector as the wave propagates and for that reason is referred to as "circularly" polarized light. If O and E components are out of phase by any other amount, when summed, the electric field vector changes in magnitude as it rotates sweeping out an ellipse; for this reason, these components are referred to as "elliptically" polarized light.

Therefore, when plane polarized light passes through a biological specimen that is "birefringent" (structurally ordered along an axis so as to treat the O and E components differently), the light will be modified to produce elliptically polarized light and change

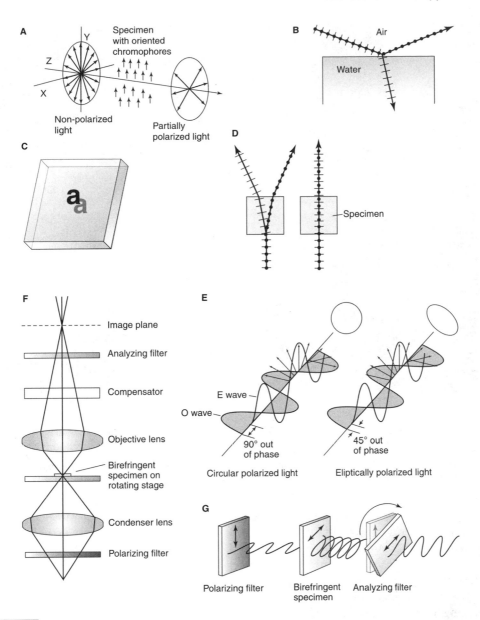

Figure 7.5

Polarized light and polarization microscopy. (A) Passage of nonpolarized light through a specimen containing orientated chromophores can produce partially polarized light. (B) Reflection/refraction of nonpolarized light at an air–water interface can produce two orthogonally polarized beams. (C) Calcite crystal showing double refraction of the letter "a." (D, left panel) Double refraction occurs when nonpolarized light is split into two polarized components each refracted at a different angle. The beam with dots represents a beam with electric vectors oriented perpendicular to the book page while the beam with slashes represents a polarized beam with electric vectors lying within the plane of the page. (D, right panel) A biological specimen with structure oriented in the horizontal axis. Nonpolarized light will be split into two orthogonally polarized beams, but typically, these beams will not be refracted at different angles. Rather, they make up the same image but become out of phase with each other as they propagate through the specimen. (E) The electric vector of light having two orthogonal components that are out of phase with each other sweeps out either a circle or an ellipse, hence the designation "circularly polarized light" and "elliptically polarized light." (F) Optical design of a polarization microscope. Light, plane polarized by the "polarizer filter," is passed through the specimen, gathered by the objective lens, and passed through the "analyzer" (a second polarizing filter) before image formation. (G) Changes in light polarization during microscopy. A birefringent specimen, if oriented optimally relative to the incident plane polarized light, will produce elliptically polarized light whose major axis has been rotated relative to the incident light. Optimal rotation of the analyzing filter will highlight the rotated light coming from birefringent structures while blocking plane polarized light coming from nonbirefringent structures.

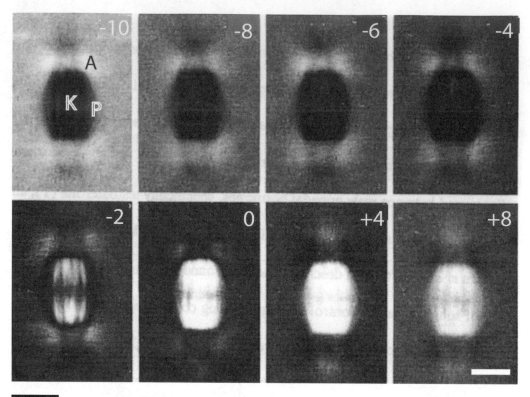

Figure 7.6
Images of a mitotic spindle from a sea star oocyte taken by polarization microscopy using a series of different analyzing filter angles. The angles in degrees, at the upper right corner of each image, are relative to zero degrees indicating that the polarizing and analyzing filters are 90 degrees to one another. Three different microtubule populations within the spindle, the astral microtubules (A), the polar microtubules (P), and the kinetichore microtubules (K) each show different birefringent properties. Bar = 2 μm.

the orientation of polarization (Figure 7.5E). This change can be selected for observation by passing the image through an analyzer that is rotated so as to optimize passage of light whose polarization has been reoriented. An optical diagram of a polarizing microscope (Figure 7.5F) shows the positioning of both polarizing and analyzing filters, as well as the rotating stage on which the specimen is placed for easy orientation. This optical design results in incident plane polarized light being rotated by the specimen and the analyzing filter (rotated correctly) selecting this rotated light for transmission while blocking out nonrotated light from the image (see Figure 7.5G).

An excellent example of a birefringent biological specimen is the isolated mitotic spindle (**Figure 7.6**). The near parallel alignment of microtubule bundles making up the metaphase mitotic spindle is efficient in rotating polarized light, thereby revealing the presence of its highly oriented structural components. As seen in the series of micrographs in Figure 7.6, the appearance of the metaphase mitotic spindle changes dramatically as the analyzing filter is rotated, its core structure appearing dark at negative rotations of the filter and bright at positive rotations demonstrating its birefringent properties. In fact, closer inspection shows that different microtubule populations within the spindle differ in this respect. Astral microtubules (A), spreading radially from each spindle pole, actually rotate polarized light in a direction opposite to that from the polar microtubules (P), which span the two poles and from kinetichore microtubules (K) that join the chromosomes at the spindle equator with the poles. Thus, the astral microtubules become brighter at negative angles of the analyzer rather than darker. Comparison of images within this series shows that while the nonpolarizing medium around the spindle is darkest at 0 degrees (crossed polarizing filters), the birefringent properties of astral, polar, and kinetichore microtubules are compensated for by analyzer angles of about $+2$, $+8$, and -6 degrees, respectively. Polarized light microscopy, therefore, not only has visualized spindle microtubule structures in unfixed and unstained specimens but also has distinguished three subpopulations of microtubules that exhibit both structural and functional differences.

Essential to polarization microscopy is complete elimination of background, nonpolarized light so as to increase the signal to noise level. This requires that the polarizing and analyzing filters used be 99.9% efficient in eliminating light that is not correctly polarized and is evidenced by the dark backgrounds in Figure 7.6. Also important is the elimination of "false-positive" images that result from mechanical strain on the lenses making up the optical system of the microscope. Strain tends to orient glass constituents that normally are randomly arranged, thereby producing birefringence; strain-free optical components are "hand picked" for use in polarizing microscopes.

■ Differential Interference Contrast Microscopy

In the 1950s, Nomarski developed an optical system that combined the use of polarized light and phase shifting by intracellular organelles to provide contrast in living cells. The principles of this technique are completely different from that of phase contrast but use some of the features of the polarization microscope discussed previously. The crux of the technique is to use the properties of polarized light even in specimens that have no birefringent properties. Plane polarized light rays are separated into very close parallel rays. The parallel rays actually travel through adjacent but slightly different parts of the specimen. Any difference in the optical path length caused by an organelle (**Figure 7.7A**, top) will slow and change the phase of one ray relative to the other. The difference in optical path length is dependent on both the refractive index and the size of the organelle. For simplicity, such a structure is referred to as a "phase object." The change in phase is directly related to the slope of the optical path length versus distance. As shown in Figure 7.7A (bottom), if the phase difference can be "read out" as image intensity, one side of the phase object will be highlighted, while the other side will be in a shadow. This readout is achieved by the fact that when the rays are precisely recombined on the other side of the specimen, the phase differences result in interference that leads to intensity differences. As a result, there is a shadowing of all organelles having an index of refraction different than that of the cytoplasm and all cells having an index of refraction different from the medium around them (Figure 7.7C, top panel).

The optical design of such a microscope is illustrated in Figure 7.7B. Light from a typical tungsten-halogen lamp is passed through a condenser lens and polarizing filter to obtain plane-polarized light. This light is then sent through a set of beveled prisms named Wollaston prisms after their early 19th century inventor. It is this set of prisms that separates each polarized ray into two wavelets, O and E, running parallel and side by side at a distance of less than a micrometer as they pass through the specimen. The electric vector of the incoming plane polarized light can be thought of as being split into two orthogonally polarized components that are physically separated by diffraction at slightly different angles. After passing through the specimen, both rays are captured by the objective lens and within the objective lens pass through another set of Wollaston prisms. This second set of prisms recombines the parallel rays that had previously been separated to once again produce polarized light rays. These recombined rays are then tested by the analyzer filter to determine whether any shift in the axis of polarization has taken place.

This critical step at the analyzer will produce image contrast (see Figure 7.7C). If both separated rays are exactly in phase, as they would be if they had passed through locations having the same index of refraction, they would recombine to form a ray of the same intensity. In addition, the ray would remain polarized in its original plane and, after passing

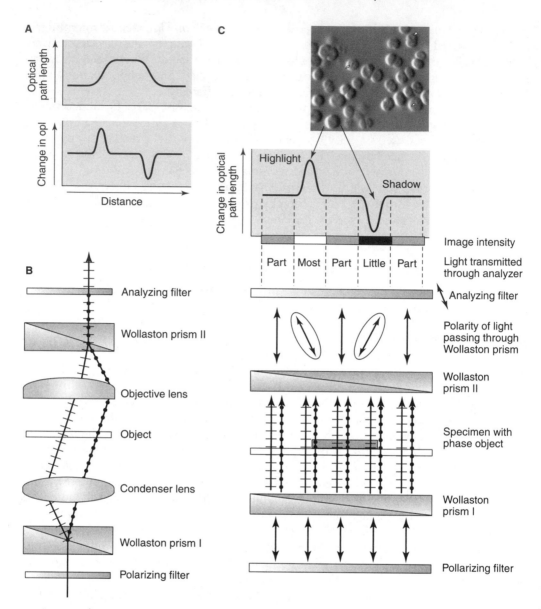

Figure 7.7

Optics of differential interference contrast (DIC) microscopy. (A, upper) Optical path length increases at the location of a phase-dense organelle caused by its increased refractive index. (A, lower left) Change in optical path length versus distance, if read out as image intensity, produces highlights on one side of the structure and shadows on the opposite side. DIC micrograph of cyanobacteria exhibiting this characteristic shading. (B) Optics of the DIC microscope. Light is polarized, split into two orthogonal beams at the first Wollaston prism, passed through the specimen, gathered by the objective lens, recombined by the second Wollaston prism, and then "analyzed." (C) Specimen effects on the DIC image. Each pair of rays passing through the specimen represents "O" and "E" waves that have electric vectors orthogonal to one another. If both components travel through either cytoplasm or the phase-dense object (e.g., an organelle), they remain in phase and when recombined at the second Wollaston prism form plane polarized light. If the two components each pass through different structures (usually occurring at the edge of an organelle), they become out of phase and produce elliptically polarized light. The analyzer blocks elliptical light rotated in one direction, giving rise to shadows and fully transmits elliptical light rotated in the other direction to give highlights.

through the analyzer rotated 45 degrees from this axis, would be seen at partial intensity (gray) in the image. Whether the two rays pass through the cytoplasm, extracellular medium, or a phase object does not matter. As long as the optical path lengths of the two rays are exactly the same, the result will be a partial attenuation—the same shade of gray.

On the other hand, if the parallel rays penetrate the specimen at the interface between a phase object and its surroundings, they would encounter different optical path lengths. As a result, the rays would be out of phase at the moment they were recombined. The recombined light would no longer be plane polarized but (depending on the phase difference) would become circularly or elliptically polarized with its new axis of polarization rotated from that of the incident polarized light; therefore, this secondary ray, after passage through the analyzer, would be seen either as a highlight (if rotated toward the analyzer axis) or a shadow (if rotated away from the analyzer axis).

In **Figure 7.8A,** differential interference contrast has been used effectively to shadow the organelles found in a fungal hyphal tip. At the distal end of the hypha (region I), the

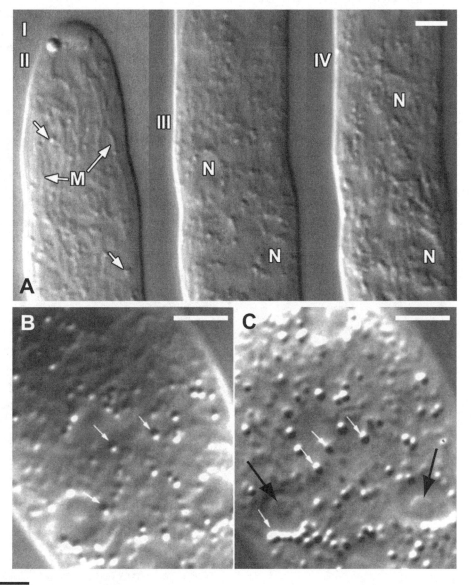

Figure 7.8

(A) Three regions of the *Neurospora crassa* hyphal tip. (A) The distal regions (I and II) contain the Spitzenkörper and underlying microfilamentous web. The proximal regions (III and IV) contain well-spaced nuclei (N). All regions contain elongated mitochondria (M). (B and C) Enlarged micrographs showing shadowed transport vesicles and nuclei exhibiting prominent nucleoli (large arrows) in a sporangium of *Allomyces macrogynus*. The same cell is shown in (B) and (C), but in (C), the Wollaston prism has been adjusted to provide better shadowing of some vesicles (small arrows). (a–c) Bars = 2 μm.

Spitzenkörper appears as a dense sphere with a second phase-dense spherical organelle to one side. In region II, numerous string-like mitochondria (M) are seen with occasional vesicles (arrows). In regions III and IV, spherical nuclei (N) are pointed out, and large numbers of transport vesicles are found. A sporangium at higher magnification (Figures 7.8B and 7.8C) exhibits individual vesicles about 1 μm in diameter (small arrows), and each nucleus has a prominent nucleolus (large arrows). Figures 7.8B and 7.8C show the same region of the sporangium, but with the Wollaston prism adjusted differently. Each prism consists of two wedges. As these wedges are slid relative to one another, the relative position of the two ray populations recombined to form an image change and the axis of polarization at each image point changes. In this way, specific vesicles (small arrows) can be highlighted differently by prism adjustments.

In addition to contrast, differential interference contrast (DIC) optics have the additional advantage that the depth of focus is relatively narrow—about 2 μm. Features just above or below this plane do not contribute to the image thus avoiding the superimposition of organelle images. Because of this feature, specimens imaged by DIC can be "optically sectioned" by small changes in distance between the specimen and the objective lens to produce a stack of images from which the specimen can be reconstructed in three dimensions.

■ Hoffman Interference Contrast

One disadvantage of DIC optics is that it does not work well with plastic tissue culture dishes that can show a high degree of birefringence. This birefringence leads to artifacts that override the delicate pattern of polarization changes produced by the specimen itself. Hoffman contrast was designed to avoid this problem by eliminating the use of polarized light.

The optical path (**Figure 7.9A**) is designed to illuminate the specimen both obliquely and asymmetrically. A slit aperture at the front focal plane of the condenser lens allows

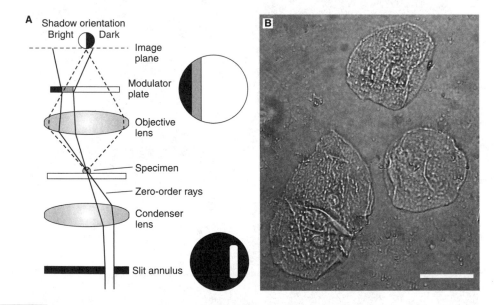

Figure 7.9
Hoffman contrast microscopy. (A) Diagram showing optics of a Hoffman contrast microscope. A slit annulus allows the specimen to be illuminated with a sheet of light at an oblique angle. The zero-order rays passing through the specimen (shaded) are captured by the objective and attenuated by passage through a half-silvered region on the modulator plate. The diffracted rays are captured by the objective but are treated asymmetrically by the modulator, being blocked on one side but not on the other side. This gives rise to a shadowing effect in the image. (B) Micrograph of human epithelial cells from the oral cavity using Hoffman contrast optics. Bar = 20 μm.

a single sheet of light to be passed through the specimen at an angle, the zero-order rays to be gathered by the objective lens and directed through a modulator plate located at the back focal plane of the objective lens. The modulator plate is completely clear except for a black, opaque stripe at one edge followed by a "gray" semitransparent strip at its inner edge. The zero-order light is directed through the semitransparent strip. Its intensity is diminished in the process and spread to all parts of the image to produce a low-level background illumination. The diffracted light, emerging from the specimen at many different angles, is collected by the entire objective lens and projected onto the modulator plate. Diffracted rays on the clear side of the plate will pass through and recombine with the zero-order light to produce destructive interference and shadows; however, diffracted rays hitting the other side of the plate will be blocked in the opaque region and will not be able to contribute to the image. As a result, features of the cell that diffract light will be shadowed asymmetrically, as shown in the image of human epithelial cells in Figure 7.9B.

References and Suggested Reading

Abramowitz M. *Contrast Methods in Microscopy: Transmitted Light* (vol. 2). New York: Olympus Corporation Publishing, 1987.

Bradbury S, Evennett PJ. *Contrast Techniques in Light Microscopy.* Oxford: BIOS Scientific Publishers, 1996.

Collett E. *Polarized Light: Fundamentals and Applications.* New York: Marcel Dekker, 1993.

Murphy DB. *Fundamentals of Light Microscopy and Electronic Imaging.* New York: Wiley-Liss, 2001.

Spencer M. *Fundamentals of Light Microscopy.* Cambridge: Cambridge University Press, 1982.

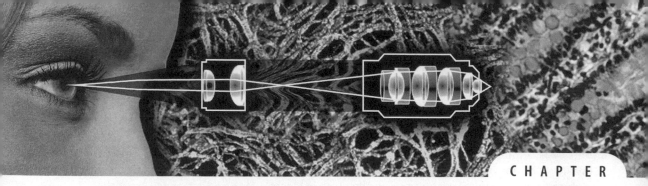

The Transmission Electron Microscope

The ability to resolve and analyze cytoplasmic detail and organization at the nanoscale provides powerful insights into the fundamental operations of cellular and molecular biology. The transmission electron microscope (TEM) (**Figure 8.1**) opens the door into the unseen world of cells and tissues for biologists while presenting great challenges in applying the most pristine preparation protocols and interpretations of data. With recent advances in TEM design, improved preparation protocols, and increased analysis capabilities, the scientific community is positioned at the beginning of a new era of ultrastructural discovery and is thus poised to make significant advancements in better understanding cell function and interaction. Indeed, the need to determine spatial cytoplasmic complexities and the elucidation of function represents one of the most challenging goals in biology today.

The TEM is a sophisticated imaging system that transmits illuminating electrons through a thin sample so that structural details can be visualized and recorded at high magnifications (up to 1 million times) and resolutions (less than 0.1 nm when using the best aberration corrected instruments). The microscope is composed of an electron emitter, a series of electromagnetic lenses and apertures, a specimen chamber and control system, vacuum pumps, and a means for viewing and acquiring images. All of these components are grouped into four integrated systems: the illumination system, the specimen manipulation system, the imaging system, and the vacuum system.

■ Illumination System

The illumination system is located at the top of the microscope column and is responsible for producing a monoenergetic electron beam, regulating its strength and directing it onto the specimen.

The Electron Gun

The electron gun is the source of the electron beam used to form an image and is made of three parts: the filament, shield, and anode (**Figures 8.2** and **8.3**).

Filament (Cathode and Emitter)   The filament is a cathode and serves as source of electrons. A typical thermionic filament is a V-shaped wire of pure tungsten with an approximate diameter of 0.1 mm (Figure 8.3E). As a metal, tungsten contains positive ions and free electrons that can be driven out of the filament. This is done in vacuo by applying

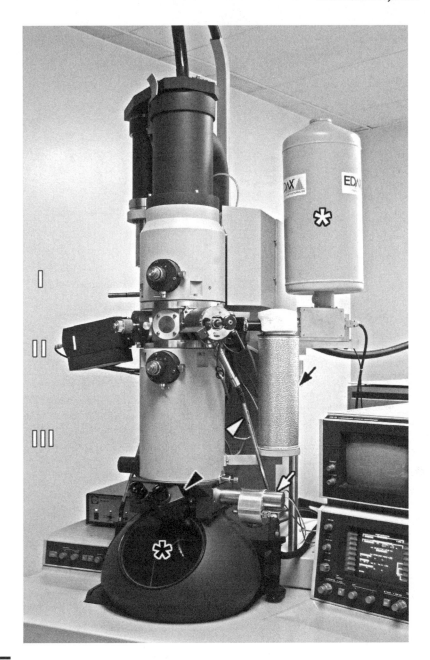

Figure 8.1
A photograph of the column of a TEM illustrating regions where the condenser lenses (I), objective lenses and specimen chamber (II), and intermediate and projector lenses (III) are positioned. Also shown are the viewing window (asterisk), binocular viewer (black arrowhead), stage manipulation rod (white arrowhead), digital camera (arrow), and liquid nitrogen Dewars for the anticontamination device (black arrow) and x-ray detector (white asterisk).

a fixed amount of negative high voltage (e.g., 80 kV) (1 kV is equivalent to 1000 V), followed by slowly heating the filament with the application of a small direct current through the filament. As the heating current is slowly increased, the number of electrons emitted off the filament will increase. The saturation point of the filament is that point at which the number of electrons emitted no longer increases as the filament is heated and is influenced by the bias setting (see later). This balancing point between optimal filament heating

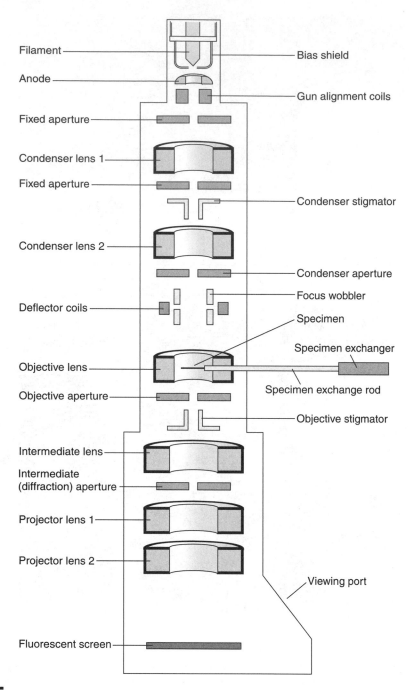

Figure 8.2
An illustration of the principal parts that make up the column of a standard TEM.

and electron emission is due to the variable self-biasing nature of the electron gun. The amount of energy required to produce sufficient electron emission from a metal is called the work function. Even though tungsten has a fairly high work function, it has an equally high melting point (3653°K) and an abundant supply of electrons that can be extracted at a temperature significantly below its melting point. Good electron emission from the

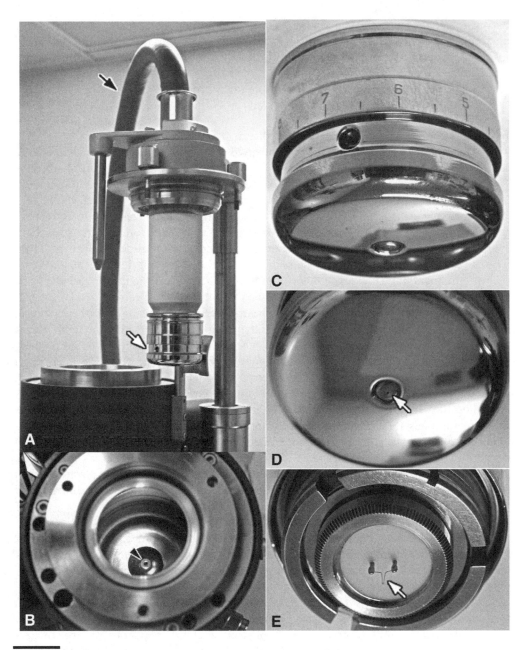

Figure 8.3
Components of the TEM electron gun. (A) The cathode of the electron gun (white arrow) raised out of the emission chamber. The high-tension cable is pointed out (black arrow). (B) Looking into the emission chamber from above the anode (arrowhead). (C) Electron gun cathode assembly. (D) A view looking up toward the shield, illustrating the aperture (arrow) that is positioned directly in front of the filament. (E) Tungsten wire filament (arrow).

tungsten filament can typically be accomplished at approximately 2600°K. A single filament can have a long and useful operating life (approximately 150 hours) if it is not used at a point of oversaturation, if the gun area is kept clean of dirt and debris, thus preventing high-voltage discharge, and if an appropriate vacuum is maintained during use.

Shield (Wehnelt Cylinder, Bias Shield, Focusing Electrode, Grip Cap)   The shield is the second part of the electron gun (Figures 8.2, 8.3A, and 8.3C) and has the primary role to direct properly the largest part of emitted electrons down the column. Without this, only a small percentage of electrons would pass through the illumination system. The shield consists of a cap-like cylinder positioned directly in front of the filament with an aperture (2- to 3-mm diameter) (Figure 8.3D) centered over the filament tip. The shield is held at a slightly higher negative potential than the filament and thus creates an environment that concentrates electrons into a cloud between the filament and shield. Only through the aperture of the shield can electrons escape to become the illuminating radiation that participates in image formation as they are directed down the column.

The difference in the negative potentials between the filament and shield is called the bias. The bias is generated and controlled by connecting the high-voltage supply directly to the shield while placing a variable bias resistor between the high voltage and filament. The degree of negative potential applied to the filament is controlled by adjusting the bias resistor, that is, increasing the bias results in a filament that is less negative (more positive) than the shield. Furthermore, each change to the bias requires appropriate adjustments to the filament saturation point. The bias setting hardly affects the number of electrons given off by the filament, but rather, it has a direct effect on the density of the electron cloud between the filament and shield, the point of saturation, and thus the number of electrons that pass through the shield aperture to become the illuminating electron beam.

The bias setting has a direct effect on the beam current (i.e., the number of electrons in the beam), which influences the brightness of the image. Working at high magnifications often requires increasing the bias in order to improve brightness. A price is paid for this, however: in decreased life of the filament and increased beam damage to the specimen. High electron beam current, sometimes called electron dose, alters the specimen by breaking bonds and by generating free radicals, which lead to structural disintegration of the sample. Embedding the sample in plastic and staining it with heavy metals protects and improves somewhat sample stability, but specimen damage can be often detected clearly. The resin-embedding medium also interacts with the electron beam. The electron beam thins out plastic where the cellular matrix does not support the sample. This thinning can reach 40% in the direction of the beam and 10% in the specimen plane. The thinning of plastic sections poses serious disadvantages when one is keenly interested in rigorous three-dimensional relationships of cytoplasmic components, such as in electron tomography.

It is important to remember and point out that additional filament heating (or current) beyond the point of saturation does not increase the beam current, the number of electrons in the beam, or the brightness of the electron beam but rather leads to a significant reduction in the life of the filament. Indeed, the filament heating should be set slightly below the point of saturation. The bias should be adjusted to a point where there is good illumination at the saturation point so as to prolong the life of the filament. In short, increase bias → increase positive potential of the filament relative to the shield → increase saturation point → increase density of electron cloud → increase number of electrons in beam (i.e., beam current) → increase brightness → increase heat and potential damage to specimen → decrease life of filament.

Anode Plate   The anode is a disc with a central aperture and is positioned beneath the gun (Figures 8.2 and 8.3B). Like the remaining parts of the microscope, the anode is grounded and is attached to the positive terminal of the high-voltage supply. With respect to the anode, the shield and filament (cathode) are held at a large negative potential, which is the force that accelerates the electrons down the space of the microscope column at a constant velocity.

Additional Electron Sources and Gun Designs   As the applications and requirements for TEM become more diverse and complex, alternatives to the standard electron gun design described above are becoming more common and in some cases mandatory. For example,

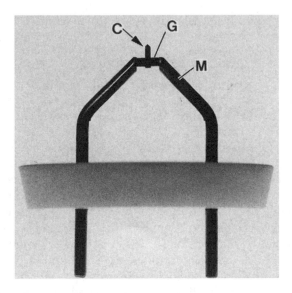

Figure 8.4
A photograph showing a lanthanum hexaboride filament. C = crystal; G = graphite holder; M = molybdenum alloy supports.

a lanthanum hexaboride (LaB_6) cathode (**Figure 8.4**) has several important characteristics that make it a highly desirable electron source for thermionic electron guns when compared with tungsten. These include a lower work function, a lower operating temperature, a higher yield of electrons and thus brighter beam even at high magnifications, and an increased lifetime (400 to 600 hours). The disadvantage of an LaB_6-based thermionic emitter is basically its higher cost (about 20 times higher than the standard tungsten filament); because of the highly reactive nature of LaB_6 when heated to temperatures required for electron emission, the TEM must be equipped with a costly vacuum system able to maintain vacuums in the gun area much greater than those typically required by standard tungsten filament emitters (10^{-7} torr or better vs. 10^{-4} torr for tungsten).

A second option for electron emitters is the field emission guns (FEGs) (**Figure 8.5**) that operate in a very different way than the thermionic emitters (i.e., the conventional tungsten and LaB_6 filaments). In the FEG design, the electron source is a precisely oriented single crystal of tungsten with a tip etched to a radius of about 20 nm. Electron emission is achieved by positioning the filament near a series of anodes held at 3 to 5 kV that act as an electrostatic lens. This design reduces the work function, allowing electrons to

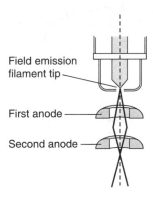

Field emission
filament tip

First anode

Second anode

Figure 8.5
A drawing of a cold FEG depicting electron extraction from a tungsten crystal by a pair of anodes.

"tunnel" through the tungsten filament and into the vacuum. Because electrons originate from a point source (thought to be 5 nm or less) on the finely tipped tungsten filament, the electron beam is highly coherent with a very narrow band width of energies, making the FEG beam very bright (high beam current) with reduced chromatic aberration. This permits the high current beam to be concentrated into smaller spot diameters, yielding higher resolutions and improved signal-to-noise ratio. Consequently, FEGs are very useful in high-resolution imaging, especially of very delicate samples such as those examined in their frozen-hydrated condition using cryo-TEM. Furthermore, the depth of field is greatly improved with FEGs because of the reduced aperture angle, which is advantageous in scanning electron microscopy (see Chapter 9). FEGs come in two designs. The first one developed was the so-called cold FEG described previously. A major factor that plagues the cold FEG is its unstable emission. This is caused either by the backscattering of ions from the anode onto the cathode tip resulting the etching of the tip, which interrupts the emission area, or by contamination of the tip when operated at ambient (i.e., cold) temperatures. Contamination of the tip can be remedied by briefly baking or "flashing" the tip periodically. Field emission guns that are operated at elevated temperatures, the Schottky FEG, can provide a virtual electron source size comparable with cold FEG source but with higher stability and no requirement to condition or "flash" the tip at intervals. Both are long-lived electron sources (more than 2000 hours), but require a very high vacuum, even greater than that for LaB_6.

Condenser Lens

Before time is spent discussing the functions of the condenser lens, some attention must be devoted to the lenses of the electron microscope in general, and we must understand how they influence electrons. All lenses in electron microscopes are electromagnets. Because electrons have such a small mass, they are easily scattered by gas molecules, and of course, glass lenses, like those in light microscopes, are of no use. Electrons, however, are charged particles and as such can be influenced by the lines of force generated by magnets. The simplest electromagnet lens is a solenoid: a coil of wire with a direct current running through it. In 1927, Gabor designed the first electromagnetic lens by partially encasing the wire coils in a soft iron shroud, which succeeded in concentrating the magnetic field through the center of the lens, or the lens bore. It was evident, however, that Gabor's design did not efficiently concentrate the magnetic field (i.e., the lens was not strong enough), and in 1931, Ruska and Knoll increased the encasement around the wire coils, leaving only a small gap in the lens bore between an upper and lower core of soft iron (i.e., the pole pieces separated by a nonmagnetic brass spacer) available for the generation of the magnetic field (**Figure 8.6**). This effectively concentrated

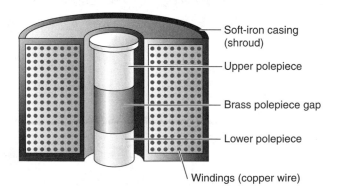

Soft-iron casing (shroud)

Upper polepiece

Brass polepiece gap

Lower polepiece

Windings (copper wire)

Figure 8.6
A drawing showing a longitudinal view through a typical electromagnetic lens. Noted are the copper wire windings, soft-iron encasement (i.e., shroud), upper and lower pole pieces, and the pole piece gap.

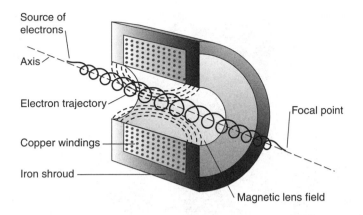

Figure 8.7
An electron is illustrated passing through a uniform magnetic field generated by an electromagnet. The electron follows a helical path through the field converging to a focal point after leaving the lens.

the strength of the lens along the short axial distance of the annular gap between the upper and lower pole pieces while allowing the lens to operate with much reduced current and consequently heat.

Under the influence of the electromagnetic lens, electrons accelerated through a vacuum take on a helical trajectory. The paths taken by electrons are not necessarily the same and are influenced by their trajectories as they enter the lens's magnetic field. For example, when a group of electrons originating from a point on the sample plane enter the objective lens with different trajectories, they will be directed along independent helical paths within the lens and will proceed eventually to converge as focused elements of the image at another point along the optical axis (**Figure 8.7**).

The function of the condenser lens, the second major component of the illumination system, is to gather the electrons from the gun and focus them onto the surface of the specimen. In doing so, this lens system regulates the intensity of the illuminating beam on the specimen. Today's TEMs have two condenser lenses. The first lens (C1) is described as a demagnifying lens that focuses or concentrates the beam to a diameter between 1 and 20 μm. This point of focus, or crossover, is called the spot size. The smaller the spot size the higher potential resolution the user may achieve, with the cost being loss in over all beam brightness. Compromised brightness for resolution can be avoided by the use of LaB_6- or FEG-based guns. The next condenser lens (C2) magnifies the spot and spreads the beam to illuminate the area of the sample being investigated. As a common rule, the illuminating beam should be a diameter large enough for the user to visualize only the region of interest. This sample region, of course, will change as the user moves from lower to higher magnifications and vice versa. Thus, a dual-condenser lens system avoids damage to the areas of the sample not being imaged and provides the user with a means for balancing the needs between magnification, resolution, and brightness.

A beam of radiation, electrons or light, can easily be altered, creating aberrations and degradation of image quality. Fixed and adjustable apertures are placed at strategic locations along the TEM column with the purpose of allaying this potential problem. Depending on the TEM, the condenser lens system may have one or more apertures. For example, just beneath the C1 lens is often an aperture of a fixed size (Figure 8.2), whereas associated with the C2 lens is a variable aperture system (**Figure 8.8**) containing apertures of several sizes. The latter are typically presented in a thin molybdenum strip containing three to four holes ranging in diameters from 100 to 500 μm. The selection, placement, and alignment of the appropriately sized C2 aperture are at the discretion of the user.

Figure 8.8
Region of the TEM column pointing out the specimen insertion chamber with specimen rod (black arrow), variable condenser aperture (white arrow), variable objective aperture (white arrowhead), variable diffraction aperture (black arrowhead), and secondary electron detector (asterisk) associated with scanning transmission electron microscopes.

The larger the C2 aperture is, the more electrons pass from the lens onto the sample yielding a brighter image; however, because peripheral beam electrons are more susceptible to heterogeneous lens strength and are not eliminated from the beam when larger C2 apertures are used, spherical aberration may result. On the other hand, a smaller C2 aperture will reduce the overall brightness of the sample, but as peripheral electrons are eliminated, spherical aberration is reduced, leading to the possibility of enhancing resolution.

Deflector Coils
Deflector coils are a series of small electromagnets located beneath the condenser lenses (Figure 8.2). This system is used for aligning the electron beam and centering it on the region of interest.

■ Specimen Chamber and Control System
The specimen chamber is typically located within the objective lens (Figures 8.1, 8.2, 8.8, and **8.9**) and is constructed with specifications that allow easy and rapid exchange of the specimen. The chamber must also have a specimen stage that is able to move samples smoothly and accurately along the x and y coordinates at right angles to the optical axis over a distance of 250 mm in increments as fine as 10 nm. These fine movements are easily regulated by the user through two external stage controls that are virtually free of backlash movements. In essence, the stage control system for TEM is a sophisticated micromanipulation apparatus. Furthermore, the system is required to maintain a stationary specimen position within 0.1 nm for up to 3 seconds for data recording purposes.

Figure 8.9
Segment of the TEM column opened for inspection in the region of the objective lens and specimen chamber. Notice the lower pole piece of the objective lens (arrow) and upper anticontamination (cold finger) blade (asterisk).

The chamber also accommodates a means for dissipating heat from the specimen and objective lens. Certain microscopes have goniometer-equipped stages, allowing the user to tilt the specimen precisely in one plane up to 60 degrees, which is used for generating electron tomography data sets and stereo pairs and moving oblique sections normal to the beam. Some stages are also able to rotate the grid within the TEM through 360 degrees, which is useful in composing images for capture and in some applications of electron tomography. Most modern TEMs have computer-controlled motorization and positioning abilities and are desirable for large-scale imaging projects. The biological samples prepared for standard TEM viewing in the form of resin-embedded thin sections are collected onto the surface of a small (approximately 3.0-mm diameter), circular, metallic screen supports called grids. Grids are analogous to the glass slides used in light microscopy. They come in a variety of designs and are made from a number of different materials (see Chapter 3, Figure 3.5F). The most common types of grids used are those made of copper because this metal is nonmagnetic and has high thermal conductivity. An often-used mesh size is 300 to 400 bars per inch, as this size provides adequate support for sectioned samples for routine work.

There are two general specimen holder designs: top-entry and side-entry. The side-entry stages (Figures 8.3 and **8.10**) are the most common in contemporary TEMs because they provide greater stability and resolution, as well as specimen position and orientation to the beam. Top-entry holders allow easier specimen manipulations within the column. Specialized stages permit a variety of options to the user, some of which are more applicable to materials science than to biological science. The tensile stage, for example, applies stretching and compression to the specimen, whereas temperature stages can apply heating or cooling. With the recent interest in viewing biological samples in their frozen hydrated states, cold stages and the technology associated with their control, sample preparation, and sample imaging and analysis have become very popular in the area of cryo-EM.

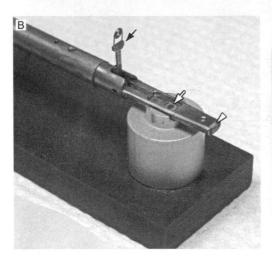

Figure 8.10
(A) Side-entry specimen holder positioned on stand. (B) Closer view of tip showing grid (white arrow) positioned in depression, grid securing arm in the open position (black arrow), and crystal (arrowhead) used for controlling holder movement. (C) Grid securing arm is shown closed. The grid is secured and ready for insertion into the TEM.

■ Imaging System

The imaging system is composed of lenses that form, focus, and magnify the final image of the specimen for viewing and image acquisition. The system includes the objective, intermediate, and projector lenses (Figures 8.1 and 8.2).

Objective Lens

The objective lens forms the initial image of the illuminated portion of the specimen that is then further enlarged by the intermediate and projector lenses. This role makes the objective lens the single most important lens in the TEM. The high resolution obtained by the objective lens is achieved by positioning the specimen close to the lens's focal plane. This is made possible by constructing the lens so that the specimen can be positioned into the heart of the lens itself, between the pole pieces (Figures 8.2, 8.8, and 8.9). The lens must also be highly energized in order to maintain very short focal lengths (1 to 2 mm). This helps in allaying chromatic and spherical aberrations. Objective lenses are of very high quality with minimal astigmatic characteristics and other aberrations. The quality of the lens, as with all electron magnetic lenses, is a direct result of the quality of the pole pieces (Figures 8.6, 8.9, and 8.11). As mentioned above, pole pieces are made from homogeneously blended soft iron and must be as symmetrical as possible to avoid

Figure 8.11
Objective upper pole piece (arrow) and encasing (asterisk) of the electromagnetic lens.

astigmatism. Although the objective lens's primary role is to form and focus the image, it also initially magnifies the image, but only to a small degree, with this task being performed by the intermediate and projector lenses (see below).

Contamination of the objective lens pole piece can result in poor image quality by distorting the power of the electromagnetic field resulting in astigmatism. The origins of most contaminating elements are from components of the specimen that volatilize during imaging and then condense on the surfaces of the objective lens pole pieces. This event can be minimized by the insertion of anticontamination devices and is mandatory for high-quality/resolution work.

Anticontaminators are positioned within the specimen chamber, close to the objective aperture (see below) and pole piece (Figures 8.1 and 8.9). They consist of two metal surfaces that are cooled via a connection to an external reservoir of liquid nitrogen and are often called "cold fingers." As contaminants volatilize from the specimen surface, most will condense on the cold metal surface of the cold finger rather than the objective pole piece. The contaminants will remain on the cold finger until it comes to ambient temperature, at which time the trapped gas molecules are released and removed from the system by the vacuum system.

An astigmator corrector is positioned beneath the objective lens (Figure 8.2). Astigmatism prevents the user from capturing high-quality images because uniform focus across the entire image is not possible. This is due to the fact that the astigmatic lens is stronger in one direction than another. Stigmators consist of electromagnetic coils, typically eight, positioned around the column that are energized to correct the asymmetry of the electron beam caused by imperfections of or contaminating materials on the objective lens or aperture. Astigmator correctors are also associated with the condenser lens (Figure 8.2).

It was noted above that a variable aperture is associated with the condenser 2 lens. A variable aperture is also associated with the objective lens, although it plays a much different role than does the condenser aperture. The objective aperture is located in the back focal plane just below the specimen and functions to enhance image contrast. To understand better how the aperture functions in forming contrast, we must understand a bit about how beam electrons interact with atoms of heavy metal stains and fixatives (e.g., osmium tetroxide, uranyl acetate), which have an affinity for and accumulate in specific areas of the specimen. As a beam electron interacts with a heavy metal atom in the specimen, it is either absorbed (i.e., giving up all its energy), inelastically scattered (i.e., giving up part of its energy with little change in direction), or elastically scattered (i.e., giving up little or no energy but changing significantly in direction). Because of their wide shift in trajectory, elastically scattered electrons are eliminated from the image beam by the objective aperture, while those beam electrons that travel directly through the specimen or have undergone inelastic scattering pass through the aperture and ultimately form the image. Thus, absorbed electrons and elastically scattered electrons are subtracted from the image, resulting in image contrast intensity and distribution based on heavy metal staining of the specimen. Objective apertures of various diameters are arranged in a metal strip and positioned with a rod, like the condenser apertures; however, the objective apertures are smaller than the condenser apertures, ranging in size from 20 to 70 μm. Thus, the smaller the objective aperture, the more elastically scattered electrons are removed, the higher the specimen contrast, and the lower the brightness.

Intermediate and Projector Lenses

Below the objective lens is a series of lenses that magnify but do not focus the specimen image formed by the objective lens (Figures 8.1 and 8.2). The first lens is called the intermediate, or diffraction, lens. The intermediate lens is able to produce an image at very low magnifications. Extremely low magnifications are useful not only for broad views of the specimen but for imaging and centering the objective aperture relative to the electron beam. In addition to serving as a magnifying lens, this lens is supplied with an aperture and can be used for obtaining low-angle electron diffraction information for a specimen. Although not often of direct interest to the biological electron microscopist, a diffraction pattern is always generated at the back focal of the objective lens. If components of a specimen make up a crystalline lattice structure, a strong diffraction pattern is formed and can be used as an analytical tool in identifying the material; however, most biological materials have a nonordered or random arrangement of atoms, which results in very poor diffraction patterns and thus few useful data.

Projector lenses follow the intermediate lens and are used to magnify further the image beyond the objective and intermediate lenses. Transmission electron microscopes today typically have two projector lenses (P1, P2) (Figure 8.2). These lenses have a broad range of focal lengths and great depth of focus and are less susceptible to spherical and chromatic aberrations. A point of interest is that each lens inverts the image it receives from a previous lens. Thus, as a lens is turned off and on through a series of magnification ranges, the image can be rotated up to 180 degrees.

Viewing System

The final image produced by the TEM is contained in the beam as variations of electron intensity. As we are not able to "see" electrons, the latent specimen image must be converted into photons. This is accomplished at the base of the column in the viewing chamber (Figures 8.1 and 8.2) where the final image is projected onto a screen containing a fine-grained coating of zinc and cadmium sulfide that act as electron phosphors. When bombarded by electrons, the coating fluoresces at intensities that are relative to the amount of beam electrons striking it. Those regions of the specimen that permit the most beam electrons to pass with little or no elastic scattering will appear bright while those regions that absorb or scatter electrons are darker. The image can be viewed on a large viewing screen or a small focusing screen that is viewed through a low-power (10$\times$) binocular microscope

(Figure 8.1). Screens are also used as meters for measuring beam current and setting the intensities for image acquisition.

Camera Systems

A system for recording images is part of all TEMs. Many TEMs will have a standard camera system made of a film canister designed to accommodate 3.25 × 4-inch sheet film and an automated image capture mechanism. These large-format negatives provide excellent resolution and should be a routine step when archiving important information. Other cameras are available, but by far the most logical addition to a TEM would be a high-quality digital camera inserted above (Figure 8.1) or below the viewing screen.

■ Vacuum System

The maintenance of a vacuum in the column of an electron microscope is basic to its operation. The primary reason for this is that electrons are easily deflected by collisions with gas molecules, rendering them useless as the source of illumination. By eliminating these molecules from the column, the vacuum increases the mean free path of an electron (i.e., the distance an electron can travel without colliding with a gas molecule). Typically, a TEM operates at a vacuum level of 10^{-6} torr (1 torr = 133 pascals, and 1.33 millibars). At these pressures, an electron has a mean free path of approximately 50 meters. A vacuum is required to avoid electrical discharge between the anode and cathode (filament/ shield), which is a common reason for a filament to fail. Also, when filaments are in use, they are particularly sensitive to oxygen; this is especially true for FEGs. Not only must a microscope have a system that will maintain high vacuum, the system must obtain workable vacuum levels within a reasonable time period. This is because many operations are performed, such as specimen insertion and removal, changing filaments, changing film, and initially turning on the TEM, that require the microscope's vacuum status to be altered. The vacuum system of a standard TEM is made of a rotary pump (also known as a mechanical, rough, or forepump), a diffusion pump, vacuum gauges, switching valves, and a network of connecting pipes. In those instruments requiring great vacuum capacity, entrapment pumps are used.

Rotary and Diffusion Pumps

Pumps function in a variety of ways and thus perform specific functions during the evacuation process and maintenance of the vacuum. The rotary pump (**Figure 8.12A**) works by compressing a specific amount of gas, concentrating its volume, and finally expelling the compressed gas out of the system. A typical two-stage rotary pump is made up of a rotor containing two spring-loaded blades that rotate within an eccentric chamber. To reduce friction, the chamber walls are lubricated by oil that collects at the base of the chamber. As the rotor spins (500 rotations per second), the walls are lubricated, and at any given time, one side of the eccentric chamber is connected to the evacuating chamber, whereas the other chamber is being exhausted. Rotary pumps are used to move large volumes of gas, taking a chamber from atmospheric pressure to approximately 10^{-3} torr. For this reason, rotary pumps are said to be low-vacuum/high-volume pumps. Rotary pumps continually expel small amounts of oil vapor and should be positioned such that they are vented to the outside and/or fitted with oil mist traps. It is critical for the upkeep of the pump that proper oil levels are maintained and that the oil is changed at appropriate intervals.

Figure 8.12
Photographs of a two-stage rotary pump (A) and oil-diffusion pump (B).

After a pressure of 10^{-3} torr is reached by the action of the rotary pump, the diffusion pump (Figure 8.12B) is used to increase the vacuum to 10^{-6} to 10^{-8} torr. The diffusion pump is referred to as an impact pump and works on a very different principle than the rotary pump. It contains no moving parts and consists of a cylindrical metal pumping chamber containing stacked, upside-down funnels within the central core. The pumping chamber is water or oil cooled via copper tubing that wraps around its exterior. At the top, the pump is connected to the chamber that is being evacuated, and near the bottom of the pump there is an exhaust port that is connected to the rotary pump or buffer tank that backs the diffusion pump. Also near the bottom is an electric heating element that boils a small volume of synthetic lubricant. The boiling lubricant forms a pressurized vapor that is forced up the central core at supersonic speeds and then is forced back down by hitting the stack of inverted funnels. The hot microscopic droplets absorb gas molecules along their path and strike the walls of the cooled pump chamber, where they condense and run down walls of the chamber to the base. At the bottom of the pump, the cooled lubricant is reheated and vaporized again and releases the trapped gas molecules. Gas molecules are removed from the diffusion pump by the action of the rotary pump. Diffusion pumps are thought of as high-vacuum, low-volume pumps and should not be used unless a chamber is first evacuated to a pressure of 10^{-3} torr. Diffusion pumps typically have liquid-nitrogen cold traps near the top of the pump just above a water-cooled baffle to prevent vapor from backfilling into the column of the TEM.

Other Vacuum Systems

Other vacuum pumps are available and may be required for those microscopes performing specialized functions. Turbomolecular pumps are impact pumps, like the oil-diffusion pump, but are oil-less systems that consist of a series of rapidly spinning rotors with vanes. Rotors spin at 20,000 to 50,000 rpm between fixed stator blades and strike gas molecules driving them down to the base of the pump. Like the oil-diffusion pump, the gas molecules are removed by a rotary pump. They are significantly better and cleaner than oil-diffusion pumps. They can also pump directly from atmospheric pressure to high vacuum unlike a diffusion pump; however, turbomolecular pumps are more expensive and demand higher maintenance.

Other options include ion and cryopumps. These are entrainment pumps and operate by trapping and holding gas molecules onto chemically reactive or extremely cold surfaces. They are used in applications requiring extremely clean, ultrahigh vacuums such as with LaB_6 guns and FEGs.

Vacuum Meters

Although each TEM manufacturer may have different names given for each vacuum monitoring system, every microscope requires vacuums to be metered at strategic locations within the system. The three common and most useful vacuum readings are:

1. High Vacuum: the vacuum within the microscope column.
2. Prevacuum Line: the vacuum detected within the camera chamber during film reloading or within the specimen airlock during specimen entry.
3. Prevacuum Column: vacuum within the microscope column during pumping from ambient pressures.

■ Practical Guide for Getting Started

After the samples have been fixed, sectioned, poststained, and stored properly, it is now time to take them to the TEM for imaging. In addition to learning the basic steps of using the TEM, it is important for the user to know that the imaging process can be a dynamic process, where he or she must be able to make adjustments to the scope either before or during the imaging session. The following section addresses this process.

The Stand-By Positions

In multiuser facilities, in which most TEMs are housed and maintained, it is extremely important that after a user has completed their imaging session that the settings of the microscope be left in a fully pumped, no-beam, no-specimen condition (i.e., stand-by position) for the next user. This is not only a measure of courteousness to the next user; it helps over the long run to prevent mistakes from occurring that could damage the instrument or result in time lost. Each facility will have its own way of doing things, but generally, the stand-by positions in which the user should leave the scope after their session are:

1. Remove specimen and return rod with empty holder to the microscope.
2. Remove objective aperture.
3. Turn down filament heat.
4. Turn off the accelerating voltage.
5. Replace any film used and bring the vacuum to working levels or (more likely) retract the digital camera.
6. Return the magnification to lowest electron microscope viewing setting.
7. Return the specimen movement controls to near center.
8. Return the computer(s) to stand-by.

Steps to Using the TEM

Preparing the Anticontamination Device (Cold Finger)   TEMs that use oil-diffusion pumps typically come equipped with an anticontamination device that is liquid nitrogen cooled and should be used routinely. This is not only necessary for the microscope but is often used to prevent contamination to the specimen. Contamination causes a loss of contrast and detail in the specimen. Care must be taken when using ultracold liquids to avoid frostbite.

Energizing the High Voltage and Obtaining a Beam   Select the appropriate accelerating voltage by turning the high tension (HT) setting knob located in prominent position on the main panel of the microscope or computer screen. As mentioned previously, most work requires 60 or 80 kV. The use of higher or lower amounts of accelerating voltage is an important decision that the user makes; the consequences should be well understood. A conventional microscope will provide a number of accelerating voltage settings, ranging from 60 to 120 kV, with 80 to 100 kV typically used. Intermediate-voltage TEMs provide a voltage range from 200 to 500 kV, while high-voltage instruments can produce accelerating voltages of 1 to 3 MV. Increased levels of accelerating voltage result in the potential of increased levels of resolution. As discussed in Chapter 5, based on the calculation of the radius of an Airy disc, or Abbe's equation,

$$d = \frac{0.612\lambda}{n(\sin\alpha)}$$

higher-energy radiation has shorter wavelengths that can lead to improved resolution. We know that electrons have mass, are negatively charged, and propagate as waves. The wavelength of electrons is expressed according to de Broglie's equations as

λ = h/mv
h = Planck's constant (6.626×10^{-23} ergs/sec)
m = mass of electrons (9.1×10^{-23} kg)
v = electron velocity

which can be expressed as

$$\lambda = \frac{1.23}{\sqrt{V}}$$

V = accelerating voltage

Thus, by merging Abbe's and de Broglie's equations, we are able to calculate the theoretical resolution of a TEM, which is 100,000 times better than that of light microscopy; however, because of extremely small aperture angles required by the TEM, spherical and chromatic aberrations, and diffraction artifacts, the theoretical resolution limit is not achievable. The realistic resolution of a modern TEM is more on the order of 0.1 nm. In addition to increased resolution, a higher voltage electron beam has the ability of greater brightness, specimen penetration, and increased depth of focus, and thus, thicker samples can be imaged. Finally, beam damage to the specimen is less at higher accelerating voltages than at lower voltages because higher energy electrons interact less with the specimen; however, with increased accelerating voltage comes a drop in specimen contrast caused by decreased inelastic electron scattering and less stability of the electron gun, which may result in high-voltage discharge. With the overall advantages of higher accelerating voltages, we are able to gather ultrastructural data from samples that have been sectioned at thickness ranging from 200 to 1000 nm, as opposed to 60 to 80 nm thick when using lower accelerating voltages. Although there are inherent disadvantages in imaging thicker sections at lower and intermediate accelerating voltages, such as increased chromatic aberration, the use of electron energy-filtering devices can all but eliminate these problems and open greater possibilities for imaging significant volumes of cytoplasmic detail and analyzing it in three dimensions.

After the accelerating voltage is turned on, one should typically wait 5 seconds for the high voltage to reach full value. Filament current is then increased to aid in electron emission. Adjust the condenser 2 lens (brightness) until a bright spot appears on the viewing screen. If the beam spot is off the center of the screen, it is centered by using the electromagnetic deflectors. Continue to increase the filament knob until an unsaturated filament image appears. If the unsaturated filament image is not symmetrical, it is adjusted by using the electronic gun tilt controls. Saturate the filament by turning and increasing the filament heating until all structure disappears or there is no change in the image when heating is applied, and then spread the beam with the brightness control to fill the screen. Self-biased filaments, which are characteristics in most microscopes, will automatically adjust the bias as the current is increased. In most modern microscopes, saturation points can be preset to help prevent the condition of oversaturation; however, the saturation point of a filament changes during its lifetime, often increasing as the filament ages; thus, the user must be trained in recognizing when saturation is not obtained and know how to adjust the system.

Beam current is a measure of the electron density of the beam and is controlled by bias, which may be read from the scope's console or computer screen. The bias may be selected before filament saturation according to beam intensity desired. Filament saturation must be readjusted whenever the bias setting is changed. For the extended life of the filament, the bias should be kept at the most usable (i.e., adequate illumination) yet minimal level. Biological samples are usually viewed with a beam current of 10 to 20 microamps (μA).

Optimizing the Focusing Binocular   In order for the operator to realize the full potential of the TEM, the focusing binoculars of the imaging system must be adjusted for each individual's eyes. For this, a small focusing screen is lowered into view, and typically an opaque wire (beam stop) is then inserted and will appear on the focusing screen. Focus each ocular of the binocular viewer on the edge of the beam stop. If this step is done correctly, the fluorescent grains on the focusing screen may be seen. Retract beam stop from column.

Inserting and Removing Specimen   Individual instructions should be given to each operator on most aspects of TEM use, but particular attention needs to be applied to this subject in detail. A few comment points will be stressed herein to cover generalities.

Never touch the specimen holder tip with bare hands; grease can be introduced into the microscope causing unstable images and increased contamination of microscope pole pieces. Always open and close the specimen holder with the special instrument provided. Place grid in depression, ensure that it lies flat, and gently close the holder. In order to avoid damage, never lay the specimen rod on the table and never set the specimen rod up on end near the edge of the table. Inserting and retracting the specimen injector rod through the airlock follows a strict sequence. Any deviations from this sequence may cause damage to the airlock, specimen tip, or objective lens stage. Refer to the specific operating instruction manual provided for the given instrument, and/or consult with the laboratory manager. For most instruments, the specimen may be inserted or retracted while the beam is on.

A summary of specimen holder insertion is:

1. Gently insert and carefully secure the specimen rod into the airlock.
2. As the specimen rod is secured, the airlock rotary pump engages.
3. When the appropriate vacuum is obtained, the rod is inserted completely into the microscope's objective lens.

Apertures   Apertures are located in three places in the microscope: the condenser lens, the objective lens, and the intermediate lens. Apertures in the intermediate lens are used for electron diffraction and are not used for routine imaging of resin-embedded thin sections. There are several selectable apertures on each aperture holder, and each aperture is selected and inserted manually or pneumatically. The optimum objective aperture size for most TEMs is generally 30 μm for a standard thin section. For the condenser aperture, 150-μm aperture gives good image quality and illumination.

Objective Aperture Alignment   For this procedure, there must be a specimen in the beam path. Bring the beam to crossover with the condenser lens controls. Adjust the microscope for diffraction mode, and adjust the beam spot with the diffraction control if needed. Center the aperture coaxially with the beam spot; using the controls located at the aperture rod entrance (**Figure 8.13**). Return to standard imaging mode.

Condenser Aperture Alignment   If removed and replaced carefully, this aperture will stay in alignment. If the condenser aperture needs alignment, obtain special instructions or consult the operations manual.

Objective Lens Stigmation   Use the holey film grid provided, or find a hole in the specimen. Stigmate at highest magnification possible for greater accuracy. Using an overfocused fresnel fringe, adjust the right and left objective stigmators until the fringe is perfectly symmetrical around the hole. The fringe must disappear equally around the hole when approaching true focus; likewise, the fringe should appear equally around the hole again when overfocusing. Phase granularity of a carbon film may also be used to stigmate for high resolution work.

Synopsis of Operations (With Microscope Turned On and Warmed Up)

1. Check that high vacuum is appropriate for instrument operation.
2. Check the liquid-nitrogen level in anticontamination device.

Figure 8.13
Close-up view of the objective aperture controls, pointing out knobs (arrows) for aperture alignment.

3. Load grid into specimen holder, and insert a sample to standby position.
4. Select accelerating voltage.
5. Turn voltage on.
6. Increase filament heat.
7. Go to crossover with the condenser, and observe the unsaturated filament image.
8. Adjust the gun tilt if needed.
9. Saturate the filament, and then spread the condenser.
10. Insert the specimen tip into the beam path.
11. Check the objective aperture size and alignment.
12. Adjust the focusing binoculars.
13. Stigmate the objective lens.

You are now ready to observe the detailed structure and beauty of your specimen!

References and Suggested Reading

Al-Amoudi A, Chang JJ, Leforestier A, McDowall A, Salamin LM, Norlen LPO, Richter K, Blanc NS, Studer D, Dubochet J. Cryo-electron microscopy of vitreous sections. *EMBO J.* 2004;23:3583–3588.

Bozzola JJ, Russell LD. *Electron Microscopy: Principles and Technigues for Biologists.* 2nd ed. Sudbury, MA: Jones and Bartlett Publishers, 1999.

Dykstra MJ. *Biological Electron Microscopy: Theory, Techniques and Troubleshooting.* New York: Plenum Press, 1992.

Hohmann-Marriott MF, Uchida M, van de Meene AM, Garret M, Hjelm BE, Kokoori S, Roberson RW. Electron tomography and its application to revealing fungal ultrastructure: Tansley review. *New Phytologist.* 2006;172:208–220.

Lucic LV, Forster F, Baumeister W. Structural studies by electron tomography: from cells to molecules. *Ann Rev Biochem.* 2005;74:833–865.

Koster AJ, Grimm R, Typke D, Hegerl R, Stoschek A, Walz J, Baumeister W. Perspectives of molecular and cellular electron tomography. *J Struct Biol.* 1997;120:276–308.

McIntosh JR. Electron microscopy of cells: a new beginning for a new century. *J Cell Biol.* 2001;153:F25–F32.

McIntosh R, Nicastro D, Mastronarde D. New views of cells in 3D: an introduction to electron tomography. *Trends Cell Biol.* 2005;15:43–51.

Subramaniam S, Milne JL. Three-dimensional electron microscopy at molecular resolution. *Annu Rev Biophys Biomol Struct.* 2004;33:141–155.

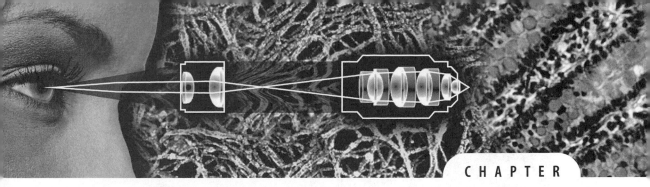

The Scanning Electron Microscope

Scanning electron microscopy is a form of electron microscopy in which a concentrated beam of electrons is made to scan a chosen region of interest over a specimen as a raster of parallel contiguous lines. In doing so, a high-resolution, three-dimensional image is built line by line of the topographical details of a specimen's surface. The scanning electron microscope (SEM) (**Figures 9.1** and **9.2**) has been likened to a high-resolution stereo dissecting microscope and, in fact, is often used at magnifications that are commonly used in light microscopy. It is unlike the transmission electron microscope (TEM), which projects electrons through a thin slice of resin-embedded sample, producing information in two dimensions. We should remember, however, that three-dimensional data can be obtained for the TEM if serial section reconstructions or electron tomography methods are employed.

The SEM got its start in 1935 when Max Knoll showed that it could actually work, and then in 1938, when the first SEM was built by von Ardenne. The actual resolutions of these early instruments were rather low, being no higher than those obtained using light microscopy. In the early 1940s, a team at RCA led by Vladimir Zworykin produced an SEM, which used a category of electrons called secondary electrons (see below) as the image-forming elements for SEM. From this work, McMullan and Oatley constructed an SEM at Cambridge University having 50 nanometer resolution in 1948, and with alterations to the electron optics in 1956 by Smith, also from Cambridge University, SEM imaging improved greatly. The next major improvement came in the early 1960s and dealt with how the secondary electron signal was collected and processed. The phosphor screen and photomultiplier used originally by Zworykin was augmented with the addition of a light pipe, which allowed direct optical coupling between a scintillator and photomultiplier. This was accomplished by the team of T. E. Everhart and R. F. M. Thornley, and the device arrangement carries their names as the Everhart-Thornley detector. With these developments the first commercial SEM came online in 1965, and with the employment of solid-state technology the SEM quickly became successful as an important imaging tool used for research and education in the biological and materials sciences. For the biologist, the SEM is particularly useful because of its large depth of field, large stage area that can accommodate bulk samples (**Figure 9.3**), and relatively easy specimen preparation methods.

Figure 9.1
Photographs of an SEM. (A) Illustrated are the column (white asterisk), stage-positioning micromanipulator (black arrow), Z-axis stage control (white arrow), image monitor (black arrowhead), system monitor (white arrowhead), and control panel (black asterisk). (B) Front view of the column (white asterisk), backscatter detector (black arrow), the port to view inside the specimen chamber (white arrowhead), the final aperture control (white arrow), photomultiplier tube (black arrowhead), and liquid nitrogen Dewar for x-ray detector (black asterisk).

■ The Systems and Principles of the SEM

As we saw with the TEM, the SEM uses electrons as a source of illumination; these are generated in a vacuum by an electron gun and are controlled by electromagnetic lenses and apertures. The SEM can be divided into discrete systems, also like the TEM, each one performing a unique function or set of functions. For the SEM, we can recognize four systems: the illumination system, the imaging system, the specimen manipulation system, and the vacuum system (Figure 9.2). It is shown later that although the SEM can be categorized into similar systems as the TEM, most of these systems are very different in their design and operation. However, the electron gun assembly and vacuum systems for the SEM are basically identical to the TEM discussed in Chapter 8 and are not addressed here.

Illumination System

Condenser lenses, deflection coils, apertures, and stigmators (Figure 9.2) regulate the primary electron beam from the time it leaves the gun to the time it impinges onto the specimen as a narrow beam of electrons. This electron beam is "demagnified" to an appropriate size (i.e., the spot size) by a series of condenser lenses and apertures before use. The effects of spot size, as well as accelerating voltage, are addressed below. The typical SEM has two condenser lenses, but for those instruments used for higher resolution, a third one is present (Figure 9.2). As we saw in the TEM, the first condenser lens (C1) forms the first crossover spot of about 50 μm. Condenser 2 performs a similar function as C1. Condenser 3 (C3) is the final condenser lens (Figures 9.2 and **9.4**) and performs the final demagnification of the electrons onto the specimen's surface and is used to focus the image by focusing this probe to a minimum diameter on the surface. Apertures are associated with the lenses that aid in regulating the spot size and allaying spherical aberration. Some apertures may be either fixed in size and location or variable and adjustable (see Final Aperture Size, below) using control knobs located on the outside of the column.

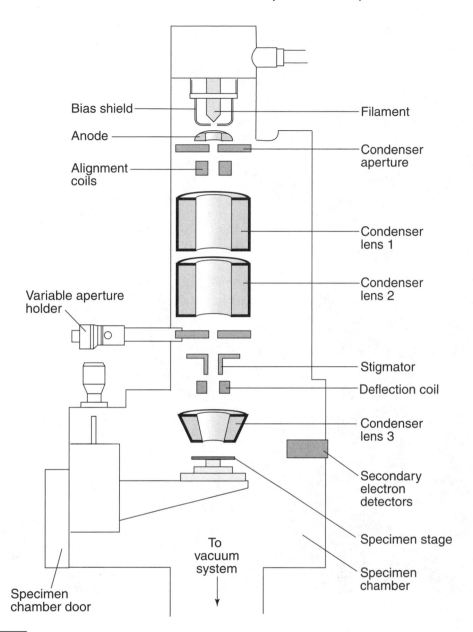

Figure 9.2
A schematic diagram of a standard SEM.

There is a set of small electromagnetic coils called the deflection coils associated with C3 (Figure 9.2). During the scanning operation, a small fluctuation of voltage is produced by a scan generator and is applied to the deflection coils, producing a magnetic field that deflects the condensed and focused spot of primary electrons back and forth across specimen's surface in a regulated, line-by-line, rastered pattern. The same fluctuating voltage from the scan generator is applied to and in perfect sync with a set of deflection coils associated with the imaging monitor and data-capturing monitor (Figure 9.1A). When the primary beam of electrons hits the sample, the transfer of beam energy into the sample results in a series of complex interactions at the surface of the specimen, giving

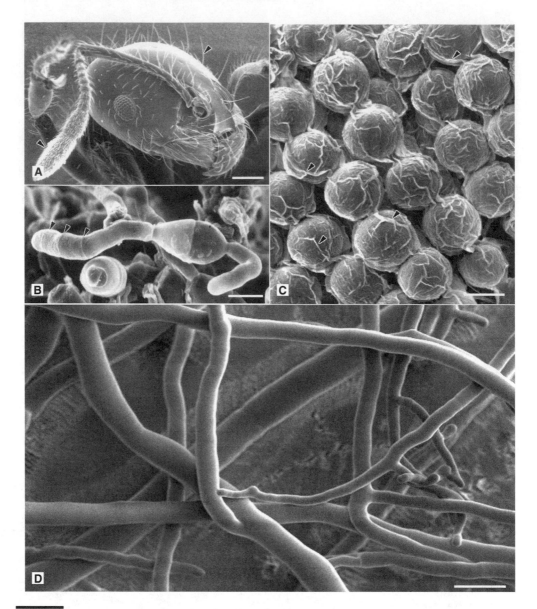

Figure 9.3

Scanning electron micrographs. (A) Head of a red fire ant, *Solenopsis invicta*, fixed in buffered glutaraldehyde, washed with water, air dried, sputter coated, and imaged at 20 kV. Bright areas pointed out by arrowheads are examples of edge effects. Bar = 30 μm. (B) Germinating teliospore of the rust fungus *Gymnosporangium clavipes* prepared using standard chemical fixation methods, critical-point dried, sputter coated, and imaged at 20 kV. Cross walls (i.e., septa) are pointed out by arrowheads. Bar = 3 μm. (C) Spores (i.e., conidia) of the common mold *Aspergillus nidulans* seen in a frozen-hydrated state with a cryo-SEM after being lightly etched, sputter coated, and imaged at 2 kV. Extracellular mucilage covering spores is pointed out with arrowheads. Bar = 2 μm. (D) Hyphal cells of the common mold *Neurospora crassa* viewed in a frozen-hydrated state with a cryo-SEM after being slightly etched, sputter coated, and imaged at 2 kV. Bar = 12 μm.

rise to electrons and photons (see below). The electrons, called secondary electrons, are those used for topographic image formation. Thus, the surface features that the user sees on the viewing monitor and captures digitally or on film are the exact features that are being scanned by the primary beam. By adjusting the area of the specimen's surface that is being scanned by the beam, one can change the magnification (i.e., larger surface areas equal lower magnification; smaller surface areas equal higher magnification). In addition

Figure 9.4
A photograph of the inside of the specimen chamber. The specimen is positioned below the final condenser lens (asterisk). The x-ray detector is indicated by the white arrow, and an arrowhead points out the secondary electron collector, or Faraday cage, of the Everhart-Thornely secondary electron detector. The backscattered electron detector is retracted in this image and is not visible.

to changes produced by the raster coils, magnifications can be controlled by varying the working distance, that is, the distance from the final condenser lens pole piece to the top of the specimen (see below) or by varying the accelerating voltage. Magnification is increased by decreasing the working distance or by increasing the accelerating voltage (**Figure 9.5**).

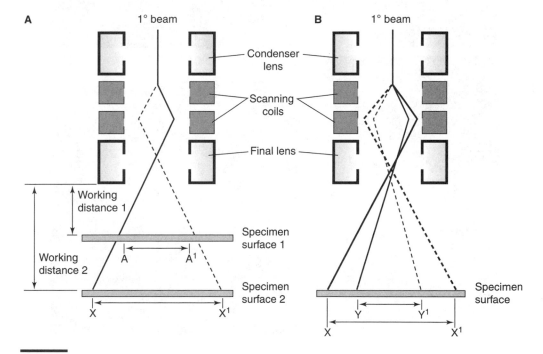

Figure 9.5
The effects of working distance and accelerating voltage on magnification. (A) Magnification increases as the working distance is reduced. (B) Higher acceleration potential results in a higher magnification.

The Specimen Stage and Manipulation

The construction of the SEM stage and specimen chamber follow two basic designs: the prepump and drawer-type design. In the former design, the major components of the stage remain in the chamber at all times, and the specimen is inserted into the chamber and mounted on the stage through an airlock. In the drawer-type design (**Figure 9.6**), the entire specimen chamber is sequestered from the rest of the column by a high-vacuum valve and backfilled with air, and the entire stage is pulled out, permitting the specimen to be secured in the stage (Figure 9.6B). Both stage designs allow for the controlled positioning of specimen in any orientation relative to the beam.

The typical biological specimen is securely attached to a small metal platform or "stub" (Figures 9.6B and **9.7**). The design of studs can vary according to the requirements of the SEM manufacturer. The stub is secured to the specimen stage (Figure 9.6B), which

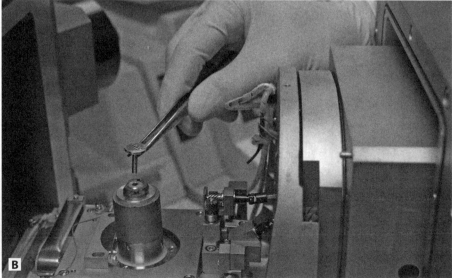

Figure 9.6
Photographs showing the opened drawer-type stage (A) and insertion of the specimen onto the stage (B). The Z-axis stage control (arrow), specimen stage (arrowhead), and the support rod for the chamber door (asterisk) are shown.

Figure 9.7
Application of carbon-based paste onto a stub (arrowhead), which is used to attach a specimen before it is sputter coated and viewed in the SEM. The three stubs on the right have double sticky-tape adhesive.

is inserted into the specimen chamber in such a way that the specimen is positioned directly below the final lens (Figure 9.4) and aperture. Stage-positioning micromanipulators (Figure 9.1A) achieve the wide flexibility of specimen positions with movement in the x, y, and z directions, rotation in 360 degrees, and tilting up to 90 degrees. All specimen manipulations mentioned above can be done while observing the specimen so that an optimal orientation can be obtained before the image is captured on film or digitally.

Imaging System

The imaging system of the SEM consists of the electron detectors, a signal processor, and an image recording system. It is appropriate, however, to start our discussion of the imaging system with an overview of the interactions of the primary electron beam and the specimen.

Specimen-Beam Interactions   When the primary electron beam of the SEM interacts with the sample, the electron beam penetrates the sample, forming the interaction volume. In most biological samples that are imaged at an accelerating voltage of 10 to 15 kV, the penetration depth is approximately 10 nm and yields signals (electrons and photons) from specific regions within the interaction zone (**Figure 9.8**). The shape and size of the

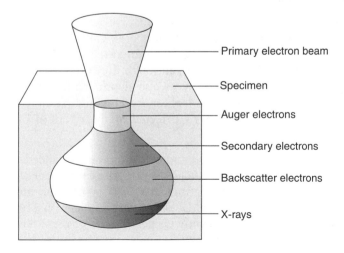

Figure 9.8
Signals emitted from the interaction zone in a typical biological specimen viewed at 10 to 15 kV accelerating voltage. According to depth, (A) Auger electrons, (B) secondary electrons, (C) backscattered electrons, and (D) x-rays.

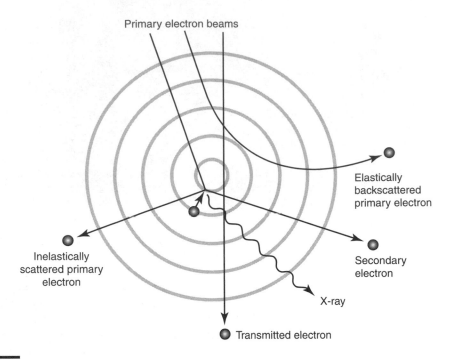

Figure 9.9
Beam-specimen interactions.

interaction volume is variable and depends on sample composition, the accelerating voltage selected, and the angle at which the primary beam impinges onto the sample. Thin sections in TEM likewise have an interaction volume, but this has much less influence on the final image or analysis of the sample. As discussed in the previous chapter, when beam electrons penetrate into the sample, there are both elastic and inelastic scattering events. These scattering events give rise to the emission of secondary electrons, backscattered electrons, characteristic x-rays, continuum x-rays (i.e., bremsstrahlung x-rays), Auger electrons, or cathodoluminescence (**Figure 9.9**). These emissions are discussed below.

Secondary Electrons   During the inelastic scattering event the energy imparted to the specimen atom may cause it to ionize and eject a so-called secondary electron. Secondary electrons have energies of less than 50 eV and are used to form the image of the specimen in the SEM. Secondary electrons, although they are produced throughout the interaction volume, have a very shallow escape depth from the specimen's surface (within 5 to 15 nm), thus making them ideal for creating high-resolution images of surface details. These electrons are efficiently detected by the standard Everhart-Thornley detector (see below).

Backscattered Electrons   Those beam electrons that have experienced single or multiple elastic scattering events and exit the specimen's surface are called backscattered electrons. These electrons have energy levels that are greater than 50 eV and, indeed, can possess energies near those of the primary beam. Unlike secondary electrons, backscattered electrons can exit the surface from significantly deeper regions of the interaction volume or may interact with the specimen's atoms several times, creating multiple interaction events within the interaction zone. Elements with higher atomic numbers will generate a greater amount of backscattered electrons. This relationship can be used to map the distribution of heavy elements on the specimen's surface visually. Furthermore, the emission of backscattered electrons can be used to detect colloidal gold particles used for cytochemical SEM investigations. These high-energy electrons can decrease image resolution and increase noise by creating secondary electrons from the specimen at significant distances away from the point where the primary electron beam impinged on the sample or if they strike parts of the microscope chamber generating secondary electrons that contribute to noise. A backscattered electron detector, either solid-state or scintillation based, located below the final lens is required to image these electrons.

Characteristic X-Rays   Characteristic x-rays are created as an atom is stabilized after being ionized by the primary electron beam as a result of an inelastic scattering event. As an electron from an inner atomic shell is ejected by a beam electron, creating an electron hole, a higher energy electron from an outer shell falls to the lower energy state, replacing the ousted electron and thus stabilizing the atom. With the fall of the electron from a higher to lower energy, an x-ray is emitted that equals the difference in energy between the two electron shells. Each shell of every atom has a characteristic energy level; thus, the x-rays that are emitted during these events have unique energy levels and wavelengths and thus represent characteristic x-rays for a particular atom. Detection of these x-rays is the basis for x-ray spectroscopy and is of great importance in elemental analysis of both biological specimens and materials (see below). With the measurement of either the wavelength or the energy of the x-ray, one is able to obtain semiquantitative and semiqualitative data of the elements present in a specimen; this may be done with a SEM or TEM. X-rays have a greater escape depth than other interaction products, come from a greater interaction volume, and, therefore, map out elemental distribution at lower resolution.

Continuum X-Rays   Also know as background or bremsstrahlung x-rays, the emission of these x-rays occurs during an elastic scattering event, as electrons from the primary beam are slowed and their trajectory is slightly altered when passing close to a specimen's nucleus. In this situation, the elastic scattering event does not result in the ionization of the specimen atom. These x-rays are not particularly useful for analytical purposes because the radiation is originating from the beam electrons, not the specimen, and because the energy levels of the continuum x-rays are random, being dependent on the distance of the primary beam electron from the nucleus of the specimen's atom. Indeed, the continuum x-ray spectrum complicates the analysis of the characteristic x-rays.

Auger Electrons   These low-energy electrons (50 to 2000 eV) are ejected from the specimen atom after absorbing the energy released after a primary ionization and stabilization event. Auger electrons are the result of a secondary ionization event. The energy of the Auger electron is equivalent to the energy difference between the orbitals and characteristic of the atom. Thus, Auger electrons are alternative to characteristic x-rays and are used for chemical analysis of the specimen's surface. They are particularly useful in the study of low atomic weight elements (their yield is higher in low atomic number elements) and materials that are within 0.5 to 2 nm of the specimen's surface.

Cathodoluminescence   Materials such as phosphors, semiconductors, and insulators emit visible, ultraviolet, or infrared photons as electrons fill holes formed by inelastic scattering events. This event is called cathodoluminescence and is not a common means for inspecting biological materials with SEM because only a few biological materials undergo this phenomenon. If an adequate signal is present, cathodoluminescence can be observed using a standard secondary electron detector that has been altered by removal of its scintillator from the front of the light pipe (see below). By doing so, secondary electrons will not be converted into photons, and only photons resulting directly from cathodoluminescence will be detected; however, as only a low percentage of photons can

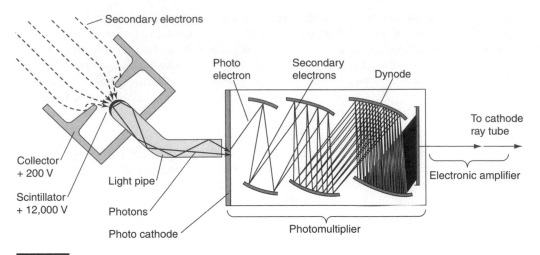

Figure 9.10
A schematic illustration of the Everhart-Thornley secondary electron detector.

be detected in this way, only preliminary studies can be conducted. Increased sensitivity can be obtained using specially designed detectors. It is a detection method used in pharmaceutical, semiconductor, and mineralogy studies.

Detection, Conversion, and Amplification of the Secondary Electron Signal   Secondary electrons that arise from the specimen as the primary electron beam impinges on its surface are used as the signal that communicates the topographic characteristics of the sample for SEM. These electrons are attracted to a detector containing a positive charge located within the sample chamber. The Everhart-Thornley secondary electron detector is the most common type used today (Figures 9.4 and **9.10**) and has undergone few changes and improvement since it was introduced in the mid 1960s. The basic parts of the detector are the secondary electron collector screen, that is, Faraday cage, scintillator, light pipe, and photomultiplier tube. The Faraday cage consists of a wire mesh or metal ring that is located at the end of the detector within the specimen chamber. It has a positive bias of up to +300 volts that draws the low-energy secondary electrons into it but does not influence the higher energy backscattered electrons or primary beam electrons; however, should backscattered electrons be scattered into the detector, they will become part of the image signal. Secondary electrons inside the Faraday cage are accelerated by a 10 to 15 kV positive charge that is applied to the scintillator (Figure 9.10). Interestingly, this positive charge has no effect on the primary electron beam because the Faraday cage confines this charge. The scintillator is a phosphor-coated plastic disc or a yttrium gallinide crystal that converts the energy of the electron impacts into photons. When secondary and backscattered electrons strike the scintillator, short-lived flashes of light (i.e., scintilla) are produced. It is thought that one photon is produced for each 3.3 eV, and thus, for every electron striking the scintillator, several photons are produced. These photons are then transmitted through a quartz or Lucite (i.e., Plexiglas) light pipe out of the specimen chamber (Figures 9.1B and 9.10) to a photocathode and photomultiplier (Figure 9.10). The photocathode has material on its surface that converts each photon into two or more electrons, or more correctly photoelectrons. These photoelectrons enter a photomultiplier, where they basically bounce between a series of dynodes and electrodes that increase the number of electrons in a cascading fashion. The bottom line is that the initial photoelectron can be amplified five to six orders of magnitude and is regulated by the voltage applied to the dynodes through adjusting the contrast control device on the microscope's console or computer screen (i.e., gain or contrast control).

Some common imaging artifacts can be caused by improper specimen preparation methods and incorrect use of the instrumentation. Charging can occur as a result of current buildup in the sample and subsequent release of these electrons in a burst, resulting in extremely bright regions of the specimen, movements of the image, and distortions of the image. Charging is caused by incomplete specimen coating (see below) or areas of the specimen that are not well grounded. If after proper grounding of the specimen to the stub and specimen coating there remains charging, then increasing the working distance, increasing the scan rate, decreasing the accelerating voltage and applying additional metal to the surface, and/or increasing the bulk conductivity (i.e., en bloc staining with heavy metals in subsequent fixations) can allay the charging phenomenon. A second imaging artifact is edge effect that can occur commonly on thin, raised, or sharp areas of the specimen (Figure 9.3A). These regions of the sample often appear much brighter than other larger scale areas of the sample. This enhanced brightness is due to the fact that the secondary electrons can exit the sample much easier from more than one surface. As this condition is inherent to the sample, it is difficult to correct, although reducing the accelerating voltage and/or repositioning the sample relative to the secondary electron detector has been reported to help in some cases.

Image Processing, Display, and Recording   Manipulations of the signal output from the SEM enhance the quality and amount of information that can be obtained from the specimen. Some of the options for processing the image are outlined here.

Contrast   The difference in contrast between the surface details of the specimen as they are viewed on the monitor provides the three-dimensional appearance of the image. The specimen generates and emits different amounts of secondary electrons viewed as differences in brightness. The amount of secondary electrons emitted is influenced by the orientation of the specimen relative to the primary beam and secondary electron detector, the angle at which the beam enters the specimen surface, the effect of specimen topography (i.e., the edge effect) locations of elements of varying atomic numbers, accelerating voltage, and charging.

Stigmation   As mentioned in the previous chapter, it is critical for high-quality imaging that the spot formed by the primary electron beam be free, as much as possible, of distortion in its roundness (i.e., a stigmatic beam). If the beam spot is not round (i.e., an astigmatic beam), then unwanted secondary electrons are generated and projected onto the monitor, making images appear smeared in one direction. Astigmatism is one of the major causes of degraded resolution in the SEM. Stigmators consist of six to eight electromagnetic coils inside the lens bore of the final condenser lens and are able to correct such distortions by introducing additional magnetic strength in the appropriate directions to counteract any astigmatic tendencies.

Brightness   As the signal leaves the photomultiplier, it is routed to a preamplifier and then is ready to be viewed on the monitor (Figure 9.1) located on the SEM's console. Increased gain on the preamplifier increases the brightness of the image overall.

Gamma   Not all signals producing the image gray scale fall within a usable range for viewing or recording; some may be very bright or very dark (**Figures 9.11A** and **9.11B**). The adjustments of the gamma control will increase the contrast range selectively of either the bright or dark areas. Gamma converts the linear input signal going to the viewing monitor into a logarithmic function—compressing the entire signal output and thus retaining the delicate details of the extreme bright and dark areas. Thus, gamma is useful when the specimen has extreme contrast.

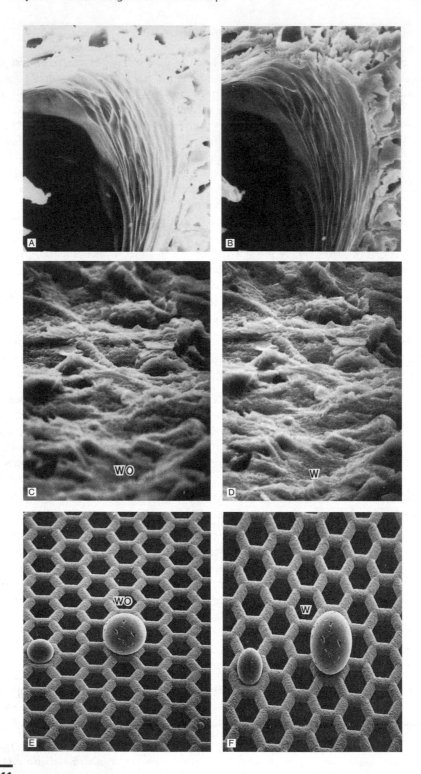

Figure 9.11
Micrographs demonstrating the effects of various signal manipulations. Gamma is off in (A) and on in (B). Dynamic focus is off in (C) and on in (D). Tilt correction is off in (E) and on in (F). Y-modulation is off in (G) and on in (H). In dual-magnification mode, two independent sets of lines are scanned over a specimen's surface, one set at a relatively longer distance (i.e., low magnification) (I) and the other set at a shorter distance (i.e., high magnification) (J). Both images are seen at the same time on the viewing monitor.

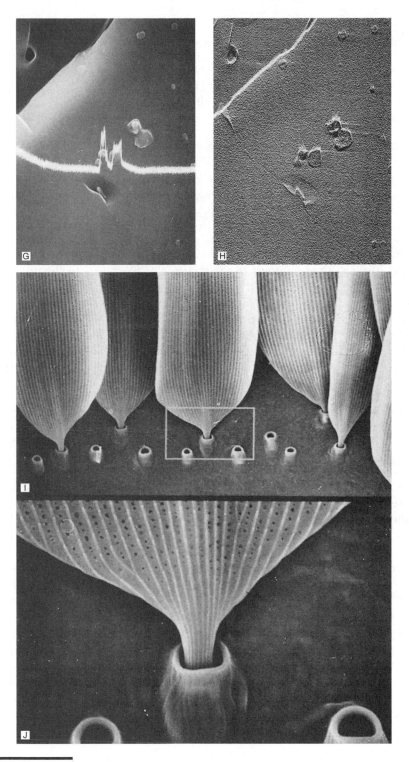

Figure 9.11   (continued)

Dynamic Focusing Also known as auto focus, dynamic focusing actively changes the focus of the final condenser lens of the SEM, compensating for the tilt angle of the specimen (Figures 9.11C and 9.11D). Without this, the electron beam spot would remain the same size as it was scanned from top to bottom of the specimen, producing an image that was in focus only where the spot was at an optimum diameter. Using dynamic focus results in a sharp micrograph of the total field of view by electronically altering the final lens in order to compensate for the out of focus regions.

Tilt Correction Tilting a specimen toward the detector will enhance the collection of secondary electrons; however, this also results in distortion of the image called the foreshortening effect, where the top of the specimen is at a higher magnification (shorter scan length) than the bottom of the specimen (longer scan length) (Figures 9.11E and 9.11F). Tilt correction compensates for this by effectively making the scans of apparent equal length at any tilt angle as the beam moves from the top to the bottom of the specimen.

Y-Modulation Y-modulation is a mode of imaging that differs from the standard brightness modulation (i.e., the brightness of the image is proportional to the number of secondary electrons emanating from the specimen) (Figures 9.11G and 9.11H). This variation in the signal strength contributes to the three dimensionality of the sample. If, however, a sample with little surface tomography is investigated, the image can be difficult to interpret, as the grayscale range is very narrow. The difficulties in interpreting specimens with few features in the z-axis can be allayed using the Y-modulation mode. Here, the specimen topography is represented in the y-axis on the CRT and not the typical z-axis.

Raster Rotation The spiral path taken by electrons as they travel down the column can result in an image shift as one adjusts the working distance of the sample from high to low. The raster rotation mode electronically compensates for this shift and allows the user to image the sample at the same orientation in all working distance settings.

Dual Magnification The dual magnification function allows for two different magnifications to be viewed simultaneously by sending alternate scans taken at different magnifications to different viewing monitors or in split screen mode on a single monitor (Figures 9.11I and 9.11J). It is a time saver because it allows the user to quickly find important areas of interest at low magnifications while viewing them at high magnifications.

Until recently, SEMs were equipped routinely with two monitors or cathode ray tubes (CRTs), one for viewing the image and one for recording the image. The viewing CRT has a long persistence with a glow lasting for seconds after the scan moves across the screen, making it easy to view. Unlike a standard television CRT, a slower scan/long-persistence glow produces a high-quality image that is easier to view. Recording images, once done with a second recording CRT, are now mostly performed using digital methods with images being recorded having millions of pixels. The advantages of digital recording over analog recording methods are that the former is quicker and less expensive to capture each image, data can be easily stored and transferred, and image quality is greatly improved by frame averaging. For further advantages and disadvantages of digital versus analog image capturing and storage, refer to Chapter 11.

■ Image Quality

As one might surmise, numerous factors influence the ultimate quality of an SEM micrograph. These factors include the microscope itself, the ability of the user to manipulate the microscope, and the method used for sample preparation. Here we discuss additional aspects of SEM imaging: beam spot size, working distance, final aperture size, and accelerating voltage and how they influence resolution, depth of field, and beam current.

Primary Beam Spot Size

Resolution of the SEM is regulated primarily by the spot size of the primary electron beam: the smaller the spot size the greater the potential resolution possible. The diameter of the beam spot on the specimen must not exceed a certain size at a given magnification: For optimal resolution at 10×, for example, the spot size must not exceed 10 μm, whereas at 100,000× a spot size of 1 nm or less is required. If the beam spot size exceeds the optimal size, secondary electrons are generated from areas outside of what is being probed by the primary electron beam, resulting in an image with less than optimal resolution. The spot size can be regulated in a number of ways. The most often used means is through the selection of the final aperture size used (see below). The inherent spot size of the electron beam can also be reduced by using an electron source with a small source diameter (e.g., the approximate source diameter of a thermionic tungsten filament is 1×10^{-6} Å, of a lanthanum hexaboride filament is 2×10^{-5} Å, and of a field emission gun is less than 1×10^{-2} Å).

The concept of resolution in the SEM is more complex than in the TEM. The complexity stems from the fact that spatial as well as temporal parameters are components in the image formation process and that there are multiple and complex interactions between the primary electron beam and atoms of the specimen that have an affect on SEM images and resolution. This is in contrast to the simpler beam/specimen interactions used to form TEM images due to the very thin (60 nm) samples imaged for general TEM applications. Thus, it is important to remember that resolution is also influenced by the fact that secondary electrons emanating from the point on the specimen that is being subjected to the primary electron beam can be additively displayed on the viewing/recording monitor over time, which results in decreased resolution. In other words, if the primary beam strikes an initial point ("A") on the specimen, resulting in the emission of secondary electrons, and these electrons are recorded and imaged essentially at the same time, the resolution is good. However, if beam/specimen interactions at point "A" result in a persistence of secondary electron production the recording of these electrons may not occur until the electron beam is positioned over a new point "B" on the specimen; then two populations of secondary electrons are additively displayed and resolution suffers.

Final Aperture Size

The size of the aperture (**Figure 9.12**) associated with the final C3 lens is typically adjustable with variable sized apertures. Reducing the size of the final aperture to between 50 and 100 μm reduces the spot size and as a result reduces the beam current and increases the depth of field (**Figure 9.13**). Conversely, a larger final aperture on the order of 200 to 300 μm will increase the spot size, resulting in a higher beam current and a decrease in the depth of field. Clearly, the appropriate selection of the final aperture is important in optimization of the SEM because of its effect on three different parameters—the spot size (and, thus, resolution), the beam current, and the depth of field. One may be surprised that smaller apertures enhance resolution because they also decrease the aperture angle of the beam. This apparent disadvantage is overcome not only by the smaller spot size but also by the fact that a smaller aperture reduces spherical aberration, which helps to improve resolution. Care should be taken when working with larger aperture settings because the increased beam current can easily damage delicate samples. Their use is appropriate, however, when the generation of characteristic x-rays is required for elemental analysis (see below).

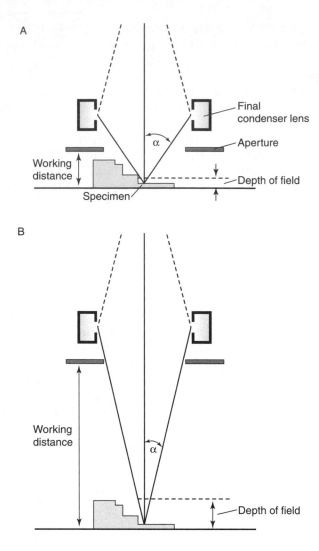

Figure 9.12
Drawing illustrating the relationship between working distance, depth of field, and aperture angle. As the working distance increases (compare A and B), the aperture angle decreases, whereas the depth of field increases.

Working Distance

The distance between the sample and the final lens is referred to as the working distance. Adjusting this distance is very useful for SEM imaging, in that it has direct influence on the depth of field of the specimen (i.e., the vertical distance within which the specimen appears in focus) (Figures 9.1A, 9.6A, and 9.12). This becomes of major importance when viewing samples with extreme three-dimensional variations in their surface features. Lowering the specimen (i.e., increasing the working distance) results in an increased depth of field, whereas positioning the sample closer to the final lens (i.e., decreasing the working distance) gives rise to a shallower field of view. As one might expect, there is an inverse relationship between working distance and resolution: Increasing the working distance compromises the resolution because of the decrease in the numerical aperture and the increased potential for chromatic aberration caused by increased focal lengths. Also, recall that working distance has an inverse relationship with magnification, mentioned above.

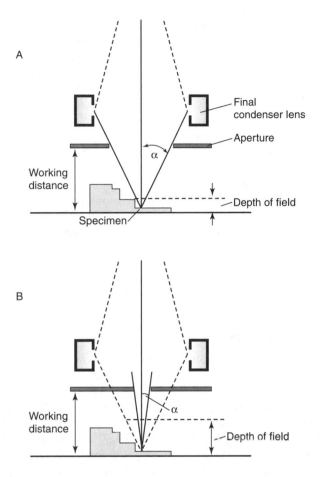

Figure 9.13
Drawing illustrating the relationship between final aperture size, depth of field, and aperture angle.

Accelerating Voltage

The accelerating voltage influences resolution in the SEM because of the fact that the higher the beam voltage the stronger or "stiffer" the beam is, and thus less susceptible to spherical aberrations and stray magnetic and electrical fields; however, it is important to note the fundamental difference in the operation of the SEM and TEM in this area. Unlike the TEM where higher accelerating voltages result in increased resolutions, higher accelerating voltages can be detrimental to resolution in the SEM. This is due to the interactions of the primary beam electrons with the sample and the path they take on entry into the sample (i.e., interaction volumes). This is discussed below, but briefly, the diameter and depth of the interaction volume increase with greater accelerating voltage because higher energy electrons penetrate deeper into the sample than do lower energy electrons. This can result in increased backscattered electrons that may increase noise by creating secondary electrons from different parts of the specimen chamber or decrease resolution by creating secondary electrons from regions of the specimen's surface outside of the primary spot. The size of the area of interaction is also influenced by the nature of the sample—the less dense the sample material the greater the interaction volume. The bottom line is that resolution in the SEM is a balance between spot size, accelerating voltage, the nature of the specimen, the signal-to-noise ratio and viewing parameters.

When using a standard SEM with a tungsten gun, most biological samples will produce the best images at accelerating voltages of between 15 to 20 kV. To obtain the best resolution and high magnification for SEM images, the microscope must use either a LaB_6 or field emission guns (see Chapter 8) in order to generate a high number of beam electrons (i.e., high-beam current), which will result in greater numbers of imaging electrons per unit area (i.e., secondary electrons), at the specimen's surface. For many biological SEM applications, the use of a standard tungsten filament is more than adequate, unless the need of true low accelerating voltages (i.e., high resolution) is required.

In summary, resolution is enhanced with reduced working distance, a smaller beam spot, and a smaller aperture diameter. The depth of field is increased with a longer working distance and smaller aperture diameters and is not influenced by beam spot size. Finally, beam current is increased with reduced working distance, increased beam spot size, and increased aperture diameter. In theory, the smaller the beam current, the higher the resolution is, and the larger the beam current, the higher the signal level is. As we have seen before, there is a downside to almost every decision made in electron microscopy, and here the steps to increase resolution and depth of field result in a decrease of beam current, and thus reduce the signal-to-noise ratio. This can be overcome to varying degrees by using lanthanum hexaboride (LaB_6) or field emission electron emitters.

■ Sample Preparation

As discussed in Chapter 3, careful sample preparation is critical for the maintenance of true, lifelike details of biological materials for ultrastructural investigations. Thus, many of the same considerations and steps that are used in TEM sample preparations are applied to SEM methods. Most biological samples need to be stabilized with aldehyde and osmium tetroxide fixation, rinsed, chemically and physically dehydrated, properly mounted on a specimen stub, and made electrically conductive before SEM imaging by coating with a thin layer of metal. Some samples are amenable to SEM viewing in their frozen hydrated state using cryo-SEM, in a hydrated state at ambient temperatures using a variable pressure or environmental SEM (ESEM, see below), or without any fixation protocols. As with TEM methods, a variety of preparation protocols are used for processing biological samples for SEM imaging. In general, the best specimens are those that are vacuum stable, unaffected by the intense electron beam, conductive (heat and electrons), and immobilized on a specimen support. A typical preparation method is described, and specialized alternatives are summarized below.

Basic Protocol
Sample Collection and Prefixation Treatments   Samples should be collected as fresh as possible and prepared rapidly. The sample size consideration is not as important as with TEM, primarily because of the fact that the tissues/cells will not be embedded in resin; however, to facilitate solution penetration and exchange the size of the specimen, it should be no larger than 2 to 4 mm in one dimension, whereas the other dimensions may be up to 15 to 20 mm, depending on the sample type. Because many biological samples may be covered by extracellular materials such a mucilage or waxes, it is possible that appropriate cleaning of the specimen's surface will be required so that structures of interest can be revealed. Likewise, the sample may be covered with debris or salts, which also may obscure important surface details. A number of methods exist for removing surface materials such as simple rinsing in water, enzymatic treatments and detergent, or mechanical cleaning.

Fixation
As discussed in previous chapters, the purpose of fixation is to preserve the fine structure of the cells and tissues in order to prevent, as much as possible, alterations or distortions

from occurring during subsequent preparations and observations. Thus, the sample will serve to represent the cell's or organism's living state. Here, as with TEM, two strategies for fixation apply to SEM: chemical and mechanical.

Chemical Fixation   The most common method of chemical fixation is immersion of the sample into an appropriate cocktail of buffered glutaraldehyde for the primary fixation, followed by a buffer rinse and secondary fixation in buffered osmium tetroxide. For the primary fixation it is particularly important that cells and tissues are fixed in physiological and environmental conditions that are similar to the living state of the sample. The concentrations of the fixatives will typically range from 1% to 4%, with buffers at a molarity of 0.1 and a pH of 6.8 to 7.2. The duration of primary fixation is typically 2 to 24 hours, and of secondary fixation about 2 hours.

An alternative to immersion fixation is vapor fixation, which can be used when delicate surface structures, such as the arrangement of chains of fungal spores associated to a fruiting body, require stabilization. Here, a specimen is confined to a small and closed chamber and positioned close to a drop or two of unbuffered osmium tetroxide (1% to 4%). Adequate fixation is often achieved after 2 hours of vapor exposure.

Mechanical Fixation   Samples that are adequately frozen in a solid state with minimal ice formation can be viewed in a frozen hydrated state directly or can be freeze etched and/or fractured (Figures 9.3C, 9.3D, and **9.14**) with an SEM that has a cryopreparation chamber and stage. Frozen samples may also be freeze dried and viewed with a standard SEM (see below). As discussed in Chapter 10, cryotechniques can stabilize samples that may be otherwise susceptible to structural and chemical modifications when standard chemical fixation and preparation methods are used. The protocols for freezing and preparing frozen samples are numerous. A common method is simply plunging the sample into liquid nitrogen slush (Figures 9.3C and 9.3D), whereas other, more involved methods, such as high-pressure freezing (Figure 9.14), are also used. Regardless of the freezing method used, the frozen samples can either be stored in liquid nitrogen or transferred to the SEM that has an appropriate cryopreparation chamber and cryostage attached. Specimens can also be infiltrated with a cryoprotectant before freezing to suppress ice crystal formation, growth, and thus cellular damage; however, their use can be limited in cryo-SEM because most cryoprotectants will produce a residue that remains

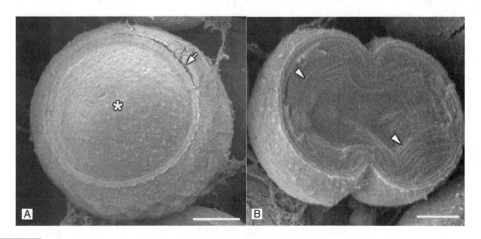

Figure 9.14
Cryoscanning electron micrographs of high-pressure frozen, freeze-fractured cells of the cyanobacterium *Synechocystis* sp. PCC 6803 after etching and sputter coating. Images are viewed at 2.0 kV. Cells fractured along various planes exposing the surface (A) and internal cytoplasmic structures (B). (A) Arrow points out artificial fracture due to overetching. Bar = 250 nm. (B) Arrowheads point out thylakoid membranes. Bar = 300 nm.

bound to the surface of the sample after drying obscuring features. Ethanol is the most useful cryoprotectant, as it is removed on evacuation. In the cryopreparation chamber, the sample may be fractured with a cold knife (i.e., razor blade), lightly etched, coated with a thin layer of metal (i.e., sputter coating), transferred to the cold stage in the specimen chamber, and examined with the cryo-SEM at temperatures as low as −150°C.

Dehydration and Critical-Point Drying

If standard chemical methods are used to fix a sample for SEM examination, it must next undergo a series of dehydration steps. A graded series of ethanol or acetone concentrations is used (e.g., 30%, 50%, 75%, 95%, and 100%) to chemically dehydrate the sample, taking it from water to 100% ethanol or acetone. Physical drying is achieved by critical point drying, which is a procedure that removes the solvent from the sample with no latent heat of vaporization or change of density. This is done by forcing the solvent to its critical point, that is, the temperature and pressure conditions under which a fluid will change directly from a liquid phase to a gaseous phase without boiling or evaporation. Thus, surface tension forces that can distort the specimen ultrastructure during solvent phase changes are entirely eliminated. Critical-point drying is the most often used means of physically drying samples, especially soft samples, for SEM. However, critical-point drying is not immune to artifact; indeed, there are instances in which extensive shrinkage of tissues can occur, and these represent the most common form of critical-point drying artifact.

The critical-point drying unit consists of a metal chamber (sometimes referred to as "the bomb") with thick walls that are able to sustain high pressures, plumbing for purging the sample of ethanol or acetone, and venting gases, and a mechanisms for temperature regulation. The basic method for critical point drying is as follows: (1) The sample in 100% ethanol or acetone is placed in a special holder and transferred to "the bomb" and cooled to about 5°C. (2) An inlet valve is opened, and liquid carbon dioxide (the transition fluid) completely replaces the ethanol or acetone in the drying chamber. (3) The sealed chamber is slowly heated, causing an increase in pressure, and the carbon dioxide eventually reaches a critical point as it transitions into the vapor phase. (4) The critical temperature is maintained to prevent condensation of the vapor back to liquid while the chamber is slowly vented to atmospheric pressure. (5) Dried specimens are removed and mounted on SEM specimen stubs.

One might ask, why bother with using carbon dioxide as a transition fluid when water itself might work? The answer is that the critical temperature and pressure of water are 375°C and 3184 psi, respectively, conditions that would cause specimen damage and present safety issues. Carbon dioxide, on the other hand, has a much lower critical temperature (31.0°C) and pressure (1073 psi), making it a more desirable and most commonly used transition fluid. Even under these conditions safety issues are present, and rupturing of the specimen chamber has been reported. This potential problem is allayed by a protective disc that will burst before dangerous pressure levels are reached. Because carbon dioxide is commonly used in the critical-point drying process, the laboratory in which the drying is done should be well ventilated, or it is recommended that the critical-point dryer (**Figure 9.15**) be used in a functioning fume hood.

Other Drying Approaches

Freeze Drying   Freeze drying can produce extremely good preservation of a sample's ultrastructure. After cryofixation, the frozen water is sublimed away at low temperatures and high vacuum in a freeze-drying machine. The time required to freeze dry a sample depends on the sample size, the temperature used, and the vacuum level. After the sample is completely dehydrated, it is returned to ambient temperature and pressure and is processed for viewing. Freeze drying can reduce the shrinkage artifact that can occur in many samples that have been critical-point dried.

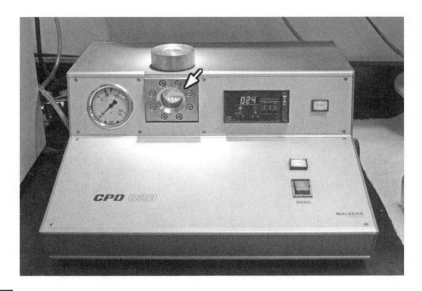

Figure 9.15
Photograph of a critical-point drying machine. Specimens are placed inside the drying chamber (arrow), and the drying process is carried out in a semiautomatic manner.

Air Drying   Some samples have strong enough surface features to withstand the tensions exerted during evaporation at ambient temperatures and pressures (Figure 9.3A). Air drying was used as the primary method of drying before the development of better preparation protocols. Most samples will suffer greatly when air drying is used and will appear extremely distorted, shrunken, cracked, and/or collapsed. It does work very well with bone, exoskeleton, wood, and shells.

Mounting the Specimen

After drying, the specimen is typically attached to an electrically conductive metal stub (Figure 9.7), which is the specimen support during the metal coating and imaging processes. The sample must be well secured to the stub so that once in the specimen chamber it does not become dislodged during routine stage movements and so that the flow of current away from the specimen produced by the electron beam is not impeded.

Before mounting the sample to the stub, one should consider the signal to be detected from the sample, the adhesive used to attach the specimen, and the orientation of the sample inside the specimen chamber. The stub most often used for secondary electron detection is made of aluminum or brass, although aluminum is generally used because it is less expensive and performs equally as well as brass. After the stub is cleaned, the specimen is attached to it using glue or tape. The mounting medium should mechanically stabilize the sample on the stub, provide an acceptable appearance that is flat and unambiguous relative to the sample, provide good electrical conductivity to help reduce problems of charge buildup over the specimen, and should be stable under the electron beam. The medium used should not contaminate the chamber by out-gassing of solvents and air when under vacuum and should not damage the sample. When backscattered or x-ray signals are of interest, nonmetallic stubs like carbon stubs, adhesives that have a carbon base, and carbon specimen coatings are suggested. During the attachment to the stubs, specimens can be easily damaged and should be handled with care while mounting. The use of fine tools, such as fine-tipped forceps, wooden applicators, needles, and a good-quality dissecting microscope, is often required. The specimen must be in close contact with the stub and not extend over the edge. Poor specimen contact can result in inadequate conductivity and

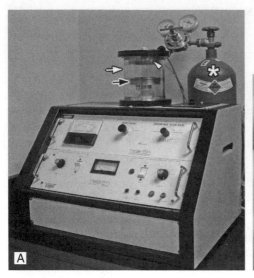

Figure 9.16
A direct current sputter coater. (A) The bell jar (white arrow), target (i.e., cathode) (arrowhead), base (i.e., anode) (black arrow), and a tank of argon gas (asterisk) are shown. (B) Insertion of the specimen (arrowhead) onto the base (arrow).

in the buildup of a negative charge, distortion of image, and/or loss of sample while imaging. After the specimen has been successfully secured to the stub, the samples should be stored in an area that is free of dust and preferably in a desiccator.

Specimen Coating

After a specimen has been mounted, it is made conductive with the application of a 20- to 30-nm-thick layer of metal such as gold, platinum, or gold/palladium. As pointed out above, a sample that is nonconductive will quickly build up a negative charge from the electron beam, producing lines on the screen and micrograph and areas on the specimen's surface of abnormally high brightness (i.e., charging). A well-mounted and metal-coated specimen will conduct the charge to the ground and help dissipate heat buildup. Additionally, the metal coating is the source for abundant secondary electrons and thus increases the signal-to-noise ratio.

The standard procedure used for applying the metal coating to the sample is the sputter-coating method. There are various types of sputter-coating machines. The most common is the direct-current sputter coater (**Figure 9.16**); it is discussed here. After samples are secured in the sputter chamber, a vacuum of about 10^{-3} torr is established using a rotary pump. The gas molecules that remain in the chamber are air or preferably inert molecules such as argon or nitrogen. When an appropriate vacuum is achieved, a negatively charged field is applied between the target (i.e., cathode) and the specimen that sets on the base (i.e., anode). The gas molecules between the target and base are ionized. The ionized gas atoms are excited to a rapidly moving state and bombard the target, which is composed of gold or palladium/gold, with sufficient force as to dislodge some of the metal atoms from the target. These atoms undergo multiple collisions with argon molecules, producing a diffuse cloud of target atoms, referred as the plasma, that falls and coats the surfaces of the specimen and stub from multiple directions. Sputtering is continued until an adequate layer of gold is deposited on the specimen's surface. At the end of the coating session, the sputtering chamber is backfilled with air. The samples are then collected and safely stored in a desiccator until imaged.

■ Analytical Microscopy

For many years the electron microscope has been used not only for imaging, but for determining a specimen's elemental and macromolecular composition. The use of electron microscopes for analytical purposes, or microanalysis, includes the methods of x-ray microanalysis, electron diffraction, and electron energy loss spectroscopy (EELS). Localization techniques, such as autoradiography, cytochemistry, and immunocytochemistry, represent additional methods that are commonly employed. Below we briefly outline the utility of x-ray microanalysis procedures relative to biological studies using SEM.

X-Ray Microanalysis

The identification of elements in SEM or TEM samples can be achieved with high sensitivity and spatial resolution by using x-ray microanalysis methods. As mentioned above, two forms of x-rays are produced continually during EM imaging: continuum or bremsstrahlung x-rays and characteristic x-rays. The latter are created as a result of inelastic scattering events, causing ionization of the specimen's atoms, resulting in the emission of x-rays that have a unique energy and wavelength signature. By analyzing characteristic x-rays, researchers can obtain data regarding the type, location, and amount of the elements that make up cells and tissues. This information can be used in studies of fundamental questions of ion distributions, toxicology, drug therapy, biomineralization studies, crime scene analyses, and disease development.

X-Ray Microanalysis Methods   Each x-ray has a characteristic energy and wavelength, which are the basis for their identification. Elemental analysis based on the measurement of energies is called energy-dispersive spectroscopy (EDS), whereas measurement based on wavelengths is known as wavelength dispersive spectroscopy (WDS).

WDS   The instrumentation and means of detecting and analyzing characteristic x-rays for WDS and EDS differ considerably. For WDS, x-rays strike a crystal with only a narrow band width of x-rays able to escape, as determined by the Bragg equation, and enter the detector (**Figure 9.17**).

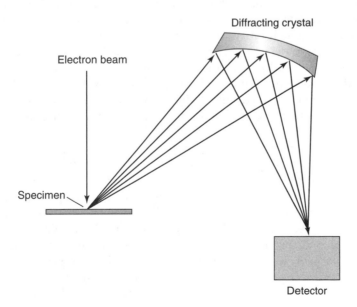

Figure 9.17
Drawing illustrating the components and operation of a WDS detector. Characteristic x-rays from the specimen enter the crystal. X-rays of a specific wavelength exit the crystalline lattice and are directed to the detector.

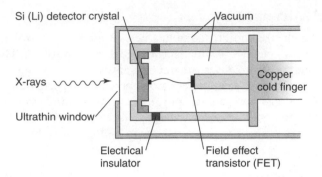

Figure 9.18
Illustration of the basic components and design of an EDS detector.

A single crystal may be used for detecting only a few elements. Different types of crystals are thus needed to analyze a range of elements. The reflected x-ray is detected as it passes through a thin window and enters the detection chamber containing a collector wire maintained at high positive voltage in an atmosphere of argon/methane gas. The x-ray causes an ionization of the gas that produces a current through the wire. The current created has a direct relationship to the energy of the x-ray and is thus used to identify it. WDS has the advantages of great spectral resolution, less detection of the continuum x-rays, an ability to detect light elements, and increased sensitivity at high beam currents. This method, however, is not often used for biological studies because the procedure often damages delicate biological samples because of the need for a larger and more energetic electron probe. A further disadvantage is that the number of elements detected simultaneously is low (typically no more than four). Thus, WDS detection has lower spatial resolution, is less efficient and is more expensive than detection by EDS systems.

EDS   The EDS system categorizes x-rays electronically and is usually sensitive to elements with atomic numbers of 11 or greater, depending on the detector characteristics. After scanning the sample for a period of time, all of the elements generating sufficient levels of x-rays will be detected. The x-rays that reach the semiconductor detector (**Figure 9.18**) first produce currents, which are then converted electronically into a voltage in which the amplitudes are directly proportional to the energy of the x-ray signals. After the voltage is converted into a digital signal, it is recorded by a multichannel analyzer. With time of sampling, the counts build up to produce an x-ray spectrum of the sample.

The EDS detectors are solid-state devices that attach to the column of the microscope and are made up of a collimator, window, lithium-drifted silicon detector, field effect transistor, Dewar, and preamp (Figure 9.18). The collimator is positioned at the front of the EDS detector and arranges the x-rays entering it into parallel paths. This functions as a means to reduce stray x-rays and backscattered electrons that do not originate from the sample from entering the detector. A thin (typically 5 to 7 μm thick) window of beryllium is located behind the collimator and establishes a protective barrier between the vacuum of the specimen chamber and the detector crystal. A problem associated with the window is that x-rays from elements below an atomic number of 11 are absorbed by beryllium and are not detected. To alleviate this disadvantage, lower density plastic or diamond windows are available. These windows have the advantage in that they absorb far fewer of the low-energy x-rays than does the beryllium window, allowing detection of elements with atomic numbers of 3 and above. Also, detectors can be designed to be used without a window, the so-called windowless detectors. Behind the window is the detector crystal, made up of a lithium impregnated silicon wafer that is maintained at cold temperatures by liquid nitrogen stored in a Dewar located on the outside of the microscope. After impacting the crystal, electron signals are generated that are directly proportional to the energy of the incident x-rays. These signals are then amplified by the field effect transistor and preamp and are processed by a multichannel analyzer that categorizes them in terms of amplitude.

Characteristic X-Ray Nomenclature The filling of electron holes by outer shell electrons results in the release of a characteristic x-ray. The name given to each x-ray is determined by the shell where the electron hole was created (i.e., K, L, M, N) and the specific electron that filled the hole (i.e., an electron originating from adjacent shell is α, from two shells away is β, from three shells away is γ). Additionally, as shells have multiple electrons in different suborbitals, different electrons spin, and therefore, different energy levels can become confusing when naming is required. A standard and somewhat simplified method of naming specific x-ray lines is as follows:

x-ray line $K\alpha_1$:

K = shell that is filled;
α_1 = the shell and orbital of the electron filling the vacancy.

Complicating matters a bit more is the fact that elements typically produce multiple types of characteristic x-rays because of numerous primary and secondary inelastic scattering events that occur. Fortunately, x-ray detectors have the ability to identify these multiple energy levels, and if enough x-rays of a given energy level, or line, are generated, they are presented as peaks in a spectrum of characteristic x-rays.

■ Additional Modes of SEM Analysis and Operations

Environmental and Variable Pressure SEM

Examining live, wet, nonfrozen, uncoated biological samples with an SEM at resolutions as good as 10 nm can be achieved in some cases when using an environmental scanning electron microscope (ESEM). This revolutionary instrument provides the means of examining hydrated samples at pressures six orders of magnitude higher than with conventional SEM because the ESEM is differentially pumped, having an aperture located just below the final lens that allows the high vacuum to be sequestered in the column while a low vacuum is maintained in the specimen chamber. The electron detector and specimen are positioned just below the aperture. Because of the close proximity of the specimen to the detector, scattering of the beam is minimal, and a large portion of the signal produced from the specimen's surface is available for collection. The detector used in the ESEM is centered on concepts of a gaseous discharge detector where the signal is propagated through the molecules within the atmosphere that surround the sample. Plants and fungi are particularly well suited for the ESEM viewing because of their rigid cell wall.

A less costly alternative to the ESEM is the variable-pressure SEM and the isolation capsule developed by Quantomix. Although they are all quite similar in imaging objects, some important differences exist. For example, both environmental and variable pressure systems can image wet or living specimens that contain water; however, ESEMs are preferable because they can operate at pressures as high as 7 torr while variable pressure systems operate at pressures no higher than 2 torr. Variable pressure SEMs can be used quite well for studies of clay, carpet fibers, forensic and archaeological samples, and metrology of advanced photolithographic masks. In contrast, the Quantomix capsule encloses wet samples in an electron transparent chamber, isolating the specimen from the high vacuum of a standard SEM specimen chamber. This allows the imaging of hydrated samples such as food, cosmetics, ink, cells, tissues, and fluids at resolutions unachievable with light microscopy. An important advantage of the Quantomix capsules is that they can be used on all existing SEMs and are not limited to newer variable pressure or ESEMs.

Clearly, progress in viewing fully hydrated samples is being made; many of the SEM systems can be optimized for viewing dynamic, delicate living specimens (e.g., fibroblast cells; in this case, the specimen must remain alive during the course of the investigation). However, such cells undergo a stress response when exposed to the electron beam, no matter what SEM method is used. Furthermore, most do not thrive at temperatures required for use in these microscopes, and unlike photons used in optical microscopy, the water of the cell is not invisible to electrons. Thus, live cell imaging is, for the time being, relegated to state of the art light microscopy methods.

■ Cryo-SEM

Another method for visualizing hydrated samples is to view them in their frozen state. Generally, a sample is cryoprotected and rapidly frozen. To examine frozen samples with an SEM, the microscope must be equipped with a cryostage and cryopreparation chamber. After freezing, the sample is transferred onto a prechilled specimen stage and inserted into the cryopreparation chamber that is attached to the SEM. Here, the sample can be fractured, lightly etched, and coated before it is transferred to the cold stage located inside the SEM specimen chamber. Frozen hydrated samples are often imaged at low accelerating voltages (e.g., 1 to 5 kV) in order to minimize charging. Well-prepared samples viewed in this state show fine details of the sample's surface whether they are fractured or not (Figures 9.3C, 9.3D, and 9.14).

■ Practical Guide for Getting Started

To obtain the best image from each specimen, the user should become comfortable with the operations of the instrument and have specimens prepared in an appropriate manner. Becoming familiar with the control panel, particularly noting the positions of the focus, magnification, and beam raster controls, is the first step. The beginning user must have written instructions, typically condensed from an operator's manual, and be comfortable in interacting with the laboratory manager if problems or questions occur. What follows is a generic protocol of basic activities surrounding the working of an SEM; these may vary considerably depending on the instrument.

Specimen Change (Based on a Drawer Design)

1. Be sure that the stage Z control is set correctly and that the stage tilt is set at 0 degrees.
2. Vent the specimen chamber by depressing the appropriate button. After a short time, you should hear the rush of air into the chamber. A minute or slightly more may pass, after which the whole chamber door should open easily. Open the door carefully so as not to damage anything as the door is opening. Gloves should be worn if there is a possibility of touching anything inside the chamber and forceps should always be used to handle specimens.
3. Insert the specimen in the stage and make sure that the way is clear for the door to close, including clearance between the uppermost portion of your specimen and the bottom of the final lens. Again, if there are questions, particularly at this point, the user must contact the facility's manager.

4. While gently holding the door closed, initiate repumping of the specimen chamber.

5. Check the vacuum status. As the vacuum in the chamber improves, the chamber valve will open, and the status of the instrument will change. Wait until the chamber vacuum reaches the appropriate vacuum and obtain a secondary electron image.

Optimizing the Image

The user now has several fundamental choices to make regarding the operating parameters for the SEM. These choices are based on the instrument used and the nature of the specimen. If the material is fragile or sensitive to the beam, you may wish to limit the heat and charge buildup by choosing a relatively low-acceleration energy and probe current. You may need to experiment to get good images. Start low and go to higher voltages in order to avoid damaging hard-to-replace specimens.

1. After the decision is made, select the appropriate acceleration voltage.

2. Select and set an appropriate spot size and beam current.

3. Align the beam so that it is centered in the column.

4. The final aperture must now be aligned around the beam.

5. The most sensitive adjustment comes last—correction of astigmatism. As mentioned above, if the beam is not perfectly round when it impinges on the specimen, it will focus at different depths along different radial planes, meaning that it cannot be focused perfectly anywhere, causing degradation of resolution. The higher the magnification, the more precise the adjustment required.

6. If your specimen is fragile or beam sensitive, begin observation at low magnification, making micrographs as required and then go to progressively higher magnifications. This will allow you to avoid making micrographs with beam damage or contamination spots.

7. If you are experiencing a poor signal-to-noise ratio (snowy picture even at slow scan rates), there are several adjustments that may be tried. First, try tilting the specimen 30 to 45 degrees toward the secondary electron detector. You may also change the Z distance, moving the specimen closer to the final lens. You may increase the probe current, particularly if you are working at low or moderate magnification. If necessary, coat or recoat the specimen with gold or platinum to increase the secondary electron signal. Finally, change the bias on the secondary electron detector to a higher positive value or try a higher accelerating voltage. Remember, if you change the accelerating voltage or the probe current, you must check the beam alignment, the final aperture centering, and the astigmatism for optimum results.

Making a Micrograph

When you are satisfied with your image, given the trade-offs necessary, it is time to document it using film or digital storage. Although most modern systems today use a digital storage format, the fundamental steps in acquiring acceptable micrographs are similar for both approaches.

1. Set the contrast and brightness using the SEM controls for optimum image recording—this will typically oversaturate the monochrome monitor.

2. Depress the appropriate buttons to begin the image capture process.

3. To control image resolution, the scan generator takes control of the SEM beam and the raster of the display and synchronizes the beam output with the image display. Higher quality images result from very slow scans, while lower quality images result from fast scans.

4. To store a satisfactory micrograph, several formats are available to the user. Consult with the instrument's user manual.

5. At the end of your SEM session burn a CD-ROM to archive your work thereby allowing the laboratory to clear the hard drive space on a regular basis.

General Rules

1. Never force a control.

2. Heed warnings or alarms given by the SEM. Seek help from laboratory personnel when needed.

3. Do not put moist specimens or poorly dried glue into the chamber of a low vacuum SEM—this is hard on the vacuum system and will greatly extend the pump down time.

4. Report any unusual sounds, smells, or sights—anything that you are not used to experiencing during the task you are doing.

5. Do not have food or drink in an SEM room.

6. Always check the stage before attempting to open the chamber.

References and Suggested Reading

Bozzola JJ, Russell LD. *Electron Microscopy: Principles and Techniques for Biologists*, 2nd ed. Sudbury, MA: Jones and Bartlett Publishers, 1999.

Crang REF, Klomparens KL. *Artifacts in Biological Electron Microscopy*. New York: Plenum, 1988.

Dykstra MJ. *Biological Electron Microscopy: Theory, Techniques and Troubleshooting*. New York: Plenum Press, 1992.

Postek MT, Howard KS, Johnson AH, McMichael KL. *Scanning Electron Microscopy: A Students Handbook*. Williston, VT: Ladd Research Industries, 1980.

Wischnitzer S. *Introduction to Electron Microscopy*, 3rd ed. New York: Pergamon Press, 1988.

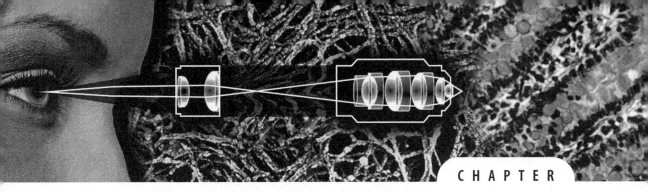

Cryogenic Techniques in Electron Microscopy

Rapid freezing constitutes a major alternative to chemical fixation for preparation of tissues for electron microscopy. The technologies used can be time consuming, more costly, and prone to lower throughput, but the rewards can be superior preservation and the elimination of artifacts associated with chemical fixation. A major hurdle in cryogenic preservation of tissues is avoiding the formation of ice crystals during the freezing process—crystals that can grow to dimensions sufficiently large to disrupt normal tissue structure completely at the electron microscopic level. Three common methods prevent ice crystal growth—the use of cryoprotectants that nucleate small crystals but slow their growth, the use of high pressure to slow the rate of ice crystal growth dramatically, and third, use of ultrarapid freezing methods that are so fast that ice crystals do not have time to grow.

After high-quality freezing has been achieved, there are four complementary techniques for further preparation of the frozen tissue. First, the tissue can be sectioned while in the frozen state or frozen in a thin layer of vitreous ice and viewed directly on a cryogenic stage in the electron microscope without thawing. Extremely thin specimens can be frozen and viewed without sectioning. Second, specimens can be "freeze substituted" by dehydration and fixation at low temperature and then embedded, sectioned and viewed at normal temperatures. Third, frozen tissues can be split open and the frozen surface replicated under vacuum using a "freeze-fracture" process; the metal replica is then viewed by normal transmission electron microscopy. Fourth, frozen tissue can be fractured and viewed directly on a cryogenic stage in a scanning electron microscope.

This chapter addresses techniques used for freezing tissues and procedures used for preparing and viewing these tissues in the electron microscope. Factors in the choice of freezing method include the type of specimen to be frozen, the allowable cost, and the instrumentation available. The pros and cons of common freezing methods are provided in **Table 10.1.** Cell suspensions and tissue culture cells are thin enough to be handled by spraying, immersion, or propane jet freezing. Thicker tissues require the use of cold metal block or high-pressure freezing.

■ Freezing Methods Using Cryoprotectants

During freezing, the formation of ice from water can occur so quickly that molecules do not have a chance to form large scale crystalline organizations. Formation of noncrystalline or "vitreous" ice, however, requires the withdrawal of heat so fast that the temperature must

Table 10.1	Comparison of Rapid Freezing Methods for Cells and Tissues		
Method	**Appropriate Specimens**	**Advantages**	**Disadvantages**
Immersion in fluorocarbon	Tissue cubes or cell pellets	Standard protocol requiring no instrumentation; freezing of consistently acceptable quality throughout tissue	Fixation and glycerol cryoprotection required
Immersion in propane or ethane	Tissue culture cells; suspensions of isolated cells, organelles, or macromolecules	Minimal instrumentation; relatively inexpensive	Thin specimens required; adequate freezing depth 5–10 μm; explosion hazard
Spray freezing	Organelles or bacteria	Excellent freezing quality for prokaryotes	Cannot be used with large cells or tissues
Propane jet freezing	Thin tissues or cell suspensions	Automated; freezing from both sides	Explosion hazard; adequate freezing depth up to 40 μm
Cold metal block freezing	Specimens of all thicknesses from organs to organelles	Automated; thick specimens accommodated	More expensive; target cells must be exposed to surface; adequate freezing depth 10 to 20 μm
High-pressure freezing	Cells and tissues as large as 0.5 mm in each dimension	Automated; adequate freezing depth relatively great (500 μm)	Costly instrumentation

drop at a rate in the vicinity of 10,000°C per second. This can be achieved only by specialized "ultrarapid" freezing devices (see below). Normally, freezing is considerably slower and involves the formation of ice crystals whose growth is "nucleated" by small groups of highly organized water molecules. Nucleation (creation of order out of chaos) is a rare event. Once nucleated, however, ice crystals grow in size extremely rapidly, much to the chagrin of microscopists seeking to preserve tissue structure. One can avoid this by making nucleation a common, non–rate-limiting step through inclusion of cryoprotectant molecules that can organize water even in the liquid state. Furthermore, once formed, these crystals grow very slowly because most of the liquid water available is hydrogen bonded to cryoprotectant molecules and cannot be incorporated into the crystal structure.

Any molecule that forms noncovalent bonds with water such as ions, sugars, alcohols, proteins, or hydrophilic polymers can act as a cryoprotectant, and examples of chemicals commonly used for this purpose are found in **Table 10.2**. Indeed, tissues may contain natural cryoprotectants because these are the same molecules that lower the freezing point of water. Some plant cells contain high concentrations of the sugar trehalose, which make them freeze resistant, whereas specialized Antarctic fish contain "antifreeze" proteins that confer the same property. Most tissues, however, require the addition of exogenous cryoprotectants in order to be frozen successfully. A common cryoprotectant used in both microscopy and preservation of cell viability during freezing is glycerol. Glycerol, having three carbon atoms each with a hydroxyl group, easily forms hydrogen bonds with surrounding water molecules, thereby forming nucleating centers.

Table 10.2 Common Cryoprotectants				
Cryoprotectant	Molecular Weight	Type of Compound	Uses	Comments
Methanol	32	Volatile liquid	Rapid freezing; windshield wash	
Glycerol	92	Nonvolatile liquid	Used for freeze fracture	Most commonly used cryo-protectant
Ethylene glycol	62	Nonvolatile liquid	Automotive antifreeze	Poisonous
Trehalose	342	Solid sugar; nonmetabolizable	In situ cryo-protectant	Found at high concentrations in freeze-tolerant plants
Polyvinyl alcohol	10,000	Solid polymer	Low osmotic strength cryoprotectant for live cells during high-pressure freezing	

Glycerol and many other cryoprotectants are used at sufficiently high concentrations (30%) to cause dehydration and other deleterious effects on tissue structure. For example, glycerol is known to incorporate into cell membranes, making them highly fluid and capable of undergoing artifactual fusion and vesicle formation; therefore, in routine procedures (e.g., preparation of specimens for freeze fracture (see below) or immunocytochemistry [see Chapter 13]), the tissue must be chemically fixed before cryoprotection, regardless of the benefits obtained by freezing. Following cryoprotection the tissue is placed on a specimen carrier and is frozen by immersion in a fluorocarbon cryogen that is cooled to liquid nitrogen temperature ($-196°C$). It might be expected that one could simply immerse a tissue in liquid nitrogen, as dermatologists do for warts. This is precluded by the fact that freezing with any fluid such as liquid nitrogen that boils easily will result in relatively slow freezing because of a layer of gas bubbles that insulates the tissue from further heat removal. Rather, a good cryogen must remain liquid at very low temperatures and be able to absorb heat rapidly without boiling; that is, it must have a high heat capacity and a relatively high boiling point. Common cryogens having these properties are listed in **Table 10.3.**

■ Ultrarapid Freezing by Immersion

The best cryogens can actually be used without cryoprotectants, providing that the specimen is thin enough to allow rapid heat withdrawal. Specimens must be only a few microns thick, with common examples being tissue culture cells, subcellular organelles, or proteins. As shown in **Figure 10.1A,** cells are often grown on dialysis or polycarbonate membrane supports and a section of appropriate size with adherent cells cut out for freezing. These membranes have the advantage that they are thin, hold little heat, and later can be easily cut during ultrathin sectioning. The cryogen of choice is either liquid propane or liquid ethane or a mixture of these. Propane is more common because it is widely available, usually being obtained in canisters used to fuel outdoor barbecue grills. One important negative of either ethane or propane is its extreme volatility

Table 10.3 Common Cryogens				
Cryogen	Molecular Weight	Freezing Point	Boiling Point	Comments
Liquid nitrogen	28	−210°C	−196°C	Boils easily; used for high-pressure freezing, cold metal block freezing in freeze fracture units, and in cryostages for EM
Liquid helium	4	−272°C	−269°C	Boils easily; used for high pressure and cold metal block freezing
Methane	16	−182°C	−161°C	Infrequently used; explosion hazard; boils easily
Ethane (R-170)	30	−183°C	−89°C	Used as an immersion cryogen; explosion hazard
Propane (R-290)	44	−188°C	−42°C	Used for immersion, spray, and jet freezing; explosion hazard
Dichlorodifluoromethane (R-12)	121	−158°C	−30°C	Freon; no longer available; formerly used for immersion freezing
Chlorotrifluoromethane (R-13)	104	−181°C	−82°C	Used for immersion freezing
Chlorodifluoromethane (R-22)	86	−256°C	−40°C	Used for immersion freezing
Tetrafluoroethane (R-134a)	100	−97°C	−26°C	Replacement for Freon; used for immersion freezing
Acetone/dry ice (CO_2)	58/44	−95°C/−79°C (sublimination)	56°C	Used for immersion freezing in biochemical work

and explosiveness! Great care must be taken to avoid any open flames or sparks nearby that could turn the process into a disaster. An absolute requirement is to carry out all procedures in a spark-proof fume hood that is vented to the outside and to do so only under experienced supervision.

The liquid propane bath is prepared by first filling a 4-liter, heavy-duty, wide-mouth dewar with liquid nitrogen and allowing boiling to subside (Figure 10.1B). Next, a nylon cover is placed over the liquid nitrogen and a copper cup placed in a central hole within the cover and allowed to cool. Finally, the hose and nozzle of the propane canister are placed obliquely against the wall of the cup, and the flow of propane is started; after contact with the cup, the propane liquefies and fills the cup two-thirds full (Figures 10.1C and 10.1D). The cup is then capped with venting until use to minimize contact with water vapor or oxygen.

The specimen to be frozen is kept in physiological buffer at an appropriate temperature before freezing to insure that the cells are as normal in function and structure

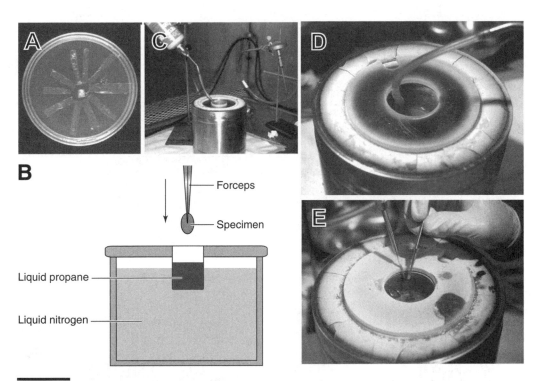

Figure 10.1
Immersion freezing. (A) Tissue culture cells being grown on dialysis membrane before freezing. (B) Diagram of an immersion freezing setup using liquid propane as a cryogen. (C) Filling the cold cup with propane. (D) Close-up of the liquid nitrogen dewar with the liquid propane cup at the center. (E) Immersion of specimen into liquid propane using forceps.

as possible. A few tens of seconds before freezing, the specimen is grasped with a watchmakers forceps, removed from buffer, excess buffer wicked away by filter paper, then immediately thrust into the liquid propane. With practice, this can be done manually (Figure 10.1E), or the specimen/forceps can be mounted on a gravity-driven plunger for rapid immersion (see Figure 15.6A). The frozen specimen is then transferred to a foam reservoir of liquid nitrogen and finally stored frozen in a liquid nitrogen dewar until further processing.

■ Propane Jet Freezing

A machine for rapidly spraying liquid propane at a specimen is illustrated in **Figures 10.2A** and **10.2B**. The specimen is housed within two thin recessed copper planchets that protect the specimen from damage by the propane jets. These planchets come in a variety of different configurations to accommodate different sample geometries (Figure 10.2C); however, the specimen itself is typically no thicker than 40 μm. The specimen is held secure in the jaws of a specimen holder that is inserted between the jets. Two jets supplied with pressurized liquid propane forcefully blast the specimen from both sides, lowering its temperature at approximately 10,000°C per second (Figures 10.2D and 10.2E). The entire process is precisely timed by a microprocessor and takes only 10 seconds. Precooled forceps are used to remove the frozen specimen and place it in liquid nitrogen for storage. The entire propane jet-freezing unit must be used in a hood to avoid possible explosion!

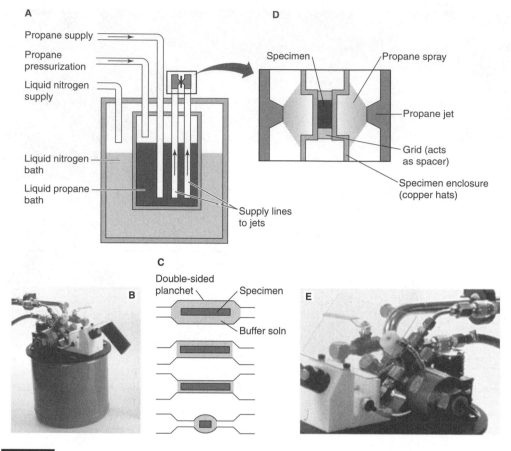

Figure 10.2
Propane jet freezing. (A) Diagram of a propane jet freezing instrument. Pressure is used to force liquid propane through the jets rapidly. (B) A commercial propane jet freezer. (C) Configurations of specimen holders for propane jet freezing. (D) Detail of specimen being frozen by liquid propane jet. (E) Detail of (B) showing plumbing for propane jets.

■ Metal Mirror Freezing

A number of instruments for pressing tissue against a cryogen-cooled polished metal block have been designed. The advantage of these techniques is that heat removal is extremely rapid because of the high conductivity and heat capacity of the copper, silver, or sapphire block. In addition, heat removal is from one side only, allowing tissue of any thickness to be used as a specimen. The disadvantage of these instruments is that only the tissue at the very surface of the specimen is rapidly frozen, typically a layer 10 to 20 μm thick. The metal mirror freezing devices used have ranged all of the way from hand held "cryopliers" to gravity-driven rods to pneumatic plungers to force the specimen against the cold metal surface (**Figure 10.3**).

The earliest successful metal mirror freezing device was developed by Dr. John Heuser and has been affectionately dubbed "the slammer." As diagrammed in Figures 10.3A and 10.3B, a copper block is cooled by a stream of liquid helium fed by a vacuum-shrouded pipe from a dewar positioned beneath the instrument. The block is housed within a machined phenolic resin compartment that is covered with a shutter to prevent atmospheric water vapor from frosting the block and preventing rapid tissue freezing. The block housing is surrounded by a circular electromagnet that is activated just as the tissue is frozen to keep the tissue in tight contact with the cold metal. The tissue specimen is

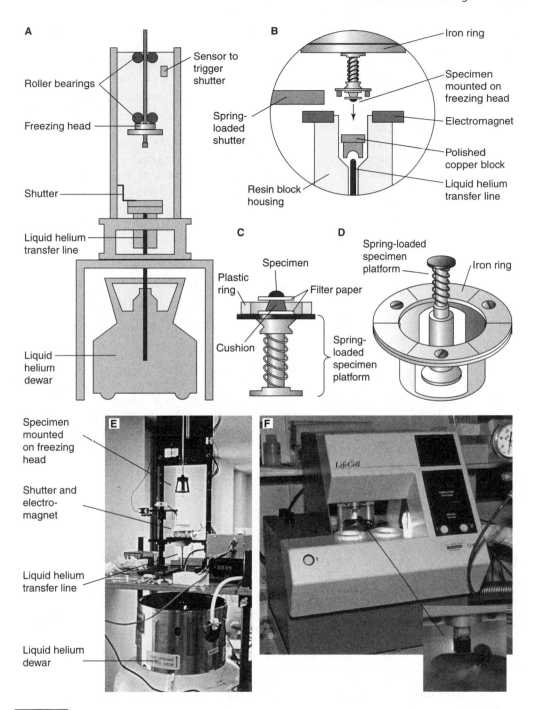

Figure 10.3

Cold metal block freezing. (A) Diagram of the cold metal block freezer designed by John Heuser. The plunger with freezing head drops by gravity pressing the tissue against a helium-cooled copper block. (B) Detail of shutter opening and specimen-block contact. (C) Detail of specimen platform. The specimen is placed on filter paper or directly on a tissue cushion that in turn rests on a filter paper surfaced aluminum planchet. A plastic ring shown in cross-section surrounds the specimen and keeps it from being overly compressed during block contact. (D) Inverted freezing head with spring-loaded specimen platform. The iron ring surrounding the specimen platform is held tight by an electromagnet during freezing. (E) A Heuser-type cold metal block freezer with liquid helium dewar below. A double-walled evacuated transfer line carries liquid helium from the dewar to the freezing machine. (F) The Life-Cell instrument for automated cold metal block freezing with liquid nitrogen. Inset: detail of Life-Cell instrument showing a pneumatically driven specimen plunger.

mounted on a thin square of liver or lung tissue sitting on a filter-paper–draped aluminum planchet (Figure 10.3C). This tissue base acts as a cushion during freezing and a platform for moving the specimen through subsequent processing steps. The specimen planchet is then mounted on a spring-loaded plunger centered within the freezing head that is surrounded by a flat iron ring that will contact the electromagnet (Figure 10.3D). The last bit of buffer is wicked away from the mounted specimen, and the freezing head/specimen is inverted and attached by a "bayonet" mount to a vertical plunger that is held in a cocked position 25 cm above the helium-cooled block. As the plunger is set into free fall, the shutter over the block is triggered to open, and the electromagnet is activated. The tissue is firmly pressed against the block, held in that position, and kept from bouncing by contact of the freezing head with the electromagnet. The frozen specimen is then immersed in liquid nitrogen for storage. A photograph of the Heuser-type freezing machine (Figure 10.3E) illustrates some of its key components.

A microprocessor-controlled version of this type of freezing machine was produced in the 1980s by Life Cell, Inc. It offered a number of improvements (Figures 10.3F). First, the cold block is held inside of a vacuum chamber covered by a pneumatically controlled shutter, thus ensuring that the block is never exposed to the atmosphere, and the buildup of frost. Second, the specimen plunger is pneumatically controlled. These features allow for automated high throughput and the use of liquid nitrogen rather than liquid helium as a cryogen.

Because heat withdrawal is from one side of the specimen, tissues frozen by metal mirror instruments exhibit a gradient of freezing quality from their surface into their interior. The superficial 10 μm of the specimen is extremely well frozen, having ice crystals that are too small to disrupt ultrastructure at the electron microscopy level. In contrast, tissues as deep as 30 or 40 μm from the surface show signs of large ice crystal growth, producing an artifactually patterned cytoplasm. At even greater depths, the holes resulting from ice crystal growth are so large that the tissue becomes unrecognizable. Clearly, either the cells to be studied must be at the surface of the tissue or the tissue must be dissected to reveal the cells of interest. Another potential problem, although minor, is that collision of tissue with the cold block produces compression and shearing forces that can distort tissue architecture in some specimens.

■ High-Pressure Freezing

Freezing at high pressure is based on the fact that ice crystal growth is dramatically slowed, allowing tissue to be frozen successfully even by "poor" cryogens such as liquid nitrogen. Because the pressure necessary is 2100 atmospheres, this represents a significant technical challenge requiring the use of a hydraulic piston to achieve this pressure within fractions of a second. The Baltec HPM 010 High-Pressure Freezer is shown in **Figure 10.4A.** The specimen is held in a double planchet that is 3 mm in diameter and somewhat similar to that used for a propane jet freezer. As shown by scanning electron microscopy, the planchets are machined from a gold-nickel alloy and "interlocked" to form a sandwich that encloses the specimen during freezing to prevent its dispersal (Figure 10.4B). The planchet is then placed at the tip of a heavy rod (Figures 10.4B and 10.4C) and inserted and locked into the ballistic chamber. The chamber is then flooded at high pressure with cold ethanol, followed milliseconds later by liquid nitrogen. At these pressures, liquid nitrogen does not boil, and freezing is accomplished within 100 msec. The loud report from pressurization is followed by venting; the specimen holder is withdrawn, and the specimen is transferred to liquid nitrogen (Figure 10.4D). High pressure is so effective at preventing ice crystal growth that specimens as large as 500 μm in diameter can be effectively frozen with no observable damage.

Figure 10.4
High-pressure freezing. (A) The Baltec HPM 010 high-pressure freezing unit. (B) Scanning electron micrographs of the high-pressure freezing specimen enclosures ("planchets") and the holder they are placed in. (C) Close-up of specimen holder with hinge cover. (D) Transfer of the specimen from the high pressure unit to liquid nitrogen after freezing.

■ Cryofixation Avoids Artifacts of Chemical Fixation But Can Cause Freezing Damage

Crucial to good cryopreservation is prevention of ice crystal growth. **Figure 10.5A** illustrates the gross disruption that can result from ice growth evidenced by large polygonal spaces from which cellular structures have been pushed aside. Ice crystal growth is often faster outside of cells than inside because cytoplasmic proteins and metabolites act as endogenous cryoprotectants to a certain extent. In fact, even in serious cases of poor freezing, cells can be dehydrated by intracellular water removal. Their cytoplasm is concentrated, and intracellular freezing damage is sometimes avoided. In many specimens, one observes a gradient in the quality of freezing, as evidenced by the absence of ice crystal growth at the surface of specimens that were in closest contact with the cryogen while ice crystals grow in numbers and size as one moves to inner regions of the specimen where heat withdrawal was slower. In some regions, ice crystal damage can be relatively subtle, as shown in Figures 10.5B and 10.5C. Small pockets in the cytoplasm of a fungal hyphal tip (arrow, Figure 10.5B) and the mottled appearance of secretory granule membranes in a mast cell (arrow, Figure 10.5C) represent clues that freezing has not been optimal.

It is seldom, however, that poor freezing can produce structures that can actually be mistaken for biological information. Rather, it simply frustrates the microscopist who must contend

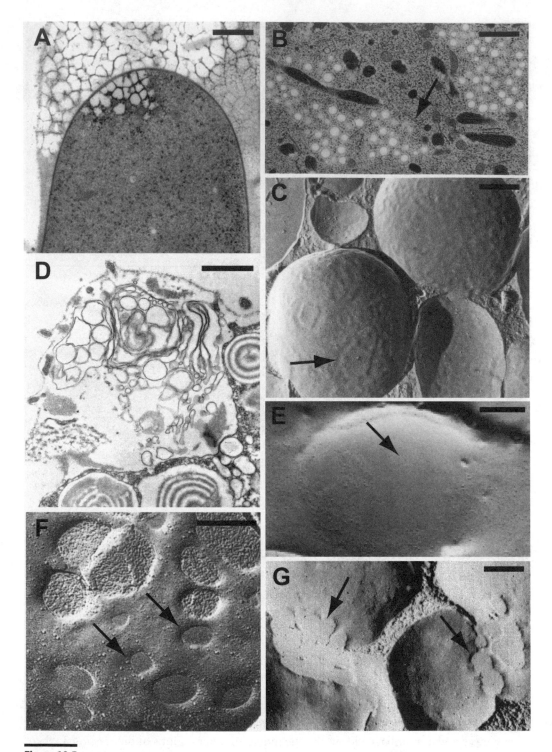

Figure 10.5
Artifacts produced by freezing and by chemical fixation. (A) Fungal hyphal tip (*Sclerotium rolfsii*) damaged by large ice crystals. (B) Cytoplasm of a fungal cell frozen by immersion freezing illustrating moderate freeze damage. (C) Freeze-fracture replica of secretory granules in a mast cell. Poor freezing has resulted in bumpy membranes having a mottled appearance. (D) Exocytosis in a sea urchin egg fixed by glutaraldehyde. Numerous large blebs are a fixation artifact. (E) Freeze-fracture replica of the plasma membrane of a mast cell showing large areas that have been cleared of IMPs (arrow). (F) Replica of the plasma membrane of a sea urchin egg undergoing exocytosis. The smooth bilayer diaphragms (arrows) represent dehydration artifacts not seen in rapidly frozen cells. (G) Artifactual membrane fusions (arrow) between secretory granules similar to those in (F). (A, B, D) Bars = 1.0 μm. (C, E−G) Bars = 0.25 μm.

with as low as a 10% yield of well-frozen material. In contrast, chemical fixation (e.g., with glutaraldehyde) can change the ultrastructure of organelles and membranes to produce artifacts that can be mistaken for genuine cellular structure. Membranes undergoing rapid reorganization during exocytosis or endocytosis are particularly susceptible to these problems because their component phospholipids are not well fixed by aldehydes. As shown in Figure 10.5D, glutaraldehyde fixation during cortical granule exocytosis in sea urchin eggs results in formation of numerous large membrane blebs and vesicles that are not seen in rapidly frozen eggs. In addition, exocytosis in many cells is preceded by large clearings of membrane particles (arrow, Figure 10.5E) and adhesions between fusing membranes (arrows, Figure 10.5F) that are not seen in cryofixed tissues. In some cases, inadequate chemical fixation is followed by artifact production during dehydration, most commonly resulting in extraction of cellular components and fissures resulting from shrinkage. Dehydration can also result in artifactual fusions between organelle membranes, as illustrated for cortical granule membranes in Figure 10.5G (arrows). Yet other problems in chemical fixation are associated with specific organelles such as mitochondria and cytoskeleton, and these are highlighted below in our discussion of freeze substitution.

■ Freeze Substitution

Freeze substitution is the first of a number of ways in which a frozen tissue can be processed for electron microscopy. **Table 10.4** provides a summary and comparison of these techniques that can be referred to as each technique is discussed in detail.

The objective of freeze substitution is to dehydrate and fix the specimen at low temperature prior to embedding in resin at room temperature. The advantages of this process are that many of the artifacts caused by traditional chemical fixation and dehydration at room temperature (e.g., protein denaturation, shrinkage, membrane wrinkling, and leaching of cell components) are entirely eliminated. As outlined by the protocol in **Figure 10.6A,** the frozen specimen is first placed in liquid acetone at $-85°C$ or alternatively on acetone solidified at $-196°C$ by liquid nitrogen and allowed to sink into the acetone as it melts. The specimen is held at $-85°C$ for 24–48 hours while the ice in the specimen is slowly replaced by acetone and dehydration is complete. The specimen vial is then warmed to $-20°C$ for 2 hours to initiate fixation of the specimen by 1% to 2% osmium tetroxide contained in the acetone. Fixation does not occur at $-85°C$ during dehydration because the reaction rate at this temperature is very slow. After fixation at $-20°C$, the specimen is warmed to 4°C briefly and then washed and placed in acetone containing 1% uranyl acetate if en bloc staining is desired. Freeze substitution protocols such as these can be carried out automatically by dedicated units such as the Baltec FSU 010 (Figure 10.6B). The specimen is then rinsed in 100% acetone, embedded in epoxy resin, sectioned, stained, and viewed, as described in Chapter 3.

In Figure 10.6C, freeze substitution has been applied to the sporangium of the fungus *Allomyces*. The plentiful organelles in these rapidly dividing cells, including mitochondria, nuclei, vacuoles, and transport vesicles, are well preserved, as indicated by their smooth membranes, dense interiors, and lack of swelling or shrinkage. In contrast, the same cells, if fixed with glutaraldehyde (primary) and osmium tetroxide (secondary), exhibit membrane distortions and extraction of materials (Figure 10.6D). In this specimen the plasma membranes are undulating and the cells have shrunk and pulled away from one another leaving a gap between cells (arrow, Figure 10.6D). By contrast, in rapidly frozen specimens, plasma membranes are smooth and without shrinkage artifacts (arrow, Figure 10.6C). Likewise, vacuoles in the chemically fixed specimen (asterisk) have been cleared of content and are irregularly shaped whereas in the freeze-substituted specimens these organelles are filled and spherical. Mitochondria, in particular, are sensitive indicators of fixation quality. In freeze-substituted specimens, mitochondria typically have smooth membranes, a dense matrix, and an array of intact cristae (Figure 10.6E), while in poorly fixed specimens they can be totally disrupted with almost no internal structure (Figure 10.6F). Indeed, these

Table 10.4	Comparison of Chemical Fixation with Methods Used to Prepare Frozen Tissues for Electron Microscopy			
Method	**Fixation**	**Advantages**	**Disadvantages**	**Comments**
Fixation at ambient temperature	Glutaraldehyde, formaldehyde, osmium tetroxide	Least labor and expense; adequate preservation in most cases	Membrane distortions and fusion; extraction of soluble cell components	Cellular cross-sections through substantial layers of tissue
Freeze substitution	Glutaraldehyde or osmium tetroxide at low temperature	Standard resin embedding and thin sectioning; excellent staining and preservation of cell structures; minimal leaching of cell constituents	Freezing damage at greater depths; process relatively lengthy	Cellular cross-sections through a few layers of cells unless high-pressure freezing is used
Freeze fracture, direct observation by SEM	No fixatives or cryo-protectants necessary	Tissue is not altered by chemicals	No permanent specimen; surface and fractured views only	Elemental analysis of fresh tissue possible
Freeze fracture and replication, standard protocol	Glutaraldehyde plus glycerol cryoprotection	Expansive membrane fracture faces; optimal for study of membrane and membrane junction architecture; three-dimensional information available in a single replica	Some cellular structures hidden by ice; best interpreted in conjunction with thin sections; expensive instrumentation required	Both cellular cross-sections and membrane panoramas through substantial layers of tissue
Freeze fracture and replication of rapidly frozen tissues	No fixatives or cryo-protectants	No artifacts from chemical fixation or chemical cryoprotection; tissue processes halted within milliseconds; three-dimensional information available in a single replica	Freezing damage at greater depths; expensive instrumentation required	Both cellular cross-sections and membrane panoramas through a few layers of tissue unless high-pressure freezing is used
Quick freeze, deep etch, and rotary shadow	Glutaraldehyde, 15% methanol wash	Beautiful exposure of cytoskeleton, extracellular matrix, protein complexes, organellar matrix and cellular "machines"; three-dimensional information available in a single replica	Freezing damage at greater depths; expensive instrumentation required; occasional chemical fixation artifacts	Both cellular cross-sections and membrane panoramas; three-dimensional relationships visualized *par excellence*

organelles are considered to be the sentinels of fixation problems because their ultrastructure is extremely sensitive to alteration.

Although freeze substitution can produce the highest quality ultrastructure available, the procedure is also used to achieve excellent immunocytochemistry results at the electron microscopy level. Because the goal of immunocytochemistry is to label specific

A. Freeze Substitution Protocol

1. Freeze tissue by ultrarapid method.
2. Place in liquid nitrogen on top of frozen acetone containing 1% osmium tetroxide.
3. Hold at –85°C for 24 to 48 hours while acetone substitutes for ice.
4. Hold at –20°C for 2 hours while osmium tetroxide fixes tissue.
5. Rinse with cold acetone; enbloc stain with 1% uranyl acetate in acetone at room temperature in the dark for 2 hours.
6. Rinse with acetone twice, then infiltrate with resin:acetone ending with two changes of 100% resin. Place in molds and cure with heat.

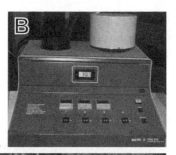

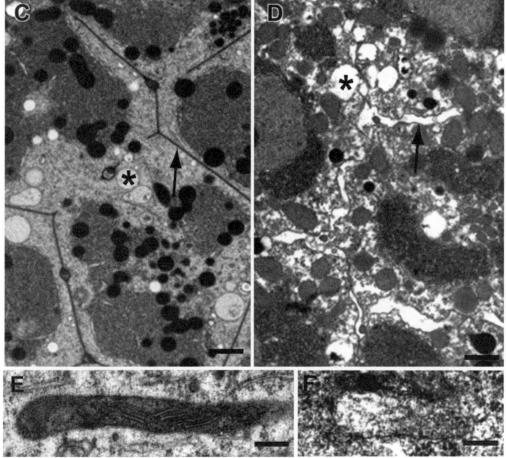

Figure 10.6

Freeze substitution technique. (A) Standard protocol for freeze substitution. (B) A Baltec freeze substitution unit providing automated processing. (C) Fungal sporangium of *Allomyces* frozen under high pressure then freeze substituted in acetone with OsO_4 fixation, then enbloc stained with uranyl acetate and embedded in Spurr's resin. Note the smooth plasma membranes (arrow) and excellent preservation of the nuclei. (D) Fungal sporangium of *Allomyces* fixed with glutaraldehyde, post-fixed with OsO_4, dehydrated in ethanol and acetone and embedded in Spurr's resin. Note the poorly preserved plasma membranes (arrow) and nuclear envelopes. Vacuole contents have been extracted and thus appear electron lucent while vacuole membranes have become disrupted (asterisk). (E) A freeze-substituted specimen exhibits mitochondria that are smoothly shaped, have a dense matrix and intact cristae. (F) Poor chemical fixation of a similar fungal sporangium leads to swollen mitochondria that are "blown out" with almost no internal structure. (C and D) Bars = 2 μm. (E and F) Bars = 0.5 μm.

proteins selectively within the cell, it is important that proteins remain in a close to native state and retain antigenicity—that is, reactivity with antibodies. This can be achieved by carrying out epoxy resin embedding and polymerization at low temperatures as well and is described fully in Chapter 13.

■ Freeze Fracture and Replica Production

The freeze-fracture technique, originally developed in the 1950s by Russell Steer and Hans Moor and practiced notably by Daniel Branton and Daniel Friend in the 1960s, was designed to visualize the architectural shapes of membranes. The reason for this is that frozen tissue, if cracked open with a razor blade, will split like a mineral through the weakest fracture plane. In cellular tissues, these fracture planes run through the hydrophobic interiors of membranes, as shown for erythrocytes in **Figure 10.7A.** Nearly all biological membranes consist of

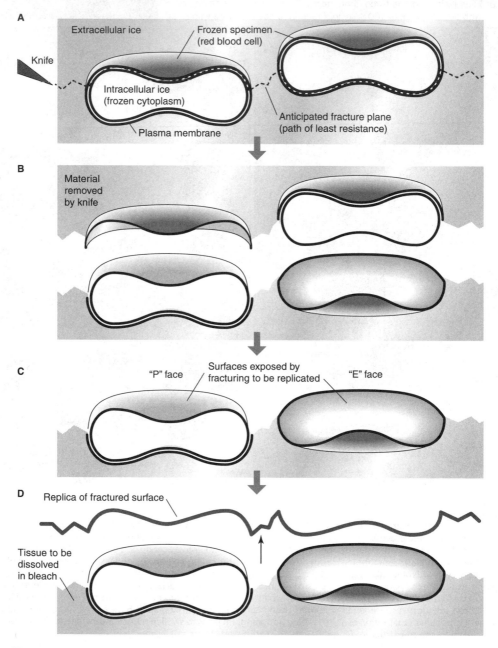

Figure 10.7
Freeze fracturing and platinum/carbon replication. (A) The plane of fracture takes the path of least resistance through the hydrophobic interior of the erythrocyte plasma membrane. (B) Material above the fracture surface is chipped off by the knife and removed during fracturing. (C) The fractured surface revealed exhibits membrane "P faces" having frozen cytoplasm behind them and "E faces" having extracellular ice behind them. (D) This surface is replicated by a thin layer of platinum/carbon, the tissue cleaned away by bleach and the replica viewed in the TEM.

a phospholipid bilayer that contains a hydrophobic interior lined by the fatty acid "tails" of phospholipids. In contrast, both outer surfaces of the membrane (extracellular and cytoplasmic in the case of the plasma membrane) are polar and are tightly bound to water (or to ice in the case of a frozen specimen). As the membrane is split down its interior, each monolayer is exposed to air and now offers a newly revealed interior surface that in contour is identical to the frozen membrane before splitting (Figure 10.7B). Usually, one of the monolayers is discarded with the ice chips that are cracked away from the specimen during fracturing; the opposing monolayer, however, is in full view waiting to be imaged. The monolayer revealed and imaged can be either an "outer" monolayer or an "inner" monolayer, and terminology has been developed to distinguish between the two. If the monolayer imaged has cytoplasm (or in older terminology "protoplasm") behind it, it is referred to as a "P face." If the monolayer has anything other than cytoplasm behind it (e.g., the extracellular space in the case of the plasma membrane), it is referred to as an "E face" (Figure 10.7C).

Examples of "P and E faces" are provided in **Figure 10.8A** and are diagrammed in Figure 10.8B to help demonstrate how this terminology is applied. This terminology is straight

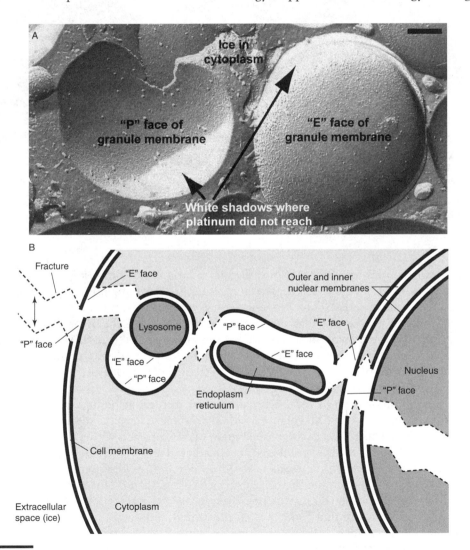

Figure 10.8
Faces of membranes in freeze fracture replicas. (A) Replica of a rat mast cell showing adjacent secretory granule membranes. Fracturing revealed the outer leaflet of one granule membrane (a P face) and the inner leaflet of the other granule membrane (an E face). Bar = 200 nm. (B) Diagram defining the membrane faces of both the plasma membrane and internal organelles.

Figure 10.9
Direct SEM observation of cryo-fractured cyanobacteria. Synechecystis pp PC 6803 was fractured, etched, and directly observed on an SEM fitted with a cryostage to keep the specimen at −120°C. In one cell, the fracture plane has torn through the cell wall and followed the cytoplasmic membrane (arrow). Another cell has been cross-fractured to reveal layers of thylakoid membranes important for photosynthesis (arrowhead). Bar = 0.5 μm.

forward for the plasma membrane but becomes a little more complex in the case of organellar membranes. For lysosome, secretory granule, Golgi apparatus, or endoplasmic reticulum membranes, the "outer" monolayer has cytoplasm behind it; thus, it is referred to as a "P face." The inner monolayer of these membranes has the organelle interior behind it (not the cytoplasm) and therefore is referred to as an "E face." Organelles that have double membranes surrounding them, such as mitochondria, chloroplasts, or nuclei, have multiple "P and E faces" possible, and these designations alternate from the cytoplasmic monolayer inward (Figure 10.8B).

One can visualize the frozen-fractured surfaces directly using high-resolution scanning electron microscopy (SEM) equipped with a cryostage to keep the specimen frozen during observation. This is best accomplished in an SEM equipped with a fracturing device and transporter such that the specimen can be fractured and immediately moved into the electron beam (Chapter 9 and **Figure 10.9**). More common is the preparation of a thin metallic "cast" or "replica" of the frozen surface that is then cleaned and visualized by normal transmission electron microscopy. Fracturing and replica formation are carried out in a "freeze-fracture unit" such as that illustrated and diagrammed in **Figures 10.10A** and **10.10B.** The instrument consists of a high vacuum chamber containing a cryostage to keep the specimen frozen, a microtome-mounted "knife" to fracture the specimen, and electron beam guns for deposition of platinum or tungsten and carbon in order to form the replica.

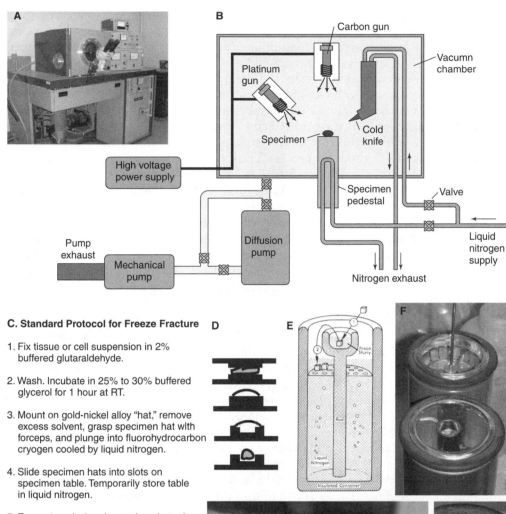

C. Standard Protocol for Freeze Fracture

1. Fix tissue or cell suspension in 2% buffered glutaraldehyde.

2. Wash. Incubate in 25% to 30% buffered glycerol for 1 hour at RT.

3. Mount on gold-nickel alloy "hat," remove excess solvent, grasp specimen hat with forceps, and plunge into fluorohydrocarbon cryogen cooled by liquid nitrogen.

4. Slide specimen hats into slots on specimen table. Temporarily store table in liquid nitrogen.

5. Evacuate unit chamber and cool specimen platform.

6. Vent chamber. Slide specimen/table onto chamber pedestal, and evacuate chamber. Cool knife. Set up electron beam guns.

7. Once specimen is at desired temperature, fracture specimen with knife, let etch if desired, and then form replica by firing electron beam guns.

8. Vent chamber, remove replica, and clean with bleach and distilled water.

Figure 10.10

The Balzer 400D freeze etch unit and sample preparation method. (A) Photograph of the unit. (B) Diagram showing the vacuum chamber, its pumping system, the electron beam guns with high voltage power supply, and the liquid nitrogen cooled knife and specimen pedestal. (C) Protocol for fixation, freezing, and fracturing of specimens. (D) Specimen "hat" geometries. (E and F) Diagram and photo of liquid cryogen bath for freezing. (G) Standard specimen hats with wells and the specimen table to which they are clamped. (H) Close up of the specimen table in (G). Leaf spring clamps hold down the "brim" of each specimen hat.

The standard protocol (Figure 10.10C) requires that 1-mm cubes of tissue or a cell suspension be fixed in buffered glutaraldehyde and then washed in buffer to remove excess fixative. The buffer and fixative conditions can be similar to that used for specimens to be embedded and sectioned. The tissue is then soaked in buffered 25% to 30% glycerol for cryoprotection. One important precaution is that tissue to be frozen and fractured must not be treated with osmium tetroxide either at a primary or secondary stage of fixation. Osmium tetroxide oxidizes membrane phospholipids, resulting in a bilayer that cannot be split easily by the freeze-fracture process.

The cryoprotected specimen is then frozen in place on a metal "hat" or planchet that is used as a carrier throughout the process. The standard "hat" is 3 mm in diameter and is made of a gold-nickel alloy (Figure 10.10D). Its diameter is the same as that of an EM grid on which the replica produced must eventually fit and its composition is designed to be resistant to chemical etching during cleaning of the replica in either sodium hypochlorite (bleach) or chromic acid. The hat incorporates a flange that will slide under a clamp to secure the specimen to a brass table during the process and a recess at the top in which the specimen is placed. Several hat designs are illustrated in Figure 10.10D, including a sandwich design (top) to crack the tissue apart without using a knife. This allows both fractured surfaces to be recovered and replicated, producing what are termed "complementary" replicas.

Before freezing the specimen, the cryogen bath must be prepared, as diagrammed and illustrated in Figures 10.10E and 10.10F. The cryogen is usually either Freon (R-12) or now (to prevent ozone depletion) a nonchlorinated refrigerant such as R-22 or R-134a. The cryogen is not flammable and therefore can be used outside of a fume hood. A gas at room temperature, the cryogen is sprayed into a 2-cm diameter well machined into a steel rod that sits in a dewar filled with liquid nitrogen. The cryogen liquifies and eventually freezes solid at liquid nitrogen temperatures ($-196°C$) but is reliquified by a brass rod just before the specimen is frozen. The hat with specimen is then thrust into the liquid cryogen, briefly swirled, and removed to a liquid nitrogen storage container.

A brass specimen table with a rotating cam securing device (arrow, Figure 10.10G) is then immersed in the liquid nitrogen bath, and the specimen hats are slid under a leaf spring clamp and secured (Figure 10.10H); the table with specimens is kept in liquid nitrogen until transfer into the freeze fracture unit. Meanwhile, the freeze fracture chamber has been prepared to accept the specimens. The chamber first is pumped to a vacuum of 10^{-5} Torr using a combination of a mechanical pump followed by an oil-diffusion pump in a manner identical to that used to evacuate a transmission electron microscope (**Figure 10.11A**). Next, the specimen pedestal within the vacuum chamber (Figures 10.11B and 10.11C) is cooled by internal plumbing carrying liquid nitrogen. The chamber is then vented and briefly brought back to atmospheric pressure, and the specimen table (with specimen hats) is slid into a slot in the top of the cold specimen pedestal and clamped down (arrow, Figure 10.11C). The chamber is then reclosed and pumped again to high vacuum.

The specimens must be brought to an exact temperature before they are fractured. The temperature of the specimen table is controlled by the supply of liquid nitrogen, which in turn is controlled by solenoid valves built into the plumbing circuit. Whenever a thermocouple located just below the specimen table detects that the temperature has risen above the set point, it triggers opening of these valves to let liquid nitrogen cool the specimen pedestal and table back down to the preset temperature. Specimen temperature is critical for good results because after the specimen is cracked open, the surface to be imaged is subjected to remodeling by both removal of ice from its surface and deposition of water vapor onto its surface. These processes must be controlled by vacuum and temperature. Ice is removed from the specimen by conversion directly to water vapor in a process called sublimation. Conversely, water

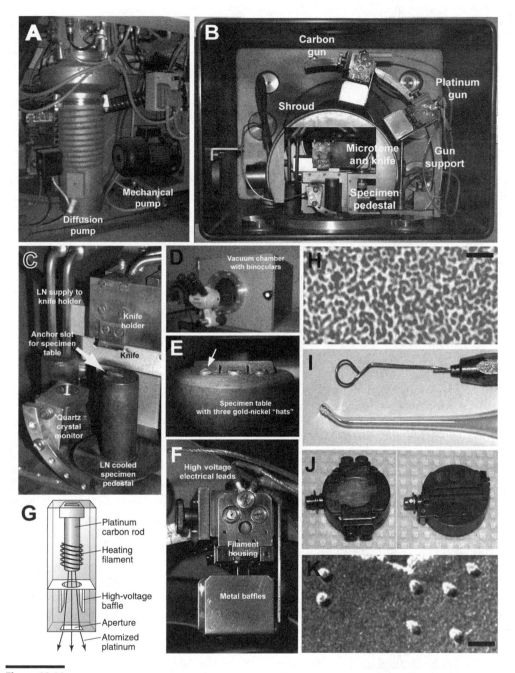

Figure 10.11
Balzer's freeze-fracture unit components. (A) The pumping system—an oil-diffusion pump backed by a mechanical pump. (B) Inside of the chamber showing relative positioning of the specimen, knife, and electron beam guns. A cold metal shroud covering the specimen pedestal has been cut away. (C) The specimen pedestal can be held at a set temperature and can be made to rotate for rotary deposition of platinum or carbon. A quartz crystal monitor is stationed next to the pedestal to measure platinum and carbon thickness during deposition. (D) The vacuum chamber with binoculars for observing the fracturing process. (E) Binocular microscope view of the specimen hats clamped onto the specimen table. A frozen droplet of isolated cells (arrow) is ready for fracturing. (F) An external view of an electron beam gun. (G) Diagram of the interior of an electron beam gun showing the heating filament, platinum/carbon rod and aperture through which the atomized platinum is shot. (H) At high magnifications, the replica exhibits microcrystalline plates of platinum. Bar = 10 nm. (I) Tools for transferring replicas during cleaning and washing. (J) Specimen tables used for rapid freezing planchets (left) and for producing "double" complementary replicas (right). (K) High-magnification micrograph showing membrane proteins seen as IMPs. Bar = 40 nm.

vapor in the "vacuum" can redeposit to form ice directly on the surface of the cold specimen. The rates of sublimation and redeposition are both temperature dependent. If the specimen is too cold (e.g., $-150°C$), redeposition wins out, and the freshly cleaved surface becomes artifactually decorated with lumps of ice. If the specimen is too warm ($-90°C$ to $-120°C$), sublimation wins out, and large amounts of ice can be removed from the specimen. Often a happy medium is achieved by setting the specimen temperature between $-120°C$ and $-130°C$. After this specimen temperature is achieved and the vacuum in the chamber has been pumped down to 2×10^{-6} Torr, fracturing can begin.

Fracturing of the specimen(s) is achieved by a knife mounted in a microtome assembly that is cooled by liquid nitrogen so that the fracturing process does not warm the specimen. The knife itself can be a razor blade or a metal blade specifically manufactured for the purpose. The objective is not to "slice" the specimen but rather to crack it open. Thus, the knife does not have to be exquisitely sharp as in thin sectioning. The fracturing process can be observed through a glass port using a binocular microscope (Figure 10.11D) that gives a view of the specimen table like that seen in Figure 10.11E. The knife microtome can be operated either manually or by motor (the former is preferred) so that the knife cracks the specimen then circles around for another pass at 50 μm deeper. Each pass should shave off only a modest layer of ice to avoid knocking the entire specimen off the hat. The last pass of the knife through the specimen should be done at moderate speed such that a nice spray of small ice chips can be seen as the knife passes.

Before making the last pass of the knife, the electron beam guns that will deposit the metal replica and back it with carbon must be made ready. As shown in Figures 10.11F and 10.11G, an electron beam gun consists of a central metal block containing a chuck that holds a carbon rod. The tip of the rod is surrounded by a coiled filament that heats the tip of the rod when a current is passed. Surrounding the rod and filament are a series of plates that accelerate atoms given off by the rod tip, which are then fired at the specimen through a baffle at the front of the gun. Because these atoms are charged and the potential difference between the rod and plates is on the order of 2000 volts, the atoms are both stripped off the rod one by one and are shot at the specimen at relatively high velocity. Adequate vacuum (less than 10^{-5} Torr) is required for this process because at atmospheric pressure, not only would the plates arc but also the atoms accelerated would collide with gas molecules before hitting the specimen. After hitting the cold fractured surface, the platinum atoms recrystallize, forming irregular plates that can be seen at high magnification (Figure 10.11H). This recrystallization process can lower the resolution of fine specimen features, and some workers use finer grained metals such as tungsten to avoid this.

The platinum gun contains a small pellet of pure metal recessed within the end of the carbon rod, and the atoms accelerated from this gun are a mixture of platinum and carbon. As shown in Figure 10.11B, this gun is usually positioned at a 45-degree angle to horizontal so that the platinum coating that forms the replica is not uniformly distributed over the fractured surface. Rather, the topology of this surface (indeed it is three dimensional!) will be highlighted by the pattern of platinum deposition—thick on the surfaces facing the gun and thin or nonexistent on surface facing away from the gun. As shown in Figure 10.8A, the curved surfaces of secretory granule membranes in a mast cell exhibit a pleasing gradient of platinum deposition that informs us of the granule shape. Because platinum is a heavy metal, it appears black in transmission electron microscopy (TEM) images. White areas in the image are "shadows"—areas receiving no platinum because surface features block the 45-degree spray of metal from the gun. In contrast, the carbon gun that provides a layer of carbon to hold the replica together is usually positioned directly over the specimen so that the entire fracture surface is coated evenly.

After the last cut is made, the platinum gun is fired for 10 to 20 seconds to apply enough platinum to coat the fractured surface to an average depth of 2 nm. This coating is thin enough not to obscure fine details of the specimen and thick enough to provide adequate contrast for TEM. The amount of platinum applied can be monitored accurately in real time by a quartz crystal monitor placed next to the specimen pedestal (Figure 10.11C). The quartz crystal changes capacitance and electrical resonance frequency as its mass is augmented by platinum deposition, which can be read out directly in units of platinum thickness. The carbon gun is fired immediately afterward to form a backing layer that is 10 times thicker than the platinum layer but that is nearly translucent in the TEM. The quartz crystal monitor is also used to record carbon deposition in real time.

Once the chamber is vented to atmospheric pressure, the replica with tissue is thawed and floated off onto the surface of a cleaning agent, usually bleach. The bleach dissolves the tissue over a period of 2 to 6 hours, and the replica, now clean, is floated on the surface of distilled water three times to remove any salts. Transfer of the replica between solutions is facilitated by a fire-polished glass bulb or wire loop (Figure 10.11I). The replica is picked up on a 100 or 200 mesh copper grid and is ready to view immediately because no staining is required. If TEM inspection reveals that tissue removal has not been complete (black residues), stronger cleaning agents, such as chromic acid or potassium permanganate, may be tried.

An alternative fracturing procedure that does not use a knife is the "double" or "complementary" replica technique. The specimen is frozen between two "sandwich" hats, and the sandwich is slid into a specimen table that operates like a book (Figure 10.11J, left). The book is closed as the sandwich is inserted under liquid nitrogen, each flange of the sandwich fitting in a slot on one side of the book. After the loaded table is in the chamber, the book is opened in order to crack the specimens. As a result, both fracture faces produced will be retained on opposite sides of the table and available for replication.

■ Imaging and Presenting Freeze Fracture Replicas

High-resolution TEM reveals not only the "shadowed" architecture of the "P and E faces" of membranes but also the fact that these faces are decorated with bumps termed "intramembrane particles" (IMPs) (see Figure 10.11K). These particles are now known to represent proteins embedded in the membrane, and the density and size of these proteins are characteristic of particular membranes and membrane specializations. Corresponding "holes" in the opposite monolayer from which these particles dislodged are not seen and are thought to have been filled in by deposition of small amounts of water vapor onto the frozen surface between fracturing and replication.

IMPs, in addition to their biological significance, can also be used to determine in which orientation a freeze fracture replica must be viewed. Our brain is very set in its way of interpreting highlighted objects three dimensionally. Objects highlighted from above, as if by a setting or rising sun, are interpreted as "sticking out," just like a landscape feature. Objects highlighted from below appear as depressions that are "recessed." IMPs, unlike any other structure, are always bumps that stick out; therefore, if a freeze-fracture image is oriented to allow proper interpretation of "innies" and "outies" IMPs will always be highlighted from above. A comparison of **Figure 10.12A** with Figure 10.12B demonstrates what misorientation can do. The images are identical except that Figure 10.12B has been inverted and is incorrectly oriented relative to the image in Figure 10.12A; many of the vesicle membranes that bulge out at the viewer in Figure 10.12A appear to recede from the viewer in Figure 10.12B. This effect is particularly obvious in the vesicle marked with an asterisk.

Freeze fracture images can be presented with shadows white and platinum black (as seen in the electron microscope) or photographically reversed wherein shadows are black and platinum deposits white (Figures 10.12C and Figure 10.12D). The later presentation requires an "internegative" (if traditional darkroom techniques are being used), but the final image has the advantage that it looks more like a SEM image, with white shades highlighting important structural features rather than information-poor shadows; however, photographic reversal to produce black shadows does not eliminate the need to orient the image

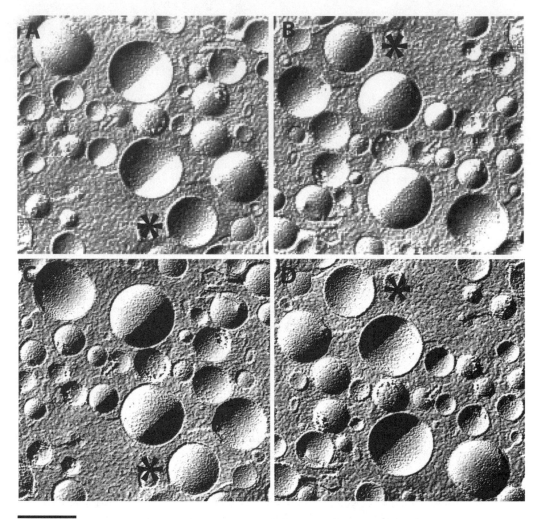

Figure 10.12
The importance of orientation and shading in freeze-fracture micrographs. (A) Freeze fracture micrograph that has been properly oriented so that illumination appears to come from the upper left. Platinum appears black and was deposited from the lower right. (B) Freeze-fracture micrograph of (A) rotated 180 degrees so that it is not properly oriented. Vesicles that appeared to be recessed in (A) now appear to bulge out (see asterisk for example). (C) Micrograph in (A) that has been photographically reversed—black for white. Platinum now appears white. Although the orientation of (C) is identical to the micrograph of (A), improper visualization of granule topography results from the photographic reversal. (D) Rotation 180 degrees of the micrograph in (C). This photographically reversed image is now oriented properly for correct topographical interpratation.

properly for correct three-dimensional interpretation. Figure 10.12D, like Figure 10.12A, is correctly oriented, whereas Figure 10.12C, like Figure 10.12B, is incorrectly oriented.

The freeze-fracture replica provides information that complements that obtainable in thin sections of the same specimen. As shown in **Figure 10.13A**, a thin section of a sea urchin egg cortex reveals microvilli at the egg surface and a single layer of secretory granules just under the plasma membrane. Also illustrated are two very different fracture planes that could be taken during the splitting of a frozen egg. The dashed line represents a "cross-fracture" that jumps from the plasma membrane to secretory granule membranes on its way into the egg interior, thereby producing a surface like that seen in the replica of Figure 10.13B. In contrast, the solid line represents a "grazing" fracture of the plasma membrane, producing surfaces like those seen in Figures 10.13C and Figure 10.13D. The "P face" in Figure 10.13C exhibits a regular array of "stumps" where microvilli were snapped off during the fracturing process.

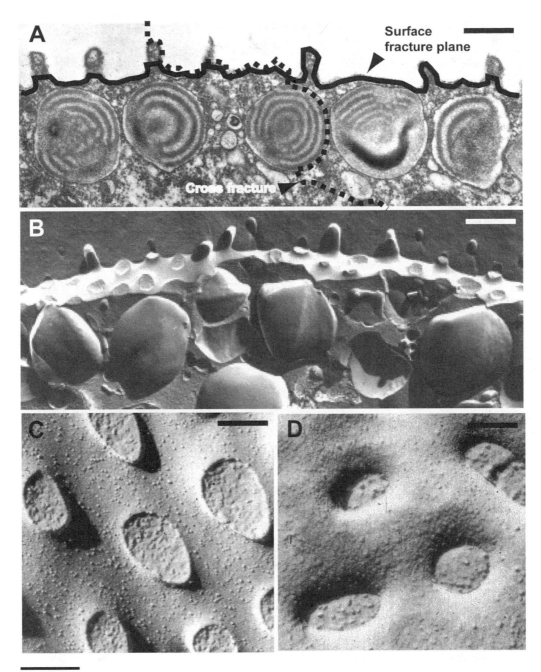

Figure 10.13
Comparison of thin sections and freeze fracture replicas of the sea urchin egg cortex. (A) Thin section of a sea urchin egg cortex reveals a plasma membrane with microvilli and secretory granules with an onion-like internal substructure. The hypothetical path of a superficial fracture through the plasma membrane (solid line) and a cross-fracture splitting into the cell interior (dashed line) are illustrated. (B) Freeze-fracture replica of the sea urchin egg cortex showing a cross-fractured specimen. (C) Superficial fracture of the plasma membrane showing an array of stumps where microvilli were removed. This is a "P face." (D) Superficial fracture of the plasma membrane showing indentations at the base of microvilli. This is an "E face." (A and B) Bars = 0.5 μm. (C and D) Bars = 0.2 μm.

The "E face" in Figure 10.13D reveals depressions wherein the fracture plane traveled a small distance into the interior of each microvillus. Thus, in the thin section, we can see the interior granule substructure not visible in replicas, whereas in replicas, we can more easily understand the three-dimensional shape of the plasma and granule membranes.

■ Live Cells Can Be Rapidly Frozen and Freeze Fractured

Rapidly frozen cells can be fractured and replicated directly with no intervening treatment by fixatives or cryoprotectants. Theoretically, such cells are still alive despite the fact that their biological processes have been halted within milliseconds by freezing. In fact, if one is rapid freezing and fracturing muscle, care must be taken that the muscle does not contract with thawing, thereby ripping the replica to shreds! Replicas of live cells, if properly frozen, reveal smooth membranes and a wealth of cytoplasmic detail, as exemplified by Figures 10.8A, 10.13B, and **10.14**. Thus, the superiority of cryopreservation demonstrated in freeze substitution (Figure 10.6C) is also apparent in freeze fracture replicas.

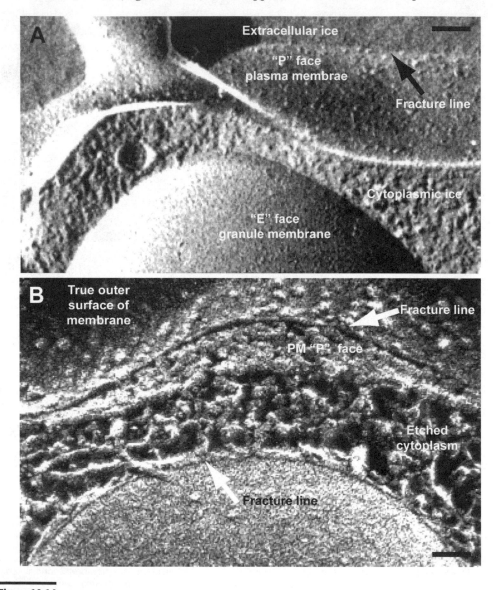

Figure 10.14
Effects of etching on freeze-fracture specimens. (A) Plasma and secretory granule membranes in a rat mast cell. This specimen has not been etched and both extracellular ice and membrane faces are smooth. No true outer surfaces of membranes are seen, only the hydrophobic interior P and E faces. (B) Plasma and secretory granule membranes in a mast cell etched extensively. Extracellular ice has been sublimated to reveal the true outer surface of the plasma membrane. The edge of the outer monolayer can be seen at the fracture line. Cytoplasmic ice has been sublimated to reveal a dense array of cytoskeletal filaments between the plasma and granule membranes. (A and B) Bars = 100 nm.

■ Quick Freezing, Deep Etching, and Rotary Shadowing

Previously, we noted that the degree of etching can be readily controlled by the specimen temperature during fracturing and, as detailed in **Table 10.5,** can range from no etching to very "deep" etching to reveal hidden structures. No etching (specimen fractured at −130°C) results in both ice and membrane surfaces appearing relatively smooth. For example, in Figure 10.14A, a secretory granule of a mast cell is separated from the plasma membrane above by a thin layer of cytoplasm. The cytoplasm is rough because of inclusions, but the extracellular ice above the plasma membrane fracture line is completely smooth and shows no sign of etching. A small amount of etching is often desirable to roughen the ice surfaces and thereby distinguish smooth membrane fracture faces from cytoplasm or extracellular ice. Such an effect is seen in Figure 10.5F and is obtained by fracturing the specimen at −110°C and allowing 20 seconds for etching before firing the electron beam guns to produce the replica. This method can be used routinely even in fixed and glycerol-cryoprotected specimens to distinguish ice from membrane.

In contrast, much deeper etching can produce views of cellular structures that would otherwise be embedded in ice and hidden. In Figure 10.14B, a mast cell specimen that has been etched extensively (3 minutes at −100°C), exhibits a cytoplasm lying between the plasma membrane and secretory granule membrane that is packed with microfilament

Table 10.5	Comparison of Etching Conditions for Freeze Fracture			
Method	**Tissue Preparation**	**Fracturing Conditions**	**Etching Conditions**	**Advantages/ Disadvantages**
Freeze fracture, no etching	Chemical or physical fixation	Temperature −130°C; vacuum 2×10^{-6} Torr	No period of etching between fracture and replication	No removal of ice; only membrane faces visible
Freeze fracture, modest etching	Chemical or physical fixation	Temperature −115°C; vacuum 2×10^{-6} Torr	20-second etching period	Ice surfaces roughened and easily distinguished from membrane faces
Freeze fracture, deep etching	Chemical fixation	Temperature −90°C to −100°C; vacuum 2×10^{-6} Torr	1- to 10-minute etching period	Removal of up to 0.5 μm of ice; reveals three-dimensional ultrastructure of cell components
Freeze drying	Chemical fixation or freezing in water	No fracturing unless to remove excess ice	5- to 15-minute etching period; temperature −85°C; vacuum 2×10^{-6} Torr	Removal of up to 1 μm of ice; reveals three-dimensional ultrastructure of cell fragments, cell cortices, isolated organelles, or macromolecular complexes

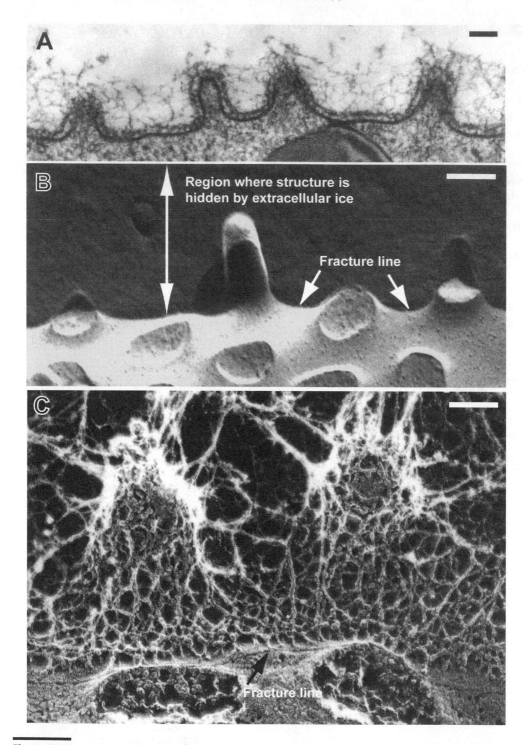

Figure 10.15
Deep etching reveals the three-dimensional structure of the sea urchin egg extracellular matrix. (A) Thin section of a freeze-substituted sea urchin plasma membrane showing microvilli that are covered with wisps of a fibrous extracellular coat. (B) Freeze fracture replica of the sea urchin egg plasma membrane showing that extracellular ice hides all biological structure on the egg surface if the specimen has not been etched. (C) The surface of a deep-etched sea urchin egg reveals a beautiful three-dimensional network of fibers appearing much like a fish net-draped over the microvilli. All structures above the fracture line (arrow) would have been covered with ice if the specimen were not deep etched. (A, B) Bars = 0.2 μm. (C) Bar = 0.1 μm.

networks. In addition, sublimation of extracellular ice has exposed the true surface of the plasma membrane that was hidden in specimens not etched (Figure 10.14A).

Deep etching can also provide dramatic results in visualizing extracellular matrix fibers. In **Figure 10.15A,** a thin section of a freeze-substituted sea urchin egg exhibits tantalizing evidence of a lacy, net-like extracellular matrix covering the plasma membrane. These wisps of fibers represent the vitelline layer, a sperm-binding structure important to fertilization physiology. The replica of Figure 10.15B shows no evidence of this structure because the fractured surface was not etched and the extracellular matrix is embedded in ice; however, in "deep-etched" specimens (Figure 10.15C), up to 0.5 μm of ice is removed before replication to expose the true glory of this fishnet-like matrix draped over each of the microvillar surfaces. Careful inspection of this micrograph reveals a subtle fracture line (arrow) that demarcates the point at which the fracture plane emerged from the hydrophobic interior of the plasma membrane to enter the extracellular ice.

Extensive sublimation of ice (e.g., 8 minutes at −95°C) can actually freeze dry specimens before replication if they are sufficiently thin. The results can be remarkable, as attested to by the images of a neutrophil sheared open by a stream of buffer to reveal the inside of the plasma membrane (**Figure 10.16A**). These replicas provide a bird's-eye view of microfilament networks and clathrin-coated vesicles (arrow and arrowheads, Figure 10.16B) and demonstrate that deep etching and freeze drying can be powerful tools by which to study the cytoskeleton.

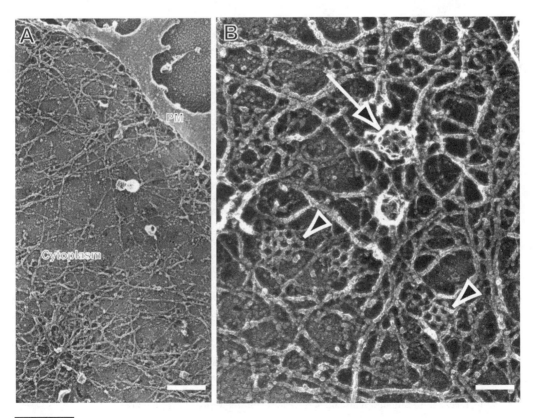

Figure 10.16
Platinum replicas of freeze-dried cell cortices from neutrophils. (A) Aerial view of the neutrophil cortex showing the inner surface (cytoplasm) and outer surface (PM) of the plasma membrane. Bar = 0.5 μm. (B) At high magnification, one can observe a network of microfilaments and clathrin baskets (arrowheads) at various stages of pinching off of plasma membrane to form endocytotic vesicles (arrow). Bar = 150 nm.

One problem that must be overcome in these procedures is that during etching, ice is easily removed, but salts contained in the ice are not. They remain behind to obscure the underlying biological structure. This problem is solved by the typical quick-freezing, deep-etching, and rotary-shadowing protocol, which calls for chemical fixation of the tissue with glutaraldehyde, washing to remove the fixative, briefly exposing the tissue to 15% methanol in distilled water to remove salts, then rapidly freezing the specimen on a metal-mirror freezing machine (see above). The frozen specimen is then fractured between −90°C and −100°C and allowed to etch 3 to 10 minutes while the liquid-nitrogen cooled knife holder is "parked" above the specimen. The cold knife holder adsorbs the water vapor produced during etching to avoid its being redeposited on the specimen surface. The "deep-etched" specimen is then rotated at 15 rpm as platinum and carbon are applied so that even deeper "nooks and crannies" of the specimen can be coated. This technique has become the "Rolls Royce" of methods by which to observe the intricate three-dimensional architecture of cellular organelles and protein complexes and its use in elucidating macromolecular structure is further described in Chapter 15.

References and Suggested Reading

Cavalier A, Spehner D, Humbel BM, eds. *Handbook of Cryopreparation Methods for Electron Microscopy (Methods in Visualization)*. Boca Raton, FL: CRC Press, 2008.

Robards AW, Sleytr WB. *Low Temperature Methods in Biological Electron Microscopy*. Amsterdam: Elsevier Science, 1985:572.

Rapid Freezing Techniques

Chandler DE. Comparison of quick-frozen and chemically fixed sea urchin eggs: structural evidence that cortical granule exocytosis is preceded by a local increase in membrane mobility. *J Cell Sci*. 1984;72:23–36.

Gilkey J, Staehelin LA. Advances in ultrarapid freezing for the preservation of cellular ultrastructure. *J Electr Microsc Tech*. 1986;3:177–210.

Jimenez N, Humbel BM, van Donselaar E, Verkleij AJ, Burger KN. Acalr discs: a versatile substrate for routine high-pressure freezing of mammalian cell monolayers. *J Microsc*. 2006;221:216–223.

Menco BM. A survey of ultra-rapid cryofixation methods with particular reference on applications to freeze-fracturing, freeze-etching, and freeze substitution. *J Elect Microsc Tech*. 1986;4:177–240.

Neuhaus EM, Horstmann H, Almers W, Mariak M, Sodati T. Ethane-freezing/methanol fixation of cell monolayers: a procedure for improved preservation of structure and antigenicity for light and electron microscopies. *J Struct Biol*. 1998;121:326–342.

Roberson RW, Chandler DE. Rapid freezing of cells and molecules. In: Spector DL, Goldman RD, Leinwand LA, eds. *Cells, A Laboratory Manual, Subcellular Localization of Genes and Their Products*. Plainville, NY: Cold Spring Harbor Laboratory Press, 1998.

Segui-Simarro JM, Austin JR, White EA, Staehelin LA. Electron tomographic analysis of somatic cell plate formation in meristematic cells of *Arabidopsis* preserved by high-pressure freezing. *Plant Cell*. 2004;16: 836–856.

Freeze Substitution

Giddings TH. Freeze substitution protocols for improved visualization of membranes in high-pressure frozen samples. *J Microsc*. 2003;212:53–61.

Matsko N, Mueller M. Epoxy resin as fixative during freeze-substitution. *J Struct Biol*. 2005;152:92–103.

McDonald K. Freeze substitution. In: Spector DL, Goldman RD, Leinwand LA, eds. *Cells, A Laboratory Manual, Subcellular Localization of Genes and Their Products*. Plainville, NY: Cold Spring Harbor Laboratory Press, 1998.

Shiurba R. Freeze substitution: origins and applications. *Int Rev Cytol*. 2001;206:45–96.

Freeze-Fracture Techniques

Branton D. Fracture faces of frozen membranes. *Proc Natl Acad Sci USA*. 1966;55:1048–1056.

Chandler DE, Heuser JE. Arrest of membrane fusion events in mast cells by quick freezing. *J Cell Biol*. 1980;86:666–674.

Gilkey JC. Freeze fracture and freeze fracture cytochemistry. In: Spector DL, Goldman RD, Leinwand LA, eds. *Cells, A Laboratory Manual, Subcellular Localization of Genes and Their Products*. Plainville, NY: Cold Spring Harbor Laboratory Press, 1998.

Orci L, Perrelet A. *Freeze-Etching Histology. A Comparison Between Thin Sections and Freeze-Etch Replicas*. New York: Springer-Verlag, 1975:168.

Rash JE, Hudson CS. *Freeze Fracture: Methods, Artifacts and Interpretations*. New York: Raven Press, 1979:204.

Robenek H, Severs NJ. Recent advances in freeze-fracture electron microscopy: the replica immunolabeling technique. *Biol. Proced. Online*. 2008;10:9–19.

Severs NJ. Freeze fracture electron microscopy. *Nat. Protoc*. 2007;2:547–576.

Steere RL. Electron microscopy of structural detail in frozen biological specimens. *J Biophys Biochem Cytol*. 1957;3:45–60.

Walther P. Recent progress in freeze-fracturing of high-pressure frozen samples. *J Microsc*. 2003;212:34–43.

Deep-Etching and Rotary Shadowing of Cells

Chandler DE. Rotary shadowing with platinum-carbon in biological electron microscopy: a review of methods and applications. *J Elect Microsc Tech*. 1986;3:305–335.

Hanson PI, Stahl PD. From the neuromuscular junction to cellular architecture and beyond: commentary on 30 years of imaging by John E. Heuser. *Eur J Cell Biol*. 2004;83:229–242.

Hawes C, Martin B. Freeze-fracture deep-etch methods. *Methods Cell Biol*. 1995;49:33–43.

Heuser J. Deep-etch EM reveals that the early poxvirus envelope is a single membrane bilayer stabilized by a geodetic "honeycomb" surface coat. *J Cell Biol*. 2005;169:269–283.

Heuser JE. Quick-freeze, deep-etch preparation of samples for 3-D electron microscopy. *Trends Biochem Sci*. 1981;6:64–68.

Severs N, Shotton D. *Rapid Freezing, Freeze Fracture and Deep Etching*. New York: Wiley-Liss, 1995:372.

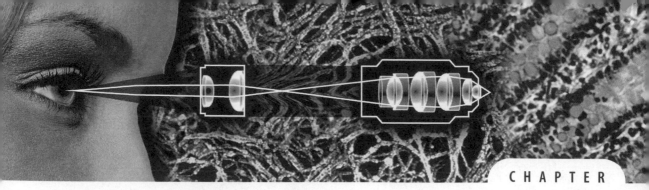

Video Microscopy and Electronic Imaging

A typical electronic imaging setup for light microscopy includes an upright or inverted microscope, a video or slow-scan charge-coupled device (CCD) camera, a storage device such as a computer or VCR, and a video/image monitor (**Figure 11.1**). Other common accessories include a camera controller, a signal processor, and a computer monitor if computer control of the camera is desired. The system may be either analog or digital, but because all image information from the camera is initially analog, there must be analog to digital conversion at some point in the system if digital data storage or analysis is desired.

The microscope has the usual function of image formation and is the first determinant of final image quality. For that reason, all of the considerations presented in Chapter 6 for obtaining the highest quality image apply. The camera is responsible for taking the light image and converting it into an analog signal by standard conventions for interrogating the image point by point. The camera may also be responsible for converting the analog signal into a digital signal in some systems. The image signal is carried to a signal-processing device or storage device using either one shielded coaxial cable for "composite" signals or three cables for color-separated (RGB) signals. Coaxial cables are used throughout the system for communication between devices. The signal processor edits the signal "on the fly" (i.e., during image acquisition in real time). Signal editing may include changes in contrast and brightness, averaging, smoothing, noise reduction and superimposition of timers, time codes, labels, or scale bars. If the analog signal is to be converted to digital at this point, an image-processing card within the computer carries out editing functions, as well as "A to D" conversion. Storage of analog signals is usually on magnetic tape while storage of digital signals can be on either magnetic tape or magnetic disk (i.e., the computer hard drive). Both analog and digital signals can be viewed by monitor either coincident with capture or at a later time.

One usually associates the digital age with advent of the personal computer around 1980, but our story really starts 40 years before with the early technology of television in the 1940s. Until that time, images, whether still or "moving," had to be placed on photographic film for playback at a later time.

Photography, although a major advance in image capture, still had a tremendous amount of labor associated with it, as well as a number of technical drawbacks. For still images, film or emulsion-coated glass plates had to be developed and stored, and then the images printed onto photographic paper. This process required a darkroom with chemicals, tanks, sinks, an enlarger, and special lighting, as well as hours of labor under red light or no light to get a print with just the right contrast and density (see Chapter 2).

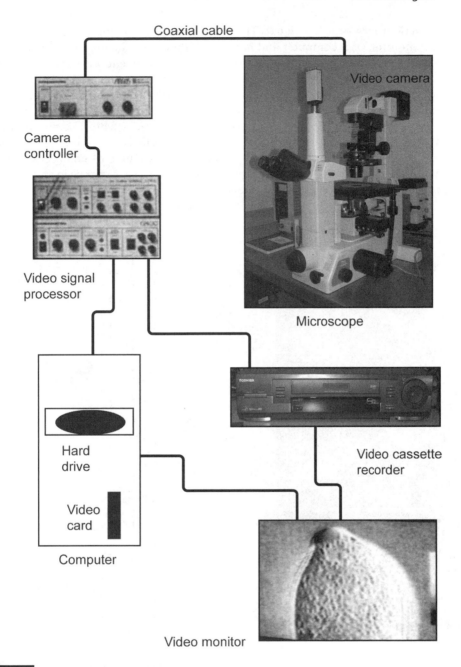

Figure 11.1
Components of a video microscopy system.

Moving pictures required a specialized high-speed motion picture camera, entire reels of film, and an appropriate projector and screen. Analysis of distance, area and shape, or pattern (morphometrics) required individual prints, regardless of whether the images were still or moving.

All of this changed when television technology developed electronic video imaging for purposes of broadcast and reception. Video cameras and now digital cameras convert light information into analog or digital electrical information, and this information is either stored on magnetic tape or disk by recorders or broadcast as long wavelength electromagnetic waves—the TV signal. The images can then be "played back" by a TV, a VCR and monitor, or a computer and monitor without ever having been printed onto film or paper. In the next section, we see that most of the conventions used in electronic imaging are those developed by the television and camera industry.

Video technology has been a tremendous step forward in being able to record the movement, shape changes, and behavior of cells and tissues. What once had to be described in words such as "cell migration to one pole of the embryo," "whip-like motion of the flagellum," and "vesicle movement along the microtubule cytoskeleton" can now be described in both moving images and in quantitative measures of change with time, including linear velocities, angular velocities, trajectories and rates of change in area, volume, and shape. An example of such studies aided by video microscopy is the swimming of sperm through the viscous jelly surrounding frog eggs (**Figure 11.2**). In Figures 11.2A–11.2C,

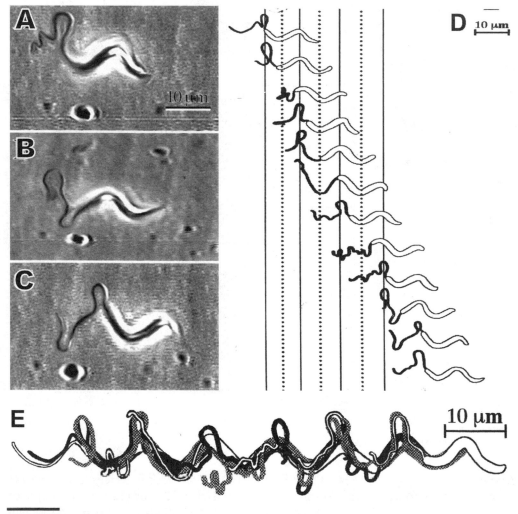

Figure 11.2
Video capture of sperm motility. (A–C) Video frames captured 66 msec apart showing the flagellar movement of a *Xenopus laevis* sperm. Note the progress of the sperm relative to a particle that is out of focus. (D) Tracings of the sperm showing that the hairpin bends formed in its flagellum propagate from base to tail as the sperm moves forward but remain stationary relative to the surrounding jelly. Full lines mark the position of upward turned bends and dashed lines mark the position of downward bends. (E) Superimposing these traces shows clearly that the sperm traces out a helical path. Bars = 10 μm.

three frames taken 66 msec apart in time, show that the sperm is propelled by the formation of hairpin bends in the flagellum that are generated at the base and propagate to the tip of the flagellum. As shown in Figure 11.2D, tracings of the sperm head and flagellum show that the position of hairpin bends remains relatively constant within the jelly and that propagation of the bend actually drives the sperm forward. Each solid vertical line is a position in the jelly corresponding to an "upward" bend, and each dotted line is a position corresponding to a "downward" bend. Superimposition of these traces (Figure 11.2E) shows clearly that the bends remain relatively constant in position while the sperm moves forward. What is not as clear from diagrams, but is very clear from the video track played in real time, is that this sperm follows a helical path in cutting through the jelly. In still images this can only be hinted at by the fact that the sperm head is helically shaped and the hairpin bends appear to arise on alternating sides of the flagellum. The flagellar movement likely contains a rotational component that may require speeds higher than "video rate" to capture. Indeed, some electronic cameras can capture up to 500 frames per second if a restricted field of view is tolerated.

■ The Video Signal

Electronic imaging requires that a two-dimensional image be turned into a one-dimensional signal. This is most easily achieved if the image is scanned point by point and the light intensity is recorded as a series of intensity values versus time. Thus, the intensity data, usually expressed as a voltage, are one dimensional. As long as the image points are scanned in the same order each time and the rate at which they are scanned is agreed on in advance, the two physical dimensions of an image, x and y, can be compressed into a single dimension in the signal—that of time. For example, let us make three "horizontal" line scans across the image of a cell, as shown in **Figure 11.3A**. In each scan, the voltage would indicate the light intensity detected at each point in the scan (black = 0, white = 0.7, gray levels in between) while distance along the x axis of the image would be proportional to time. The position of each scan on the y axis would simply be denoted by the order in which the scan appeared. A number of conventions would have to be agreed on in advance. For a start, these conventions might include (1) the black and white voltage levels, (2) the order and duration of horizontal scans, and (3) the number of horizontal scans constituting an image. These and all other conventions together would be referred to as a "format."

In the early days of broadcast TV, a standard scanning pattern was adopted: A single image, having a width to height ratio of 4 to 3, would be scanned left to right using 525 horizontal lines (Figure 11.3B). Between each horizontal line scan there would be a short period in which no image data were collected to allow the scanning mechanism to get from the end of one line scan to the beginning of the next. This was referred to as the horizontal flyback period. After the 525th line scan, the image data were complete, and one could go ahead and acquire the next image. To do this, the scanning mechanism would have to jump from the bottom of the image to the top of the next image. This data-free period was called the vertical flyback period.

In this manner, a nonstop series of images could be transmitted by a single voltage versus time signal. It was agreed that because the TV transmission region of the electromagnetic spectrum is not infinitely broad and the simultaneous transmission of numerous signals (each representing a TV station) might be desirable, each signal would be allowed a total frequency range (bandwidth) of 6 MHz (6 million cycles per second) and no more. (As you know, it is really a mess to listen to two radio or TV stations at the same time). This finite bandwidth represents a limit on the rate of information transfer for any one signal and is equivalent to 30 standard images or frames per second. The 30-frames-per-second standard is now referred to as the "video rate" or "real-time video" because it has been

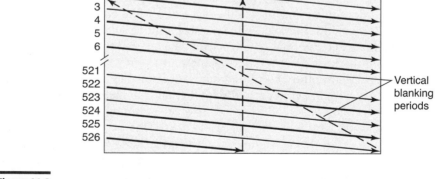

Figure 11.3
The scanning pattern of an NTSC video signal. (A) Three consecutive horizontal scans of an image result in a voltage versus time signal, with white represented by a high voltage and grays or black represented by lower voltages. These scans can be joined together with a horizontal blanking period between each scan to accommodate for the time it takes to get from the end of one scan to the beginning of the next (the "flyback" period). (B) Scanning pattern of an NTSC video signal. Every 1/30th of a second, a frame consisting of a series of 525 horizontal scans is performed by the camera. Each horizontal scan (solid arrows) is followed by a horizontal "flyback" period (dashed lines). At the end of each frame, there is a vertical "flyback" period (dashed arrow). The scan data from each frame are divided into odd fields containing odd numbered scans and even fields containing even numbered scans to make the NTSC signal. The playback device (usually a television), by "reading" the NTSC signal, projects 60 fields per second (thereby eliminating flicker) that, if properly positioned (interlaced), contain the same information as 30 frames per second.

used as a standard for 60 years. However, it is actually an arbitrary number that was arrived at because of a limitation placed on bandwidth by the U.S. government.

Surprisingly, the 30-frames-per-second standard does not match up with human physiology. If our brain is subjected to 30 frames per second, the picture appears to flicker like the old movies that had to be speeded up to make them look acceptable. In order to avoid this, the frame rate has to be increased to greater than 45 frames per second; at these speeds, the brain fuses one image with the next without the perception of flicker. How was this problem solved? The world of cinema solved it in a very clever way. Although motion picture cameras were designed to capture only 24 frames per second on film, the image stream was expanded to repeat each frame twice for a total of 48 frames per second! The television industry took a different route. Because broadcasting more than 30 frames per second was not permitted, it was decided that the information in each frame would be divided up into two "fields"; therefore, our brain would be presented with 60 fields per second, and flicker would be eliminated. Each field would contain only half of the image information, but each half would have to be gathered from the entire image without duplication. The simplest solution was to present an "odd field" and then an "even field." The odd field would contain all of the odd numbered horizontal scans (1 to 526), whereas the even field would contain all of the even numbered horizontal scans (2 to 526) (Figure 11.3B).

Things happen fairly fast in a video signal. Because there are 30 frames per second and 525 horizontal scans each frame, there is a total of 15,750 horizontal scans each second. This means that each horizontal scan only lasts 63.4 microseconds! As shown in **Figure 11.4A,** the voltage in each scan ranges from 0.7 volt, which represents white, to 0 volts, which represents black, with all voltages in between representing various shades of gray. (Color did not exist for the first 10 years of broadcast TV.)

Between each horizontal scan there is a "blanking period," in which the signal is arbitrarily set to just less than 0 volts. This period obliterates the horizontal flyback as well as the last few scans of the previous frame and the first few scans of the next frame. This is just as well because the image data during this period are either noisy or entirely lacking. Another advantage of including a blanking period between each scan is that additional technical signals can be edited in. A critical signal during the blanking period is the presence of a synchronization pulse that is seen as a square wave during which the voltage drops to −0.29V. This pulse marks the beginning of each horizontal scan; in this manner, the TV is able to resynchronize its "clock" every 64 microseconds to that of the camera that created the video signal. Just imagine if these synchronization pulses did not exist. Over the course of a 2-hour movie, the TV clock would have to keep time with an accuracy of 30 microseconds or face the disaster of having its picture halfway off the screen. In reality, the needed accuracy of 1 microsecond in 2 hours is found only in the most expensive clocks that would far out price the cost of the TV! Synchronization pulses avoid the need for every single piece of video equipment to have such an accurate clock, whether it is a camera, a monitor, or a VCR. Instead, each piece of equipment has a clock of moderate price that suffices to keep everything within an error of a few seconds every hour. Another important calibration signal that occurs during the blanking period is the color burst. Color was added to the standard video signal as an afterthought because its technical feasibility came along years after the standard video signal had been invented. Color information was "added on" as a high-frequency subsignal that is out of phase and interspersed with the main signal frequencies containing intensity information. It is very easy to separate the out-of-phase, high-frequency color signal using comb and capacitance filters. Thus, separation of the color and luminance information is carried out in real time, and these are processed separately in televisions and VCRs, as described later in the chapter.

A. Composite monochrome video signal format

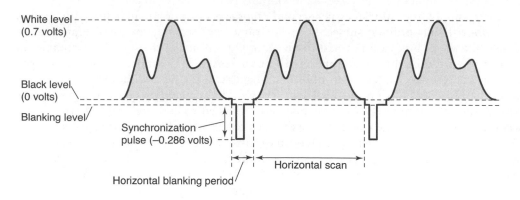

B. Composite color video signal format and decoding

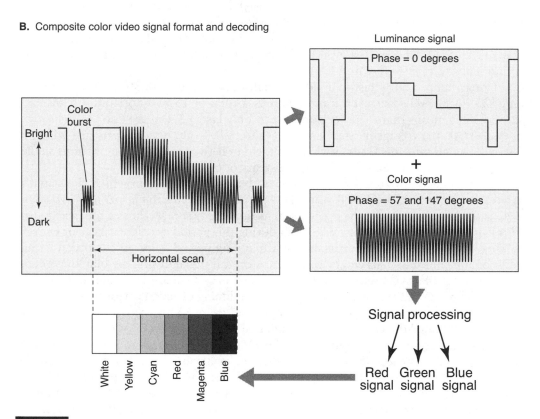

Figure 11.4
Black and white and color NTSC video signals. (A) A composite black and white video NTSC signal. Two horizontal scans are each preceded by a blanking period, during which the signal voltage is set to zero. Also present during each horizontal blanking period is a square wave synchronization pulse of −0.29 volt. During each horizontal scan, black is represented by 0 volt, white by 0.7 volt, and gray levels in between. (B) The color NTSC signal. The NTSC color signal has a high-frequency color information carrier that is out of phase with the luminance (intensity) signal (jagged lines in horizontal scan). Also present is a color burst during each horizontal blanking period, a standardization signal for color frequencies and phase. The television receiver separates the luminance and color signals using a comb filter. The color and luminance signals are then reprocessed to produce three separate signals—red, green, and blue—that contain both color and luminance information. Each color signal drives a separate electron beam gun that excites only screen phosphors of the correct color.

For a test pattern that contains vertical zones of light intensity ranging from white to black, a horizontal scan in a standard video signal will appear as a series of stair steps (the luminance signal in Figure 11.4B). If one adds color to these zones without altering the intensity differences, a high-frequency color signal is superimposed on this stair-step pattern. The frequency of this color signal is a code for the color to be displayed at that particular point in the horizontal scan, as seen in our test pattern that is divided into zones of different colors (the composite signal in Figure 11.4B). Calibration of these frequencies and their phase difference is essential to correct separation and interpretation of color by the receiver. The color burst during the blanking period before each horizontal scan is a test pattern laid down by the camera to inform and calibrate the receiver. None of the synchronization and calibration information transmitted during the blanking period is displayed on the screen and is for internal use only.

The standard video signal then is composed of intensity information, color information, blanking periods, and synchronization signals all integrated into a single voltage versus time format that can be carried by a single cable. For this reason, it is referred to as a "composite" video signal. As we see later here, the color information in this signal is relatively low in spatial resolution, while the luminance information has a relatively high spatial resolution. This usually does not matter because it matches the way our brain works.

Our brain uses luminance information to define objects, their detail, and the movement and depth of objects in our visual field. Indeed, the rod photoreceptors in our retina that detect luminance (light intensity) far outnumber the cone photoreceptors that are specialized for detecting color. An everyday application of this fact is that children's drawings look far more recognizable in coloring books. Because the child has colored outside the black lines really makes little difference, as it is the black lines, and not the color, on which our brain places priority. Nevertheless, in some applications, high-resolution color can be important. If that is the case, a video signal can be generated separately for each primary color so that each color has spatial resolution equivalent to the luminance information. This necessitates a separate cable for each color—red, green, and blue—and for that reason is referred to as an RGB signal or a "component" signal. RGB signals are typically used in professional-quality video equipment, especially in digital formats, including the HDTV format. RGB component signals are also used in high-quality, consumer-grade equipment, including that used for bioimaging applications.

Because the standard video signal format has been largely created and imposed on the broadcast industry by government regulation, it should not be surprising that the "standard" video signal is different from one part of the world to another. The signal that we have discussed to this point (Figure 11.4A) is that developed by the National Television Standards Committee of the U.S. government. For short, it is referred to as an NTSC signal. For 60 years, this signal has been used exclusively in the United States and in countries with their communications industries integrated with that of the United States (e.g., Canada, Mexico, and Japan). Other parts of the world use different standards, as shown in **Table 11.1**. The PAL signal has been used as a standard throughout Europe, China, and Australia, while the SEACAM signal has been used in France, former Soviet Union countries, and parts of Africa. These standards use about the same bandwidth. For example, the PAL and SECAM signals contain fewer frames per second (25) than the NTSC but a greater number of horizontal scans per frame (625 versus 525 in NTSC).

Equipment designed to handle NTSC formatted video cannot handle the PAL or SEACAM format and vice versa. Whenever equipment is purchased, it is always worth checking to make sure that the video format served is compatible with that handled by your other equipment. This point is also important in regard to video formats used by VCRs in recording onto magnetic tape that are entirely different from broadcast video formats (see later).

Table 11.1	Video/Television Formats	
Format	**Location**	**Characteristics**
NTSC (National Television Standards Committee)	North America, Japan	30 frames and 60 fields per second; 525 horizontal scans; 4.2 MHz bandwidth, analog, composite; aspect ratio 4:3; color encoded by vectors 147° and 57° out of phase
PAL (Phase Alternating Line)	United Kingdom, most of continental Europe	25 frames and 50 fields per second; 625 horizontal scans; 5.5 MHz bandwidth, analog, composite; phase of color carrier is shifted 180° for each image line
SECAM (Sequential Color and Memory)	Russia, France, many African and Asian countries	25 frames and 50 fields per second; 625 horizontal scans; 6.0 MHz bandwidth, analog, composite. Two separate color signals from sequential lines utilized; memory is required to obtain both color difference signals for decoding at the receiver
HDTV, progressive scanning	United States	60 frames per second; 760 horizontal scans; 1280 pixels per horizontal scan; bandwidth 6 MHz; aspect ratio 16:9
HDTV, interlaced scanning	United States	30 frames and 60 fields per second; 1125 horizontal scans; 1920 pixels per horizontal scan; bandwidth 6 MHz; aspect ratio 16:9
RS 170 (EIA)	United States, Japan	Standard black and white (monochrome) format used as a basis for the NTSC color format; developed by the Electronics Industry Association (EIA)
RS 330	United States, Japan	Closed circuit format (nonbroadcast) recommended by the EIA using 525 horizontal scans and 60 full frames per second
CCIR (Comite Consultatif International des Radio Communications)	Europe	Closed circuit format (nonbroadcast)

Of further interest is the fact that what was "standard" for 55 years is now being phased out in many parts of the world! One reason for this is that the old standard video signals are analog signals. That is, intensity is a function of voltage and can be varied continuously in the old signal through many gray levels. The new signals are digital, whereby voltages are expressed as a finite set of gray levels represented by numerical data. In fact, all of the data needed for a video signal, including intensity, color, image position, and timing, can be coded into numbers (digitized) and broadcast in that form. This allows a large amount of image processing both before transmission and after reception that greatly improves resolution. Such processing produces "high-definition" TV without increasing the bandwidth of the signal! As a result, high-definition TV allows the use of "big-screen" presentations without excessive empty magnification. In order to appreciate this advance, however, we need to discuss the matter of resolution as it applies to video images.

In all types of imaging, resolution is related to the amount of detail that can be obtained in an image. Although the amount of detail in a video image can be limited by diffraction, as discussed earlier (Chapter 4), it is usually not. Rather, the limitation is in squeezing enough image information into a video format of set bandwidth. Video resolution is usually determined empirically on the monitor screen by the use of a test pattern. The test pattern consists of alternating black and white stripes (lines) running horizontally (to test vertical resolution) or vertically (to test horizontal resolution). Resolution can be gauged by how many lines can fit on the screen and still be seen as separate lines. Thus, on screen, a video image might be said to have a vertical resolution of 200 TV lines (200 TVL) and a horizontal resolution of 300 TVL. In this way, video resolution is expressed as a relative value; it cannot be expressed as an absolute distance because monitor screens come in many sizes.

Vertical resolution is limited by the number of horizontal scans in a frame, a number that is set by the format and cannot be changed. In theory, the NTSC format should be capable of producing a 525 TVL resolution, one line for each horizontal scan. In practice, this number is dramatically reduced by the manner in which these scans are projected and whether the monitor is adjusted for optimal reception and display. First, about 50 of the horizontal scans are obliterated by the vertical blanking period. Second, not all of the remaining scans actually get onto the viewing area of the picture tube, and about 20 scans are eliminated in this manner. We are down to 460 TVL. As if this is not bad enough, we have to consider the fact that any given TV line may not register exactly on a single horizontal scan but might fall between two horizontal scans. The extent to which this reduces resolution varies but is estimated, on the average, to reduce vertical TV lines by a factor of 0.7. This is called the Kell factor and is named after a television pioneer who studied this problem. Although we are now down to a vertical resolution of 320 TVL because of geometrical and format considerations, we have to consider the performance of the cathode ray tube itself. First, recall that the scans of each frame are actually projected as two fields. To obtain the best resolution, these fields must be projected in perfect register one after the other. The two fields are said to be "interlaced" not only because odd and even fields alternate but also because in the fused image of each frame, adjacent horizontal scans alternate in which field they came from. Another factor is whether the scanned beam of electrons that produces the image on the screen is focused to the smallest possible diameter. If these adjustments are out of whack, the vertical resolution is lower still and can represent a further 40% to 50% drop in TV lines visualized to a value of 180 lines. No wonder high-definition TV was needed!

Horizontal resolution is dependent on a different set of factors. Because the test lines are now vertical, the ability to distinguish between lines comes down to two major limitations: (1) how fast can voltage changes of a horizontal scan be faithfully followed by the electronics of the camera, transmitter, or receiver and (2) the extent to which fast voltage changes can be accurately transmitted by a signal of limited bandwidth. The fidelity of fast voltage changes can be limited by both the circuit design and the quality of the electronic components within any unit handling the video signal. The circuitry in any part of the system, whether it is the camera, a signal processing unit, a transmitter or receiver, the VCR, or the monitor, can degrade the horizontal resolution achievable by the system. For this reason, any video component of quality will have its horizontal resolution specified; at this point, 400 TVL is considered very good; 500 TVL is excellent, and 700 to 1000 TVL is exceptional.

The bandwidth limitation is most severe in the case of broadcast TV. If one considers that to get 800 TVL of resolution one needs at the very least a sine wave that fluctuates from maximum to minimum 800 hundred times (400 complete cycles), each maximum representing a black line and each minimum representing a white line. This would have to occur in about 40 μsec—the time it takes to do 75% of a horizontal scan (to be comparable, horizontal resolution is measured over the same relative distance as the vertical resolution is measured; this is about 75% of a horizontal scan because the standard image dimensions (aspect ratio) are three vertical to four horizontal); therefore, to get 800 TVL, we would need a sine wave that has a frequency of $(400/40) \times 10^6$ or

10 MHz. Thus, a rule of thumb is that for every 80 TVL of horizontal resolution, we need a bandwidth of 1 MHz for the signal. Thus, TV broadcast X signals that are limited to 4.2 MHz in bandwidth for luminance information can support a maximum horizontal resolution of 350 TVL. In your living room, the horizontal resolution of the NTSC signal is frequently as low as 200 to 250 TVL.

These resolution limits of about 180 TVL vertical and 350 TVL horizontal are for an analog NTSC signal. Higher resolution can be achieved by creating a different format having a greater number of horizontal scans per frame. Such a new "high-definition" format would need to use either a larger bandwidth or digital coding and file compression. File compression is commonly carried out by MPEG-2 standards developed during the 1990s from previous JPEG and MPEG-1 standards. Digital coding and broadcasting are done in different formats in different countries. Common to all formats, however, is the use of file compression to fit information more efficiently into bandwidths that are about the same as analog transmission bandwidths (6 MHz in the United States). File compression of digital data is extremely effective for the following reason: Each frame of a video is different from the preceding frame, but these differences may be slight. Although a moving object may change radically in position and light intensity from one frame to the next, the background may undergo no changes whatsoever; therefore, if each frame can be coded simply as a collection of the changes that have occurred since the last frame, essentially an update, we can discard most of the information in each frame that is not changing. Using this approach, digital video files can be compressed by 90% with little loss of information and as much as 98% with noticeable losses in information! This efficiency allows not only a higher resolution, but also the option of using progressive scanning without interlacing fields and even three separately digitized signals, one for each primary color.

In the United States, the new high-definition digital format for broadcast TV creates images that are 1080 picture elements vertical by 1920 horizontal to achieve a maximal effective resolution of approximately 724 TVL by more than 1000 TVL. As you have probably already experienced, the new format has an aspect ratio of 9 vertical to 16 horizontal and is much wider than the old format. The term "home theatre" TV is probably appropriate.

■ Video Cameras: The Old Tube Type

For many years, vacuum tubes were at the heart of electronics because within a vacuum streams of electrons can be guided or organized by electric and magnetic fields without interference from collisions with gas molecules. For this reason, the picture tubes in many of our televisions and monitors were, until recently, giant vacuum tubes having considerably more bulk than the new flat, liquid crystal screens. Early video cameras were also based on these evacuated "video" tubes.

The first problem for these cameras was conversion of a photon signal to an electron signal. Such a conversion can be accomplished by using either photoemission or photoconduction. Photoemission, using the photoelectric effect, occurs when a specialized light sensitive material is bombarded with photons, thereby knocking loosely held electrons out of their normal orbit and making the electrons available to move as a current. In a photomultiplier tube, as diagrammed in **Figure 11.5A,** these electrons are abstracted from the coating on the front window and accelerated to a positively charged anode that acts as a collector. The dish-shaped collector, termed a dynode, after being struck with one electron, emits up to 10 electrons, thereby amplifying the signal. If a number of dynodes are positioned sequentially, the electron signal can be amplified an order of magnitude for each dynode present. Often amplifications of up to 10^{10} can be achieved, and for this reason, the photomultiplier tube can be an extremely sensitive detector of light. The magnified signal is then detected as a current in which magnitude is measured by the voltage drop across a resistor.

A

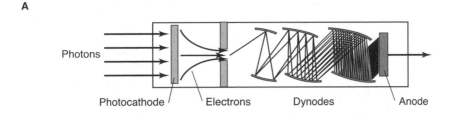

Photons

Photocathode / Electrons Dynodes Anode

B

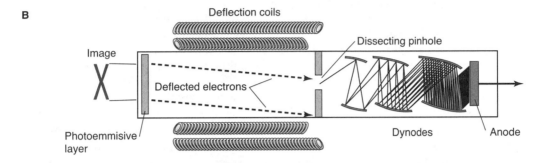

Deflection coils

Dissecting pinhole

Image

Deflected electrons

Photoemmisive layer

Dynodes Anode

C

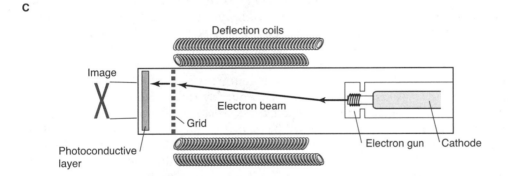

Deflection coils

Image

Electron beam

Grid

Photoconductive layer

Electron gun Cathode

Figure 11.5
Photomultiplier and video tubes. (A) A photomultiplier tube. A photon signal at the front window of the tube generates an electron signal via the photoelectric effect and the electrons are guided to and amplified by a series of dynodes. (B) Early video tube using the photoelectric effect. A photonic image is focused on the front window of the tube, the signal converted to electrons, and the electrons guided to a partition by deflector coils. Only electrons from one point in the image pass through the pinhole dissecting aperture to reach the dynode amplifier. Commands to the deflector coils allow point image information to be scanned in a set order. (C) A modern photoconductive video tube. An electron beam gun fires a stream of electrons at the front window in a set pattern determined by the deflecting coils. At each point on the front window, the electrons provided traverse a semiconductor material at a rate determined by the local conductance of this layer. The conductance of the layer is increased in proportion to the number of photons hitting it since the last readout. Thus, the magnitude of the current leaving the front window, as measured by the voltage across a resister, is a measure of the light intensity at each point in the image.

The high sensitivity of the photomultiplier tube makes it ideal for applications in astronomy, spectroscopy, and scanning confocal microscopy (see Chapter 12). Its drawback for electronic imaging, however, is that it sums the entire light intensity falling on the front window without regard to the patterns of light distribution that make up an image. Theoretically, one could use a massive array of photomultipliers, one for each image point, but this would be very expensive, and miniaturization would be impossible.

The solution was to allow the photomultiplier tube to receive electrons from only one image point at a time and to scan the image point by point. The earlier video tubes did this by having a set of scanning electrodes that, based on their voltage, could direct the electrons emitted from a single location (image point) on the front window to a pin-hole aperture through which there was access to the first dynode of the photomultiplier (Figure 11.5B). Through systematic changes in the voltage applied to these deflection coils, each point on the front window could be assayed for electrons being knocked out by photon bombardment. The amplified current from each point would be measured by passage through a resistor and the voltage drop across the resistor versus time used to form the video signal. The circuit was completed when the current measured was fed back to the coating on the front window to replace the electrons knocked out by light. As soon as all points were scanned, the process started over for the next frame.

Video tubes incorporating this technology worked but required much more light than subsequent designs. The reason for this lack of sensitivity is clear. As shown in Figure 11.5B, at any given moment, electrons coming from each image point on the front window were directed toward the barrier containing the pinhole access to the photomulti-plier, forming a "ghost" image on the barrier. Nevertheless, less than 1% of the electrons hitting were allowed through the pinhole for recording: 99% of the signal was being thrown away!

The next generation of video tubes used a different technology—photoconduction—so that virtually every photon hitting the front window became part of the signal. As shown in Figure 11.5C, the front window of these phototubes is backed by a coating of Sb_2S_3 or $ZnSe$, which when bombarded with light has electrons knocked out leaving pos-itively charged holes within a negatively charged layer. The presence of these fixed posi-tive charges greatly increases the local electrical conductivity of this layer. The more photons that strike over a period of time, the greater are the number of positive charges and the change in conductivity. If one then interrogates this layer using a beam of elec-trons, those image points having high photoconductivity will allow many of these electrons to pass to a metal electrode and subsequently to be "read out" by a resister (Figure 11.5C). The signal read will reflect the increase in photoconductivity at that point in the image, and its magnitude will be related to the number of photons striking that image point during the preceding period of "exposure." The electrons used for in-terrogation are pulled off a filament by a high-voltage electric field and then accelerated and focused by electrodes. This focused beam is scanned in a raster pattern across the front window to interrogate each image point in order, thereby providing information for an entire frame. Just before the beam hits the photosensitive coating, it is slowed to a velocity of nearly zero by a field set up by a negatively charged grid (dashed line, Figure 11.5C). In this manner, the electrons requesting permission to cross this layer are of very low energy and do not in their own right change the number of "holes" in the coating. In addition to creating a "photoconductivity" current to be read out, these elec-trons fill up the holes generated previously and reset the image point to zero so that new photon collisions can be recorded before the next interrogation. Thus, every photon, wherever it strikes the front window, will leave a hole that will be sensed the next time the interrogation beam passes by. No photons are wasted.

The efficiency and detection spectrum of such a video tube depend mainly on the type of photosensitive coating that is used. There are approximately six coatings in general use; these range from heavy metal oxides and sulfides to metal amalgams. The efficiency of a coating can be measured as a quantum efficiency—that is, what is the probability that an electron will be knocked out by the capture of a single photon? This quantum efficiency is 1.0 if an electron is knocked out every time a photon hits. A much lower quantum effi-ciency would be 0.15, where, on average, an electron is knocked out only once every seven photon collisions. Most of these coatings are efficient at wavelengths between 400 and 700 nm, although Newvicon and Silicon Vidicon coatings, used frequently for scien-tific applications, have high efficiency in both the visible and IR wavelengths.

■ Solid-State Cameras

The disadvantages of tube cameras are that they are bulky, they can be burned by too much light, they are somewhat fragile, and they are slower than the fastest solid-state cameras. Solid-state electronics have no vacuum tubes because electrons are handled completely within the solid state, not as beams to be manipulated. Although comparatively bulky in their original design, solid-state circuits can now be fabricated on small silicon chips using thin-film technology. Included in this technology are arrays of photodiodes referred to as CCD chips. Each photodiode consists of three layers of material, rather like a capacitor (**Figure 11.6**). At the heart of a diode is a nonconducting "depletion layer" that is photo sensitive. When a photon strikes this layer, the energy of the photon produces charge separation (i.e., an electron that can migrate is removed from a positively charged "hole" that can also migrate, but in the opposite direction). The electrons migrate to a p-type silicon conducting layer above that acts as a storage site and is referred to as a "well." The holes migrate to an n-type silicon conducting layer below. The potential energy well for electron storage is created by the presence of a positively charged electrode overlying the p-type silicon layer but separated from it by an insulator—silicon dioxide. The voltage and area occupied by the "gating" electrode determine how many electrons can be stored before the well is full.

In solid-state cameras, each recordable image point, often referred to as a "picture element" or "pixel," is represented by a diode that gathers light independently of adjacent diodes in the CCD chip. The diodes are arranged in orthogonal rows and columns like a checkerboard. After a period of light gathering, the electrons stored within the wells of each diode on the chip are read out in order, column by column. In **Figures 11.7A** and **11.7B**, this process is illustrated by an array of cups, each cup representing a diode with its well of accumulated electrons. Within a row, the electrons in each cup are poured repeatedly from one cup to the next. As each cupful reaches the edge of the chip, it is moved as a current to a readout device that measures its magnitude. This process reads out the electron groups in a predetermined order, as illustrated in Figure 11.7C.

A critical feature of the process is that each group of electrons originating from a single diode is kept separate and distinct from other groups as they are passed from one cup to the next all of the way to the readout device. This is accomplished by dynamic control of the voltages on each gating electrode. As shown in **Figure 11.8A**, electrodes at neighboring wells held at a positive voltage are separated from each other by one or more electrodes held at 0 volts. This pattern isolates one well from another and is maintained during read out. By shifting electrode voltages in a specific temporal pattern, each electron group is swept toward the edge of the chip for readout. As the chip diodes are read

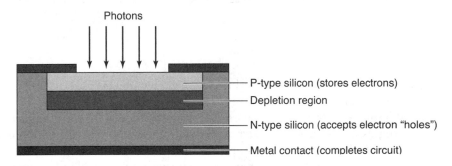

Figure 11.6
Design of a silicon diode photodetector. Photons striking the depletion layer result in charge separation—negative electrons are free to move to a storage site in the p-type silicon layer until readout. Positive "holes" remain stationary or move to the n-type silicon layer but are refilled during readout of the photodiode.

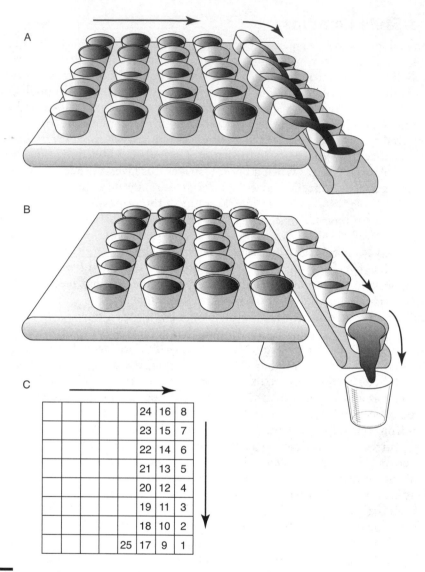

Figure 11.7
Readout pathways in a solid-state chip. (A) If electrons stored by each photodiode in the solid-state chip array are likened to differing fluid levels in an array of cups, the aliquots in each cup are shifted in order to the side of the chip. (B) They are then read by a "serial register" one column at a time. (C) The order in which photodiodes are read out by the scheme in (A) and (B).

out, they are reset to their initial state and begin to gather the next set of photons that will provide data for the next image.

Control of gating electrode voltage patterns allows readout of the chip at set intervals, thus serving as an electronic shutter that determines the exposure time for each image. A mechanical shutter is not necessary on some CCD chips, except to protect the chip from dust and mechanical injury. Electronic shuttering is capable of exposures as short as 1/50,000 of a second—much faster than mechanical shutters can achieve. In addition, the pattern of readout can be changed such that the output of neighboring diodes is no longer kept separate but is grouped together—a process referred to as "binning."

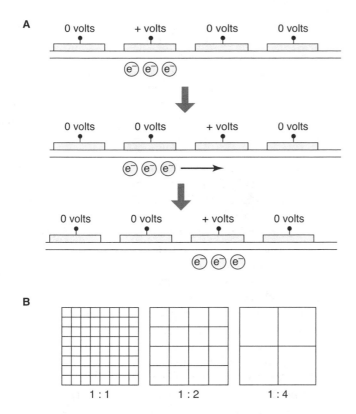

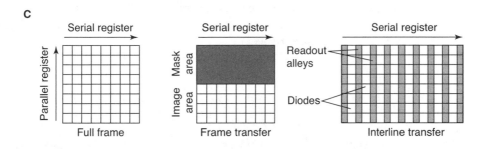

Figure 11.8

Gating electrodes control readout. (A) The potential well for storage of electrons harvested from each photodiode is created by positive voltage on an overlying gating electrode. Each potential well is separated from neighbors by gating electrodes set at zero volts. Coordinated temporal control of these voltages allows the electron groups in neighboring wells to be swept to the side of the chip simultaneously but in order and separate from each other for readout. (B) Readout speed can be increased by "binning"—creating larger pixels from groups of smaller pixels. (C) Readout strategies in solid-state chips. Full-frame readout of all pixels is slowest but preserves spatial resolution (left). Frame transfer uses half of the chip as a cache during readout. Frame rate is increased twofold, but spatial resolution is reduced twofold (middle). Interline transfer chips have dedicated readout alleys, allowing high-speed continuous readout (right).

For example, as shown in Figure 11.8B, the output from each square of four diodes can be read rather than each diode individually. This will result in a drop in image resolution but will allow the chip to be read out in one quarter of the time!

The advantages of binning when a higher frame rate is desired points out that readout of the chip takes a significant amount of time—in most cases, longer than the exposure time to collect the photons. For this reason, and especially to serve the needs of video imaging, chip design has focused on obtaining as high a speed in readout as possible. Figure 11.8C depicts the design of three types of chips that differ in how they are read out—full frame readout, frame transfer readout, and interline transfer readout. Full frame readout is the slowest. Between each exposure period, every pixel on the chip is read. The exposure might take only one hundredth of a second, but the readout period afterward takes 0.20 seconds—20 times as long! Thus, one can expose and read only about two or three frames per second. (This is rather like the situation in a manual film camera: The exposure may be short but it takes much longer to wind the film.) To get around this, frame transfer designs use one half of the chip for photon capture and the other half of the chip for storing the data while it is read out. This allows one image to be read out while the next image is being acquired, shortening the overall time per frame. Frame transfer is faster but has one major drawback: The resolution of the image is half that of the full frame chip because only half of the chip is used for image acquisition. In addition, frame transfer designs do not get us up to real-time video speed—30 frames per second. They more commonly achieve rates of 5 to 10 frames per second. Both full frame and frame transfer chips usually require a mechanical shutter to avoid blurring the image caused by photon capture and readout occurring simultaneously.

The interline transfer chip does reach video rate because its design includes "alley-ways" that are dedicated for electron transfer. Thus, transfer of electron groups to the readout device is occurring continuously during image acquisition. Some advanced chips even combine interline and frame transfer architectures to gain a further increase in speed. An "aerial" view of the diodes in the interline transfer chip shows that up to 50% of the surface area of the chip is dedicated to electron transfer, and as little as 40% of the chip is used for photon capture. Although resolution (the number of diodes on the chip) is maintained, the interline transfer design suffers from a loss of light sensitivity because of the fact that only 40% of the incident photons are being used. One way in which this problem has been minimized is by addition of a layer above the diodes that actually contains an array of miniaturized lenses that focus up to 90% of the photons onto the light harvesting regions of the chip diodes (**Figure 11.9**).

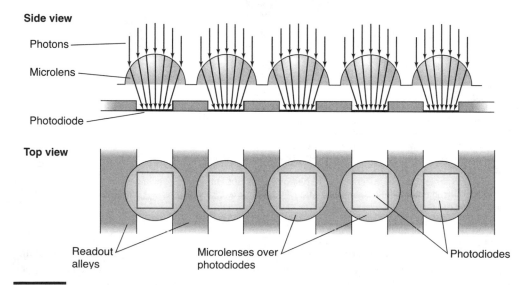

Figure 11.9
Microlenses increase sensitivity of solid-state chips. Microlenses fabricated on-chip focus incoming photons at light sensitive surfaces and away from readout alleys, thus increasing chip sensitivity.

Table 11.2 Comparison of Video and Slow-Scan Cameras		
Feature	**Slow-Scan Camera**	**Video Camera**
Chip type	Full Frame Transfer	Interline transfer
Pixel size	10×10 to 25×25 µm	6×6 µm
Pixel array size	512×512 to 4000×4000	780×480
Well depth/pixel	200,000 electrons	10,000 electrons
Gray levels	65,000	256
Full frames/second	2–4	30
Dynamic range	Up to 35,000:1	100:1
Bits/pixel	10 to 16 monochrome	8 monochrome
Binning	Yes	No
Sensitivity	Can be photon counting	2–4 Lux
Cooled	Yes, water or air	Uncommon
Dark noise	1 electron/pixel/second	
Read noise	6 electrons	
Signal-to-noise ratio	>60 dB	45–55 dB
Subarrays (regions of interest)	Yes	No

Other characteristics that affect readout speed include pixel array size, diode size, and well size. In solid-state cameras, there is a tradeoff between speed, resolution, and sensitivity, just as there is in various types of film used in traditional photography. Cameras that are optimized for resolution and sensitivity are referred to as "slow-scan" cameras and are used when time resolution is not a factor. Used for still images in light microscopy and for electron microscopy, these cameras achieve frame rates of two per second. Video or "high-speed" solid-state cameras provide image rates between 30 and 500 frames per second but at the cost of resolution and sensitivity.

Table 11.2 compares the features of these two classes of cameras. Slow-scan cameras can afford to have larger diode arrays that take longer to read out. The size of these arrays in commercial solid-state cameras increases every year and currently stands at about 6 million pixels arranged in a 3000 by 2000 raster in high-end consumer models. A fourfold increase in pixel number is equivalent to a twofold increase in image resolution; this is approximately the increase in resolution seen in current cameras compared with the original consumer-grade cameras having 1.4 million pixels. Slow-scan cameras can also afford to have physically large diodes—about 25 by 25 µm—compared with the small diodes of video cameras—about 7 by 7 µm square. Large diodes enable a higher number of electrons to be stored before readout, typically up to 600,000 per diode. These diodes are said to have "deep wells," the advantage of these being that a much greater range of light intensities can be distinguished. A CCD chip having large diodes can often provide adequate signal over a 300- to 3000-fold range of light intensity and for that reason is said to have a large "dynamic range." A large dynamic range is important if images having both strong and weak light intensities are to be recorded accurately and is particularly important in applications such as fluorescence microscopy.

In contrast, video cameras have typical pixel arrays of 768 by 488 that provide video resolutions that are compatible with that of broadcast TV—200 TVL vertical by 300 TVL horizontal. This standard array of 375,000 pixels is quite meager in comparison with slow-scan cameras, and indeed, many video cameras are now designed to do better by taking advantage of video-recording formats that have higher resolutions, such as S-VHS and DV. The smaller pixel size of video camera chips leads to a relatively limited dynamic range, usually no more than one order of magnitude (10-fold) and a relatively low sensitivity.

■ Solid-State Color Cameras

CCD chip diodes are sensitive to visible light over a broad energy range between 400 and 700 nm and therefore do not discriminate between colors. In order to construct a color image, cameras use color filters to obtain a separate blue, green, and red signal. These three signals can be acquired by either different regions of the same chip, simultaneously or sequentially on the same chip, or each collected on a separate chip. Cameras that collect simultaneously on one chip (such as typical consumer-grade color video cameras) use a Bayer mask placed directly on top of the CCD chip that guides red, green, and blue wavelengths to different but adjacent pixels. This solution costs less but comes with a loss of resolution because usually three or four pixels are used to represent each image point rather than a single pixel. Cameras that use three separate sensors are referred to as "three-chip" cameras and typically achieve a horizontal resolution of 800 TVL rather than the 350 TVL obtained by single-chip color video cameras. An optical diagram of one (**Figure 11.10**) shows that the optical train is separated into three by dichroic mirrors or wedge-shaped prisms that act as dichroic mirrors to separate the red, green, and blue components of the image so that each can be recorded by one of the three CCD chips. The output obtained from each chip is then combined to form a composite video signal, or alternatively, the output of each is used to form one channel of a three-channel RGB signal. In general, color cameras require a higher level of light than do black and white cameras and for this reason are usually not used for low-light-level work, such as fluorescence microscopy.

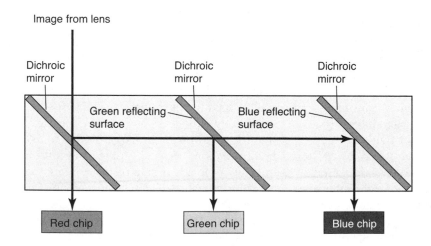

Figure 11.10
Light paths in a three-chip color camera. Dichroic mirrors at prism interfaces are used as wavelength-dependent beam splitters so that red, green, and blue components of the image can be detected by separate chips.

■ Digital Image Processing

The signal that originates in a CCD camera is analog, not digital. Electron flow from the chip versus time is converted to voltage changes versus time to produce an analog video signal. The analog signal can then be digitized by an analog-to-digital converter at one of three sites. The A to D converter could be in the camera itself in which case the camera is referred to as a "digital camera." Alternatively, the analog camera signal could be fed into a digital video cassette recorder and digitized before it is recorded onto magnetic tape. Third, the analog signal could be sent to the A to D converter on a computer card and the digitized signal stored on the hard drive of the computer. The latter is referred to as a "frame grabber" because it can not only convert a stream of data from analog to digital "in real time" but also "grab" (isolate) one particular frame from the signal to display on a monitor or to store as a digital image file.

During digitization, the analog signal is sampled at discrete time intervals, and the value for voltage sampled is converted to the nearest voltage level (gray level) that can be digitally coded. The frequency and type of sampling are important to the quality of the image, and these vary from one digitization standard to another. In general, the luminance signal is sampled with twice the frequency of color signals because it is the luminance signal that is most important for image resolution. Typical sampling protocols include 4:2:2, 4:1:1; and 4:2:0, meaning that for every period in which the luminance signal is sampled four times, the blue and red color signals are each sampled one or two times. The number of gray levels available (and hence the ability to encode a large dynamic range) depends on how much digitized information (significant places) is allowed in each sample. Common formats include 8, 10, 12, and 16 bits, with each bit being equivalent to one digit in binary expressions. An 8-bit format allows 256 gray levels, 12-bit 4096 gray levels, and 16-bit 65,535 gray levels. Eight bit formats are standard, while 12- and 16-bit formats are handled only by specialized software and hardware. Some programs such as Photoshop will convert 16-bit files into 8-bit files for easier manipulation but do so at a loss of dynamic range in the image. Digitized color signals can have each of the three color channels encoded as 8-, 12-, or 16-bit signals leading to a total signal having 24, 36, or 48 bits.

The most common formats for single image files are JPEG (Joint Photographic Experts Group) and TIFF (Tagged Indexed File Format) (**Table 11.3**). The TIFF format

Table 11.3 Digital Image File Formats				
Format	**Characteristics**	**Use**	**Advantages**	**Disadvantages**
Bitmap (general use)	Uncompressed	File transfer	Full resolution and bit depth	Large files
TIFF	Uncompressed	Archival and publication	Full resolution and bit depth	Large files
JPEG	Variable compression	Photographic image format	Smaller files; user-defined quality	File information lost each time the file is opened
MPEG-2 MPEG-3 MPEG-4	Standard for digital video; compressed	Digital video transfer; HDTV	Compression does not degrade image noticeably	Compression artifacts near moving objects

contains separate and distinct information on the properties (intensity, color, position) of each pixel in an image. These files are large but contain a compete set of properties for every image point. In contrast, the JPEG format is a compressed format, which means that some of the information is missing. Instead of each pixel being associated with its own set of attributes independent of other pixels, attributes are now assigned to groups of pixels, instead of each separately. The advantage is that the number of bits used per pixel on the average can be set by the user to produce either a large file that has a large dynamic range or a small file that is restricted in its dynamic range in regard to both color and intensity. The disadvantage is not only the loss of information but also the fact that each time a JPEG file is opened, modified, and saved it undergoes another cycle of compression, resulting in a further loss of original data.

Many software packages handle JPEG and TIFF files differently. Microsoft Power-Point, for example, assumes that with a TIFF file, all pixels should be processed independently and should remain separate and distinguishable. In contrast, using JPEG files, it assumes that pixel properties are not to be kept separate and that image smoothing and pixel averaging can be done automatically to make the image "look good," even if it results in loss of data; therefore, it is recommended that all raw image data should be saved for archival purposes as TIFF-formatted files but that subsequent processing of the image can use either TIFF or JPEG formats. Pixel properties often are repetitive: For a white sheet of paper, one only has to specify that the first pixel is white and that all other pixels are the same. Thus, compression can make sense. On average, image files can be compressed to one tenth of their full size before any loss of quality is noticeable. Indeed, digitized video formats such as MPEG can be compressed even further because there is tremendous duplication of information between one frame and the next (see below).

■ Signal-to-Noise Ratio

Successful electronic imaging requires that the signal containing image information be considerably stronger than the random electron noise present. The signal-to-noise ratio therefore is an important parameter used to judge whether an imaging component such as a camera, VCR, or monitor can process image information without significant degradation by noise. This ratio is usually expressed in decibels, an exponential scale that is related to the ratio by the following equation:

$$dB = 20 \log_{10} [\text{signal/noise}]$$

Thus, a signal-to-noise ratio of 10 is equivalent to 20 dB, a ratio of 100 equivalent to 40 dB and a ratio of 1000 equivalent to 60 dB. A rough guide when looking at the equipment specifications is the following: more than 65 dB, highest quality, used for the most demanding professional applications; 60 to 65, excellent, will perform demanding applications at a high level; 50 to 60 dB, very good, will perform all but the most demanding applications well; 40 to 50 dB, acceptable, adequate for routine imaging; and less than 40, unacceptable, will perform routine tasks with a noticeably lower quality result. **Figure 11.11** provides simulated results obtained for an array of squares processed at different signal-to-noise levels.

The signal-to-noise ratio can be increased dramatically by the reduction of electronic noise. One important source of noise in both video tubes and CCD chips is the "dark current"—output caused by random kinetic energy in the absence of light. Because this

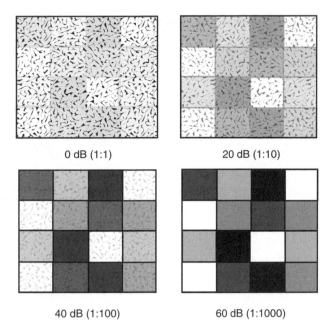

Figure 11.11
Simulated image recognition at various signal to noise levels.

type of noise is temperature dependent, it can be reduced to low levels by cooling the tube or the chip. Cooling to 20°C below ambient (typically 5°C to 10°C) can be achieved by Peltier devices that use bimetallic junctions to move heat from the tube or chip housing to external radiator flanges. Cooling to 0°C to 5°C can also be obtained by forced cold air systems—a miniaturized version of the air conditioning system in your house. More dramatic cooling to 2°C can be achieved by water circulating through plumbing in the camera, whereas temperatures as low as −30°C can be achieved by circulating cold antifreeze through the camera. Most low-light-level CCD cameras are capable of being cooled in one of these ways. Cooling is particularly useful in fluorescence microscopy, where low light levels can require exposure times of many seconds if not minutes. These long exposure times not only summate (integrate) the desired signal but also integrate the random noise to unacceptable levels if cooling is not used.

■ Storage of Electronic Signals

Creating a formatted electronic signal that represents an image is of no use if it cannot be stored or displayed. Typical storage media include magnetic tape, as used in video cassette recorders, and magnetic disks, CDs, DVDs, and solid-state chips used in computers (**Table 11.4**). Each of these varies in its properties and technology, and some of these basic differences are addressed here.

The earliest technology used to store electronic data was the phonograph record, invented by Edison in 1901. Initially, sound itself and later an electrical signal from a microphone were used as a mechanical signal to form the geometric variations in the

Table 11.4 Characteristics of Image-Storing Media

Medium	Type	Size	Speed	Cost	Stability	Comments
Black/white photograph or negative	Silver grains	10 Mb	1 Mb/min	$100/Gb	>150 years	Highest resolution available
Color photograph	Organic dyes	30 Mb	1 Mb/min	$500/Gb	20 years	Degradation of color dyes limits stability
Ink Jet print (color)	Organic dyes	2 Mb	5 Mb/min	$300/Gb	1–20 years	Degradation of color dyes limits stability
Laser Jet print	Carbon	5 Mb	50 Mb/min	$100/Gb	>20 years?	Screening degrades resolution
Floppy disk	Magnetic	1.4 Mb	0.2 Mb/min	$150/Gb	20 years	Can be corrupted or erased by magnetic fields; format can be made obsolete by technology advances
Zip disk	Magnetic	250 Mb	20 Mb/min	$40/Gb	20 years	Can be corrupted or erased by magnetic fields; format can be made obsolete by technology advances
Hard disk (internal)	Magnetic	50–500 Gb	1000 Mb/min	$1/Gb	20 years	Can be corrupted or erased by magnetic fields; format can be made obsolete by technology advances
Magnetic tape	Magnetic	120 Gb	1000 Mb/min	$0.02/Gb	40 years	Can be corrupted or erased by magnetic fields; format can be made obsolete by technology advances
Flash drive	Solid state	64 Mb to 4 Gb	100 Mb/min	$30/Gb	20 years?	Format can be made obsolete by technology advances
RAM	Solid state	128 Mb to 4 Gb	1000 Mb/min	$100/Gb	Temporary	Format can be made obsolete by technology advances
Compact disc (recordable)	Organic dyes	700 Mb	200 Mb/min	$0.30/Gb	20 years?	Subject to optical dye degradation; format can be made obsolete
Compact disc (pressed)	Pressed plastic	700 Mb	N/A	$15/Gb	>40 years	Format can be made obsolete
DVD (recordable)	Organic dyes	4.7 Gb	200 Mb/min	$0.50/Gb	20 years?	Subject to optical dye degradation; format can be made obsolete

groove of a wax master and reproduced in plastic or metal for storage. When played on a phonograph, the geometric data of the record groove were converted to mechanical vibrations in the phonograph needle that in turn, through a piezoelectric device, were converted to a voltage versus time signal that could be amplified and used to drive acoustic speakers. In the 1940s, it was found that one could store an electrical acoustic signal on magnetic tape. Basically, the fluctuations in voltage were used to drive an electromagnet (the recording head). The resulting fluctuations in magnetic field at the head produced localized oriented patterns of ferric oxide particles adhering to a plastic tape that was drawn past the head. In this manner, the "recording tape" became magnetized in very specific micropatterns, and it was in this form that the electronic signal was stored. Playing back the tape simply reversed the process by changing the magnetic field pattern into a voltage versus time electronic signal.

An acoustic signal is a very slow stream of data compared with that of a video signal. A common observation that demonstrates this is that up to 80 minutes of music can be stored on a compact disc, but only one-half minute of real-time video can be stored on the same CD. Thus, mechanical storage of information, suitable for low-density acoustic signals, is impractical for storing information-rich video signals. On the other hand, magnetic tape and other magnetic media can store data at high densities. One 2-hour video tape can store about 120 Gb of digital information—equivalent to 180 CDs or 25 DVDs—at less than $1/10$th the price. (Movies can be stored on DVD only in the form of highly compressed digital files.) Thus, magnetic tape is still the medium of choice for storing large amounts of image or video information.

Extensions of magnetic tape technology are the floppy, zip, and hard disks used in computers for data storage. High-speed rotation of hard disks is very advantageous for recording and accessing data quickly. Many hard drives have data access times in the microsecond range, and capturing or reading of digitized video data in real time is feasible. This rapid access time is allowed by the fact that all data are available to a magnetic detecting head that floats above the rotating disk on a thin layer of air without actually contacting the disk (**Figure 11.12A**); however, the cost of these storage devices, although dropping rapidly, is still much greater than that of magnetic tape.

As mentioned above there are other important alternatives—optical media familiar in the form of compact discs (CDs) and digital versatile discs (DVDs). CDs use tracks of long and short optical signals that are interpreted as 1s and 0s. Commercial CDs are "pressed" (i.e., the tracks of long and short microscopic indentations are made by pressing them into plastic much like manufacture of a phonograph record). Alternatively, these long and short signals can be "burned" into the CD optically by laser-light–induced chemical alterations in a dye layer on the disc. The latter type of CD is that made by "CD burners" commonly found in computer systems. Both types of CDs are read out by a drive that uses laser light to detect the sequence of dots and dashes in the track (Figure 11.12B). The laser light is scattered or absorbed momentarily whenever an indentation or dye mark is passed and the precise flickering pattern of this light converted into a digitized electronic signal. DVDs are used in essentially the same way as CDs, except that they often contain multiple dye layers to make room for larger amounts of data.

Finally, digital electronic data can be stored in solid-state devices similar to the RAM memory chips in your computer. Solid-state storage devices are now being made with capacities as high as 4 Gb, and capacities are likely to increase with time. Common examples include the flash cards used with digital cameras and portable USB drives generally having capacities from 256 Mb to 2 Gb. Although these devices are too expensive

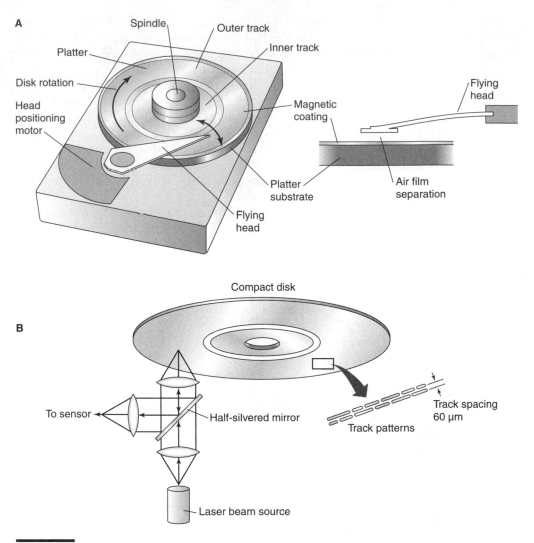

Figure 11.12
Storage of digital data. (A) Top and side views of a computer hard drive. (B) Oblique view of a compact disc drive.

for permanent storage, they are extremely useful for transferring data between camera and computer or between other imaging instruments.

One important factor that we have not discussed is the permanence of information storage. How long can we expect to preserve important image data? What are the best methods for archiving image information? First, a developed photographic emulsion on paper, glass, or metal is the standard by which to compare when we discuss archiving images. Black and white photographs have stood the test of time and are known to last greater than 150 years if stored properly. They are directly accessible by eye and can have extremely high resolution, equal or greater than the resolution obtained in electronic images.

In comparison, magnetic media are not very permanent. The oriented magnetic fields of ferric oxide particles gradually deteriorate with time, particularly at high temperatures, and are thought to retain their fidelity for only 30 to 40 years. In addition, there is always danger that a strong but transient magnetic field can reorient the ferric oxide and "erase" the information present at any time. Optical media have lifetimes that are as yet uncertain. CDs that have indentations pressed into them are likely to be long lived, providing they (like records) are kept away from heat. In the past, there has been some speculation that the reflective metal coating placed on CDs or DVDs might degrade with time because of oxidation, but evidence for this is lacking. It seems likely that such CDs might last 100 years or more. CDs and DVDs that rely on dye marking rather than indentations for information storage are likely to last a shorter period of time (20 to 30 years?) because of eventual photo degradation of the dyes.

There is, however, a very important consideration in archiving electronic images that is not related to physical stability of the medium: the problem of outdated formatting and technology. The format in which either analog or digital data is stored may have changed completely long before the medium physically degrades. One needs only to think of the Atari and Commodore video games of the 1970s that can no longer be played because the computer operating systems to read them are no longer manufactured; it is difficult to find them, even in antique shops. For this reason, important data to be maintained in an archival state should be transferred to the most recent stable format and medium about once every 10 years.

Table 11.4 provides a quick comparison of the common media presently available in regard to size, cost, speed, resolution, and permanence. Video data, because of their potentially massive amounts, are likely best stored in two stages. Magnetic tape still provides a fast, cheap way to store large amounts of raw data at the initial stage, but the format in which it is stored should be at as high a resolution as possible. At the second stage, a smaller volume of edited data is best stored on indentation-coded CDs or DVDs.

■ Video Cassette Recorders

Because magnetic tape is still an important and probably the most common storage device for video data, it is useful to consider what these devices do. In a typical analog recorder, the electronic video signal is passed through a tiny electromagnet called a "head" that produces a very minute magnetic field as small as 20 μm wide. As shown in **Figure 11.13A,** the ferric oxide–coated tape is moved past the head such that the magnetic field of the head, as it changes with time, lays down a pattern of oriented ferric oxide that will remain oriented for years, unless disturbed by another magnetic field. Generally, there are two heads, both of which record onto the tape, producing diagonal stripes of information. If one could visualize the geometrical pattern of information laid down for an NTSC signal, it would appear as shown in Figure 11.13B. Each diagonal stripe would correspond to a single horizontal scan. The set of scans representing the odd field would be laid down by one head (A), while the even fields of the same frame would be laid down by the other head (B), and this pattern would repeat for each frame. A stereo audio signal to go along with the video signal would be laid down by separate heads as two straight tracts at the side of the tape or in the diagonal tracks within a designated region.

Interestingly, the signal that is recorded onto tape is not the NTSC video signal as supplied to the recorder but, rather, a version of the signal that has been transformed into a recording format such as VHS or S-VHS. Conversion to a VHS format requires several steps, as illustrated in **Figure 11.14A.** First, all of the color information is separated from the luminance information using a comb filter. This is easy to do because the color

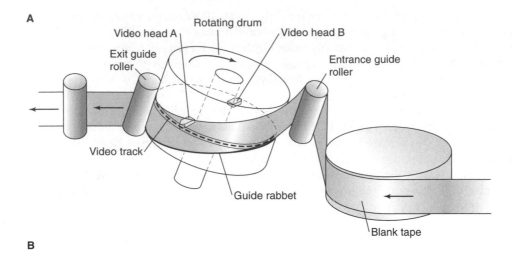

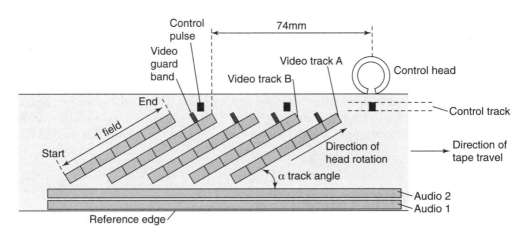

Figure 11.13
Video cassette recording technology. (A) The electromagnetic recording heads on a rotating drum lay down tracks as the tape is moved past the drum. (B) Organization of video information as diagonal tracks on magnetic tape. Audio tracks, containing less information, run straight on one side of the tape.

information is both out of phase and encoded in a very restricted set of frequencies. The luminance (intensity) information is then converted from an amplitude-modulated signal to a frequency-modulated signal. What this means is that changes in light intensity that were previously expressed as changes in voltage amplitude are now encoded as changes in frequency. This conversion allows better separation of signal from noise during playback, just as a radio FM signal is relatively free from static compared with a radio AM signal. Because the luminance information is more information rich and requires a wider bandwidth, it is contained in frequencies from 1.3 to 4.5 MHz. The color information, which is already frequency modulated, is carried at a frequency range of 0.2 to 1.0 MHz in order to separate it completely from the luminance information. Formation of an S-VHS signal is similar, except that the bandwidth used for luminance information is increased from a total of 3.2 to 5.5 MHz (Figure 11.14B). This increase in bandwidth dramatically increases the horizontal resolution from 240 TVL in VHS to 400 TVL in S-VHS.

A. VHS

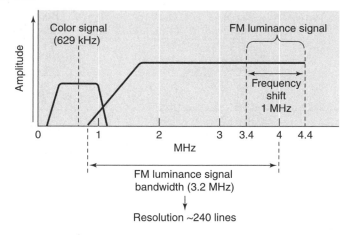

Resolution ~240 lines

B. S-VHS

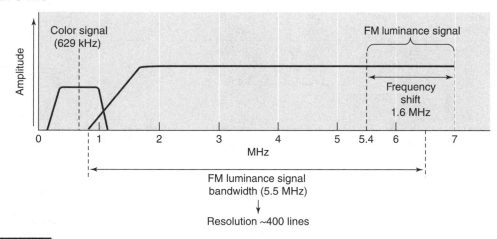

Resolution ~400 lines

Figure 11.14
Specifications of common recording formats. (A) In the VHS format, luminance information is converted to a frequency modulated (FM) signal with a 3.2-MHz bandwidth. The color information is coded as a low-frequency signal that is entirely separate from the luminance frequency band. (B) The S-VHS format is similar to the VHS format, except that enhanced luminance information is provided by a 5.5-MHz bandwidth.

VHS, S-VHS, and other formats less frequently used (e.g., the "Betamax" format used by Sony in the 1970s) are analog formats. More recently, digital recording formats have been used in both consumer grade and professional grade recorders, allowing an even greater increase in horizontal resolution. This resolution ranges from 500 TVL in the DV (DVC) format to greater than 1000 TVL in professional broadcast formats such as D-5 HD, as shown in **Table 11.5.** This increased horizontal resolution is obtained not just from higher quality electronic components and faster data transmission rates, but also from digital file compression, which for the DV format is about 5:1. By eliminating repetitious information, compression leaves more space for larger pixel arrays, such as those used by digital high-definition TV. Before high-definition TV, the higher resolution of S-VHS and the digital recording formats were unable to be broadcast and found use in only closed circuit systems such as those used for video microscopy. Now, with digital TV at hand, the higher resolution of these formats can be appreciated even in the course of mass media entertainment.

Table 11.5 Examples of Video Recording Formats

Format (Year Developed)	Type	Luminance Bandwidth	Color Bandwidth	Total Data Rate	Signal/ Noise	Horizontal Resolution
Type C (reel to reel; 1976)	Analog composite					
Betacam SP	Analog component	4.1	1.5		47 dB	400
VHS (1976)	Analog composite	3.2	0.63			240
S-VHS (1987)	Analog composite	5.5 MHz	0.63			400
Hi-8 (1993)	Analog	4.2	0.74			
M-II (1986)	Analog; professional	4.5	1.5		47 dB	
DV25; DVC; mini-DV (1996)	Digital 8-bit intraframe compression	13.5 MHz (sampling rate)	3.4 MHz (sampling rate)	25 Mbps		460–500
D-1 (1986)	Digital; component; 8-bit; uncompressed professional	13.5 MHz (sampling rate)	2.75 MHz (sampling rate)		56 dB	
D-6; D-7; DVC100; DVCPROHD	Digital; HDTV; 10-bit component; intraframe compression; professional	30 MHz (sampling rate)	15 MHz (sampling rate)	100 Mbps	54 dB	869 (limited by bandwidth, not pixel array)
D-5HD (1994)	Digital; HDTV: 10-bit uncompressed professional	120 MHz (approx.)		235–330 Mbps		>1000
HDV (2003)	Digital; MPEG-2 interframe compression; consumer/ professional	55–75 MHz (sampling rate before compression)		25–30 Mbps		

■ Playback of Video and Electronic Images

The grand daddy of video playback is the black and white TV. At the heart of its technology was the cathode ray tube (CRT), more informally named the "picture tube." The oldest examples of this genre were not-so-large, hand-blown vessels that by necessity were rounded and circular when seen from the viewing end. These tubes were evacuated and operated in a manner quite similar to the photoconductive video tubes described earlier in this chapter. Opposite to the viewing end was a centered electron beam gun that, by heating of a filament to generate free electrons and a series of high voltage plates to accelerate and focus these electrons, could produce a high-density, narrow beam. This

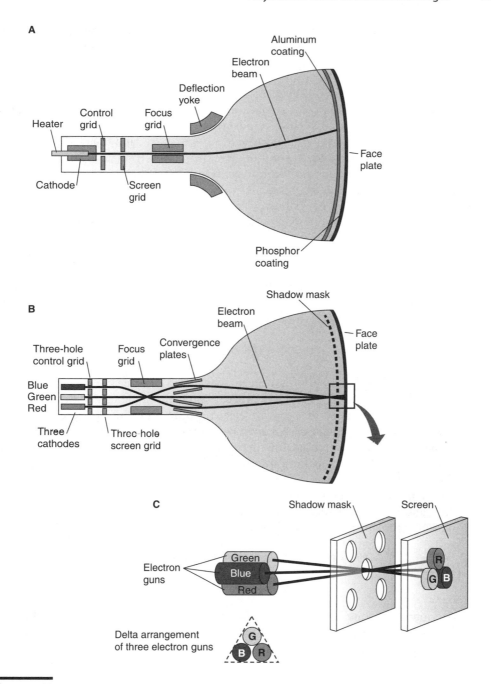

Figure 11.15
The cathode ray tube. (A) A black-and-white picture tube having one electron gun whose beam is scanned across the screen, exciting the phosphor coating to emit white light. Scanning of the beam is carried out by the deflection yoke coils. (B) A color picture tube having three electron beam guns, one for each primary color, arraigned in a triangular configuration. The three separate beams are scanned together, but after passing through a mask close to the screen, each beam strikes a different and appropriately colored phosphor. (C) Geometrical relationship between the electron guns, the shadow mask, and screen phosphors in a color tube.

beam was then moved across the screen (the viewing end of the picture tube) in a set pattern identical to the standard video raster used by the camera to create the video signal. Control of the direction and pattern of the electron beam is managed by the scanning coils through which the beam passes on its way to the screen (**Figure 11.15A**). After the beam reaches the screen, it interacts with a phosphorescent coating on the inside surface, causing the phosphor to absorb the electrons and emit photons. The photons emitted make up the actual image seen by the viewer.

Color TVs and computer monitors basically work on similar principles with some minor modifications. Instead of one electron beam gun, there are three separate guns for the red, green, and blue primary color signals (Figure 11.15B). The three electron beams are simultaneously rastered across the screen in synchrony. The screen, instead of having a single phosphor, now has three phosphors emitting red, green, and blue light. They are arranged as an array of spots, each group of three forming a triangle (Figure 11.15C). The key in producing an appropriate color image is first to be able to separate the NTSC signal into red, green, and blue color signals, with each signal being sent to a separate gun. Second, there must be a mechanism insuring that electrons from the "red" gun strike only the red phosphor spots. This is achieved by placing a mask with an array of holes that center on each triangle of phosphor spots. The three electron beams, one from each gun, go through these holes at different angles, thereby striking the three different phosphor targets independently.

Dissection of the NTSC signal into three separate color channels takes place as follows. As pointed out in Chapter 5, light intensity or luminance is not independent of color but is given by the following equation:

$$Y = 0.3\ R + 0.59\ G + 0.11\ B \tag{11.1}$$

Thus, both luminance and color information can be carried by three independent parameters, for example, R, G, and B. Thus, both types of information, intensity and color, can be produced by three and not four separate electron beam guns in a picture tube. The NTSC signal, however, groups this information differently, albeit as three independent signals that can be dissected electronically then reformatted into R, G, and B signals. One parameter of the NTSC signal is pure luminance (Y), virtually identical to the earlier black and white signal. This enables black and white receivers to use the NTSC signal with no problem; they just ignore the color information. The second and third parameters are the I and Q waves, which carry color information and represent the following transformations of the primary color signals:

$$I = 0.60R - 0.28G - 0.32B \tag{11.2}$$

$$Q = 0.21R - 0.52G + 0.31B \tag{11.3}$$

These equations show that I carries red and yellow-green information while Q carries blue and blue-green information. Based on the relative density of cone types in our retina, the Q signal can afford to have much less detail (equivalent to a bandwidth of 0.5 MHz) than the I signal, which requires the 1.5-MHz bandwidth, and both have much less detail than the luminance signal, which is allotted the 4.2-MHz bandwidth. Thus, by grouping color information in this way, bandwidth can be spared in the Q wave. The relative frequency range of each wave is illustrated in **Figure 11.16A**. Both color signals modulate a carrier wave centered at 3.58 MHz above the carrier frequency, whereas the audio signal is a narrow band centered 4.5 MHz above carrier frequency.

Given that frequency range for the color waves is superimposed on the frequency range for the luminance information, how are these two kept separate during transmission, and how are they distinguished from one another at the TV receiver? There are two mechanisms at work: phase separation and frequency pattern separation. First, I and Q

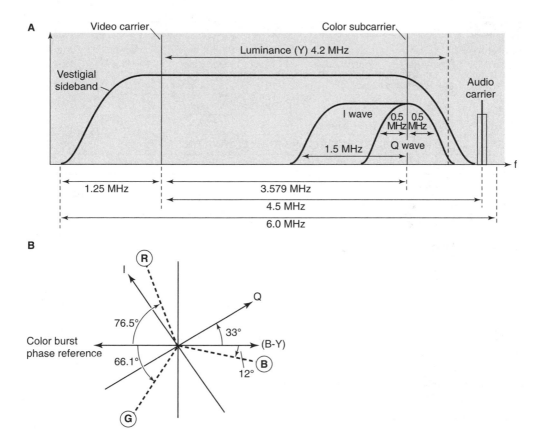

Figure 11.16
The commercial television bandwidth. (A) The 4.2-MHz band on the upper side of the carrier frequency carries luminance information. The color information is contained in the I and Q waves that have smaller bandwidths of 0.5 and 1.5 MHz, respectively. The audio signal is carried in a narrow band above the image information. (B) Phase relationships between color and luminance signals in the NTSC signal. By definition, the color-burst phase reference signal has a phase of 0 degrees. In comparison, the theoretical phases of the three primary colors are designated by dashed lines. In practice, linear combinations of color and luminance information are broadcast with a phase of 180 degrees (B-Y wave), 57 degrees (I wave), and 147 degrees (Q wave).

waves are out of phase with each other and with the luminance signal. Relative to the color burst at the beginning of each horizontal scan, the Y wave is in phase. The I wave is 57 degrees out of phase, and the Q wave is 147 degrees out of phase (Figure 11.16B). Capacitance filters can distinguish between the phase of these three components and separate them.

Second, luminance information is largely clustered around the horizontal scan frequency (15.75 kHz) and its harmonics. This makes sense because any object that is stationary and of a similar gray level will be a pattern that will be repeated in each horizontal scan at the scan rate frequency or in multiples of horizontal scans at a harmonic frequency. In contrast, changing detail during movement will briefly force the luminance information to frequencies between harmonics. Thus, the frequencies between scan rate harmonics are seldom used and a perfect place to stash away the color information. At the TV set, a comb filter dissects out these color-rich, luminance-poor regions of the frequency spectrum rather cleanly. The separated Y, Q, and I waves can then be remixed properly to produce the R, G, and B signals used by the electron guns.

Monitor Type	Pixel Array	Color Mode	Number of Colors	Frames Per Second	Lines Per Frame	Data Rate Mb/Sec
VGA	640 × 480	4 bpp	16	60	525	9.2
SVGA	640 × 480	8 bpp	256	60	525	18.4
SVGA	640 × 480	16 bpp RGB	65,536	60	525	36.8
SVGA	640 × 480	24 bpp RGB	16,777,216	70	525	64.5
SVGA	1024 × 768	8 bpp	256	70	840	55.0
SVGA	1280 × 1024	4 bpp	16	70	1120	45.9
SVGA	1600 × 1200	8 bpp	256	70	1310	
HDTV	1920 × 1080	24 bpp RGB	16,777,216	60	1180	

Table 11.6 Standard Computer Monitor Arrays

Data are from Table 1.2 of Inglis and Luther and Table 10.3 of Luther.

Computers and their monitors are allowed much more flexibility because their communications are not limited to a relatively narrow bandwidth. Indeed, bandwidth for CPU and monitor communications has increased over the years and is used in three different ways. First, the number of pixels in the screen array has increased. Standard arrays are listed in **Table 11.6**, including VGA, SVGA, and HDTV classes along with their size. Second, the amount of luminance and chrominance information dedicated to each pixel has increased from 8 bit to 3 channels of 8 bit. Eight-bit encoding typically can provide 256 levels of luminance or hues of colors while 24-bit encoding supplies thousands of luminance levels and millions of colors. Third, the rate at which the screen images are presented is high (60 or 70 frames per second compared with 30 frames per second for real-time video rate) and getting higher. We are likely to experience further increases in the near future, allowing not only better luminance resolution and color rendition, but also capabilities for high-speed video analysis on consumer equipment that currently is only provided by expensive professional broadcast equipment.

Ultimately, any CRT accepting a digital signal, whether it be a broadcast signal or from a hardwired digital source, such as a computer or digital video cassette recorder, must convert its input back into an analog signal that can power the electron beam guns and scanning coils or alternatively send a voltage to each pixel on a liquid crystal display.

■ Digital Video Processing

The most efficient video capture and processing are achieved by converting the initial analog signal to digital immediately, that is, within the camera itself. As explained above, how the analog signal is sampled is of prime importance. Digitization by standard NTSC conventions samples luminance 13.5 million times per second, 8 bits at a time while the two chrominance channels are sampled at half that rate (a 4:2:2 ratio). This rate of sampling for luminance is adequate to provide the standard of 720 pixels horizontal by 486 pixels vertical in each frame, with corresponding lower pixel numbers for color rendition. The standard DV format (DV25) produces a data stream of 25 megabytes per second and uses a 4:1:1 ratio—a lower rate of color sampling to conserve on data rate and the storage space required. Despite this, horizontal resolution is higher at about 460–500 TVL than S-VHS or Betacam SP, the best analog formats commonly available (Table 16.5). In addition, there are a number of other reasons for using a digital format for video, including (1) the ease of file compression to achieve high quality images with less bandwidth, (2) no generational loss (i.e., a digital copy is exactly the same as the original), and (3) the ease of postproduction and editing by computer as opposed to old fashion erasing, cutting, and pasting carried out on magnetic tape.

Compression

Digital video formats can take advantage of at least five different types of compression: sampling rate, frame size, frame rate, intraframe compression, and interframe compression. Sampling rate compression uses sampling at lower than standard rates ("undersampling") and is used in the color channels in virtually all digital video formats. Some formats may compress by reducing the number of luminance pixels per frame, resulting in a smaller frame size and/or by reducing the frame rate; these types of compression are common for web video and digital still camera video where levels of compression as high as 50:1 are needed to reduce either the data rate or amount of storage space used. Intraframe compression refers to elimination of spatial detail within a single frame, usually color detail, and this can be acceptable provided it is not too severe. Interframe compression offers the most "bang for the buck." Because each frame of a video is set within the context of earlier or later frames, adjacent frames may be very similar to one another, in which case, one could avoid repetitious data from one frame to the next. For example, if nothing has moved in the scene between two frames, the two frames are identical. A more normal situation is that a few objects in the scene may be moving against a still background, and thus, new frames need only revised data about the moving objects. To do this, three types of frames can be created by the software: "I" frames, which have complete information and are used as key or reference frames; "P" frames, which are predicted frames based on the content of earlier frames; and "B" frames, which are interpolated from both prior frames and later frames that follow. Because P and B frames contain only information in those portions of the frame that are changing, P frames can be one tenth the size of I frames and B frames even smaller in size. Interframe compressed video contains all three types of frames, with more I frames being required during periods of fast action (e.g., car chases) and very few I frames being needed when motion is at a minimum (e.g., someone talking but standing still).

Digital video formats such as DV25 or MPEG-2 consist of a set of standards to which the video stream must conform; however, the information needed can be compressed or reorganized in many different ways by different types of hardware or software as long as afterward the data stream can be "decoded" to conform to the format standards. These processes of coding and decoding, represented by sets of computer algorithms, are referred to as "codecs," and there may be a number of codecs in use for any given format. Thus, software that handles the MPEG-2 format may or may not be able to decode a MPEG-2 file, depending on which of the dozen or more MPEG-2 codecs the software has available to it. Not all codecs are equal. Some are more effective (higher image quality per unit storage space) or faster than others.

Generational Loss

Whenever one copies an analog signal any deviation in voltage caused by instrument or environment will be thought to be part of the signal and will be superimposed on it (**Figure 11.17A**); therefore, each new generation (copy) of the signal will have more noise

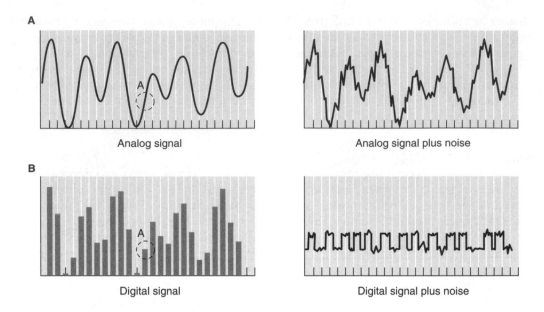

Figure 11.17
Electrical noise degrades analog but not digital signals. (A) Analog signals consist of a continuous voltage versus time stream (left panel). The addition of noise creates voltage fluctuations that cannot be separated from the signal (right panel). (B) Digital signals are not continuous and consist of a sequence of intensity magnitudes (numbers) sampled at standard time intervals (left panel). Sampling of the analog signal at A is represented by the bar labeled A in the number stream. The number sequence is coded into 0s and 1s, represented by sequences of square wave excursions of voltage. Even with the addition of noise, there is little difficulty in discerning the signal because noise does not come in square wave pulses (right panel).

than the previous; this often results in analog signals that are remarkably degraded after three or four generations. This does not happen to digital video because the information is encoded in ones and zeroes and therefore represented by trains of square wave voltage pulses that are almost never mistaken for noise (Figure 11.17B). Editing and image processing can involve many signal generations, hence the importance of this feature.

Image Processing
Compression is one example of image processing but is only a beginning. We have all craved the computer-generated "special effects" that we see in movies, and more subtle changes in brightness, contrast, and color range that are usually noticed only if they are *not* done. Wholesale video image processing is now taken for granted in commercial products and certainly could not be achieved without digital signals. Today, video image processing at a reasonable level is available even to consumers and scientists; however, one does need a computer and software that are up to the task.

First, assuming that a high-quality digital video camera is being used, one must get the digital signal to the computer. The DV25 and DV50 formats are a good choice and these formats use data packaging that exactly matches the IEEE 1394 serial transfer protocols. This protocol, called "firewire" by Apple (the developer) and I-link by Sony, can currently handle data streams up to 400 megabits per second (Mbps). The compatibility of the DV format and IEEE 1394 transfer is not a mystery—they were developed together. The data stream coming into your computer via an IEEE 1394 card at up to 50 Mbps (in the case of DV50) is sent directly to your hard drive for storage. This translates to a storage space of 3 gigabytes per minute or 180 gigabytes per hour; thus, you had better order a computer that has a large hard drive. Acquisition speed of the hard drive is also important. Because hard drives vary from moment to moment in their acquisition rate, a large buffer is needed to temporarily store data while it is being entered into your hard drive during capture or being extracted from your hard drive during editing or processing. A lot of RAM will do the trick, with 2 or more gigabytes being desirable; leave room for more memory expansion as your demands on the system escalate. During later editing and processing, assuming that you would like to do it in real time, one will have to perform calculations on at least 32 Mb of video data each second; thus, your CPU speed had better be lightning. Three GHz is a good minimum to start with; thus, you are not going to get away cheap. Even though you wish to make digital signals your standard fare, you will undoubtedly want to import and digitize analog signals at some point so you should not be without a video card that can do the job. Finally, you will probably want more storage space on either tape (a DV digital VCR would be nice) or swapable hard drives and playback on a digital monitor; however, this is all doable for less than $5000 and is often within a scientist's budget! Do not forget video-editing software, such as Adobe Premier Pro, that can handle a variety of digital formats.

References and Suggested Reading

Video Microscopy and Television Technology

Benoit H. *Digital Television, MPEG-1, MPEG-2 and Principles of the DVB System,* 2nd ed. Oxford: Focal Press (Elsevier), 2002.

Inglis AF, Luther AC. *Video Engineering,* 2nd ed. New York: McGraw-Hill, 1996.

Inoue S, Spring K. *Video Microscopy,* 2nd ed. New York: Plenum Press, 1997.

Murphy DB. *Fundamentals of Light Microscopy and Electronic Imaging.* New York: Wiley-Liss, 2001.

Luther AC. *Principles of Digital Audio and Video.* Boston: Artech House, 1997.

Noll AM. *Television Technology: Fundamentals and Future Prospects.* Norwood, MA: Artech House, 1988.

Roberts RS, ed. *Television Engineering: Broadcast, Cable and Satellite, Part 1.* London: Pentech Press, 1985.

Shotten D, ed. *Electronic Light Microscopy Techniques in Modern Biomedical Microscopy.* New York: Wiley-Liss, 1993.

Tancock M. *Broadcast Television Fundamentals.* London: Pentech Press, 1991.

Wayne RO. *Light and Video Microscopy.* New York: Academic Press, 2008, p. 354.

Digital Cameras and Microscopy

Spring KR. Scientific imaging with digital cameras. *BioTechniques,* 2000;29:70–76.

Sluder G, Wolf DE, eds. *Digital Microscopy,* 3rd ed. (A second edition of "Video Microscopy," *Methods in Cell Biology,* Vol. 72), Academic Press, New York, 2007, p. 632.

Digital Video and Image Processing

A Digital Video Primer: An Introduction to DV Production, Post-Production, and Delivery. Adobe Corporation, 2006:54. Available in PDF format at www.Adobe.com.

The DV, DVCAM and DVPRO Formats. 2006:23. Available in HTML format at www.adamwilt.com.

Gonzales RC, Woods RE. *Digital Image Processing,* 3rd ed., Upper Saddle River: NJ: Prentice Hall, 2007, p. 976.

Wu Q, Merchant F, Castleman K. *Microscope Image Processing.* New York: Academic Press, 2008, p. 576.

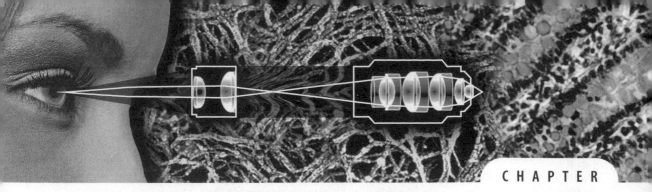

Fluorescence Microscopy

Because of its sensitivity and ability to detect even single molecules, fluorescence microscopy has become the most widely used technique in light microscopy. This chapter presents the fundamentals of wide field fluorescence microscopy and laser-scanning confocal microscopy.

■ The Molecular Basis of Fluorescence

Molecules absorb light by moving their electrons from a lower energy state to a higher energy state, a process described in Chapter 5. Only a subset of these molecules reuses this energy to emit light—the phenomenon referred to as fluorescence. The high-energy electrons in most molecules simply lose their energy by transitions through a series of rotational and vibrational states that are much too close to each other in energy level to result in emission of visible light. Most of this kinetic motion is transferred to the solvent and dissipated as heat. Fluorescent molecules, however, take the short way down. After releasing a certain amount of kinetic energy, the electrons involved fall directly to the ground state, thereby emitting light.

These events can be represented by a Joblonski diagram, as shown in **Figure 12.1.** Both ground and excited states for an electron are represented as an energy trough. Within each trough is a ladder of energy levels closely spaced that represent different degrees of kinetic energy available within that orbital. Absorption of a photon provides energy to move an electron from the ground state to the excited state (upward arrow, Figure 12.1). After being in the excited state, the electron loses potential energy by conversion to kinetic energy and then drops all of the way down to ground state as it is relieved of its energy by emission of a photon (downward arrow, Figure 12.1). All of these events take place within 10^{-12} seconds, which is referred to as the "fluorescent lifetime" of the excited state. Also, the energy retrieved during photon emission is less than that absorbed by the electron a moment earlier. This results in the emitted light having a longer wavelength (lower energy) than that of the absorbed light. This difference in energy, referred to as the "Stokes shift," is shown in **Figure 12.2A.** The emission spectrum, a probability envelope for light emission by a dye, is shifted to longer wavelengths compared with the excitation spectrum, a probability envelope for light absorption. These spectra represent a population response by millions of dye molecules. Although the excitation spectrum is broad, representing the influence of numerous rotational, vibrational, and solvent states, these all lead to the lowest energy level of the excited state from which the emission

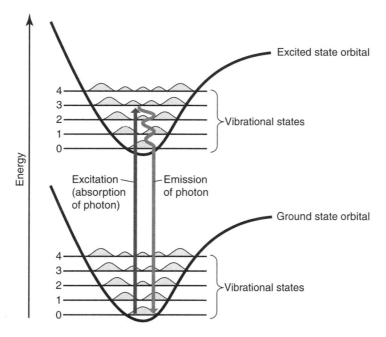

Figure 12.1
Electronic transitions during fluorescence. The Jablonski diagram showing energy levels of the ground state and excited state as well as the vibrational energy sublevels in each. Absorption of a photon is accompanied by transition of an electron from the ground state to an energetic vibrational level in the excited state. After decay of the electron energy to the lowest vibrational level in the excited state, a photon of lower energy is emitted as the electron drops down to the ground state.

transition is launched. For this reason, the peak emission wavelength (2, Figure 12.2A) is the same regardless of the excitation wavelength (e.g., 1, 3, or 4 in Figure 12.2A). Only the probability of emission as indicated by emission intensity is affected. The electronic transition that powers emission is also subject to a large number of kinetic energy states that broaden the emission spectrum. Enough time exists between the excitation and emission transitions that the kinetic energy states experienced in one are independent of the energy states in the other. For this reason, excitation at any wavelength within the excitation spectrum will lead to a normal and full distribution of emission energies. Figure 12.2B illustrates the excitation and emission spectra of two common fluorescent dyes—tetramethylrhodamine isothiocyanate and fluorescein isothiocyanate. The Stokes' shift of both dyes is about 25 nm, and their emission spectra show a characteristic tail toward longer wavelengths because of the electronic transition ending in multiple rotational and vibrational energy levels within the ground state.

The Stoke's shift allows one to separate the exciting light from the emitted light using filters. This process is essential for fluorescence microscopy. As shown in **Figure 12.3A**, the key optical components of a fluorescence microscope include a mercury or xenon lamp whose light is brought in at 90 degrees to the optical axis between the objective lens and eyepiece. The light is reflected downward through the objective lens onto the specimen. This arrangement is referred to as epifluorescence illumination. Fluorescent dye molecules in the specimen absorb the light and emit light of a longer wavelength. The emitted light is gathered by the objective lens to form an intermediate image that is then magnified by the eyepiece. An important fact is that the emitted, image-forming light has to

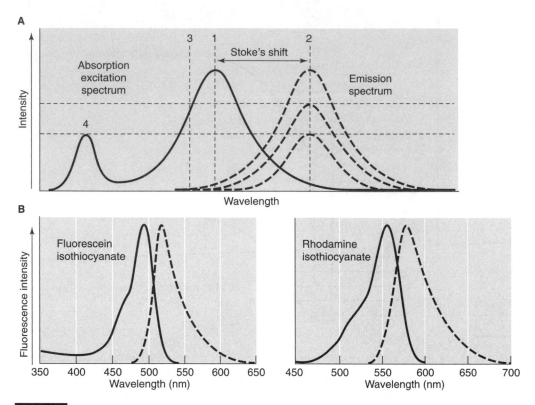

Figure 12.2

Fluorescence excitation and emission spectra. (A) Theoretical excitation/absorption and emission spectra showing that the intensity of emission but not its wavelength is affected by the excitation spectrum. (B) Excitation and emission spectra for the common fluorophores fluorescein and rhodamine.

pass straight through the mirror that reflected the incoming excitation light! Fortunately, the emitted light is of a longer wavelength, and the "dichroic" mirror used is able to transmit the longer wavelength while reflecting the shorter wavelength exciting light.

Positioned at the junction where illuminating light enters the optical axis is a filter cube containing three different components (Figure 12.3B). The component through which the exciting light first passes is referred to as excitation filter; it lets through light of the correct wavelength for exciting the fluorescent dye while blocking other wavelengths. The dichroic mirror (sometimes referred to as a beamsplitter) reflects this exiting light 90 degrees to illuminate the specimen. Emitted light coming back up the tube travels straight through the dichroic mirror and then passes through the third component of the filter cube—the barrier filter. The barrier filter allows the emitted light to pass on to the eyepiece but blocks any remaining exciting light that may have bounced off the specimen. Thus, excitation light is discarded in three ways after it excites the specimen: (1) It is transmitted through or scattered by the specimen and only a fraction of it is regathered by the objective, (2) the dichroic mirror reflects it back into the lamp housing, and (3) the barrier filter blocks its progress to the eyepiece. Efficient removal of incident light from the fluorescence image requires all three mechanisms because the incident light intensity is often three orders of magnitude greater than that of the light emitted by fluorescence.

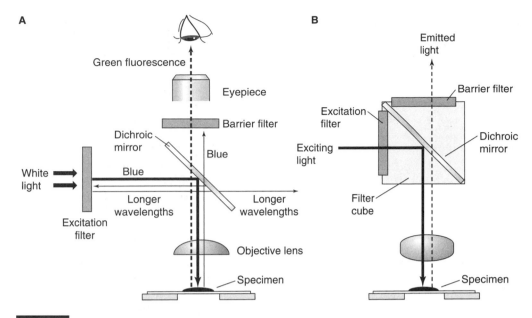

Figure 12.3
Location of filters in a fluorescence microscope. (A) Design of an epifluorescence microscope. White light from a full spectrum light source is passed through an excitation filter to select the color to be used for excitation (e.g., blue). The exciting light is reflected 90 degrees by a dichroic mirror and focused onto the specimen by the objective lens. Light emitted by the fluorophore at a longer wavelength (e.g., green) is gathered by the objective lens, transmitted through the dichroic mirror and passed through a barrier filter before image is seen. Blue exciting light scattered by the specimen is eliminated from the image both by reflection at the dichroic mirror and by absorption at the barrier filter. (B) The excitation filter, barrier filter, and dichroic mirror sets needed for specific dyes are components of interchangeable filter cubes.

■ Optical Filters

The filters used in transmission and fluorescence microscopy are of two major types: absorption and interference. Absorption filters efficiently remove selected wavelengths of light by incorporation of dyes (**Figure 12.4A**). **Table 12.1** lists a series of standard absorption filters used in different regions of the visible spectrum. The absorption spectrum for the RG-630 filter shows that red light is transmitted 95%, while only 1% of the blue light is transmitted. In contrast, the BG-28 filter transmits blue light efficiently (65%), while red light is almost completely blocked. For both filters, it should be noted that the cutoff between absorption and transmission is not sharp but occurs over 20 to 50 nm in wavelength, rather like the broad peaks of dyes in solution. In contrast, interference filters, based on structural features rather than dye properties, exhibit a much sharper cutoff between transmission and nontransmission. As shown in Figure 12.4B, interference filters are made of series of layered materials. The distance between layers determines which wavelengths will undergo positive interference and be transmitted. The distinction between transmitted and nontransmitted wavelengths can be very sharp—this can change over a range of 5 nm.

Filters are also classified based on the nature of their transmission spectrum. A shortpass filter blocks out all wavelengths of light longer than a specified cutoff, but allows shorter wavelengths to be transmitted (**Figure 12.5A**). A long-pass filter does the opposite. It blocks all wavelengths shorter than a specified cutoff, but allows longer wavelengths to go through (Figure 12.5B). Bandpass filters block passage of light at all wavelengths except for a narrow band of wavelengths (Figure 12.5C). These filters are the most selective, usually of the interference type, and are particularly useful when one is trying to excite two or more dyes simultaneously. Selection of wavelength by interference properties can also be used to produce dichroic mirrors that have a sharp wavelength cutoff in their

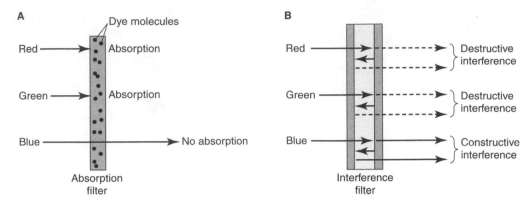

Figure 12.4
Construction of two types of filters. (A) Absorption filters contain plastic- or glass-embedded dyes that absorb unneeded wavelengths of light. (B) Interference filters are constructed of multiple layers of semi-reflective material and pass only those wavelengths that exhibit constructive interference between the transmitted and reflected rays. For example, the filter illustrated passes only blue light because the distance between its two reflective surfaces exactly equals a multiple number of wavelengths for blue light.

optical properties. These mirrors act as beamsplitters because they reflect light at wavelengths below and transmit light at wavelengths above the cutoff (Figure 12.5D). As discussed earlier, these mirrors serve as an important component of the filter cube used for epifluorescent illumination (Figure 12.3B). Selection of limited regions of the spectrum

Table 12.1	Terminology for Schott Standard Dye-Based Absorption Filters		
Transmission Range	**Range Designation**	**Examples**	**Description**
Ultraviolet	UG	UG-1	Wide pass, 300–400 nm
		UG-11	Wide pass, 250–400 nm
Blue/blue-green	BG	BG-1	Wide pass, 300–450, 700–1100 nm
		BG-28	Wide pass, 350–500 nm
Blue/green	VG	VG-6	Wide pass, 450–600, 600–800 nm (attenuated)
		VG-9	Wide pass, 450–600 nm
Visible, IR but not ultraviolet	WG	WG-295	Long pass, 295 nm cut off
		WG-360	Long pass, 360 nm cut off
Green, yellow, red	GG	GG-400	Long pass, 400 nm cut off
		GG-500	Long pass, 500 nm cut off
Yellow, orange, red	OG	OG-550	Long pass, 550 nm cut off
		OG-590	Long pass, 590 nm cut off
Red, Infrared	RG	RG-630	Long pass, 630 nm cut off
		RG-715	Long pass, 715 nm cut off
Visible but not IR (heat filters)	KG	KG-1	Wide pass, 300–750 nm
		KG-5	Wide pass, 350–700 nm
Visible, uniform attenuation (neutral density filters)	NG	NG-5	Wide pass, 400–1000 nm, 30% transmission
		NG-11	Wide pass, 350–1000 nm, 70% transmission

Information is from Schott, as posted at www.optical-filters.com.

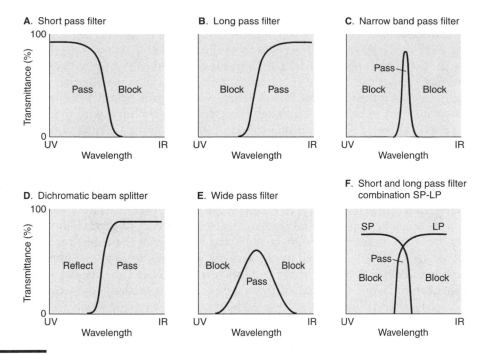

Figure 12.5
Types of filters used in fluorescence microscopy. Short-pass, long-pass, and wide-pass filters (A, B, E, F) are typically absorption filters, while dichroic beam splitters and narrow bandpass filters (C, D) are typically interference filters.

can also be achieved by using a single wide-pass absorption filter (albeit with low precision) (Figure 12.5E) or by a short-pass/long-pass combination (Figure 12.5F).

Spectral properties of typical filter sets found in filter cubes are shown in **Figure 12.6.** The filter set in Figure 12.6A is that used for dyes that absorb blue light and emit green

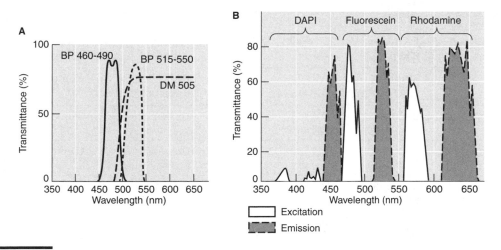

Figure 12.6
Component filter sets match the spectral properties of the dye(s) used. (A) Components of a green filter cube suitable for visualizing fluorescein emission. (B) Spectral characteristics of a triple excitation and barrier filter suitable for visualizing DAPI, fluorescein, and rhodamine fluorescence simultaneously.

light, the most common being fluorescein. The exciter filter is a narrow-bandpass filter having a transmission window between 460 and 495 nm. Blue light needed to excite the dye will pass through this filter. Green light from the source (510 to 540 nm) that might be confused with light emitted from the dye is not transmitted by this filter. The blue light passed through strikes the dichroic mirror and is reflected through the objective and onto the specimen. This mirror (DM 505) will reflect wavelengths below 505 nm but not those above 505 nm. Again, any green light that might be confused with dye emission will not be reflected onto the specimen; instead, it will pass straight through the dichroic mirror and be absorbed by the blackened wall of the cube behind the mirror.

The reflected blue light will then be focused onto the specimen by the objective lens. A small amount of the light will be absorbed by the dye in the specimen and then re-emitted by fluorescence, as described above. A vast majority of the exciting light, however, will be scattered off the specimen or will pass right through the specimen and most of it lost. Scattering of blue exciting light and the fluorescence emission of green light from the specimen occurs in all directions. The portion that falls within the acceptance angle of the objective lens will be regathered by the lens and sent back through the filter cube to form an image. Before the image is formed, however, the filter cube removes any blue exciting light that might be confused with fluorescence. This occurs in two steps as previously described. First, blue light is reflected by the dichroic mirror 90 degrees back to the light source. Second, the blue light is blocked by the "emission" or "barrier" filter. This filter is a bandpass filter that transmits light between 505 and 545 nm (including the fluorescein emission) but blocks any light having a shorter wavelength.

Table 12.2 lists the filter specifications for common general use filter cubes in fluorescence microscopes, each containing an appropriate filter set for one region of the visible spectrum. If the microscope is used for a specific dye, only one of these (or alternatively a custom filter set) might be in use. If the microscope is a multiuse instrument, it would likely have three or more of the filter cubes listed. In some cases, when multiple dyes are in use, a single filter cube can be designed to keep the signals separate for each. An example of such a filter cube, illustrated in Figure 12.6B, shows that the excitation spectra (in white) and emission spectra (in gray) for each of three dyes (4',6-diamidino-2-phenylindole or DAPI, fluorescein, and Texas red) are kept separate by a judicious choice of filters and

Table 12.2	Specifications for Standard Filter Sets Used in Fluorescence Microscopy				
Excitation Light	Excitation Filters	Dichroic Mirror	Barrier Filter	Typically Used for	Emission Color
UV	BP 340–380 BP 350–410	DM 400 DM 410	LP 430	DAPI, Hoescht 33258	Blue, green
Violet	BP 350–460 BP 420–490	DM 455 DM 460	LP 470	Lucifer Yellow	Blue, green
Blue	BP 450–490 BP 470–490	DM 500 DM 510	LP 520	Fluorescein isothiocyanate	Green, yellow
Green	BP 515–560 BP 530–560	DM 565	LP 575	Tetramethyl-rhodamine isothiocyanate	Yellow, orange, red
Yellow	BP 550–570	DM 580	LP 590	Lissamine R	Orange, red
Orange	BP 580–600	DM 605	LP 610	Texas red	Red, infrared
Red	BP 610–650	DM 660	LP 670	Cy-5	Infrared

dichroic mirrors. Even when dyes are chosen so as to have wide spectral separation, the filters required are usually not those optimal for each single dye alone. Thus, the advantage of being able to obtain multisignal images in color (without the registration problems of superimposing three or more separate images) comes at the cost of lower sensitivity for each signal.

■ Fluorescent Dyes

Very few techniques use materials made solely for the purposes of science. Fluorescent dyes were first synthesized in the late 19th century for use in the textile industry, and today, this still remains their common use. **Table 12.3** provides a short list of the most common dyes and their spectral characteristics. Dyes such as fluorescein and rhodamine continue to be popular because of their history of use for over a century; however, a number of new dyes have been recently synthesized that have superior qualities for microscopy. These qualities include photostability, high-quantum efficiency, low toxicity, and spectral suitability.

All dyes undergo photodegradation (commonly referred to as bleaching) and lose intensity if they are illuminated continuously. Susceptibility to photodegradation varies markedly from one dye to the next. Fluorescein, for example, is easily degraded and rapidly loses intensity; this fact is readily observed when one moves a fluorescein-labeled specimen from one field to another—the previously illuminated field is now very dim compared with the emission of the new field. Bleaching can actually prevent photography of weak fluorescence signals. In contrast, newer dyes such as those in the Alexa series exhibit little or no bleaching despite the fact that several members exhibit spectra almost identical to that of fluorescein.

Quantum efficiency, the yield of photons emitted relative to the number of photons absorbed, can range anywhere from 0.1 to 0.9 for typical dyes. Higher quantum efficiency allows the dye to be detected at lower concentrations. One example of an early dye that had a low efficiency was Quin-2, a first-generation calcium chelator whose fluorescence signal was used inside secretory cells to monitor calcium-mediated exocytosis (see Chapter 14 for details). The quantum efficiency was so low that the high concentrations of probe required altered the calcium signals being detected! Subsequent generations of calcium sensing dyes such as fura-2 and indo-1 had much higher efficiencies, resulting in signals that had not been attenuated by the dye. Toxicity of dyes can be a problem whether it is directed toward cells in the specimen or toward the microscopist. Well-known examples are DNA stains such as ethidium bromide and Hoescht dyes that are designed to intercalate into the minor grove of the DNA double helix. This fact accounts both for their selective staining ability and for their highly carcinogenic properties. Newly improved DNA stains such as SYBR green exhibit lower toxicity and increased brightness and selectivity.

Fluorescence dyes, like other light absorbing organic molecules, have extended systems of conjugated double bonds, often in the form of multiple aromatic rings, some having heterocyclic N and O atoms (e.g., see the structure of fluorescein in Figure 5.16C). As described in Chapter 5, these systems of conjugated double bonds create molecular orbitals in which the energy levels are separated by amounts equivalent to that of visible photons. This feature is essential for both the absorption and emission of photons during fluorescence. Also present are several charged groups that help make the dye more soluble in aqueous buffers. These charged groups also determine what structures will be labeled by the dye; basic dyes that are positively charged will bind to negatively charged molecules such as DNA, RNA, and acidic sugars, while acidic dyes that are negatively charged bind to positively charged molecules such as proteins and organelles such as

Table 12.3 Common Fluorescent Dyes

Fluorescent Dye	Peak Excitation Wavelength	Peak Emission Wavelength	Use, Comments
Indo-1	346/330	475/401	Calcium imaging
Fura-2	363/335	512/505	Calcium imaging
DANS	340	525	Protein conformation changes
DAPI	358	461	DNA/nuclear stain
Hoechst 33258	352	461	DNA/nuclear stain
Cascade blue	400	420	Immunocytochemistry
Lucifer yellow	428	536	Cell filling/tracing
Acridine orange	500/470	526/650	DNA, acidic organelles
$DiOC_{18}$	484	501	Membrane/lipophilic probe
FM-1-43	479	598	Membrane/lipophilic probe
Fluorescein isothiocyanate	494	518	Immunocytochemistry; bleaches relatively quickly
Cy-2	492	510	Immunocytochemistry
Alexa 488	495	519	Immunocytochemistry; very photostable
BODIPY fluorescein	505	513	Immunocytochemistry; relatively photostable
Fluo-3	506	526	Calcium imaging
Oregon green	496	524	Protein labeling
Propidium iodide	536	617	DNA/nuclear stain, flow cytometry
Tetramethyl rhodamine isothiocyanate	555	580	Immunocytochemistry; relatively photostable
BODIPY tetramethyl-rhodamine	542	574	Immunocytochemistry
SNARF-1	548	587	Proton imaging
Cy-3	550	570	Immunocytochemistry
Alexa 546	556	573	Immunocytochemistry
Phycoerythrin R	565	575	Immunocytochemistry
$DiIC_{18}$	549	565	Membrane/lipophilic probe
Texas red	595	615	Immunocytochemistry, relatively photostable
Alexa 594	590	617	Immunocytochemistry
Cy-5	650	670	Immunocytochemistry
BCECF	482/503	520/528	Proton imaging

Spectral values are for most probes in aqueous solution bound or linked to their target. Most values are taken from the handbook at probes.invitrogen.com.

mitochondria and secretory granules. Some dyes are pH sensitive and undergo changes in fluorescence as the pH is raised or lowered—a fact that is used to advantage in detection of basic or acidic organelles. Other fluorescent dyes have few or no charged moieties, indicating that they are "hydrophobic" or "lipophilic." These dyes tend to stain structures containing lipids such as lipid droplets and membranes.

■ Quantum Dots

Quantum dots are an alternative to conventional fluorescent dyes. These multiatomic clusters of semiconducting elements absorb and emit light much as traditional fluorescence dyes do, but they have a number of useful properties that dyes do not have. Quantum dots are extremely bright, have broad excitation spectra and narrow emission spectra, and are extremely photostable. These properties extend directly from the fact that these dots consist of rigid ensembles of semiconducting atoms—essentially nanocrystals that are protected from the aqueous solvent by a relatively inert and invariant matrix.

As shown in **Figure 12.7A,** a quantum dot consists of a core of semiconducting atoms (black), an insulating shell of atoms (white), and a final coating that renders the outer surface hydrophilic and easily suspended in aqueous buffers. The core represents the

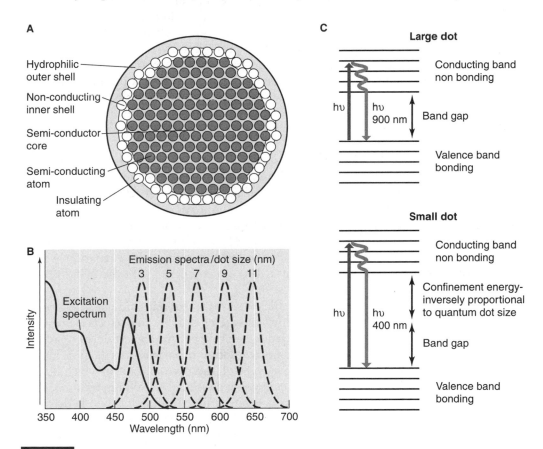

Figure 12.7
Structure and spectral properties of quantum dots. (A) Substructure of a quantum dot. (B) Excitation and emission spectra of quantum dots of varying diameters. (C) Electronic energy levels in a quantum dot. There is an energy difference between the conduction band and valence band, and hence, the wavelength of light emitted can be tailored by controlling the size of the dot.

fluorophore and consists of a mixed semiconductor containing several thousand atoms arraigned in an invariant geometry. Typical core compositions include CdSe, CdS, and CeSe, all being elements that can form coordination bonds as well as both extended and localized electronic orbitals. The electronic and optical properties of these nanocrystals are unique. Like metals, they form extended electronic orbitals that can involve an almost infinite number of atomic nuclei, as in copper or silver wire. In metals, nearly all outer-shell electrons circulate freely in these orbitals and can be directly measured as a current when a voltage gradient is applied. These numerous orbitals have energy levels so close to one another that they form an energy continuum referred to as a "conduction band." These electrons are delocalized and do not participate in bonds between nuclei. In con-trast, the inner-shell electrons are held locally by specific atomic nuclei. They participate in chemical bonds, and because of this, they are referred to as "valence" electrons.

In metals, the jump in energy from a valence electron to a conducting band electron is so small as to present almost no barrier—a tiny voltage will suffice. In contrast, the energy jump from valence to conducting band in a nonmetal insulator is so high that it never happens—hence its insulting properties. Semiconducting elements are in between. Usually the semiconductor combines atoms such as selenium and sulfur that are willing to give up ground-state electrons with atoms such as cadmium that readily form con-ducting bands. The jump between valence and conducting band usually occurs only at fairly high voltages; alternatively, just enough energy can be provided by absorption of a photon. In fact, if these atoms make up an ensemble as they do in quantum dots, there will be thousands of different but closely spaced energy levels available in the con-ducting band. For this reason, these semiconductor nanocrystals are powerful ab-sorbers of light—100 to 1000 times more powerful than the best dyes available. In addition, this also explains why the light absorbed can range over a broad expanse of energies. Dye molecules have only a few excited energy levels available, and therefore, they have absorption/excitation spectra that encompass only a limited range of wave-lengths; if it were not for solvent interactions, the emission spectra of dyes would be even narrower.

In contrast, the quantum-dot excitation spectrum can span hundreds of nanometers, as shown in Figure 12.7B. This is of great advantage in fluorescence microscopy for three reasons. First, this broad range of wavelengths makes continuous spectrum light sources more efficient than expensive monochromatic laser illumination. Second, excitation over a broad range of wavelengths increases fluorophore absorption by manyfold, thereby providing for more intense fluorescence emission. Third, the efficiency of excitation ac-tually increases at shorter wavelengths. Thus, one can separate the excitation wave-lengths from emission wavelengths by at least 100 nm compared with the 20 to 40 nm achieved with dyes.

The core of the quantum dot, having efficiently absorbed light and elevated one or more electrons to the conducting band, is now ready to emit light. Fluorescence emission occurs as an electron drops down into a ground state from the lowest energy excited state—just as it would in a dye molecule; however, two properties of emission in the quantum dot are very different from that in a dye. Unlike a dye molecule in which bonds between atoms and solvent molecules are undergoing continual stretching, rotating, and bending motions, these nanocrystals are held together by coordination bonds that are fixed in orientation and length. Multiple vibrational states in dyes often result in broad-ening of the emission spectrum because the spectrum is actually a sum of spectra each of which represents an electron transition to a different vibrational state within the ground state orbital. The result is that the emission spectrum consists of not one but several peaks that when added together produce a substantial tail extending from the longer wavelength end of the main emission peak (for examples, see Figure 12.2B). In comparison, the ground state of a quantum dot does not include widely spaced vibrational states as do dye molecules. Another factor is that dye molecules are relatively small and exhibit random whole molecule motions that we describe as kinetic energy. In contrast, quantum dots are larger in size and exhibit considerably reduced velocities compared with individual dye molecules; therefore, the emission spectrum of a quantum dot is very narrow, a feature that is of great advantage when using multiple fluorophores as discussed in the next section (Figure 12.7B).

Another important feature of quantum dots is that the conduction band energy from which the electron drops to ground state, and thus the wavelength of light emitted, depends on the size of the core. There are two size effects involved. First, quantum dots must contain enough semiconductor atoms to ensure that the conduction band and ground-state energy levels are sufficiently close to be in the right range for emission of visible light. Second, now that we are in the visible light range, there is an opposite effect. The physical size of the dot is actually smaller than the volume required by quantum mechanics for a "proper" low-energy orbital to be set up! Thus, these orbitals are confined to an abnormally small volume, and the smaller the dot the more confining it is. Confinement of the molecular orbital raises its energy—the electron is forced to spend time closer to oppositely charged nuclei without actually being able to jump down to ground state and get as close as it wants (Figure 12.7C). To accomplish this, the electron must have more kinetic energy, and one might think of it as an added energy of confinement. Thus, decreasing the size of the dot increases confinement and the overall energy of the conduction-band orbitals. As a result, smaller dots emit more energetic, shorter wavelength photons when their electrons do drop down to ground state than do large dots (Figure 12.7B). Thus, quantum dots made out of CeSe can be "tuned" to emit light at any wavelength between 525 and 655 nm simply by changing their size. CdSe cores about 3 nm in diameter emit blue-green light, while CdSe cores 10 nm in diameter emit red light. Currently, the production of quantum dots having a very strict size distribution leads to emission spectra much narrower than that of dyes. Further improvements in producing dots of uniform size may lead to even better properties in the future.

Finally, in order for the semiconductor core to perform optimally, it must be coated by two outer layers. The first layer, referred to as the "shell," consists of material much like the semiconductor core, except that it does not participate in forming the conduction bands (open circles, Figure 12.7A). Its function is to minimize the effect of irregularities in the core surface that could otherwise serve as electron traps that de-energize electrons without the emission of photons. The second and outermost layer is to provide a hydrophilic coating. Without this coating, the dot would be hydrophobic and completely incapable of interacting with water. As a result, the dots would aggregate and form a solid phase incapable of functioning. The coating provides therefore a hydrophobic interior that bonds strongly with the core and shell and a hydrophilic exterior that interacts with water so as to produce a hydrated particle. The hydrophilic surface is covered with carboxyl groups that produce a high density of negative charge and serve as a useful chemical handle for conjugating quantum dots to proteins or other biological molecules. Quantum dots are too big to go into solution like dye molecules do. Rather, they are colloidal particles that are small enough to remain suspended by normal kinetic energy at room temperature, provided that they have an incentive not to aggregate. In the case of quantum dots, the negative surface charge repels nearby particles, thereby stabilizing the colloidal suspension.

Quantum dots have been used not only for immunocytochemistry at the cellular level but also for study of fluid flow in tissues and for macroscopic labeling of protein and nucleic acid blots.

■ Using Multiple Dyes

Microscopy using "double-label" and "triple-label" techniques carries with it its own set of technical requirements. The most important is to use dyes that can be easily distinguished by their spectrum. Although this seems obvious, in practice, this can be a challenge. The main reason for this is that dyes in solution exhibit broad excitation and emission spectra that frequently have long tails. These tails frequently overlap with the spectra of other common dyes. Consider a double-label experiment in which fluorescein and rhodamine are being used. **Figure 12.8A** provides the excitation and emission spectra for both dyes, and it can be seen immediately that the fluorescein emission spectrum forms a long tail at longer wavelengths. This tail overlaps with the emission spectrum for rhodamine. This means that the spectral "window" for detecting rhodamine emission, typically from 570 to 650 nm, will also be detecting some of the fluorescein emission. This is referred to as "bleedthrough" or "cross-talk"—some of the fluorescein signal is bleeding into the rhodamine signal. The effects of crosstalk can be seen in Figures 12.8B and 12.8C. In Figure 12.8C, rhodamine staining of mitochondria in plant cells is superimposed on a fluorescein signal coming from the cell walls of these cells (arrows). In Figure 12.8C, when cross-talk is eliminated, the signal is seen from only the mitochondria.

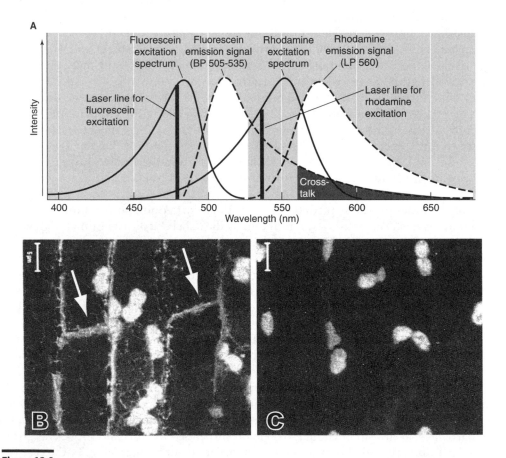

Figure 12.8
Bleedthrough caused by the spectral overlap of two dyes. (A) The fluorescein and rhodamine emission spectra overlap because they have long tails. The region of overlap leads to "cross-talk" or "bleedthrough," in which part of the fluorescein signal is mistaken as rhodamine signal if both dyes are excited simultaneously. (B) Rhodamine-stained mitochondria in plant cells. A fluorescein signal from the plant cell walls (arrows) has "bled through" and is seen in the rhodamine channel. (C) Rhodamine-stained mitochondria in the same tissue in the absence of fluorescein channel crosstalk. Bars = 5 μm.

The presence of bleedthrough can be evaluated by labeling the specimen with fluorescein only and then visualizing the specimen using the rhodamine filter set. Any fluorescence seen is either autofluorescence or bleedthrough fluorescence from the fluorescein. These two possibilities are distinguished by (1) observing an unstained specimen with the rhodamine filter set to detect autofluorescence and (2) observing the fluorescein-labeled specimen using the fluorescein filter set to determine whether the bleedthrough pattern is identical (but at lower intensity) to the actual fluorescein emission pattern. Similar controls can be done to evaluate whether the rhodamine signal is being detected by the fluorescein filter set. In this case, a tail extending from the rhodamine emission spectrum into the shorter wavelengths reserved for fluorescein detection (typically 500 to 540 nm) would be the problem. As we can see from Figure 12.8A, this does not appear to be a likely problem because the rhodamine emission spectrum falls very sharply on the short wavelength side. If such a problem arose, it would more likely be caused by use of a wide-pass or long-pass filter for fluorescein emission, rather than a narrow-bandpass filter appropriate to a double-label experiment.

There are several ways to minimize fluorescein bleedthrough. First, by the use of appropriate filters, one can narrow the rhodamine detection window to eliminate more of the fluorescein tail. This solution comes at the cost of losing sensitivity in the rhodamine channel. Second, one could reduce the intensity of the fluorescence signal by either using less fluorescein dye or reducing the intensity of fluorescein excitation through appropriate filter changes. This solution is at the cost of reducing sensitivity in the fluorescein signal. The third solution is to change one of the dyes being used; in this case, one would want a fluorescein substitute that either emits at shorter wavelengths and/or exhibits less of a long-wavelength tail. Alternatively, one could use a rhodamine substitute such as Texas red that emits at even longer wavelengths.

In order to minimize these problems from the onset, one should use dyes that are reasonably well separated in spectrum. For double labeling, the most likely combination is a fluorescein mimic, such as Alexa 488 and tetramethylrhodamine. Both dyes are relatively stable to photobleaching. For triple labeling, one might use a combination of an ultraviolet-excitable dye such as cascade blue that emits in the blue, Alexa 488 that emits in the green, and Texas red that emits in the red. Finally, quadruple labeling requires restricted filter bandwidths and dyes ranging all of the way from ultraviolet excitable to red excitable. Cascade blue, combined with fluorescein-like Alexa 488, yellow emitting tetramethylrhodamine, and the far red-emitting Cy 5 will work if the filer sets are designed correctly. Of even greater use for multidye labeling is the use of multiwavelength, full-spectrum determinations that use computer software to find the linear combination of dye spectra corresponding to the experimentally observed spectrum at each image point. This technology is now available in some laser-scanning confocal microscopes (this chapter), interferometry-based microscopes (see Chapter 13), and monochronometer/line scan camera equipped microscopes (this chapter).

Second, we must optimize image registration if we are using a multi-image approach to record our fluorescence. One must use a high-quality objective lens that has low chromatic aberration that would otherwise cause color registration problems. Generally, the best objective lenses for fluorescence microscopy are high numerical aperture fluorite (e.g., Plan NeoFluor) or apochromates that are not flat field corrected to increase light throughput. Furthermore, recording of the image requires either color photography or registration of multiple black and white images, one for each wavelength. To achieve registration, one can use a mixture of small fluorescent latex beads emitting at three or four different wavelengths as a test specimen; beads about 1 micron in diameter or less are available from a number of microscopy supply houses. Alternatively, one can use quantum dots that are smaller and brighter than latex spheres. A comparison of the fluorescence image for each wavelength with the brightfield image will not only determine whether there is a registration problem but will also indicate how to correct for the error at each wavelength. In some cases, it is possible to use beads as an internal standard by mixing them with the cells to be visualized.

■ Autofluorescence and Background Fluorescence

Tissues and cells themselves often contain fluorophores, common examples of which are listed in **Table 12.4.** Some tissue fluorophores may be subjects of study, while others are simply a nuisance that get in the way of visualizing the "real" data.

An example of an endogenous fluorophore subject of much biophysical study is chlorophyll, the light-harvesting pigment found associated with photo-system proteins of organisms ranging from photosynthetic bacteria to higher plants. If one uses the plant chloroplast as an example, chlorophyll is found as a two-dimensional array of molecules arraigned on a protein substrate, the assembly referred to as an "antennae complex." In essence, the chlorophyll array acts as an organic photosensitive semiconductor. Absorption of a photon bumps an electron up to an excited energy state that is transferred from chlorophyll to chlorophyll until the electron is handed off to a photo-system protein complex for conduction through the membrane and donation to oxygen. Along this electronic pathway, there are several points at which a drop in electron energy can result in either fluorescence or transport of hydrogen ions to form a hydrogen ion gradient to be used as an energy source for ATP synthesis. The fluorescence emission spectrum of chlorophyll represents an important probe of an electron pathway that is protein conformation dependent. The fluorescence detected microscopically can be used both to determine photo-system distribution and density as well as to evaluate the functional state of photo-system proteins in vivo. Nevertheless, the strong fluorescence of chlorophyll prevents the use of many exogenous probes in tissues such as leaves that contain a large amount of this light-harvesting pigment.

Endogenous fluorescence can also form the basis for semiquantitative distribution of a number of vitamins and cofactors important to the synthetic process in many cell types; however, they can also act as background fluorescence that makes detection of exogenous fluorescent dyes difficult. In some cases, fluorescent products are actually formed during chemical fixation of the tissue. A common example occurs in tissues that have been fixed with glutaraldehyde to cross-link protein constituents. The protein polymers that result contribute to significant autofluorescence that is very noticeable in imaging fluorescein-like dyes. Fortunately, this can easily be avoided by fixing tissues with formaldehyde rather than glutaraldehyde so that cross-linked protein polymers are not formed.

Table 12.4 Endogenous Fluorophores		
Fluorophore	**Source**	**Emitted Light**
Oxidized riboflavin	Milk	Blue
Fatty acids	Margarine	Blue
Catecholamines	Brain, adrenal medulla	Blue
Riboflavin (Vitamin B$_2$)	Yeast, liver, butter	Green, yellow
Carotenes	Carrots, leaves	Yellow
Chlorophylls	Leaves of green vegetables	Red
Porphyrins	Brown egg shells, liver, bile	Red

Autofluorescence must be evaluated as a potential complication in any study using exogenous fluorescent probes. Its presence can readily be detected by observing unstained tissue using the same filter set that will be used for the exogenous dye. In many cases, one can work around autofluorescence by either using dyes that fluoresce in a different spectral region or accounting for autofluorescence during data analysis.

■ Laser Scanning Confocal Microscopy

Wide-field microscopy, that is, microscopy in which images are formed by lenses gathering light from all specimen points simultaneously, has been the cornerstone of light microscopy, including the vast majority of fluorescence microscopy up to 20 years ago. One disadvantage of wide-field microscopy is that the image is formed by both in-focus and out-of-focus light. Objective lenses form useful images of thick specimens over only a limited range of depth. This range, referred to as depth of focus, is relatively large for low-magnification lenses but relatively narrow for high-magnification lenses. As a result, one can bring different layers of the specimen into focus sequentially by continuous movement of the focusing knob to change the distance between the surface of the specimen and the objective lens. Some techniques such as brightfield and phase contrast have greater depth of focus, while other techniques such as differential interference contrast have a relatively narrow depth of focus.

We usually rely on the focused information of the image and ignore the light that is being gathered from regions of the specimen either deeper than or more superficial than the region "in focus." The out-of-focus light presents little problem in transmission microscopy because it just adds to the general brightness of the background; however, for techniques that require a black background, such as fluorescence and darkfield microscopy, out-of-focus light is distracting and decreases the viewer's ability to recognize in-focus features.

Several solutions are available for this problem. The simplest is to avoid thick specimens by using sections of tissue. Understanding of the three-dimensional aspects of the sample, however, is limited to trying to combine the information present in a series of sections cut in sequence from an embedded sample. The number of hours spent doing "serial sectioning" is tremendous, but it is the old "tried and true" method for reconstructing the three-dimensional features of a specimen.

Much more palatable is using the out-of-focus light as well as the in-focus light to model the structure of a specimen using computational methods. These computational methods are referred to as "deconvolution" techniques, and their attractive feature is that most of the work is done by computer rather than the operator themselves. In a typical deconvolution experiment, a series of images are recorded at specific depths within the tissue, usually by electronic imaging. In order for the deconvolution algorithms to use the out-of-focus information, they must be provided with information on the geometry of out-of-focus light that is very much dependent on the lens design. What is needed is the "point-spread function"—how does the light emitted from a single point in the specimen spread out as it becomes "out of focus?" What is its intensity distribution in other optical planes? This can be determined by using tiny fluorescence beads—the closest thing we have to "point" sources of fluorescence.

The main problem with deconvolution methods is the substantial computation time involved, which can range from minutes to hours for a single image. Indeed, it is always a matter of hours to deconvolute a "stack" of optical images used to reconstruct the structure of a specimen. Fortunately, there is a very simple solution to the out-of-focus light problem—get rid of it! The easiest way to do this is to place a pinhole aperture just at the position where the image-forming train of light is focused to a "cross-over point." Here, the light rays emitted from a single optical plane within the specimen will pass

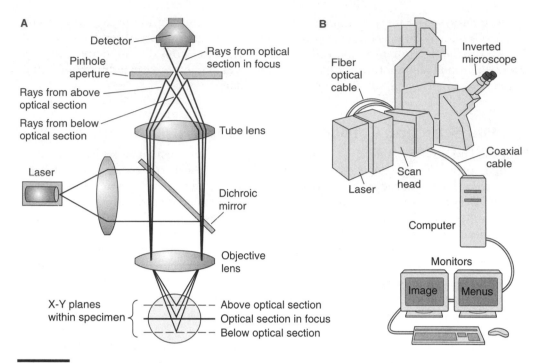

Figure 12.9
Confocal microscope design. (A) Confocal microscopy uses a pinhole aperture in front of the detector in order to eliminate out-of-focus light. The image recorded therefore comes from one thin optical plane within the specimen (solid line). Rays coming from other planes (dashed lines) do not go through the pinhole and therefore do not reach the detector. (B) Typical components of a laser-scanning confocal microscope workstation. Often multiple lasers are used each to service a different spectral region.

directly through the center of the aperture. In contrast, light rays coming from planes below or above will be brought to crossover either before or after the aperture, respectively, as shown in **Figure 12.9A.** For this reason, the out-of focus light will not pass through the center of the aperture but at some distance from it. Providing that the aperture is small enough, this out-of-focus light will be blocked by the aperture and never reach the light/image detecting device. The size of the aperture therefore will determine the thickness of specimen plane that is being visualized, and this thickness is usually much smaller than the depth of focus for the objective lens.

This means that images of specific planes within the specimen—usually referred to as "optical sections"—can be isolated and recorded, and entire "stacks" of these images used to reconstruct the structure of the specimen without ever having to mechanically cut the specimen apart. (You did not really want to use that microtome did you?)

Producing and recording optical sections in this manner is referred to as "confocal microscopy." As shown in **Figure 12.10A,** the image of a single optical section is much sharper than that obtained with nonconfocal microscopy in Figure 12.10B. Even if all the out-of-focus light is removed from a standard micrograph, one still sees in-focus information coming from many depths in the tissue superimposed (referred to as a projection image); spatial information in the z-axis (the optical axis) is absent. In contrast, a systematic stack of optical sections, as shown in Figure 12.10C, provides information along x-, y-, and z-axes. Information along the z-axis can have a resolution that is almost as great as that obtained in the x- and y-axes—approximately 0.3 μm. Computer reconstruction of this information can be displayed in three-dimensional or pseudo three-dimensional representations.

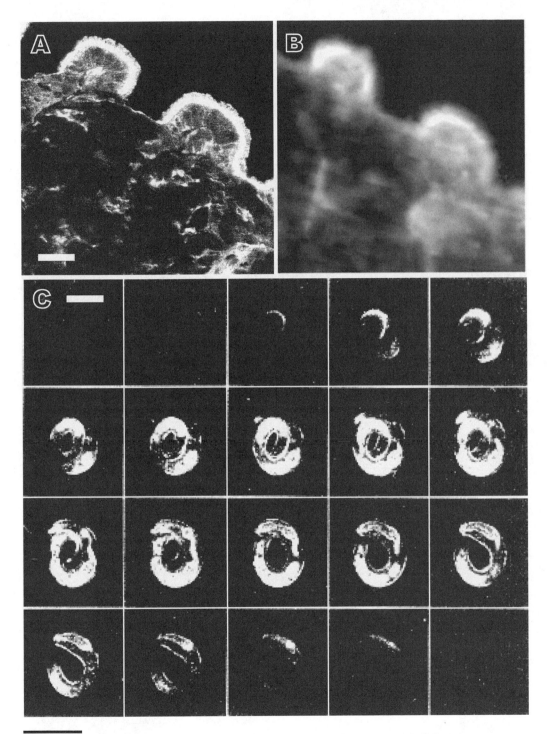

Figure 12.10
Comparison of confocal and nonconfocal images. (A) A single optical section from confocal microscopy of an epithelial cell layer of a frog oviduct is crisp and detailed. Bar = 10 μm. (B) Immunofluorescence from a similar specimen by nonconfocal microscopy is fuzzy because of the preponderance of out-of-focus emitted light. (C) A series of optical sections at different Z levels used to image a *C. elegans* embryo. Such a "stack" of images can be processed to construct a three-dimensional model of the embryo. Bar = 200 μm.

Thus, what is the trick? What technologies are required for confocal microscopy? First, to capture and process the information contained in an entire stack of images efficiently, we have to use electronic imaging and computer processing. Second, in order to provide the high-intensity illumination required for efficient fluorescence emission, we need lasers. Third, we need a scanning device that can focus the laser beam onto one specimen point at a time and raster the beam across the specimen in an exact and reproducible pattern. Fourth, we need a perfectly aligned pinhole aperture that can be changed in size along with a sensitive photomultiplier or charge-coupled device CCD chip for detection of light. All of these technologies are assembled in one integrated package in most current instruments, as illustrated in Figure 12.9B. Although the principle of confocal microscopy was reported over 50 years ago, it was not until the 1980s that the availability of lasers and computers made such an instrument feasible. We now examine in more detail each component of a laser-scanning confocal microscope (LSCM).

■ Lasers

Lasers use light-amplified light emission from solids and gases to provide highly coherent and intense sources. The workhorse for much laser-driven microscopy in the past has been the gas laser, as diagrammed in **Figure 12.11A.** Lasers operate on the principle of stimulated light emission. Stimulated emission results when an incident photon interacts with an atom having an electron in an excited state and triggers the drop of the electron to the ground state accompanied by emission of a second photon. The second photon has a

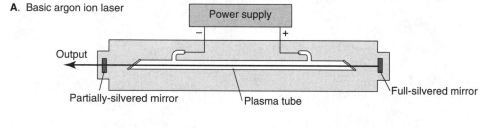

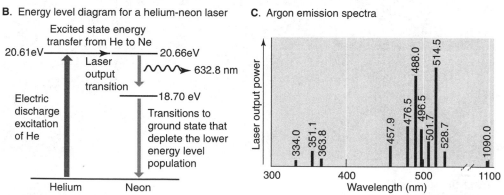

Figure 12.11
Gas laser design and principles. (A) The gas laser consists of a noble gas filled tube excited by high voltage discharge. One end of the tube contains a full-silvered mirror, while the other end contains a partially silvered mirror. Light emitted will ricochet from one end of the tube to the other to stimulate further emission. A small portion of the light will pass through the partially silvered end to be output as a coherent beam. (B) Electronic energy diagram of a helium-neon laser. A key feature is that the lower level energy state of the photon-emitting transition is continuously depleted so as to maintain a population inversion. (C) Typical line spectrum of a gas laser.

direction, polarization, and phase identical to the stimulating photon. This "two-for-one" amplification process can result in a positive feedback cycle that produces an extremely intense beam of collimated, polarized, in-phase light of a specific wavelength—the desired result whenever laser light is used. To achieve this effect, a number of special conditions need to be met.

First, there must be vastly more atoms with electrons in the high-energy state than there are atoms with electrons in the ground state. This condition is referred to as a population inversion and is essential so that photons gained through stimulated emission will not simply be reabsorbed by neighboring atoms in the ground state. At normal temperatures, over 99.99% of electrons are in the ground state so that a substantial effort must be made in order to change that. A common mechanism for generating a population inversion is to use high voltages to ionize noble gases—a mechanism similar to that of a xenon arc lamp. An important difference is that in an arc lamp only a tiny portion of the gas atoms need to be in an excited state, while in a laser, almost all of them have to be. Second, the lower energy state to which an electron drops during stimulated emission must be depopulated rapidly to avoid reversing the process by absorption.

A common scheme is to use four different electronic states to accomplish a population inversion, as illustrated in Figure 12.11B for a HeNe laser. A helium electron is first raised to an excited state using high-voltage discharge. Because the ground state and the excited state consist of a set of vibrational states, the energy absorbed in this process can be any one of a close set of energies, which makes this process robust. Second, this excited state is transferred to neon. The neon excited state then decays through release of heat and vibration to a somewhat less energetic excited state, which will serve as the upper state for stimulated emission. Third, a photon of proper wavelength stimulates the electron to drop from the upper to the lower state, thereby emitting a second photon. Fourth, the electron is removed from the lower state by a rapid, spontaneous decay to the ground state of neon. The important features of this scheme are that (1) the four electronic energy states used all have associated vibrational states so that the energy required at each transition is represented by an envelope of energies, which provides for greater throughput; (2) the upper energy level for stimulated emission is highly populated because electrons from an entire series of excited vibrational states in both helium and neon decay to this one level; (3) the lower energy level that receives electrons during stimulated emission is extremely sparsely populated because any electron entering this level is immediately siphoned off to the ground state below; and (4) the energy transition during stimulated emission is different in magnitude from all other electronic transitions. This four-level scheme sets up a situation in which any new photon generated by stimulated emission is more likely to encounter an atom that is "cocked" and ready to emit a new photon than one that is going to absorb the photon already produced. These conditions lead to a stimulated emission chain reaction.

Even in the best lasers, stimulated emission does not occur with high probability, and a photon will often have to travel several centimeters between each emission event. To highly amplify stimulated emission, therefore, the path length must be meters long! Because this is impractical, a resonant cavity is built into the device (Figure 12.11A). On each end of the laser are mirrors that reflect the emitted photon beam back and forth numerous times, with more photons being picked up each time. Although the mirror on the back end of the laser reflects with nearly 100% efficiency, the mirror at the front end reflects only 98% of the light that hits it. The remaining 1% to 2% is transmitted through the mirror and constitutes the laser beam output. Stimulated emission requires light input; provision for this can range from an incandescent lamp to light of a very defined wavelength provided by a LED. Although lasers produce intense beams of coherent, highly culminated light of very narrow bandwidth, they are not very efficient—only 1%

Table 12.5	Lasers Commonly Used for Confocal Microscopy	
Laser	Emission Lines	Examples of Dyes Excited
Ultraviolet-argon	351, 364	DAPI, Indo-1, Fura-2
Argon	457, 488, 514	Fluorescein isothiocyanate, Alexa 488, Oregon green, Cy 2
Krypton	488, 568, 647	Tetramethylrhodamine isothiocyanate, Alexa 546
Green He/Ne	543	BODIPY-R, Cy-3
Red He/Ne	633	Cy-5
Red laser diode	638	Cy-5
Titanium/sapphire	700–1000	Two-photon excitation

to 10% of the energy invested actually comes out as light. The remainder is given off as heat, and for this reason, most lasers of reasonable power (including those used in microscopy) must be cooled by forced air or forced water convection.

Lasers can operate either continuously (if the amplifying effect is great enough) or in pulses, with each pulse being triggered by light input. Most lasers used for confocal microscopy are of the pulsed variety, with the pulse frequency being in the picosecond to femtosecond range. Because this frequency is determined by the input of stimulating light, these lasers are said to be "pumped" by the external light source.

Most lasers emit at a number of "lines," as illustrated for an argon laser in Figure 12.11C. **Table 12.5** lists the most commonly used lasers and their most useful emission wavelengths. Because of their cost, durability, and wavelengths and because they are air cooled, the argon gas laser and the red and green helium-neon lasers are usually the starting point for laser-equipped microscopy. These lasers can last for 5 to 10 years of normal use and are not extremely expensive ($10,000). Krypton gas lasers are common but less durable; they often need to be replaced every year or two even with normal use. Ultraviolet argon lasers are quite expensive ($50,000 to $100,000) and often last only 3 or 4 years; however, these lasers do allow the use of ultraviolet excitable dyes, some of which (e.g., DAPI, Fura-2) are quite popular.

Light-emitting diodes (LEDs) are now replacing gas lasers as the light source of choice in confocal microscopy. An LED essentially operates like a light-sensitive diode in reverse: A silicon diode sandwich containing a layer of n-type silicon and a layer of p-type silicon is charged using plate electrodes so as to produce and maintain an excess of electrons and electron holes. These electrons and electron holes meet up at the interface between the two silicon types, and the electrons fall back into the holes, thereby emitting their energy as photons. The light at the interface is focused by reflection from polished and silvered surfaces to produce a beam. The major advantage that LEDs have over gas lasers is their long life. Most last two to four times as long as their gas laser counterparts and cost about the same.

■ Delivery of Laser Light to the Scanning Head

For many years, optical devices have been designed and tested with each component securely fastened to a solid platform so that after optical alignment was achieved (not a trivial task) it could be maintained for long periods of time. In addition, the entire instrument could become a rigid whole and be relatively immune to external vibration.

With the perfection of fiber optics—flexible solid glass strands that can carry light—optical components can now be coupled without being bolted to the same optical table. The coupling of multiple lasers to the scanning head of a confocal microscope is now done in this manner, allowing the lasers to sit on the floor under or beside the microscope. Fiber optic cables actually consist of two layers of glass, a central core through which the light is piped and an outer coat that is made of fused silicon glass that has a refractive index different from that of the core. Providing that light enters the fiber core at a proper angle, it will be propagated efficiently by total internal reflection off of the interface between the core and the coating. The intensity of the light during passage is reduced by 10% to 40%. Coupling of the laser to the fiber optic cable is managed by first spreading the laser beam and then focusing it with a culminating lens into the core of the cable.

A bundle of fiber optic cables delivers the laser light to a wavelength selection device (**Figure 12.12A**). Because lasers emit only at selected wavelengths, narrow-band excitation filters are generally not necessary. Wide-band excitation filters positioned in a computer-controlled filter wheel can be used if sequential scans using one laser at a time are desired, but this type of approach has several drawbacks. First, it takes more time for each scan to be collected separately and up to 0.5 seconds for changing filters between scans. Second, different filters for different laser lines can introduce image-registration problems. Instead, excitation of multiple dyes using multiple laser lines are typically carried out by simultaneous scanning. Light output from two to four lasers is bundled using dichroic mirrors and a culminating lens that produces one small-diameter beam used for scanning.

Better yet, recently manufactured LSCMs use a single acoustic-optical tunable filter (AOTF) that can change wavelengths and the percentage of transmission within microseconds. The filter consists of a tellurium oxide crystal whose structure can be compressed by acoustic waves (Figure 12.12B). These waves of compression, created by a radiofrequency piezoelectric device, propagate through the crystal much like sound waves propagate through air or water. These denser, compressed surfaces within the crystal have a high index of refraction compared with the crystal structure in between them and act as interfaces at which light waves are refracted. The light refracted from multiple compressions undergoes constructive interference, providing that the distance between compression waves is appropriately related to an integral number of wavelengths. This relationship holds for only a very narrow band of wavelengths. Thus, the compression waves of an AOTF act exactly like an interference filter with two important differences. First, the frequency at which compression waves are generated, and therefore the distance between them, is continuously adjustable over the entire visible light spectrum. This means that the same crystal, given the proper compression frequency, can isolate any wavelength within a few nanometers. Second, even at the permissible wavelength, the amount of light that is diffracted is proportional to the refractive index of the compressions. This too is continuously adjustable from 0% to 100% transmittance by changing the power of the piezoelectric device. Of even greater value is the fact that both wavelength and transmittance can be changed within microseconds!

This means that the laser line scanning the specimen can be changed from one horizontal scan to the next or even within the same horizontal scan. For example, "regions of interest" can be scanned or laser irradiated regardless of their shape simply by turning the incident light on or off via the AOTF at the appropriate points within each horizontal scan. A second important use is to eliminate the effects of bleedthrough from one channel to another that occurs during simultaneous scanning by two or more lasers. With AOTF control in a single horizontal scan, one channel can be collected in the

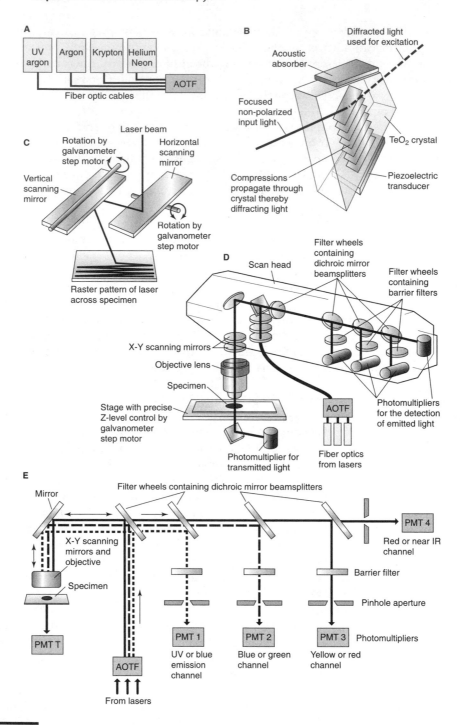

Figure 12.12

Components of the laser scanning confocal microscope. (A) Light output from multiple-lasers can be delivered to one AOTF by fiber optic cables. The AOTF selects which spectrum line will be furnished to the scan head of the LSCM. (B) A tellurium oxide crystal is subjected to waves of compression by a radiofrequency transducer. These compressions act as a tunable interference filter in which the frequency of compressions determines the wavelength of light selected and the magnitude of the compressions determines intensity of the light selected. (C) The scanning system of a confocal microscope includes precise step-motor–controlled mirrors. (D) Physical design of the scan head in a Leica NT-SC system. (Some components have been omitted for simplicity.) (E) Optical design of the scan head. Selectable dichroic mirrors and barrier filters contained in filter wheels dissect the emitted light into different spectral channels, each to be detected by a separate photomultiplier (PMT) tube. An additional PMT below the stage (PMT T) can be used to acquire a brightfield image.

forward direction, and another channel can be separately recorded in the reverse direction. The fast AOTF switching of lasers occurs precisely at the "turnaround" point between forward and reverse scanning. In this manner, only one dye is being excited on each horizontal scan, making bleedthrough impossible.

■ Scanning the Beam Onto the Specimen

The culminated excitation beam is scanned onto the specimen using a set of scanning mirrors and the objective lens. The position of each of the two scanning mirrors is controlled by a sensitive and rapidly responding galvanometer step motor that is computer controlled. The x-axis mirror moves as the beam is swept horizontally across the specimen, while the y-axis mirror moves only between horizontal scans so as to change the level of the scan. In this manner, a complete raster pattern can be achieved with the excitation beam (Figure 12.12C). Just as precise synchronization has to be achieved in video signal production and readout, the data stream from a laser scanning microscope has to be synchronized with the movements of the scanning mirrors. As the rastered beam passes through the objective lens and hits the specimen, fluorescence occurs, and the emitted light is gathered by the objective lens, deflected off the scanning mirrors, and then sent to the detectors using a beamsplitter. This occurs so quickly that the mirrors have scarcely changed position between passage of the exciting beam and the emitted beam.

In order to produce stacks of optical sections, the distance between the objective lens and the specimen surface must change each time a new "section" is to be acquired. This requires extremely precise movements along the z-axis, as little as 50 nm. This movement can be produced in two ways: The specimen itself or the objective lens can be moved up or down. The first method, often used on older instruments, requires that the specimen be light in weight—a tissue on a slide or small Petri dish is okay, but a large organ or an entire organism might not be okay. The second method, moving the objective lens turret, is used on newer instruments having infinity corrected optics and has the advantage that the stage is stationary and robust—able to carry the weight of rather large specimens. In both cases, the stage or objective lens turret is moved by delicate computer-controlled galvanometer motors.

■ Light Collection and Data Analysis

The emitted light is sent to a series of photomultiplier detectors, as shown in Figure 12.12D and E. At each detector, light of a certain wavelength range is stripped off by a dichroic mirror. After a 90-degree reflection off the dichroic mirror, the light is sent through a barrier filter, through the confocal pinhole, and to the detector. Most LSCMs have between two and five detectors. In order to fine tune the system to a specific set of dyes, the dichroic mirror and the barrier filter at each position can be selected from those on a filter wheel, usually by computer-controlled robotics. Typical choices for mirrors and filters are listed in Table 12.2. The usual strategy is to strip off the shortest wavelength first (ultraviolet or blue) and then strip off longer wavelengths in sequence working from green to yellow to red or near infrared.

The signal from each photomultiplier is separately amplified by dynode voltages that are preset by the operator. In order to maintain consistency from one experiment to the next and between experimental samples and controls, the magnifications used and the photomultiplier presets should be consistent. To reduce background noise, the photomultiplier tubes on some instruments can be cooled. At the computer, the signals from each phototube are digitized and then mapped by synchronization with the scanning mirrors. The magnitude of the signal at each pixel is pseudocolored according to channel or in some cases according to magnitude. Maps from several channels are often displayed simultaneously for comparison.

Because position data, as well as intensity data, can be digitized in a consistent manner for every optical section, it is relatively straight forward for appropriate software to create three-dimensional reconstructions of the specimen and to determine exactly how signals from different channels relate to one another in three dimensional space. For such data analysis, it is often useful to refer to the term "volume element" or "voxel"—a rectangular "cube" that has the dimensions of a pixel in the x- and y-axes and the thickness of the optical section in the z-axis. The rendering of voxel data so as to present three-dimensional models is treated later, but here we discuss "colocalization," a term used to compare multichannel data in a single optical section.

Colocalization refers to the process of identifying which voxels display significant signals in each of two channels. Colocalization data are often more important than the spatial pattern of each channel alone. For example, if one is trying to demonstrate that a certain protein is found in only the Golgi apparatus, one would look for voxels in which both the signal for the protein and signal for a marker of the Golgi coincide. After signal thresholds are set, computer analysis can easily locate those voxels and display them in a different pseudocolor.

■ Line Scanning and Spinning Disc Confocal Microscopy

The data-acquisition pattern described previously is referred to as "point scanning" because each image point is acquired and analyzed separately. Its advantages include easy use of binning, regions of interest, and image analysis to produce robust colocalization, three-dimensional modeling, fluorescence recovery after photobleaching (FRAP) and fluorescence resonance energy transfer (FRET) data (see Chapter 13). Its disadvantage is that it is slower than two other techniques available. One technique that improves speed is use of a CCD chip rather than a photomultiplier tube as the detector. Photomultiplier tubes are sensitive; however, their signal contains no spatial information, and their use requires point scanning. A CCD chip, however, allows acquisition of an entire horizontal line of image data at one time. Scanning is now accomplished with a single mirror used to position each horizontal line in sequence, as illustrated in **Figure 12.13A**. Currently, line scanning allows capture of 512×512 pixel images at the rate of 120 frames per second, suitable for resolving very rapid cellular events in time. This technology may become even faster in the future.

Another technique allowing rapid data acquisition is use of a spinning disc with an appropriate pattern of holes to interrogate the specimen point by point. As shown in Figure 12.13B, the laser beam is adequately spread to illuminate a specific pattern of holes in the spinning disc such that illumination will flash briefly onto each specimen point of a raster. The use of a patterned disc spinning at a rate of 6000 to 8000 rpm was pioneered by the Japanese microscopist Nipkow as a method to rapidly point scan a specimen. Although the method allows frame acquisition rates as high as 360 per second, it suffers from the fact that diffraction-induced spread of the beam during passage through these pinholes lowers excitation intensity and spatial resolution in the image. This was subsequently ameliorated by adding a second disc containing microlenses, one for each pinhole, to focus the illuminating light into the pinhole. This combination microlens/pinhole disc is referred to as a Yokogawa-Nipkow disc (to recognize both inventors) and solves this problem. This technique is not only fast, but also can be used with

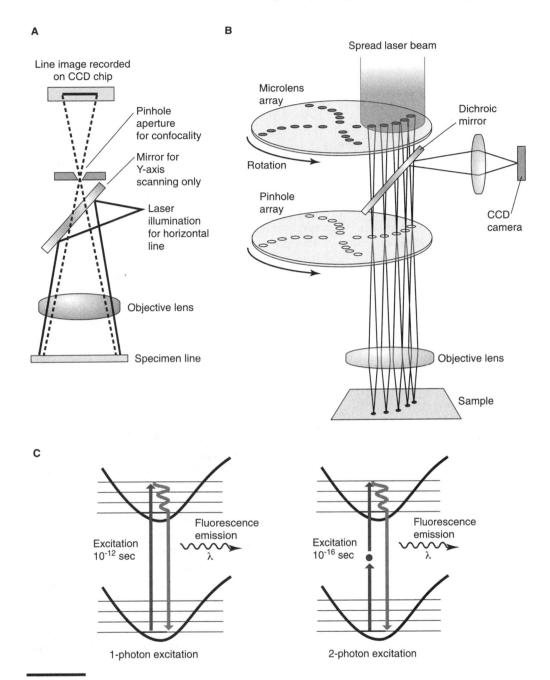

Figure 12.13
Alternative techniques in confocal imaging. (A) Acquisition of image line-by-line rather than point-by-point increases the scanning rate dramatically. (B) "Spinning disc" confocal microscopy uses an array of pinholes in a rapidly rotating disc to illuminate the specimen; very high scanning rates are achieved. (C) "Multiphoton confocal microscopy" uses excitation of fluorophores with two photons of low energy light rather than one photon of high energy light.

either lasers or vapor arc lamps as light sources. As each spot on the specimen is illuminated, the light emitted from that specimen point is gathered by the objective lens, passed back through the same pinhole in the disc, reflected 90 degrees off the optical axis, and focused onto a CCD chip for detection.

■ Multiphoton Confocal Microscopy

An electronic transition from ground state to excited state must absorb a relatively precise amount of energy from a photon. This need for a predetermined amount of energy, however, does not preclude the energy being supplied by two or more photons as long as the serial absorption events occur within an extremely short time—about 10^{-16} seconds (Figure 12.13C). This can be accomplished providing that the photon intensity is extremely high at the site of absorption. Indeed, even when laser light is focused by an objective lens, adequate photon fluxes are achieved only very close to the absorption site—photon density drops off with the sixth power of the distance from the focal point. As a result, the light emitted after a multiphoton absorption event is confined to a precise distance from the objective lens, and the image formed with this light will automatically be from one thin plane of the specimen. In other words, confocality is achieved without the image light passing through a pinhole aperture on the way to the detector.

Although excitation with three or four photons is theoretically possible, two-photon excitation is the most common. In this case, if a fluorophore normally absorbs a single photon having a wavelength of 400 nm, then it will also absorb two photons each having a wavelength of 800 nm. For this reason, the light source used for two-photon microscopy is a near-infrared (IR) laser typically supplying wavelengths between 700 and 1100 nm. As shown in Table 12.5, this is often a titanium-sapphire laser that is pumped by LEDs; the wavelength of laser light can be tuned within the range of 700 to 1000 nm. The remainder of the instrument is virtually identical to a standard laser scanning confocal workstation.

Illumination with IR light has several technical advantages that microscopists pay large sums of money to acquire. First, IR light, being of longer wavelength, is not as subject to scattering, as is visible light. This is born out by the fact that scattering of light is related to the fourth power of the wavelength with short wavelengths scattered to a much greater extent. A reduction of scattering allows deeper penetration of the specimen by the IR light, with the result that specimens as thick as 500 μm can be imaged throughout their depth. In contrast, one-photon confocal microscopy is usually limited to a depth of approximately 100 μm.

Second, IR light, because of its lower energy, is not capable of producing free radicals, as is energetic ultraviolet light. A serious limitation to viewing live tissues in a single-photon confocal microscope is the toxicity of free radicals formed during illumination. Continuous imaging of live tissue in a healthy state is possible for only 15 to 30 minutes, and this amount of time may be reduced in thicker specimens requiring higher intensity illumination. In contrast, IR excitation does not produce high levels of free radicals and therefore can be used to image live cells continuously for hours. This means that this technology is particularly useful in studying developmental or pathological processes that may occur over a period of hours or days. In some tissues, however, the IR illumination is absorbed, leading to heat that can damage the specimen.

■ Formats for Presentation of Three-Dimensional Data Sets

The ability of confocal microscopy to provide a series of optical sections that can be used to create a virtual three-dimensional model requires us to examine the best methods for presenting this information. First, we should understand how our brain creates three-dimensional space. Our visual processing of two-dimensional images is clear: Images having width and height along x- and y-axes are easily mapped onto our retina as a two-dimensional representation. The photoreceptor array in our retina, laid out in two dimensions, transmits this information to the brain by the optic nerve—an axon bundle that effectively acts like a huge parallel cable, with each "pixel" generated by a photoreceptor sent by its own axon. The data from each eye are mapped onto the visual cortex and processed by one side of the brain in essentially two dimensions.

These two-dimensional data are then evaluated for clues as to how to assign depth—that is, relationships between objects in the z-axis. There are four major ways in which this is done. First, the brain looks at object size and coloring. We are all familiar with the fact that more distant objects appear smaller and that their colors are muted, while closer objects appear larger and more brightly colored. Second, the brain looks for shadows. As described in our discussion of freeze-fracture images (Chapter 10), the positioning and shape of shadows indirectly provide information on object depth and texture. No skilled artist is without the ability to provide visual cues through appropriate shading. Third, the brain can store and analyze changing two-dimensional information as an object is rotated. The power of these two-dimensional analyses in providing three-dimensional perception is no better illustrated than in our legendary ability to recognize people's faces.

The final and most important perception of depth (assuming no other senses are used) comes from the brain comparing what the left and right eyes see. Because the left and right eyes are separated by some 60 to 90 mm, they have slightly different views of the world; more precisely, they view the world from a different angle. Small differences in object orientation in the two visual fields are analyzed by cross-talk between the two sides of the brain and give rise to what we commonly refer to as "depth perception." This is the "3-D" that we talk about when referring to comic books, movies, and "stereoscopic" vision. Clearly, in order to create this sensation of depth, a slightly different view of the world must be recorded by each eye, and for illustration purposes, two separate images must be presented, one to each eye. These two separate and slightly different images are referred to as "stereo pairs."

Based on these features of our visual system, there are a number of ways in which one can present three-dimensional image data. First, one can create an image that has perspective—that is, closer image features are larger, and deeper image features are smaller. This has the disadvantage that the deeper layers cannot be presented to our visual field at their full resolution—we must eliminate some information, and this is undesirable. For that reason, microscopic models are rarely presented in this manner. In some cases, pseudocoloring can be used to provide depth information, with hot colors (white, red, and yellow) representing closer features and cool colors (green and blue) or muted colors representing more distant colors. Color coding for depth, however, is less effective than shading, rotating, or stereoscopic presentation.

Shading to impart depth can be very effective. **Figure 12.14A** presents a neuron with numerous processes as an example. Rotation of the model is often combined with shading. Model rotation is simulated by constructing two-dimensional representations of the model every two degrees of rotation from -45 to $+45$ degrees and constructing a movie clip from these frames played repetitively. Most three-dimensional reconstruction programs have the ability to present shaded and rotating models. Another technique that is used routinely to present confocal microscopy data is a movie clip presentation of sequential optical sections from top to bottom that produces the perception of "flying through" the specimen.

Presentation of true stereoscopic images requires more effort but is well worthwhile. Presentation of separate still images to the left and right eyes first requires that two-dimensional representations of the model be constructed at rotations of $+6$ to $+12$ degrees and -6 to -12 degrees. These two images (at the same magnification) can then be placed side by side to form a stereo pair that when presented through a stereoscopic viewer allows each eye to see only one of the pair. Even today, stereoscopes for this purpose

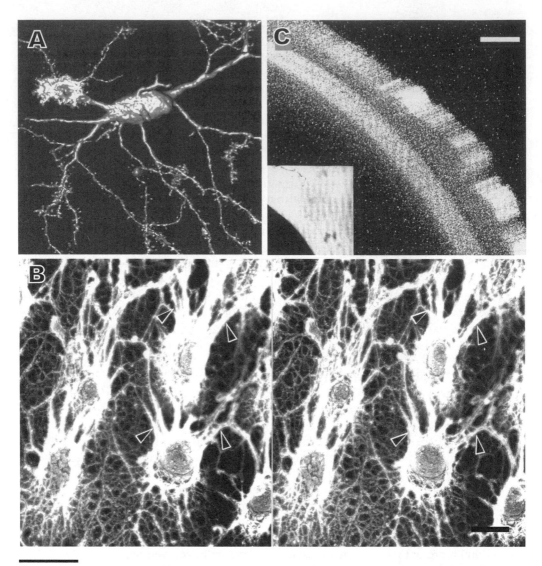

Figure 12.14
Presentation of three-dimensional images. (A) Contours of a three-dimensional model represented by shading. (B) "Stereo pair" electron micrographs of the surface of a sea urchin egg. The two images viewed separately through a stereo viewer can be fused by your brain to produce a three-dimensional image. Bar = 0.2 μm. (C) Stack of images superimposed to produce fluorescent surfaces in the jelly layers of a *Xenopus laevis* egg. In a red-blue "anaglyph" of this image, these surfaces could be perceived as three-dimensional through use of red-blue glasses. Inset: transmission micrograph of the jelly layers. Bar = 100 μm.

function much the same as the old parlor versions 100 years ago. It is important that the two images of a stereo pair be placed about 65 mm apart—the same as the average distance between our left and right eyes. It is also important to determine which image should be on the left and which should be on the right. If the images are exchanged right to left, features that should appear protruding toward the viewer will instead appear to be receding from the viewer. An example of a three-dimensional model presented as a stereo pair is shown in Figure 12.14B.

Photographs of actual objects, microscopic or macroscopic, can also be presented as stereo pairs, providing that the object is photographed from two different appropriate angles. If the object is microscopic, it must be tilted to obtain photographs from different angles. Tilting is defined as rotation around an axis that is perpendicular to the optical axis—either the x-axis or the y-axis. Specialized holders for tilting specimens, referred to as "goniometer" stages, are available for all research grade electron microscopes. The specimens tilted must be three-dimensional to start with, and good examples include freeze fracture replicas (Figure 12.14B) and thick resin-embedded sections. At the light microscopy level, tilting stages are uncommon because one usually uses specialized stereoscopic light microscopes that provide differently angled views to the left and right eyes simultaneously without specimen tilting.

Whenever one does use specimen tilting to obtain a stereo pair of images, the angle of tilt should be appropriate for the magnification. At high magnification the angle of tilt is usually less (±6 degrees), while at lower magnifications, the angle of tilt is usually greater (±10 to 20 degrees). The greater the angle of tilt between the two stereo pair images, the greater is the effect of depth, provided that the angle is not so great that ones eyes are strained to superimpose the two images.

Stereo-paired photographs can also be obtained of macroscopic objects, as demonstrated by the hundreds of thousands of stereocards published in the late 19th century. You can make your own stereo pairs with your regular camera. One simply takes a picture of the scene then steps to the right or left a few feet and takes another picture. One item to pay close attention to is to make sure that the picture frame is not moved in the vertical direction. To avoid this, one usually lines up a prominent feature in the scene with a frame edge or a focusing area edge. For the best effect, inclusion of objects at a series of depths, such as telephone poles, buildings, rocks, or trees, is advantageous.

Presentation of stereo images that are superimposed is also common, and these superimposed images are referred to as "anaglyphs." One technique is to superimpose images of different colors (e.g., blue and red). In that case, if one wears glasses that cover the left eye with a blue lens and the right eye with a red lens, then the left eye will see only the red image, and the right eye will see only the blue image. In this manner, each eye sees only one of the images, despite the fact that they are superimposed. A stack of optical sections prepared as an anaglyph, albeit without color, is shown in Figure 12.14C. The one disadvantage of color anaglyphs is that the stereoscopic image appears in black and white; however, one can project colored stereoscopic images during lectures by providing the audience with glasses that have polarizing lenses. In this case, each of the images making up a stereo pair must be projected by a separate polarizing projector, one image polarized horizontally and the other image polarized vertically. After reflection off a screen that maintains light polarization, the two images are seen separately by your right and left eyes because of the polarizing lenses in the glasses (see Chapter 16).

This technology can be extended further to include three-dimensional movies such as those produced and distributed by IMAX. In this case, the goggles that you wear have liquid crystal lenses that fluctuate in their light polarizing ability at video rate and in synch with the frames being shown on the screen through radio control from the projection equipment. For those of you interested in image technology (and you all are because you are microscopists!), any one of the three-dimensional IMAX movies is a must see! Similar technology is available for use with computers and monitors running specialized software.

References and Suggested Reading

Fluorescence Microscopy and Spectroscopy

Abramowitz M. *Fluorescence Microscopy: The Essentials.* New York: Olympus-America, 1993.

Haugland RP. *Handbook of Fluorescent Probes and Research Chemicals,* 6th ed. Eugene OR: Molecular Probes, 1996.

Herman B, Tanke HJ. *Fluorescence Microscopy (Royal Society Microscopy Handbooks),* 2nd ed. New York: Springer Verlag, 1998.

Horobin RW, Kiernan JA, eds. *Conn's Biological Stains: A Handbook of Dyes, Stains and Fluorochromes for Use in Biology and Medicine*. Oxfordshire, England: BIOS Scientific, 2002.

Jaiswal JK, Goldman ER, Mattoussi H, Simon S. Use of quantum dots for live cell imaging. *Nat Methods*. 2004;1:73–78.

Lakowicz JR. *Principles of Fluorescence Spectroscopy*, 3rd ed. New York: Springer, 2006, p. 954.

Mason WT, ed. *Fluorescent and Luminescent Probes for Biological Activity*, 2nd ed. London: Academic Press, 1998.

Miyawaki A, Sawano A, Kogure T. Lighting up cells: labeling proteins with fluorophores. *Nat Rev Mol Cell Biol*. 2003;5:S1–S7.

Molecular Probes Website. Molecular Probes, Inc. www.probes.invitrogen.com. Largest collection of information on fluorescence probes anywhere, including Haugland R. *Handbook of Fluorescent Probes and Research Chemicals*.

Periasamy A, ed. *Methods in Cellular Imaging*. Oxford: Oxford University Press, 2001.

Quantum Dot Technology Website. Quantum Dot Corporation. www.qdots.com.

Sharma A, Schulman SG. *Introduction to Fluorescence Spectroscopy*. New York: John Wiley, 1999.

Spector DL, Goldman RD, eds. *Basic Methods in Microscopy: Protocols and Concepts from Cells: A Laboratory Manual*. Woodbury, NY: Cold Spring Harbor Laboratory Press, 2005, p. 375.

Laser Scanning Confocal Microscopy

Abelson JN, Simon MI, Conn PM, eds. *Confocal Microscopy. Methods in Enzymology*, Vol. 307. New York: Academic Press, 1999:696.

Gratton E, Barry NP, Beretta S, Celli A. Multiphoton fluorescence microscopy. *Methods* 2001;25:103–110.

Hibbs AR. *Confocal Microscopy for Biologists*. New York: Springer, 2004, p. 474.

Larson DR, Zipfel WR, Williams RM, Clark SW, Bruchez MP, Wise FW, Webb WW. Water-soluble quantum dots for multiphoton fluorescence imaging in vivo. *Science* 2003;300:1434–1436.

Masters BR. *Confocal Microscopy and Multiphoton Excitation Microscopy: The Genesis of Live Cell Imaging (SPIE Press Monograph Vol. PM161)*. Bellingham, Washington: SPIE Press, 2006, p. 230.

Matsumoto B. *Cell Biological Applications of Confocal Microscopy*, 2nd ed. *Methods in Cell Biology*, Vol. 70. New York: Academic Press, 2002, p. 507.

Paddock S, ed. *Confocal Microscopy. Methods in Molecular Biology*, Vol. 122. Philadelphia: Humana Press, 1998.

Pawley JB, ed. *Handbook of Biological Confocal Microscopy*, 3rd ed. New York: Plenum Publishing, 2006, p. 988.

Sheppard CJR, Hotton DM, Sheppard C, Shotton D. *Confocal Laser Scanning Microscopy. (Royal Society Microscopy Handbook)*. Oxfordshire, England: BIOS Scientific, 1997.

White JG, Squirrell JM, Eliceiri KW. Applying multiphoton imaging to the study of membrane dynamics in living cells. *Traffic* 2001;2:775–780.

Presentation of Three-Dimensional Images

Chandler DE, Kazilek CJ. Extracellular coats on the surface of *Strongylocentrotus purpuratus* eggs: stereo electron microscopy of quick-frozen and deep-etched specimens. *Cell Tissue Res*. 1986;246:153–161.

Heuser J. Protocol for 3-D visualization of molecules on mica via the quick-freeze, deep-etch technique. *J Electron Microsc Tech*. 1989;13:244–263.

Russ JC, Dehoff RT. *Practical Stereology*, 2nd ed. New York: Plenum, 2000.

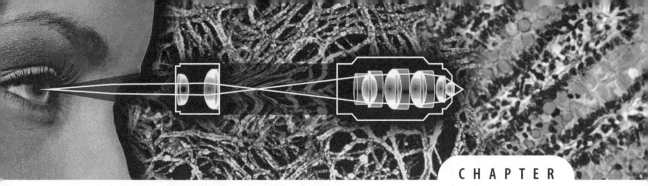

Microscopic Localization and Dynamics of Biological Molecules

By the nature of the technique, microscopy has always had as its primary objective the visualization of structure; however, the discovery of cells, subcellular organelles, and molecular complexes can be attributed not only to technical advances in image formation but also to advances in the "labeling" of structures based on their chemical composition, the earliest of which was the use the colored textile stains to visualize subcellular organelles. Details of structure inevitably lead to questions of physiological function that is dependent on specific molecules being at the right place at the right time. The ability to detect the distribution and dynamics of specific molecules therefore was and is the key linking the work of microscopists to that of biochemists and physiologists. In fact, this interdisciplinary link led to the formation of the field of cell biology in the 1950s and 1960s. This chapter addresses those tools that microscopists use to understand the dynamic distribution of macromolecules in cells.

■ Cytochemistry

Cytochemistry represents the earliest technique used for localizing specific macromolecules, usually enzymes, by microscopy. The enzyme to be localized is usually either a marker enzyme used to identify a specific class of subcellular organelle or an enzyme of functional importance. The technique is based on conducting an assay for the enzyme on a microscopic scale so as to produce a product that can be seen in either the light or electron microscope. Because the reagents needed for the assay must penetrate the cell membrane and possibly organelle membranes, the specimen for cytochemistry must be prepared in a manner that permeabilizes these membranes but at the same time does not redistribute or eliminate the enzyme's activity. One common method is to fix the tissue in formaldehyde, carry out the cytochemical procedure, then embed and section. Alternatively, frozen sections can work extremely well as long as the tissue structure is adequately preserved.

During cytochemistry, the tissue must be incubated in a buffer appropriate for optimal activity of the enzyme to be localized, taking into consideration osmolality, ion composition, pH, and temperature. Length of incubation is usually determined empirically, but 1 to 3 hours is common. Included in the buffer is an appropriate substrate that can be acted on by the enzyme at the concentrations required for maximal activity. The substrate

General Procedure for Enzyme Cytochemistry

1. Cut tissue into small cubes.

2. Fix tissue with formaldehyde.

3. Wash out fixative.

4. Incubate tissue with enzyme substrate using optimal temperature and buffer conditions.

5. Precipitate substrate (if needed).

6. Postfix with osmium and/or glutaraldehyde.

7. Embed and section tissue.

8. Observe areas of precipitated product by light or electron microscopy.

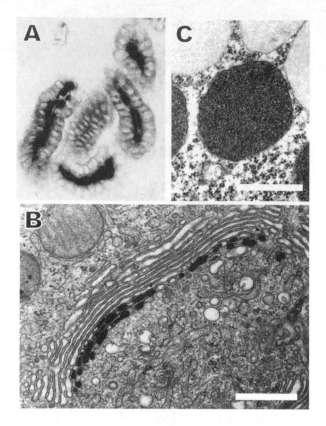

Figure 13.1

Cytochemistry: protocol and results. (A) Kidney epithelial cells from a zebrafish exhibit dark staining resulting from an alkaline phosphatase-catalyzed reaction. (B) Thiamine pyrophosphatase cytochemistry stains specific cisternae on the trans (maturing) face of the Golgi apparatus. Bar = 0.5 μm. (C) A peroxisome labeled by catalase cytochemistry using diaminobenzidine as a substrate. Bar = 0.5 μm.

must be designed such that the product is a precipitate that is either highly colored or fluorescent for light microscopy or electron dense for electron microscopy.

A classic example is the localization of alkaline phosphatase in tissues such as intestine or kidney using light microscopy. As outlined in the protocol of **Figure 13.1,** tissue is fixed with formaldehyde, then washed with TBS containing 50-mM Tris HCl at pH 9.5. Next, tissue cubes are incubated at room temperature with the chromogenic substrate 5-bromo-4-chloro-3-indolyl phosphate, which can be hydrolyzed by alkaline phosphatase and then oxidized by nitroblue tetrazolium to form a blue dye that precipitates. The darkly stained regions in Figure 13.1A identify the location of the enzyme in the zebrafish kidney. Similar cytochemistry for thiamine pyrophosphatase can be carried out at the electron microscopy level, as illustrated in Figure 13.1B. In this case, the pyrophosphate produced as a product is precipitated with lead, a heavy metal that is readily visualized in the electron microscope. The compartments of the Golgi apparatus labeled with this product represent specific cisternae on the trans (maturing) face of this organelle that face the plasma membrane.

Another classic example of cytochemistry is the identification of peroxisomes at the electron microscopy level using catalase activity. Catalase is an enzyme that oxidizes substrates using hydrogen peroxide. One substrate for this reaction is 3,3-diaminobenzidine,

Table 13.1	Examples of Cytochemical Procedures	
Organelle	**Marker Enzyme**	**Visualization**
Lysosome	Acid phosphatase	Lead phosphate precipitation
Plasma membrane	Sodium/potassium ATPase	Lead phosphate precipitation
	Alkaline phosphatase	Lead phosphate precipitation
	Adenylate cyclase	Pyrophosphate precipitation
Peroxisome	Catalase	Diaminobenzidine oxidation
	Peroxidase	Diaminobenzidine oxidation
Golgi	Galactosyl transferase	
	Nucleoside diphosphatase	Lead phosphate precipitation
	Thiamine pyrophosphatase	Pyrophosphatase precipitation
Mitochondria	Cytochrome oxidase	Oxidized dye
	Succinate dehydrogenase	Oxidized dye
Smooth endoplasmic reticulum, Golgi	Acyltransferase	Potassium ferricyanide precipitation
Specific granules, neutrophils	Myeloperoxidase	Diaminobenzidine oxidation

which when oxidized forms a polymeric precipitate that is darkly stained with osmium tetroxide. The reaction carried out by catalase is the following:

$$\text{Diaminobenzidine} + H_2O_2 \rightarrow \text{Oxidized Diaminobenzidine Polymer} \quad (13.1)$$

First the tissue is fixed with paraformaldehyde; it is then washed and incubated with 2% diaminobenzidine and 0.15% H_2O_2 at a pH of 10.5 and at 37°C. The tissue is then postfixed with osmium tetroxide (which turns the polymeric product a dark black) and dehydrated, embedded, and sectioned. After counterstaining lightly with lead citrate, electron-dense diaminobenzidine products are searched for by electron microscopy. Figure 13.1C illustrates a typical result showing a densely stained peroxisome from a germinating plant seed. This technique is robust and is frequently used to trace neuronal cell processes by microinjection of exogenous horseradish peroxidase into a neuron, allowing the enzyme to diffuse throughout the cell, and then staining the tissue by diaminobenzidine cytochemistry to visualize the complex dendritic and axonal processes of the peroxidase-loaded neuron.

Table 13.1 lists other examples of cytochemical assays that are robust and have been frequently used. Despite the widespread use of cytochemistry for localization of characteristic marker enzymes, this technique is limited in its application because it can be applied directly only to enzymes; however, localization of nonenzymatic proteins and other macromolecules can be done by immunocytochemistry.

■ Autoradiography

This technique is used to localize radioactive macromolecules or metabolites in cells or tissues. The basic method, outlined in **Figure 13.2**, consists of incubating the tissue in a radioactive precursor of the macromolecule, allowing the macromolecule to be synthesized and delivered to its site of action and then fixing, embedding, and sectioning the tissue. The sections, with radioactive macromolecules embedded, are coated with a photographic emulsion and stored in the dark for a few days or several weeks. The emulsion

Procedure for Autoradiography

1. Incubate tissue with radioactive precursor (e.g., H^3-leucine) for a short period (the "pulse").

2. Wash away radioactive precursor and add high concentration of nonradioactive precursor (e.g., leucine).

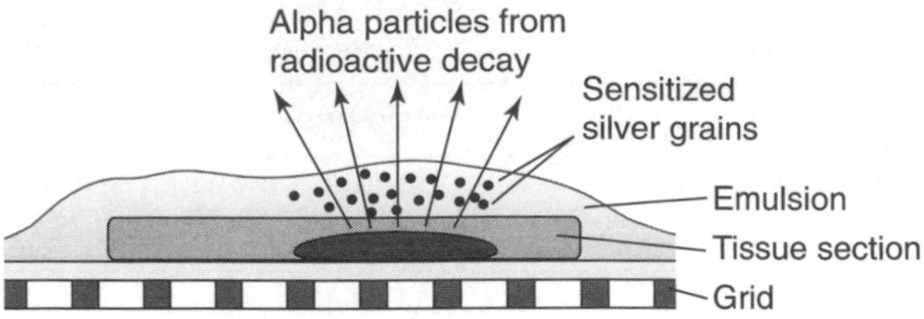

Alpha particles from radioactive decay
Sensitized silver grains
Emulsion
Tissue section
Grid

3. Continue incubating tissue in physiological buffer and remove and fix specimens at scheduled times (e.g., every 30 minutes). Embed and section each specimen.

4. Apply photographic emulsion to sections and store in the dark for several weeks.

5. Develop emulsion-coated sections and observe by microscopy.

6. Quantitate silver grains lying over each class of organelle.

Figure 13.2
Autoradiography protocol. Silver grains in the photographic emulsion are sensitized by passing radiation (α particles); these grains form reduced silver when developed that can be seen in the microscope.

is then developed and fixed, and the section is viewed in the light or electron microscope. The location of radioactive products is pinpointed by developed silver grains that result from the emission of alpha particles from the radioactive source and their passage through the emulsion coating above. These particles, like the photons that interact with photographic emulsions, will sensitize nearby silver nitrate crystals that can then be reduced to black metallic silver during development.

In the microscope, the silver grains lie directly over or very close to the organelles containing the radioactive products, thus, revealing their location. The use of a radioactive isotope that emits alpha particles is essential because high-energy electrons will pass right through the emulsion without interacting with the silver nitrate crystals.

A classic example of the use of autoradiography may be found in Jameison and Palade's work on the exocrine pancreas, the first demonstrating that secreted proteins are synthesized in the rough endoplasmic reticulum, moved to the Golgi where sugars are added, then packaged into secretory granules, and finally released by exocytosis into the extracellular medium (**Figure 13.3A**). Subsequently, Castle, Jamieson, and Palade carried out similar experiments on salivary glands. Their approach was to incubate the pancreas or salivary gland in vitro in medium containing ^{3}H-leucine, a common amino acid that would be incorporated into all proteins made. This incubation lasted only 5 minutes so that proteins labeled during a short "pulse" period could be followed as they traveled through the cell during a "chase" period in which only nonradioactive proteins would be made. At a series of times thereafter, ranging from 5 to 120 minutes, tissue specimens were fixed, sectioned, and then processed by autoradiography. Several weeks later, the emulsion-coated sections were developed and observed by electron microscopy. Figures 13.3B and 13.3C show that tissues fixed within 5 to 10 minutes of the radioactive leucine "pulse" exhibited silver grains over the endoplasmic reticulum, while tissue fixed at 60 minutes after the pulse showed silver grains over the newly formed secretory granules.

If one counts the silver grains above each type of organelle in each tissue specimen and plots these values versus time (Figure 13.3D), one finds that the wave of radioactive protein is first seen in the endoplasmic reticulum where it is synthesized, later passes through the Golgi apparatus where sugars are added, and finally is packaged into secretory granules. This elegant demonstration began the field of intracellular protein trafficking and garnered a share of the Nobel Prize for Dr. Palade.

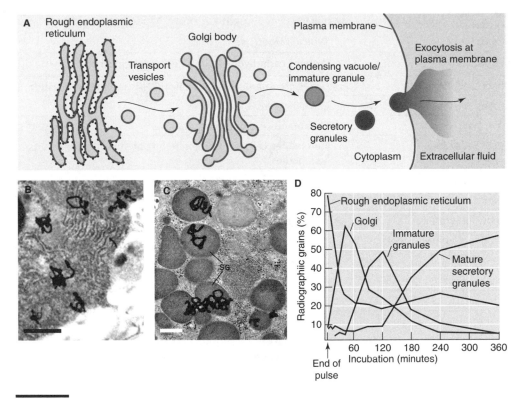

Figure 13.3

The secretory pathway as studied by autoradiography. (A) Both pancreatic and parotid salivary gland secretory proteins are synthesized in the rough endoplasmic reticulum, transported to the Golgi apparatus for glycosylation, and then packaged into immature condensing vacuoles. Further packaging produces mature secretory granules that are later released by exocytosis. (B) Electron autoradiograph showing developed silver grains over the rough endoplasmic reticulum. Bar = 0.5 μm. (C) Electron autoradiograph showing silver grains over secretory granules. Bar = 0.5 μm. (D) Quantitative time course of silver grain appearance over organelles in the parotid secretory cells.

Since then, autoradiography has been used extensively to study the synthesis and routing of proteins, nucleic acids, and sugar polymers within the cell. Its use has dropped dramatically in recent years, however, because of the ease with which green fluorescent proteins (GFPs) can be used to reveal protein dynamics in living cells.

■ Use of Fluorescent Probes

The resurgence of light microscopy is largely due to the ability of fluorescence microscopy to locate specific molecules. Specific proteins, carbohydrate polymers, and lipids can be localized by the use of antibodies. Specific DNA and RNA molecules can be localized by the use of oligonucleotide probes of complementary nucleotide sequence. A summary of useful probes for localization is found in **Table 13.2**. Critical to the use of these probes is their covalent link (conjugation) to an appropriate fluorescent dye either directly or indirectly. Frequently, one can purchase proteins or oligonucleotides already conjugated to the appropriate dye molecule. In some cases, one must carry out conjugation reactions oneself; hence, a brief discussion of how this is usually done follows.

Attaching small organic dyes to proteins is carried out by linking them to reactive chemical groups that attack amino groups in proteins. For this purpose, isothiocyanate and succinimidyl ester derivatives of dyes are the most commonly used, and these can be purchased. As shown in **Figures 13.4A** and **13.4B**, the dye derivative, on reaction with an

Table 13.2 Classes of Probes for Fluorescence Localization of Macromolecules	
Probe	**Typical Target/Use**
Antibodies	Proteins, carbohydrates, immunocytochemistry
Oligonucleotides	Genes, chromatin, gene-specific mRNAs; in situ hybridization
Lectins	Carbohydrates
Phospholipids	Membranes, membrane fluidity
Lipophilic probes	Membranes, cell tracing, membrane potential measurements, lipid bodies
Ion chelators	Free metal ion quantitation, organelle localization
Ligands and inhibitors	Receptors
GFP fusion proteins	Protein trafficking
Nucleic acid-binding molecules	DNA, RNA
Alkaloids	Cytoskeletal proteins
Toxins	Ion channels, neurotransmitter release sites
Substrates	Enzymes, enzyme products
Dextrans	Endocytosis
Microspheres	Cell movements, membrane movement, molecular movements, optical parameter determination

amino group, is covalently linked to protein. The reaction is stopped by adding an excess of free amino groups in the form of lysine, and the conjugated protein is dialyzed to remove the unlinked dye. The fluorescence of the conjugated protein, compared with dye standards, can be used to calculate how many dye molecules, on average, are linked to the protein. The most common use of dye conjugation is for linkage to primary or secondary antibodies for use in immunocytochemistry (see below).

Linkage of dyes to oligonucleotides can be carried out chemically or by the addition of derivatized nucleotides during synthesis. One chemical means of linking dyes to previously synthesized nucleic acids is to use platinum–dye conjugates that react directly with the purine bases of the nucleic acid polymer (Figure 13.4C). Much more common is to use specialized nucleotide triphosphate (NTP) derivatives for the actual synthesis of oligonucleotide probes. These derivatives, however, must be capable of being recognized as a substrate by the DNA or RNA polymerase that actually carries out the synthesis. One approach is to use amino derivatives of NTPs, which after synthesis can be conjugated to dyes exactly as proteins would be (Figure 13.4D). A second approach is to use NTPs that already have chromophores, such as BODIPY or fluorescein linked to them before oligonucleotide synthesis (Figure 13.4E). A third approach is to use NTPs that have linked to them a distinctive chemical group that is readily recognized by antibodies (Figure 13.4F, upper panel). One such chemical group is digoxigenin, an alkaloid isolated from the digitalis plant that occurs nowhere else in nature. Oligonucleotides having digoxigenin incorporated into their structure are then localized using immunocytochemistry with antidigoxigenin antibodies. A similar approach is to use biotin derivatives of NTPs for oligonucleotide probe synthesis (Figure 13.4F, lower panel). The biotin-containing probe can then be localized by dyes conjugated to avidin, a protein from egg white that binds to biotin very strongly.

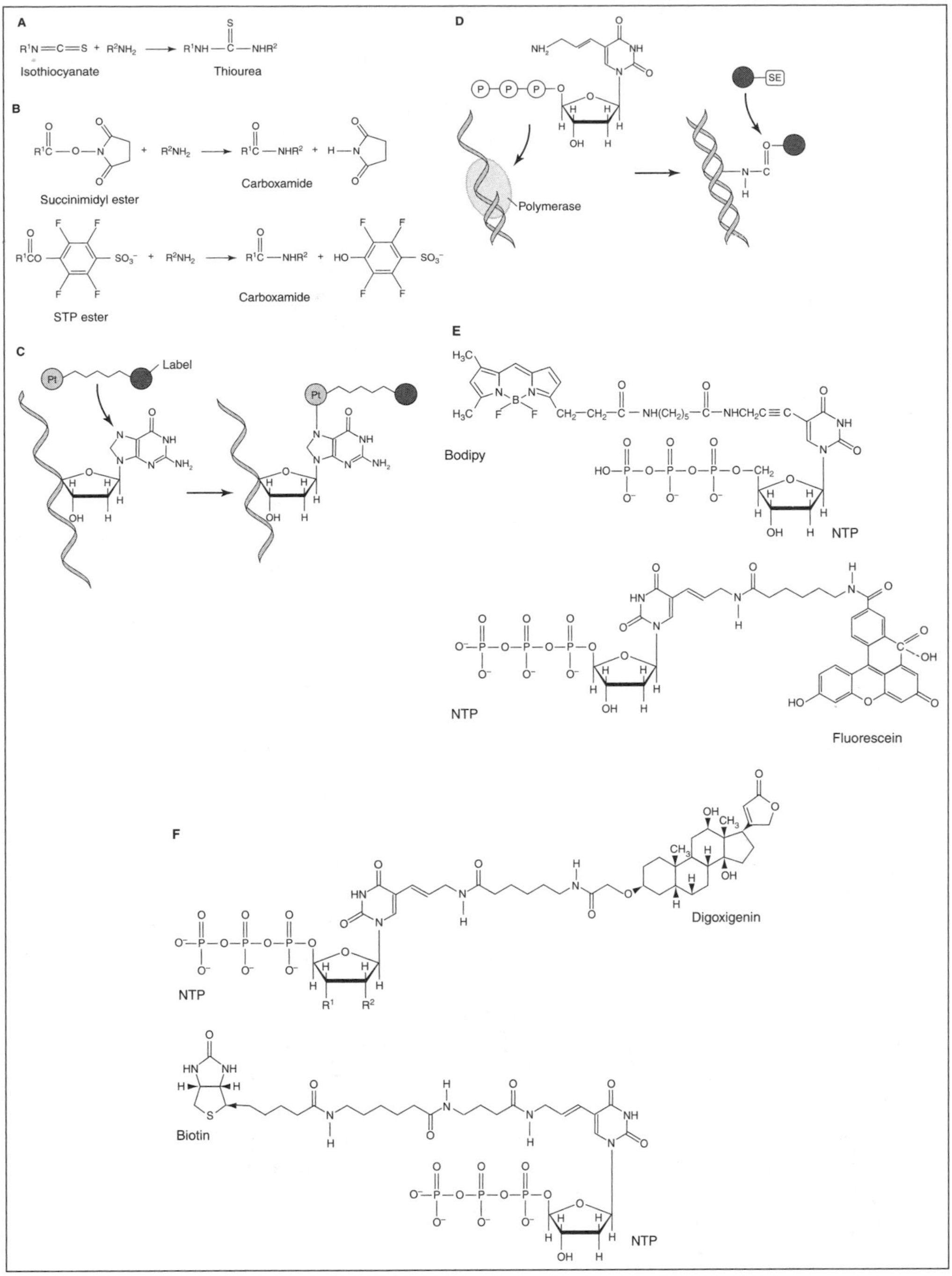

Figure 13.4
Conjugation of fluorescent dyes to proteins and nucleic acids. (A and B) Covalent linkage of dyes to amino groups on proteins. R^1 is the dye; R^2 is the protein. (C) Direct labeling of guanine bases in nucleic acids using platinum-coordinated dye derivatives. (D) Use of amino-derivatized nucleotides to synthesize nucleic acids that can be labeled directly with succinamidyl ester (SE) derivatives of dyes. (E) Bodipy- and fluorescein-conjugated nucleotides used to synthesize fluorescent nucleic acid probes. (F) Digoxigenin- and biotin-conjugated nucleotides used to synthesize nucleic acid probes that can then be localized by immunocytochemistry.

■ Immunocytochemistry

Immunocytochemistry localizes specific macromolecules by the use of antibodies that bind to architectural features that are relatively unique to that protein or macromolecule. Although there are five different classes of antibodies, the IgG class is most frequently used in this technique. IgG molecules consist of two heavy (large) chains and two light (small) chains that are linked by disulfide bonds to form a "Y"-shaped molecule (**Figure 13.5A**). The stem of the Y consists of amino acid sequences that are common to all IgGs, and for this reason, it is referred to as the "constant region" or by the abbreviation "Fc." The two arms of the Y, formed by both the heavy and light chains, contain variable amino acid sequences and for this reason are referred to as "variable regions" or the abbreviation "Fab." These regions are variable because they form the antigen binding sites that need to be chemically and topologically complementary to the antigen itself.

Mammals are thought to be capable of producing over 100,000 different antibodies, each having a differently shaped antigen-binding site. Each antibody is produced by a different set of "B" lymphocytes. The B lymphocytes are continually produced in the bone marrow, circulate throughout the circulatory and lymphatic systems of mammals, and take up residence in secondary lymphoid organs such as the spleen, tonsils, Peyer's patches in the intestines, lymphoid nodules in the lungs, and lymph nodes throughout the body. The B lymphocytes represent a mixture of sets of cells called clones. The members of each clone only produce one antibody, but the presence of thousands of clones gives the immune system a vast ability to produce many different antibodies. After an antigen is introduced into a mammal and it comes in contact with B lymphocytes, those lymphocyte clones able to bind the antigen through display of their antibody on the cell surface are stimulated to divide rapidly. In this way, the effective clones are amplified, and the production of antigen-binding antibodies by these cells is increased manyfold. The animal can then be bled to obtain serum having a high concentration of antibodies capable of binding to the antigen (see protocol, Figure 13.5A). By this procedure, an antibody mixture is obtained having antibodies that recognize a number of structural features or epitopes of the antigen. For this reason, these are referred to "polyclonal" antibodies because multiple clones of lymphocytes were activated to produce antibodies that bind to multiple epitopes on the antigen.

Alternatively, the spleen of the animal can be harvested as a source of the antibody-producing lymphocytes, the lymphocytes "immortalized" by fusion with a cancer cell, and then the hybrid cells grown in culture to produce antibodies almost indefinitely (Figure 13.5B). In order for these antibodies produced in vitro to be useful, however, one must identify the cells that are producing antibodies that actually bind to the antigen; therefore, one must use assays at the cellular level to identify individual cells producing a desirable antibody, a process that is labor intensive and referred to as "screening." After being identified and separated, these individual cells naturally divide to form expanded clones that produce only one antibody. Antibodies produced in this manner are termed "monoclonal" and typically recognize only one epitope on the antigen.

Antibodies can be raised to almost any macromolecule, although some macromolecules are better antigens than others. What is needed first is a purified antigen that can be injected into an animal in order to raise an immune response to the antigen. If the antigen is not purified, the immune response will likely be directed toward every component in the mixture. In the case of polyclonal antibodies, this will lead to a lack of specificity. It is not as great a problem if one is producing a monoclonal antibody because the hybridoma cells will be screened to find the right antibody-producing cell anyway. The animal immunized is typically a mouse for monoclonal antibody production or a rat, rabbit, chicken, goat, or horse for polyclonal antibody production. The choice of animal depends partly on how much immune serum is needed (the larger the better) and partly on cost or feasibility of maintaining the animal (the smaller the better).

A. Protocol for preparation of polyclonal antibodies

1. Purify antigen.

2. Inject antigen with adjuvent into host animal at multiple sites to immunize.

3. Boost immune response with one to three further antigen injections at one to two week intervals.

4. Bleed animal periodically (about once per month).

5. Separate plasma from blood cells, allow to clot, and remove serum.

6. Test serum for antibodies by enzyme-linked immunoassay (ELISA).

B. Protocol for preparation of monoclonal anitbodies

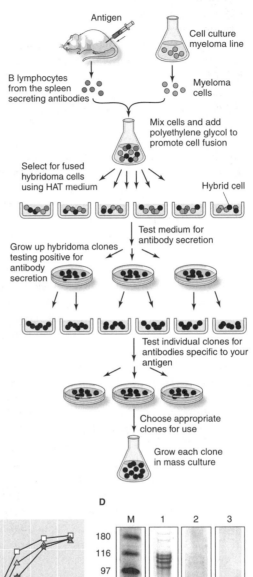

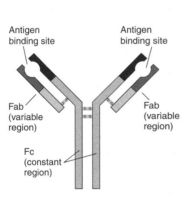

C. Antibody characterization

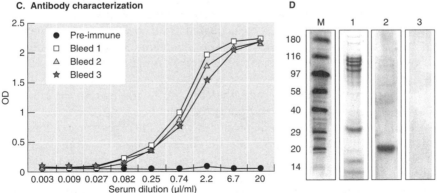

Figure 13.5
Antibody production and characterization. (A) Production of polyclonal antibodies by immunization of a host animal. (B) Production of monoclonal antibodies from mice. (C) Results from a typical ELISA. Serum harvested in three different bleeds from an immunized rabbit contained detectable antibodies at titers as high as 1:4000, while preimmune serum showed no evidence of antibodies at titers of 40. (D) Western blot analysis of antibody binding. Lane M: marker proteins with molecular weight in kilodaltons at left. Lane 1: Coomassie Blue-stained sodium dodecyl sulfate polyacrylamide gel of proteins extracted from the jelly of frog eggs. "Allurin," a sperm chemoattractant protein, appears at a relative mobility of 23 kD. Lane 2: Western blot of lane 1 using an antiallurin polyclonal serum at a dilution of 1:20,000. Only allurin is labeled. Lane 3: Western blot of lane 1 using preimmune serum.

Presentation of the antigen is important in mounting a strong immune response. For this reason, antigens are commonly mixed with adjuvants containing inflammatory agents and injected subcutaneously at a number of sites. Inflammation at the site of injection attracts macrophages that play an important role in presenting fragments of the antigen to B lymphocytes and thereby stimulating their multiplication. Often, only 0.1 to 0.5 mg of antigen is required to obtain a substantial response provided that it is divided between a number of injection sites. The number and location of sites are often trade secrets among commercial antibody-producing companies. These days, it is usually time and cost-effective to hire a company to produce the needed antibodies rather than doing it yourself.

After the antibody-containing serum or culture medium is available, the specificity of the antibody must be documented. Binding of the antibody to the antigen can be determined by enzyme-linked immunoabsorbent assay (ELISA). The antigen is adsorbed to the bottom of a microtiter plate, and the antibody or serum is added to multiple wells using serial dilution to provide trials that have a wide range of antibody concentrations. The amount of antibody bound in each well is then visualized by the use of a second antibody to which is linked an enzyme. The enzyme substrate is then added, followed by incubation to produce a product that is easily quantitated by either absorption or fluorescence. An example of such an ELISA for an immune rabbit serum produced by immunization with two 15-amino acid peptides is shown in Figure 13.5C. Plotting enzyme activity versus dilution gives an indication of the amount of antibody present and its affinity for the antigen. The amino acid sequence (and presumably structure) of these immunizing peptides was identical to certain regions of the protein to be localized; this approach of using synthetic peptides fragments for immunization is often used when the actual protein cannot be purified in adequate amounts for immunization.

Specificity of the antibody for a particular protein in a tissue or biological specimen to be examined is carried out by Western blotting (Figure 13.5D). A tissue extract is separated by sodium dodecyl sulfate-polyacrylamide electrophoresis. Staining of the gel for protein using Coomassie blue reveals five major bands and several minor bands (lane 1), while staining of a similar gel by silver staining reveals more than 20 proteins in this mixture. A gel with identical protein bands is electrophoretically transferred to a nylon or polyvinyl fluoride membrane to produce a duplicate array of proteins on the surface of the membrane. The membrane is then processed by incubating it in the antibody to be characterized and washing and applying a second antibody linked to an enzyme. The membrane with attached enzyme is then incubated in substrate to produce a product that is either fluorescent or chemiluminescent. The light emitted from the product is then detected by either exposure to film or by a laser-scanning fluorescence detector. The result in Figure 13.5D (lane 2) shows that product formation (and therefore binding of the antibody) occurred at only one band, thereby demonstrating that the immune serum studied contained antibodies directed against only one protein in the tissue extract.

Finally, to verify that the antibody produced was made in response to immunization, serum taken from the animal before it was immunized is tested in a similar way. This is referred to as preimmune serum. The desired result (shown in Figure 13.5D, lane 3) is that none of the antibodies present in the preimmune serum is able to bind to the proteins in the tissue extract. The preimmune serum also contains no antibodies against the peptides used for immunization, as demonstrated by the ELISA assay (Figure 13.5C). The immune serum is now ready to be used for immunocytochemistry.

Immunocytochemistry procedures are nearly identical to those used during ELISA and Western blotting. Briefly outlined, the tissue to be examined is fixed, permeabilized, incubated with primary antibody to locate the protein or macromolecule of interest, washed, and incubated with a second antibody that locates and binds to the first, and then the dye, enzyme, quantum dot, or metal particle linked to the second antibody is visualized by microscopy (**Figure 13.6A**). Because there are many subtleties along the way, we discuss each step in detail.

Specimens can be prepared for immunocytochemistry as either whole mounts (i.e., unsectioned) or sections. In either case, the cells must first be fixed in an antigen-friendly manner. If the antigen is a protein, the protein can be fixed chemically using formaldehyde (for light or electron microscopy) or by coagulation using cold methanol (for light microscopy only). In the first case, fresh formaldehyde is prepared from paraformaldehyde and diluted in phosphate buffer to reach a final concentration of 4% wt/vol. The tissue is fixed for 1 to 2 hours, usually at room temperature, and is then washed free of formaldehyde using phosphate buffer. To maintain structural integrity, especially in specimens being prepared for electron microscopy, one can add 0.1% glutaraldehyde to the formaldehyde fix. A higher concentration of glutaraldehyde runs the danger of cross-linking the protein antigen so severely that it is no longer recognized by the antibody. For the same reason, the specimen must not be treated with osmium tetroxide. If one is using 100% cold methanol, the solvent is usually cooled to –20°C in a refrigerator freezer before use. After initial fixation, the specimen can be brought to room temperature and passed through several changes of 100% methanol before embedding or immunolabeling. If sections are to be prepared, the specimen is dehydrated with a graded series of ethanols and embedded in either paraffin for light microscopy or epoxy resin for electron microscopy. Sections are then cut and immunolabeled. If a whole mount is being used, one can immunolabel immediately after fixation.

In order to discuss the details of the procedure, we use paraffin sections as an example. The protocol outlined in Figure 13.6A is referred to as "indirect" immunolabeling because a secondary antibody with conjugated fluorophore is used in addition to a primary antibody. Early immunocytochemistry protocols used a "direct" method in which the primary antibody itself was conjugated to a dye. This method is rarely if ever used now because indirect methods offer two important advantages: (1) Binding of multiple second antibodies to a primary antibody results in amplification of the signal and greater sensitivity, and (2) second antibodies with almost any type of label are available commercially at relatively low cost. **Table 13.3** provides examples of common types of secondary antibody labels and their uses.

Immunolabeling of paraffin sections begins with removal of the paraffin so that antibodies will have access to the tissue. Strong organic solvents such as xylene must be used to dissolve the paraffin, despite the fact that these can denature proteins and lower antigenicity. After paraffin removal, the specimen is rehydrated and washed with Tris-buffered saline (TBS), the preferred buffer for immunolabeling. The specimen is then incubated in a blocking agent that contains relatively high concentrations of protein. Cells and tissue specimens contain many sticky macromolecules that will bind proteins of any kind including antibodies. For this reason, these sites are referred to as "nonspecific" binding sites. If one incubated the tissue without blocking these sites, the antibodies would stick even though the sites contained none of the antigen that the antibody is looking for. This would result in a false-positive signal. Instead, the protein in the blocking agent sticks to and fills these sites before the antibody is applied. Common blocking agents include bovine serum albumin, calf or lamb serum, or nonfat milk solids containing large amounts of casein. The latter, affectionally called "blotto" in laboratory lingo, is cheap because it can be bought at the grocery store and is often used during Western blotting. On the other end of the cost spectrum are proprietary blocking agents sold in ready-to-use form but really containing nothing more than the usual buffer and ingredients.

The first or primary antibody is then diluted in blocking buffer and applied to the specimen in as low a volume as possible to conserve antibody. The dilution used must

A. Typical Protocol for Immunocytochemistry at the LM Level

1. Fix tissue or cells with 1% formaldehyde, then embed, section, and remove paraffin.

2. Expose to tris-buffered saline-tween 20 (TBS-T) with blocking protein at 20°C for 30 minutes.

3. Wash three times with TBS-T, 15 minutes each wash.

4. Incubate with TBS-T and blocker and primary antibody at 4°C overnight.

5. Wash three times with TBS-T, 15 minutes each wash.

6. Incubate with TBS-T and blocker and secondary antibody at 20°C for 2 hours.

7. Wash three times with TBS-T, 15 minutes each wash.

8. Observe by fluorescence/transmission light microscopy.

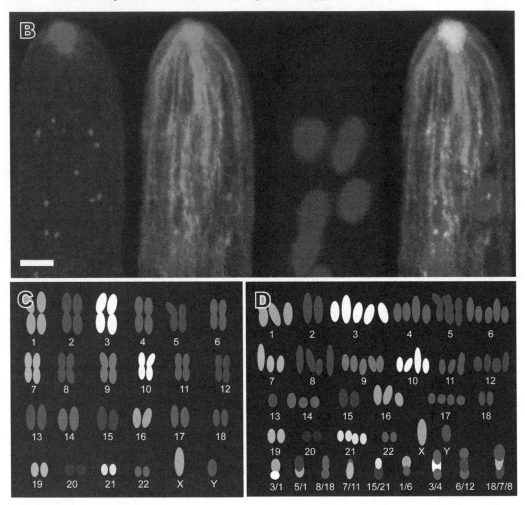

Figure 13.6 (See Plate 6 for the color version.)
Immunocytochemistry at the light microscopy level. (A) A typical protocol. (B) Laser scanning confocal microscopy illustrating a single hypha of the fungus *Allomyces macrogynus* labeled with antibodies for γ-tubulin (left, red) and β-tubulin (green) and DAPI for DNA (blue). γ-Tubulin identifies microtubule organizing centers. β-Tubulin identifies cytoplasmic microtubules, and DAPI identifies nuclei. An overlay of all three images is presented at the right. Bar — 2 μm. (C) Simulation of human chromosomes identified by "painting" with five oligonucleotide probes to repetitive nucleotide sequences. Chromosome numbering follows a standard convention from largest to smallest. The use of interferometric microscopy for this purpose is explained in the text and in Figure 13.11. (D) Simulation of human chromosomes from a malignant tumor. Chromosome "painting" indicates the presence of many chromosomal abnormalities including deletions, fusions, and multicopy regions. Chromosomes are numbered according to the origin of their chromatin as indicated by the painting process.

Table 13.3 Commonly Available Secondary Probes		
Probe	**Target**	**Visualization Method**
Dye-conjugated antibodies	Primary IgGs	Fluorescence microscopy
Dye-conjugated oligonucleotides	Gene-specific DNA or RNA sequences; repetitive DNA sequences	Fluorescence microscopy
Dye-conjugated lectins	Oligosaccharides	Fluorescence microscopy
Quantum dot-conjugated antibodies	Primary IgGs	Fluorescence microscopy
Dye-conjugated Fab fragments	Primary IgGs	Fluorescence microscopy
Horse radish peroxidase-conjugated antibodies	Primary IgGs	TEM or LM localization of polymerized HRP product
Alkaline phosphatase-conjugated antibodies	Primary IgGs	LM localization of colored or fluorescent alkaline phosphatase product
Dye-conjugated avidin	Biotinylated antibodies or proteins	Fluorescence microscopy
Alkaline phosphatase-conjugated avidin	Biotinylated antibodies or proteins	LM localization of colored or fluorescent AP product
Colloidal gold-conjugated antibodies	Primary IgGs	TEM or SEM
Colloidal gold-conjugated protein A or protein G	Primary IgGs	TEM or SEM
Ferritin-conjugated antibodies	Primary IgGs	TEM localization of ferritin iron core
Colloidal gold-conjugated avidin	Biotinylated antibodies or proteins	TEM or SEM

be determined empirically, with 1:100 to 1:2000 being common. If the antibody was purchased commercially, the provider will usually suggest what dilution to try first. If the antibody was custom made, the dilution that you found optimal for Western blotting will give an indication of what to use. Generally, for immunocytochemistry, the primary antibody is used at a concentration that is 10 to 20 times greater than that optimal for Western blotting. Incubation of the primary antibody with the specimen is usually carried out for 6 to 24 hours, sometimes at room temperature, but more often at 4°C.

The specimen is then washed thoroughly with TBS—at least three times for 10 minutes each wash. The second antibody is then diluted in TBS and applied to the specimen, typically with mild mixing for 1 to 2 hours at room temperature. Again, the dilution of secondary antibody is optimized empirically, but one usually starts with the dilution recommended by the distributor. Finally, the specimen is washed thoroughly to remove the second antibody (three times minimum) and then soaked in distilled water containing glycerol (to prevent evaporation), and an antifade reagent such as n-propyl gallate and a cover glass are applied. The antifade agent prevents light-induced breakdown of the dye either during storage or microscopy by scavenging free radicals. Microscopy is best carried out within a few days, but specimens can remain usable for weeks provided that they are stored in the dark at 4°C.

Because the purpose of the second antibody is to bind selectively to the first antibody, one must use an antibody that is directed toward the constant region (Fc) of an IgG of the correct species. For example, if the primary antibody was made in rabbit, one might use a second antibody such as antirabbit IgG made in goat. The animal used to make the second antibody must be a different species than the animal used to make the first antibody because in general animals do not make antibodies against their own proteins. If one is doing double labeling, even more care must be exercised in the selection of antibodies to avoid confusion between the two signals. First, the two primary antibodies have to have been raised in different animals (e.g., rabbit and chicken). Second, the second antibodies used have to have been raised in a third species; one might use an antirabbit IgG antibody and an antichicken IgG antibody both raised in goats. Finally, the two secondary antibodies being used must be conjugated to different dyes so that their signals can be optically separated; therefore, we might choose an antirabbit IgG conjugated to fluorescein (or a fluorescein-like dye) and an antichicken IgG antibody conjugated to tetramethylrhodamine.

An example of a "triple label" experiment is shown in Figure 13.6B (see Plate 6 in the Color Addendum). In the first panel, we see a hyphal tip of the fungus *Allomyces macrogynus*. The primary antibody used was a mouse monoclonal antibody raised against γ-tubulin, a protein found at microtubule-organizing centers. The secondary antibody was an anti-mouse IgG raised in goat and conjugated to Texas red. The second panel shows the same hyphal tip imaged in the green channel. This signal is localized to β-tubulin found in microtubules. The primary antibody was a polyclonal anti-β-tubulin raised in rabbit, and the secondary antibody was an anti-rabbit IgG raised in goat and conjugated to Bodipy FL. In the third panel, nuclei are imaged using the DNA-binding dye 4', 6-diamidino-2-phenylindole (DAPI), which fluoresces blue. An overlay of all three channels is shown in the final panel. The labeling procedure conforms to the need for a different species source for each primary antibody and for distinguishable fluorophores for each secondary antibody. Minimal cross-talk exists between channels.

Similar considerations are required when choosing antibodies for immunoelectron microscopy. In this case, however, double labeling is usually accomplished by using secondary antibodies that each have a different size of gold particle linked to it. For example, one second antibody might be attached to 10-nm diameter gold particles, while another is attached to 30-nm diameter gold particles. Table 13.3 illustrates the variety of second antibody detection methods for both light microscopy (LM) and transmission electron microscopy (TEM) as well as other labeling methods that do not require antibodies.

Immunocytochemistry always requires proper controls. First, one should examine a specimen prepared for microscopy by the same procedure, except for omission of immunolabeling. Signal from this control indicates that there are components that autofluoresce or that the filters chosen are incorrect and incident light scattered off the sample is being observed. If autofluorescence is so strong that it has to be reduced, consider using a different dye in a spectral range in which autofluorescence is less.

A second control should omit the primary antibody. Any staining present in this control is due to binding of the secondary antibody; if present, one should first make sure that washing protocols to remove unbound antibody are adequate. Alternatively, the antibody may be binding nonspecifically; in this case, one needs to improve blocking by either changing the blocking agent or increasing its concentration. In addition, one may need to decrease the concentration of second antibody used if that is consistent with maintaining a reasonable signal. Although less likely, the second antibody may be binding specifically to some tissue component, in which case one should use a different second antibody, possibly from a different species.

Third, for polyclonal antibodies one should carry out a control using preimmune serum as the primary antibody. Again, there should be little or no staining. If there is staining, several possibilities exist. The staining may be nonspecific; in which case, improving the blocking agent and making sure that careful washing is performed at each step should take care of it. Alternatively, the staining may be specific. This means that the animal used to raise antibodies was already immune to some component in your specimen even before exposure to your chosen antigen. If this is the case, one can note the pattern of staining in the preimmune control and ignore this pattern in the specimens stained with immune serum. One can also go to the trouble of trying to identify the component staining with preimmune serum and then using the material to clear the immune serum of antibodies directed against it—a process referred to as "preabsorption." If preimmune serum is not available, one can use an irrelevant primary antibody that is directed against a component known to be absent from the specimen or use a primary antibody from a species other than that recognized by the second antibody.

The protocol summarized in Figure 13.6A, although typical, contains many steps in which macromolecules can be denatured and antigenicity of the specimen can be reduced, leading to less than optimal results. This problem is particularly common in electron microscopy and can occur during fixation because of excessive cross-linking, during dehydration because of water removal, during embedding because of inaccessibility, or during curing because of heat.

For example, the usual embedding resins for electron microscopy (e.g., Embed) are quite hydrophobic, and usually harsh solvents such as acetone or propylene oxide need to be used. In addition, such resins are cured at 60°C for 24 hours. Both the solvents and the high temperature diminish or eliminate antigenicity. Epoxy resins are not penetrated by aqueous antibody solutions so that at best only a surface labeling of sections are obtained. Occasionally, electron microscopists have even used resin etching to increase accessibility, but these methods are chemically harsh.

To avoid these problems, electron microscopists normally use relatively hydrophilic resins that allow macromolecules to retain a more antigenic structure. Currently, the best resins for electron microscopy immunocytochemistry are LR White, LR Gold, and Lowicryl. These acrylic resins require that the specimen be dehydrated; however, small traces of water can remain, and polymerization will not be affected. Although these resins can be cured with heat (LR White) or with the aid of a catalyst at 4°C (LR White and LR Gold), these resins are often cured by ultraviolet light at low temperatures (0 to −25°C for LR White and LR Gold; −35°C to −80°C for Lowicryl) when used for EM immunocytochemistry.

Another approach to preserving antigenicity is to use frozen sectioning to avoid dehydration of the tissue before immunolabeling. A very successful protocol for this, developed by Tokuyasu and others, is to chemically fix the tissue with formaldehyde, cryoprotect the tissue with sucrose, freeze the tissue, and cut frozen sections (**Figure 13.7**). The sections are then immunolabeled with a primary antibody, washed, labeled with a second antibody, and washed. After labeling, the sections are embedded in resin and then counterstained lightly with heavy metals. Tissues prepared in this manner usually show robust labeling with colloidal gold or ferritin-conjugated antibodies (arrows and inset, Figure 13.7).

Another approach is to use ultrarapid freezing, combined with freeze substitution, low temperature embedding with LR White or Lowicryl resin, cutting of ultrathin sections, and then immunolabeling the sections by primary and second antibodies. This protocol is outlined, and typical results are shown in **Figure 13.8**. The advantage of using this protocol is that it avoids chemical fixation until the tissue is in a frozen state, allowing water-soluble structures to be retained and avoiding structural artifacts. Both protocols avoid heat before labeling and provide fairly good structural preservation (see Chapter 10 on cryopreservation).

Typical Protocol for Immunocytochemistry at the Electron Microscopy Level Using Frozen Sections

1. Fix tissue or cells with 1% formaldehyde, infiltrate with 2.3 m sucrose, freeze and cut frozen ultrathin sections.

2. Thaw sections and wash with tris-buffered saline-tween 20 (TBS-T) with blocking protein at 20°C for 30 minutes.

3. Wash three times with TBS-T, 15 minutes each wash.

4. Incubate with TBS-T and blocker plus primary antibody at 4°C overnight.

5. Wash three times with TBS-T, 15 minutes each wash.

6. Incubate with TBS-T and blocker plus colloidal gold-conjugated secondary antibody at 20°C for 2 hours.

7. Wash three times with TBS-T, 15 minutes each wash.

8. Fix with glutaraldehyde, embed in epoxy resin, and observe by TEM.

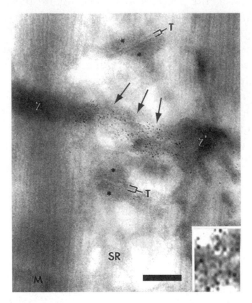

Figure 13.7
Immunocytochemistry at the TEM level using frozen sections. Ferritin-conjugated antibodies localizing the scaffold protein titan in rat skeletal muscle. Myofibrils (M), sarcoplasmic reticulum (SR), the Z-line (Z), and transverse tubules (T) are noted. Bar = 0.2 μm. Inset: The iron-containing cores of the ferritin label at high magnification.

Typical Protocol for Immunocytochemistry at the TEM Level Using Rapidly Frozen, Low Temperature Embeded Tissue

1. Freeze tissue by ultrarapid techniques (see Chapter 10).

2. Freeze substitute tissue in acetone containing 0.5% glutaraldehyde at −80°C for 48 hours; warm to −20°C for 2 hours, then to 4°C for 1 hour, and finally to room temperature.

3. Embed in acrylic resin; cure using ultraviolet light at low temperature for 24 hours; cut sections and pick up on gold or nickel grids.

4. Expose sections on grids to tris-buffered saline-tween 20 (TBS-T) with blocking protein at 20°C for 30 minutes.

5. Wash three times with TBS-T, 15 minutes each wash.

6. Incubate with TBS-T and blocker and primary antibody at 4°C overnight.

7. Wash three times with TBS-T, 15 minutes each wash.

8. Incubate with TBS-T and blocker and secondary antibody at 20°C for 2 hours.

9. Wash three times with TBS-T, 15 minutes each wash, then stain.

10. Observe by TEM.

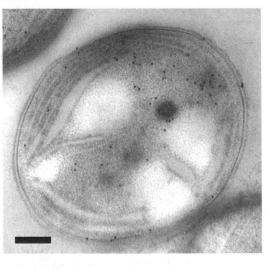

Figure 13.8
Protein localization at the ultrastructural level in rapidly frozen and freeze substituted *Synechocystis* sp. PCC 6803. Colloidal gold markers conjugated to the secondary antibody are seen distributed over regions of the cytoplasm. Bar = 0.5 μm.

■ Fluorescent Proteins (Green and Otherwise)

Some of the most elegant experiments in protein localization and dynamics are done in live cells. A typical approach is to conjugate a dye directly to the protein and microinject the protein into a live cell. Notable successes have been achieved using labeled-tubulin and labeled actin to visualize dynamics of microtubule and microfilament polymerization and depolymerization during live cell behavior.

More recently, this field has been revolutionized by letting cells make their own fluorescent proteins. Jellyfish of the genus *Aqueoria* produce a protein that when excited with blue light fluoresces green. The protein is used naturally as means for these creatures to identify one another and to locate prey. The excitation and emission spectra of the native protein are shown in **Figure 13.9A.** Sequencing and molecular cloning of this protein led to the production of its recombinant form in prokaryotic as well as eukaryotic cells,

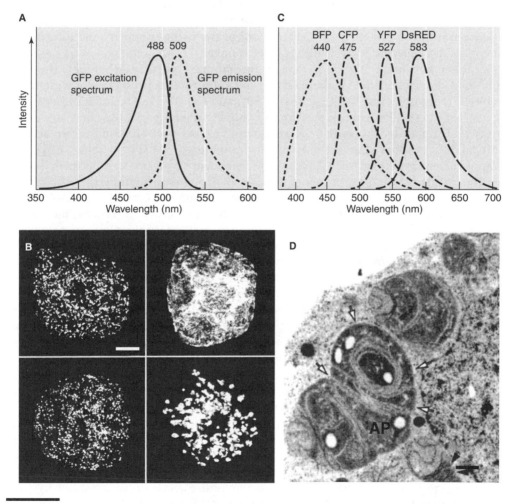

Figure 13.9

Spectral properties of fluorescent proteins. (A) Excitation and emission spectra of GFP. (B) Localization of GFP constructs targeted to peroxisomes (upper left), endoplasmic reticulum (upper right), Golgi bodies (lower left), or plastids (lower right) in cultured BY-2 tobacco cells. Bar = 10 μm. (C) Emission spectra of other members of the fluorescent protein family. (D) GFP constructs of a peroxisomal membrane protein has resulted in aggregation of peroxisomes (AP). Bar = 0.5 μm.

including mammalian. A mammalian cell that has been transfected with a vector coding for GFP will become fluorescent because of manufacture and incorporation of the GFP. GFP is relatively small in size (22 kD), and fusion proteins containing a GFP domain linked to a normal cellular protein can often be expressed with no loss of function on the part of the cellular protein domain.

How the cell traffics and localizes GFP fusion proteins will depend on how the cell normally handles the cellular protein to which it is attached. Thus, by endogenous fluorescence of GFP, the dynamics of protein movement and localization can be followed in living cells. An example of such a study, shown in Figure 13.9B, is targeting of GFP to four different organelles in tobacco tissue culture cells. In each case, a vector coding for GFP linked to either a protein or a localization peptide sequence characteristic of that organelle was prepared by standard genetic engineering techniques. The cells were then transfected and allowed to express transiently each protein construct and each construct, followed by fluorescence microscopy to determine its final location in the cell. In each case, the GFP construct translocated to the correct destination, either peroxisomes (upper left), endoplasmic reticulum (upper right), Golgi apparatus (lower left), or plastids (lower right). In a similar manner, studies of gene expression can use the fluorescence of fluorescent green protein constructs to monitor the tissue specificity and time course of expression of specific proteins during development.

Another advantage of GFP is that by recombinant protein technology, the fluorophore of the protein can be altered to produce fluorescent proteins of a wide range of colors, including cyan (CFP), yellow (YFP), and red (RFP). Although the emission spectra for these FP variants overlap, as shown in Figure 13.9C, they are sufficiently well separated to be distinguished with narrow pass emission filters or with confocal microscopes that use a monochronometer or AOTF to select a narrow band of the emitted light (see Chapter 12). This means that one can perform double and even triple labeling experiments with these proteins, providing that the microscope has the appropriate optics to separate these channels. Recombinant DNA technology has also made possible improvement of the fluorophore over what nature has provided, including increases in its absorbance and quantum yield to provide a brighter signal.

Fluorescent proteins are not foolproof and do have some drawbacks. Not all cells are efficient in their rate of transfection and their expression of the protein. Even if the protein is expressed in adequate quantities, there is no assurance that the fusion protein with its GFP domain will act exactly like the parent cellular protein. One nasty habit that crops up is that these fluorescent proteins have a certain propensity for dimerization, and some fusion proteins having the FP domain "inherit" this trait. Again, recombinant technology has come to the rescue in providing new GPs that show less tendency for polymerization. One usually takes care to include controls that help ascertain whether the FP fusion protein is acting normally. First, the localization of the fusion protein should be identical to the native cellular protein, as determined by immunocytochemistry. Second, the structure of the cell and the positioning of its organelles should be similar in both wild-type cells and FP-transfected cells. An example of a dimerization problem is seen in Figure 13.9D. Immunocytochemistry of wild-type tobacco cells stained for a peroxisomal membrane protein shows a typical localization to these organelles scattered throughout the cell. In contrast, when these cells are transfected with a recombinant fusion protein that had GFP linked to the peroxisomal protein and imaged in thin sections, the fusion protein correctly localized to peroxisomes, but the peroxisomes are now clustered because of dimerization of the GFP domains.

The advantages of FP labeling, however, far outweigh the occasional artifacts encountered. Furthermore, many FPs have now been genetically engineered to reduce dimerization and boost quantum yield resulting in a new generation of "enhanced" fluorescent proteins, for example, enhanced green fluorescent protein (EGFP). The alternative to fluorescent proteins, microinjection of dye-conjugated proteins, is feasible only in large cells; one is limited to observing a small number of cells, and specialized equipment is necessary. Furthermore, FPs are extremely useful in monitoring the interactions of proteins in vivo by fluorescence resonance energy transfer (FRET), as we shall see later.

■ In Situ Hybridization

Localization of specific DNA and RNA sequences in cells is extremely useful in the study of gene expression and for detecting chromosomal abnormalities characteristic of cancer and of genetic defects during development. The kinds of questions that are often answered using these techniques are as follows: Which cells within a complex tissue are producing mRNA coding for a specific protein? Is there more than one cell type expressing this mRNA? When are cells doing this and for how long? What transcription factors are controlling expression of this specific gene? Does cell position or cell–cell interaction lead to expression? On what chromosome is a specific gene located? Do chromosomes contain DNA sequences that are missing or exhibit an abnormally high copy number? Are their chromosome deletions, repeats, or disjunctions that are too small to be detected by chromosome morphology alone?

The principle underlying these techniques is the ability to produce oligonucleotides that are labeled either with fluorescent dyes or with derivatives that are detectable by immunocytochemistry. The oligonucleotides must be of a specific nucleotide sequence that is complementary to the DNA or RNA being localized (i.e., having an "antisense" sequence). Hybridization of the probe with the target under appropriate conditions will lead to visualization.

Localization of mRNA is somewhat more demanding than localization of DNA because RNA is so easily hydrolyzed by RNAases that are found almost everywhere, including our fingers. Tissue prepared for in situ hybridization is usually processed in buffers that contain inhibitors of RNAase, such as diethylpyrocarbonate (DEPC). A typical protocol is provided in **Figure 13.10** to provide an understanding of the process. First, either DNA or RNA probes are produced by in vitro polymerization of labeled nucleotides, as described earlier (Figure 13.4). RNA probes, although subject to degradation, are considered to hybridize better with mRNA and are recommended over DNA probes by some workers in gene expression studies. Epithelial cells of the frog oviduct labeled with an antisense oligonucleotide that will hybridize with mRNA coding for a specific secretory protein appear darkly stained with the alkaline phosphatase product (Figure 13.10A). In contrast, the same epithelium, labeled with a sense probe that should not bind to the mRNA shows no darkly stained product (Figure 13.10B).

A similar procedure can be applied to locating specific genes in the nucleus. In Figure 13.10C, a nucleus that has been labeled with a fluorescent oligonucleotide having a gene-specific sequence exhibits bright spots of fluorescent localization (arrows). The remainder of the DNA in the nucleus was stained nonspecifically with DAPI. The same approach can be used for locating genes (Figure 13.10D) or identifying chromosomal regions (Figure 13.10E) in chromosome squashes prepared by cytogeneticists.

Typical Protocol for in Situ Hybridization

1. Prepare RNA or DNA oligonucleotide probes using in vitro polymerization, PCR amplification, and digoxigenin-labeled nucleotides. Hydrolyze probe into 100 to 200 base segments. Synthesize both sense and antisense probes.

2. Prepare deparaffinized tissue sections.

3. Incubate sections with either antisense probe or (as a control) sense probe.

4. Wash sections with TBS-T and blocker, then inclubate with antidigoxigenin antibody.

5. Wash with TBS-T three times.

6. Incubate with alkaline phosphatase-linked second antibody.

7. Wash with TBS-T three times.

8. Incubate sections with dye-linked substrate.

9. Visualize by brightfield or fluorescence microscopy.

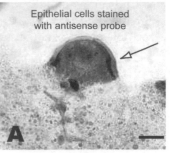

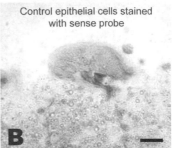

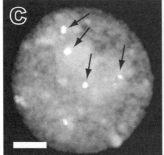
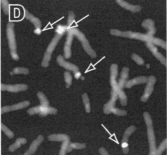
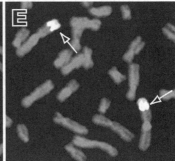

Figure 13.10
In situ hybridization. (A) Staining of frog epithelial cells with a digoxigenin-labeled oligonucleotide anti-sense probe. Bar = 10 μm. (B) Control epithelial cells stained with an oligonucleotide sense probe. Bar = 10 μm. (C) A nucleus stained with DAPI for DNA and for specific chromosomal regions (arrows) using fluorescence in situ hybridization (FISH). Bar = 2 μm. (D) A chromosome "squash" labeled by FISH using a gene-specific oligonucleotide probe (arrows). (E) A chromosome "squash" labeled by FISH using a region-specific oligonucleotide probe (arrows). DAPI was used as a sequence-independent stain in (D) and (E) to visualize all chromosomes.

■ Multispectral Imaging Microscopy and Chromosome Painting

In our discussion of immunocytochemistry at the light level, we noted that the use of multiple fluorescence probes is often limited by the number of available spectral windows that do not overlap. Standard data acquisition (one intensity value per channel per pixel) places a restraint on the number and type of dyes that can be used simultaneously and on the sensitivity that can be achieved; however, applications such as "chromosome painting" in which chromosomal abnormalities can be plotted by use of five to eight dye-linked oligonucleotide probes, motivated the development of systems that can scan complex emission spectra and use computer-driven data analysis to create maps that identify the chromosomal source of each pixel in the image. These systems were first developed by NASA and the Jet Propulsion Laboratory for spectral imaging in space missions, but subsequently, these methods have been applied to microscopy with great success.

The main goal is to collect the entire emission spectrum from each pixel of the image. Granted, one has to block out those portions of the spectrum that coincide with exciting light wavelengths because their high intensity would overpower any emitted light information in these regions. There are three basic ways to collect spectral data. The first is by using a monochronometer prism or diffraction grating on the emission side of a fluorescence microscope to isolate each wavelength and to rotate this prism/grating to obtain a spectral scan. This is exactly the technology that has been used for years in spectrophotometers and fluorometers and is used today on laser scanning confocal microscopes for the purpose of separating the emissions of two spectrally similar fluorophores (e.g., GFP variants). This technology has several drawbacks. First, it is very insensitive and slow because information at only one wavelength can be collected at any one time. Second, its lack of sensitivity is further compounded by the fact that typically spectral data for only one set of pixels is collected at a time. The second method is to use fiberoptic coupling to convert the pixels of a two-dimensional fluorescence image into a linear array that is delivered to a diffraction grating. As the grating is rotated, its spectral output for each pixel is scanned onto a CCD chip, which is then read out in such a manner that each column of diodes represents the spectrum of one image pixel.

The third and fastest method uses interferometry to collect spectral data from all pixels simultaneously. As shown in **Figure 13.11A**, a normal fluorescence microscope equipped with vapor arc lamp illumination and multiwindow filter cubes is used. The emitted light image, rather than being sent directly to the eyepiece or camera, is passed through an interferometer and is then focused onto the front window of a CCD camera. The interferometer, by splitting the emitted light into two equal trains, reflecting one of them off of a vibrating mirror, and recombining them to allow interference to occur, is able to provide the entire emission spectrum within one cycle of vibration. Let us take a closer look at how this works.

As the image light enters the interferometer (Figure 13.11A), it is split into two equal trains in a non–wavelength-dependent manner using a half-silvered beam splitter. The two trains are each reflected off of a mirror and are precisely recombined at the exit port for projection onto the CCD camera chip. The paths of the two trains during their separation are virtually identical except for one critical difference. One of the mirrors can be made to vibrate at about 1 Hz, thereby changing the path length of one train relative to the other in a predicable manner. During one cycle of vibration, the mirror (and thus path length of the affected light train) is changed by several micrometers—equivalent to several wavelengths of the visible spectrum. As the path length changes, every wavelength of light within the affected train will move in and out of positive interference with the other train. This is, in effect, like scanning the emission spectrum; however, the information recorded by the CCD chip is very different from a traditional spectrum. It is encoded as an intensity versus path length distribution, as shown in Figure 13.11B, rather than as an intensity versus wavelength distribution, as illustrated in Figure 13.11C. The major central peak of this distribution represents the strong positive interference experienced when the path of one train is different from that of the other by an exact integral number of wavelengths. The information peripheral to the central peak is essentially a series of correlation coefficients that relate each data point in the emission spectrum to intensities at another wavelength in the spectrum. The two wavelengths whose intensities are being related differ exactly by the difference in path length between the two trains and therefore experience positive interference. For example, if the path length of the two trains differed by an integer multiple of 200 nm, then every data point in the spectrum would positively interfere with light that was either 200 nm less or 200 nm greater in wavelength. Data that are a function of path length difference

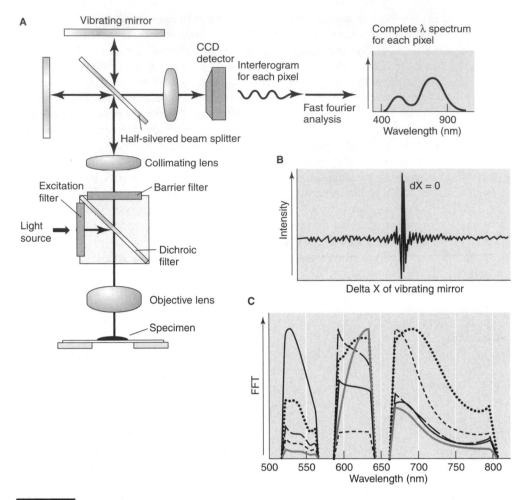

Figure 13.11
Spectral imaging using interferometry. (A) The optical path in an interferometric microscope. An emission image from a specimen labeled with multiple dyes is passed through a three-windowed barrier filter into an interferometer. Output from the interferometer is captured by the CCD chip of a sensitive black and white video camera. Within the interferometer, the image rays are split by a beamsplitter and then later recombined by interference. Changing the phase retardation of one of the split beams results in an intensity versus time interferogram for each pixel much like that seen in (B). Fast Fourier transform (FFT) mathematics converts the interferogram into a spectrum which can then be analyzed by curve stripping to determine what linear combination of dyes is present at each pixel. (B) A typical interferogram. (C) Emission spectra of five different dyes as seen in three widows. The linear combination of these spectra that best fits the pixel spectrum is used to identify the source of the fluorescence. The pixel is then given an appropriate pseudocolor indicating the source.

can be mathematically transformed by Fourier analysis to a typical intensity versus wavelength plot.

To do this, the path length difference must be calibrated, and usually interference of laser light at a specific known wavelength is used as an internal standard. Because the frequency/pathway-dependent data stream is organized spatially as an image on the CCD chip, the spectral data can be analyzed separately for each pixel. The emission spectrum for each pixel is then compared with reference spectra for each of the fluorophores being used; reference spectra for each of 5 dyes are shown in Figure 13.11C to illustrate the fact that given these spectral data one can distinguish each dye from the others. Curve stripping and fitting algorithms are then used to model the experimental spectrum using a linear combination of the reference spectra. The coefficients arrived at are an estimate of the relative concentrations of each probe at that pixel. The coefficients are then compared with typical coefficients found in each chromosome. The oligonucleotide probes used hybridize with repetitive DNA sequences that are found at different levels in different chromosomes. Thus, the relative values of these coefficients represent a fingerprint that identifies the source of chromatin in each pixel. Based on the coefficient profile, each pixel is then pseudocolored. The resulting array of chromosomes is not only aesthetically pleasing (simulated in Figure 13.6C) but is also very informative. Even if a small bit of chromatin from chromosome 1 has been translocated to another chromosome, it will be detected. Such translocations as well as chromosome deletions, fusions, and duplications are frequently seen in cancer cells and in fetuses with genetic abnormalities that can generate birth defects (compare the simulation in Figure 13.6C with that in Figure 13.6D).

Interferometric microscopy with its ability to gather robust spectral data in a reasonable amount of time (10 to 20 seconds) will soon find applications far beyond the realm of cytogenetics.

■ Localization of Lipids, Carbohydrates, and Other Molecules Based on Chemical Affinity

Although antibody probes for proteins and oligonucleotide probes for DNA and RNA have extremely high specificity, localization of other molecules can be achieved based on bulk chemical properties rather than interaction with a specific molecular structure. Good examples include the detection of lipids in membranes, carbohydrate polymers in the extracellular matrix, and functional features of specific organelles.

Probes of Membrane Environment

Membranes consist of a lipid bilayer that is hydrophobic on the inside and hydrophilic on both exterior surfaces. Most probes of membrane structure therefore are amphipathic molecules having a polar end and a nonpolar end, just like phospholipids themselves. Typically, an amphipathic membrane probe will be linked to a fluorophore at its polar end because most fluorophores are themselves polar. Examples of such probes are listed in **Table 13.4**. Some probes, such as $DiIC_{18}$ and FM4-64, are used as general membrane markers. Their carbon-rich fatty acid tails nicely anchor these probes in the membrane bilayer; a single aggregate of this type of dye in contact with a membrane can supply probe molecules that diffuse throughout connected membranes regardless of their architectural complexity. They are often used to "light up" endoplasmic reticulum membranes (**Figure 13.12A**, green label in Plate 7 in the Color Addendum) or to label newly formed membranes, as demonstrated in Figure 13.12B.

Other probes that resemble phospholipids are used to detect fluid and gel-like microdomains within membranes. The phospholipid and cholesterol constituents of membranes can undergo fairly dramatic phase changes as temperature is changed, with the lipids being in a more fluid, rapidly mobile state at higher temperatures and a relatively immobile gel-like state as lower temperatures. The transition between the two states can occur over a relatively narrow range of temperature. In recent years, it has also become apparent that at stable physiologic temperatures membranes can contain regional domains that differ in fluidity properties—specifically the formation gel-like "rafts" surrounded by fluid seas of phospholipids. The phospholipid and cholesterol compositions of these two phases are quite different. For that reason, phospholipids that have long hydrophobic tails such as $DiIC_{20}$ interact more favorably with constituents of

Table 13.4 Lipophilic Probes Used to Study Membrane Properties		
Probe	Spectral Parameters	Membrane Property/Use
BODIPY FL C_5-HPC	Ex 488 Em 501	Fluidity and continuity
$DiOC_{20}$	Ex 484 Em 501	Nonfluid or raft-like membrane microdomains; neuronal tracer
$DiOC_{18}$	Ex 484 Em 501	Neuronal tracer; endoplasmic reticulum tracer; cell migration, lineage and circulation
$DiAC_{18}$	Ex 456 Em 590	Neuronal tracer; endoplasmic reticulum tracer; cell migration, lineage and circulation
$DiIC_{18}$	Ex 549 Em 565	Neuronal tracer; endoplasmic reticulum tracer; cell migration, lineage and circulation
$DiDC_{18}$	Ex 644 Em 665	Neuronal tracer; endoplasmic reticulum tracer; cell migration, lineage and circulation
$DiRC_{18}$	Ex 750 Em 780	Neuronal tracer; endoplasmic reticulum tracer; cell migration, lineage and circulation
FM 1-43	Ex 479 Em 598	Plasma membrane tracer
FM 4-64	Ex 506 Em 750	Plasma membrane tracer
RH 414	Ex 500 Em 635	Plasma membrane tracer
Octyldecylrhodamine B		Liposome and membrane fusion
Filipin		Tracer for cholesterol and membrane rafts

lipid rafts and tend to accumulate in these gel-like domains. Probes that have short fatty acid tails and double bonds such as BODIPY FL C_5 prefer to remain within the fluid domains of the bilayer. Figure 13.12C (see Plate 7) demonstrates these properties in phospholipid vesicles. The gel-like islands that criss-cross the vesicle are localized by $DiIC_{20}$ and appear red, while the larger fluid areas of the membrane localized by BODIPY FL C_5-HPC appear green.

Carbohydrate Polymers and Glycoproteins

For those interested in carbohydrate polymers, the tools of choice are lectins—proteins produced by plants that bind to specific sugar residues. As shown in **Table 13.5**, there are a wide variety of lectins available having very different specificities usually directed toward a single sugar residue or to two or three residues in a specific sequence. Sugar residues are present as large polymers in the extracellular matrix of many tissues either as networks or linked to proteoglycans. For this reason, amplification in the labeling process is usually not necessary, and direct labeling—attaching the fluorophore or gold particle directly to the lectin—is used. An example of such labeling is shown in **Figures 13.13A** and **13.13B** in which the sugar polymers in the extracellular jelly layers of a frog egg have been incubated with fluorescein-conjugated wheat germ agglutinin, a lectin that binds to N-acetylglucosamine residues. This technique reveals that the middle jelly layer is selectively labeled at two neighboring shells, indicating the presence of N-acetylglucosamine-containing glycan layers at this position. At the electron microscopy level, rapid freezing and deep etching (see Chapter 10) reveal lectin-conjugated gold particles attached to the polymer strands of this layer (Figure 13.13C).

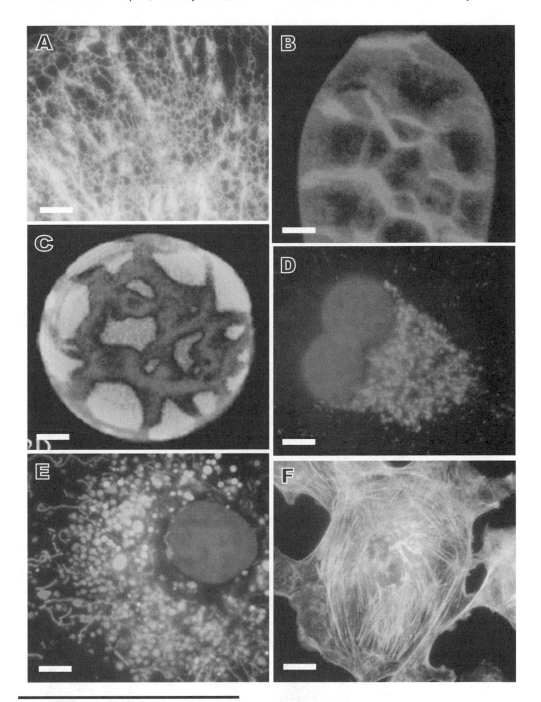

Figure 13.12 (See Plate 7 for the color version.)
Fluorescent probes for cell membranes and organelles. (A) Staining of mitochondria and the endoplasmic reticulum network in a bovine pulmonary artery endothelial cell using Mito-Tracker red and ER-Tracker Blue-White DPX (green). Bar = 1 μm. (B) FM-64 staining of the plasma membrane delineates newly formed fungal zoospores after septation of the fruiting body. Bar = 5 μm. (C) A giant unilamellar liposome stained with DilC$_{20}$ (red) and BODIPY FL C$_5$-HPC (green) to localize gel and fluid regions of the membrane respectively. Bar = 5 μm. (D) A BODIPY TR C$_5$-ceramide probe (red) is used in a bovine pulmonary artery endothelial cell to selectively stain the Golgi apparatus. The nucleus is counterstained blue with Hoescht 33342. Bar = 1 μm. (E) Selective staining of mitochondria and lysosomes in a bovine pulmonary endothelial cell using Mito-Fluor Far Red 680 and Lyso-Tracker Green DND-26. The nucleus is counterstained blue with Hoescht 33342. Bar = 2 μm. (F) Alexa 488-phalloidin staining of microfilaments (green) and antitubulin/Fab-Alexa 594 staining of microtubules (orange) in bovine pulmonary endothelial cells. Bar = 2 μm.

Table 13.5	Examples of Lectins Commonly Used as Probes for Carbohydrates	
Lectin	**Source**	**Saccharide Target**
BS-1	*B. simplicfolia*	α-Galactose
Concanavalin A	*C. ensiformis*	α-Mannose, α-glucose
Snail	*H. pomatia*	N-acetyl-galactosamine
Lentil	*L. culinaris*	α-Mannose
Winged bean	*P. tetragonolobus*	Galactose
Caster bean	*R. communis*	β-Galactose
Lotus	*T. purpureas*	α-L-fucose
Wheat germ agglutinin	*T. vulgaris*	N-acetyl-glucosamine
Furze	*U. europaeus*	α-L-fucose
Hairy vetch	*V. villosa*	N-acetyl-galactosamine
Examples are taken from the Sigma-Aldrich catalogue.		

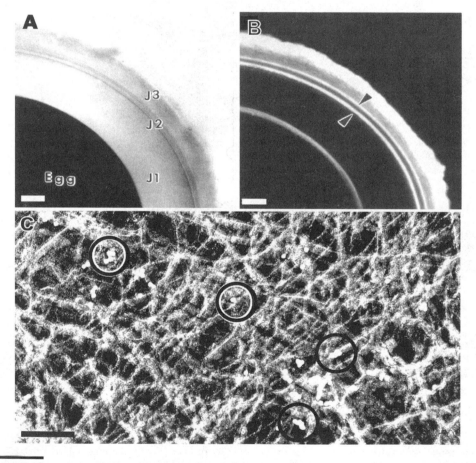

Figure 13.13

Xenopus laevis egg jelly layers stained by wheat germ agglutinin lectin. (A) Brightfield micrograph showing the three jelly layers (J1, J2, and J3) surrounding the *X. laevis* egg. Bar = 100 μm. (B) Laser-scanning confocal micrograph of the jelly layers stained with fluorescein-conjugated wheat germ agglutinin (WGA) to localize N-acetylglucosamine residues. J2 contains two layers of lectin labeling (arrowheads), and J3 is labeled in its outer region. Bar = 100 μm. (C) Electron micrograph of the J2 jelly layer prepared by immunocytochemistry followed by quick freezing and deep etching. Circles point out WGA-colloidal gold labeling of extremely fine fibrils that are likely to be N-acetylglucosamine polymers. Bar = 100 nm.

Table 13.6 Examples of Probes for Organelle Function

Organelle	Dye	Principle
Lysosomes	Lyso-tracker Green; acridine orange	Detects low pH
Mitochondria	Mito-tracker Red; Rhodamine 123	Detects membrane potential
Golgi	BODIPY-Ceramide	Golgi-specific membrane probe
Endoplasmic reticulum	$DiIC_{18}$	Intercalates into membranes
Nucleus	DAPI, Hoechst 33258	Intercalates into DNA
Actin cytoskeleton	Rhodamine Phalloidin	Binds to filamentous actin
Endosomes	Lucifer Yellow	Taken up by endocytosis

Organelle-Specific Probes

Table 13.6 lists a variety of other probes that are used to identify cellular organelles. Most of these probes either identify a structural feature of the target or detect a functional property of an organelle. For example, DAPI and Hoescht dyes bind to the major groove of DNA and serve as general stains for chromosomes and nuclei, as illustrated in Figures 13.12D and 13.12E (blue label). BODIPY-ceramide specifically localizes the Golgi apparatus— the exclusive location of ceramide synthesizing enzymes—as shown in Figure 13.12D (red label). Potential sensitive dyes such as carbocyanines, rhodamine 123, and "Mito-tracker" bind preferentially to mitochondria that exhibit a very substantial membrane potential (–180 mV), higher than that of the plasma membrane in any cell. For this reason, they are effective in localizing these organelles, as shown in Figures 13.12A (orange label) and 13.12E (red label). Weak acidic dyes, such as acridine orange or "Lysotracker," accumulate and fluoresce brightly in acidic environments and are used to identify lysosomes and endosomes (Figure 13.12E, green label). Dye-conjugated toxins that bind specifically to a single protein target are also very useful. A good example is the fungal toxin phalloidin, which binds to filamentous actin and is used routinely to identify microfilament networks (Figure 13.12F, green label).

■ Single Molecule Fluorescence

Fluorescence microscopy is usually carried out on large populations of macromolecules labeled with multiple dye molecules. The study of individual macromolecules, however, can reveal both biophysical properties and dynamics that cannot be determined by a population approach. In order to monitor the fluorescence of individual dye molecules linked to individual macromolecules, one must be able to (1) use high-sensitivity detection techniques, (2) limit the number of molecules in view at any one time, and (3) lower background fluorescence contribution to a minimum. A standard way of doing this is to excite a very thin layer of molecules using an evanescent wave of light and to detect emitted light using a silicon intensifying unit coupled to a CCD chip or a photomultiplier detector for counting of single photons.

Central to the design of a workstation for this purpose is a microscope equipped for fluorescence microscopy but having (1) an intense vapor lamp or laser light source, (2) a piezo-electric–controlled precision stage, (3) an objective or prism for total internal reflection of incident light, (4) a silicon-intensified CCD camera, and (5) computer-based control of stage, camera, and data acquisition.

The illumination system can consist of either a vapor arc lamp or a laser. In both cases, the illuminating light must be focused by lenses into a narrow beam of high intensity and in the case of an arc lamp must be passed through an excitation filter. The incident beam of light is not directed at the specimen itself but rather is reflected off a glass surface that is either directly in contact with the specimen or is extremely close to the specimen. One method for doing this to place a prism in contact with the specimen through which the beam of exciting light is guided at a low angle so as to reflect off the inside surface of the coverglass completely (**Figure 13.14A**). Alternatively, the bottom surface of the specimen may be coated with immersion oil and a prism placed in contact with the oil layer. The beam of exciting light will pass through the prism to enter the oil/coverglass layer at low angle and by multiple reflections fill this layer with reflected light (Figure 13.14B). Both of these methods use an inverted microscope. A second method uses epifluorescence optics to direct the incident beam down the periphery of a high numerical aperature objective lens (Figure 13.14C). This beam is totally reflected off the front surface of the lens at a low angle and re-enters the objective lens going away from the specimen. This method is compatible with either upright or inverted microscope design. In all three cases, total internal reflection takes place at the interface between aqueous specimen and glass surfaces because the critical (Brewster's) angle of reflection has been exceeded. This means that the specimen receives no direct light from this beam, hence the term "total internal reflection fluorescence." Instead, reflection of high-intensity beams is accompanied by an evanescent wave of radiation that extends from the reflection interface into the specimen for a very short distance, effectively about 100 nm (Figure 13.14D). The evanescent wave excites a very thin layer of fluorescent molecules without background contributions from molecules at greater distances. The specimen, usually an array of molecules or a cell plasma membrane, must be extremely close to this surface.

The light emitted by these molecules is gathered by the objective lens, and the intermediate image formed is applied to the front window of an intensified CCD chip camera. Intensification refers to the fact that the photon or electron signal to the CCD is amplified manyfold before actual detection, thereby allowing the camera to image at very low light levels. These types of cameras were originally developed for the military for use at night but have been adopted with excellent results for fluorescence microscopy of single molecules and weak-emitting fluorescent probes.

Intensified cameras come in three basic designs. Older units typically use a silicon intensification target within a vacuum tube and are referred to as silicon intensified target (SIT) cameras. In these devices, the electrons knocked out of the photocathode by photons are accelerated, and the image pattern is formed by the electrons focused onto a silicon target. The high energy of each electron, after collision with the target, results in the formation of numerous charge separation events, thereby amplifying the signal several orders of magnitude. An electron beam then integrates the silicon target to read out the charge distribution in a manner like that of the photoconductive video tube cameras described in Chapter 11. A second method using an electron-bombarded back-thinned CCD chip for intensification is essentially the solid state equivalent of the SIT tube. Electrons produced at the photocathode are accelerated and focused by high-voltage coils onto the back surface of a CCD chip that has been thinned by ion milling until it is relatively transparent to light or electrons (Figure 13.14E). Now each energetic electron produces several hundred-charge separations within a photodiode, thereby amplifying the signal. The CCD chip is then read out as described in Chapter 11 for solid-state cameras.

State-of-the-art photon-counting cameras use a third device—the proximity-focused image intensifier—that provides a much higher amplification of the signal (Figure 13.14F). In this design, photo electrons from the cathode are channeled through tunnels as small as 6 μm in diameter inside a microchannel plate. Multiple collisions of electrons with the walls of these tunnels result in manyfold amplification of the electron signal, much as would happen on a macroscopic scale in a photomultiplier tube. The signal, now consisting of up to 100,000 times the number of electrons that were emitted by the photocathode,

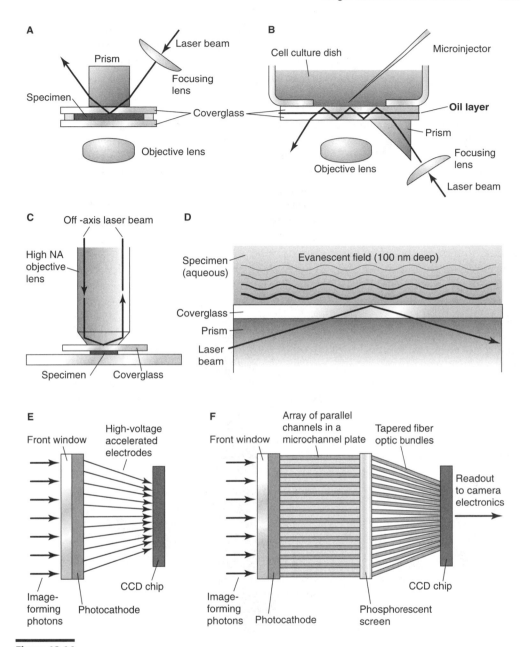

Figure 13.14
Design of a single molecule fluorescence workstation. (A) A prism is used to direct a laser beam to reflect off a coverglass specimen enclosure. (B) Laser beam reflection off of the coverglass serving as the bottom of a culture dish. This arraignment allows access to the specimen for such purposes as microinjection or hormone addition. (C) An off-axis laser beam directed down the barrel of an objective lens will be reflected off the coverglass of the specimen. (D) All methods previously mentioned result in coherent laser light being internally reflected off the internal surface of the coverglass or prism. No light rays pass through this boundary into the specimen but the radiation sets up a thin evanescent electromagnetic field that can excite single dye molecules within 100 nm of the glass surface. (E) Electron-bombardment thinned CCD detection of low light level images. (F) Microchannel plate intensifier for low light level imaging.

hits a phosphorescent screen on the back side of the microchannel plate. This screen converts the electron signal into emitted photons that are then guided by fiberoptics directly to the surface of a CCD chip. The image pattern at the front photocathode is maintained all of the way to the CCD chip surface because of the presence of a single microchannel and a single optic fiber for each pixel of the image. The resolution of the tunnel array in the microchannel plate is the limiting factor in resolution of the image because both the fiber optic bundle and the diode array on the CCD chip generally have more tightly packed arrays. In this manner, extremely low-level signals consisting of single photons can be detected by the chip. Adding to the sensitivity of the device is the possibility of binning the diode elements of the CCD chip. The result is that under the best circumstances, the entire life history of a single fluorescent dye molecule can be recorded.

All dye molecules are subject to light-induced destruction (bleaching), although this varies from dye to dye. The life history of a typical dye molecule might consist of 50 to 100 photons being emitted over a period of 30 to 60 seconds before the dye is extinguished, never to fluoresce again. Although this history is finite, it is sufficient in certain cases to follow the biological behavior of the molecule to which it is attached. Imagine, for example, a dye-conjugated DNA polymerase molecule that is bound to a single DNA strand being used as a template for DNA synthesis. As the polymerase synthesizes a complementary strand of DNA from nucleotide triphosphates, it will move relative to the template. By monitoring the dye attached, the instantaneous velocity of polymerization can be followed. By using templates of known nucleotide sequence, one can determine whether A, T, C, and G are each added at the same or different rates, something that would be impossible if one were monitoring a population of enzymes.

Similarly striking experiments can visualize the dynamics of events at the cell surface. Exocytosis and endocytosis of small membrane bound vesicles can be monitored with precision using the total internal reflection approach to fluorescence excitation. Likewise, fusion of artificial phospholipid vesicles with a planer bilayer membrane can be imaged in real time (**Figure 13.15**). If one prepares vesicles that are loaded with a

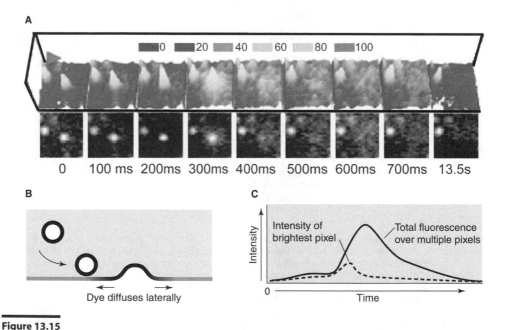

Figure 13.15
Liposome fusion imaged by single molecule fluorescence. (A) Three-dimensional (upper panel) and two-dimensional (lower panel) intensity maps of images captured by an intensified CCD camera during liposome fusion. In each frame, the upper peak/spot represents a liposome that approached the target membrane but did not fuse, while the lower peak/spot represents a liposome that did fuse. (B) During fusion, lipophilic dye molecules in the liposome membrane diffused into the target membrane, resulting in the dequenching of dye fluorescence. (C) As a result of dye dequenching, fusion is accompanied by an immediate rise in peak intensity and followed by a longer lasting rise and spread of total emission intensity.

fluorescent phospholipid probe at high density, these probe molecules will self-quench their own emission because they are packaged so close to one another. After the vesicle membrane fuses with the planar target membrane, the probe molecules will diffuse laterally; as a result, quenching will be reduced, and fluorescence efficiency should rapidly increase. This is exactly what is observed by total internal reflection fluorescence. In the upper panel of Figure 13.15A, two vesicles seen as peaks are docked at the target membrane; one vesicle fuses with the membrane at $t = 300$ ms, and one vesicle (the upper one) does not. Fusion of the lower vesicle results in a rapid increase in fluorescence intensity, as well as a rapid spread in fluorescence because of lateral diffusion of the probe within the target membrane (Figure 13.15B). Consistent with this model is the fact that peak fluorescence at the site of fusion undergoes a rapid but transient increase, while total fluorescence summed over the entire area of probe diffusion undergoes a large and long-lasting increase (Figure 13.15C). Although the steps in this process were previously known, the exact time course was not. Granted, in these experiments, many dye molecules are being visualized at once, and this is not a single molecule experiment; however, the advantages of this technique in the study of vesicle trafficking in live cells are undisputed.

■ Detection of Molecular Motions and Interactions: Fluorescence Resonance Energy Transfer, Fluorescence Recovery after Photobleaching, Fluorescence Correlation Spectroscopy, Fluorescence Polarization, and Speckle Microscopy

Some molecular motions can be detected on either a single molecule or population basis. Molecular rotation can be studied by fluorescence polarization; translational diffusion behavior can be studied by either fluorescence recovery after photobleaching (FRAP) or fluorescence correlation spectroscopy (FCS). Interactions of proteins with other proteins can be studied by fluorescence resonance energy transfer (FRET) or speckle microscopy.

Fluorescence Polarization

In order for a chromophore to absorb light, the electronic transition must be oriented in the same plane as the oscillating electric field vector of the photon. If polarized light is used, only a fraction of the chromophores in a population will be oriented correctly. Subsequent emission of light a few nanoseconds later will produce a photon having polarization identical to the absorbed light provided that the chromophore has not changed its orientation; however, if the dye molecule has reoriented during that time, the emitted light will become depolarized—the direction of its electric field vector will have become randomized. A change in polarization or randomization can be detected by using a polarizing filter as an analyzer of the emitted light. Rotation of a small dye molecule is rapid, and depolarization of the emitted light is expected; however, if the dye is attached to a macromolecule, the rotation may be considerably slower (**Figure 13.16**). In this case, the degree of depolarization observed will be related to the speed with which the macromolecule rotates. For example, labeled phospholipids in a bilayer membrane may rotate very quickly at high temperature when the membrane is in a "fluid" state but rotate very slowly at low temperatures when the membrane is in a "gel" state. For this reason, structural transitions in membranes are often measured by fluorescence depolarization.

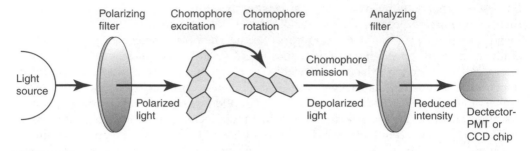

Figure 13.16
Fluorescence polarization detection of molecular movement. Polarization of emitted light, as analyzed by a polarizing filter, is decreased if the chromophore is moving fast enough to change orientation during its fluorescence lifetime.

Fluorescence Correlation Spectroscopy (FCS)

Translational diffusion of dye-labeled macromolecules can also measured by microscopic spectroscopy. As shown in **Figure 13.17A,** laser illumination of molecules diffusing through the beam and photon counting of the emitted light leads to a train of peaks that

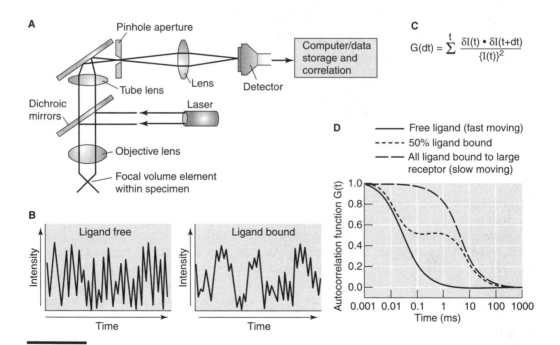

Figure 13.17
Workstation for FCS. (A) Laser light is focused into a small volume within the specimen to excite any dye-conjugated molecules moving through the beam. The photons emitted by these dyes are captured by the objective lens and counted individually by a sensitive detector. (B) Detector current versus time is recorded by a computer and analyzed by correlation algorithms. Each photon detected is represented by a peak. Fast-moving fluorophores are represented by well-separated peaks (left panel), while slow moving fluorophores are characterized by groups of peaks (right panel). (C) A typical correlation function used for detection of grouped photon counts. (D) The correlation function drops quickly with increasing correlation times for fast moving fluorophores such as one attached to a small ligand (solid line). Because of photon peaks being grouped, the correlation function of a slow moving fluorophore remains high until larger correlation times are reached (dashed line). Such data might be expected if the fluorescent ligand had bound to a large receptor protein. A mixture of bound and free ligand would result in the correlation coefficient exhibiting two separate decreases (dotted line).

appear to be noise but, in fact, are not. Mathematical analysis of this train by computer shows that large, slowly moving molecules lead to a series of closely spaced peaks caused by multiple photon counts that track the macromolecule through the beam. In contrast, the rapid passage of small molecules produces single, isolated peaks (Figure 13.17B). These patterns can be evaluated on a population basis by a peak correlation approach, the correlation equation being given in Figure 13.17C. First, the average intensity of all data points over time, {I(t)}, is calculated. Next, the amount that each data point differs from the average, $\delta I(t)$, is calculated. One then correlates the intensities of all pairs of points in the data stream that differ in time by dt. The degree of correlation, as denoted by the correlation coefficient G(dt), is then plotted versus dt. One obtains a curve that is near a value of 1.0 at very low time differences but that at longer time intervals drops to 0. The value of 1 at very short times reflects the fact that we are comparing points in the same photon count, and thus, their intensity is highly correlated. The value of 0 at long times reflects the fact that there is no correlation between the two data points if they are far apart in time because the average $\delta I(dt)$ summed over all data points is zero. The important result is the time interval dt over which the correlation drops from 1 to near 0. If this interval is short, the labeled molecules zoomed right through the beam and are probably small; if the interval is long, the labeled molecules are relatively large. They passed leisurely through the beam, giving rise to many correlated photon counts that are reflected in G(dt) remaining high for a longer period.

From these data, one can derive not only the diffusion coefficient of the labeled molecule but also its interaction with other molecules to form larger complexes. FCS has been applied very nicely to the binding of ligands to receptors. In Figure 13.17D, from left to right, the first trace depicts data for a ligand, a small molecule. Its correlation coefficient rapidly drops to low values at time intervals of 0.01 msec, and from this, a very high diffusion coefficient can be calculated. The third trace (dashed) illustrates data for the receptor-bound ligand; the correlation coefficient does not drop until the time interval reaches 10 msec, suggesting a diffusion coefficient that is much lower for the ligand–receptor complex. At intermediate concentrations of ligand that lead to half-maximal binding of the ligand to the receptor (middle trace, dotted line), the correlation curve reveals two populations of ligands—fast-moving ligands that are unbound and slow-moving ligands that are bound to the receptor.

Fluorescence Recovery after Photobleaching (FRAP)

Molecular movements on a population basis can also be measured by photobleaching dye-linked molecules in a specific area and visualizing the return of fluorescence to that area due to molecular movement (**Figure 13.18A**, left). Typically, strong laser illumination is used for rapid photobleaching, while either weaker laser illumination or vapor arc lamp illumination is used to detect fluorescence rebound, a period during which photobleaching is not desired. Figure 13.18B depicts such an experiment using GFP tagged-galactosyl transferace (GalTase-GFP), an enzyme that is typically restricted to the Golgi apparatus in interphase cells but which temporarily becomes part of a generalized membrane reticulum during mitosis. Photobleaching of GalTase-GFP in a mitotic cell, using a high intensity horizontal laser scan (arrow) followed by imaging over the following 120 seconds demonstrates redistribution of the tagged protein to the bleached region. Fluorescence recovery is exponential and its quantitation allows calculation of an observed diffusion coefficient (Figure 13.18C).

A related technique, fluorescence loss in photobleaching (FLIP), takes the opposite approach. A selected area is photobleached continuously, thereby acting as a continuous site for dye destruction (Figure 13.18A, right). As dye molecules move to their "death," the surrounding areas become less fluorescent, and this drop in fluorescence can be measured.

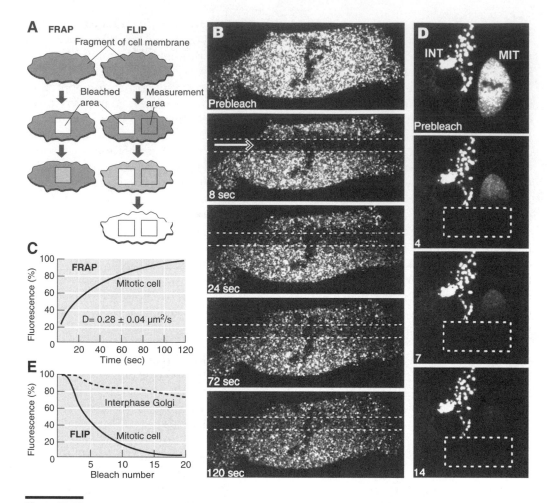

Figure 13.18

Molecular movement detected by fluorescence recovery after photobleaching (FRAP) and by fluorescence loss in photobleaching (FLIP) methods. (A) Left panel. Laser-induced photobleaching of a restricted membrane region and return of fluorescent lipids to the region during a FRAP experiment. Right panel. Repeated laser-induced photobleaching of a specific membrane region during a FLIP experiment leads to depletion of chromophores from other regions providing they are conjugated to a mobile molecule. (B) FRAP detection of GFP-galactose transferase dynamics in PTK1 cells. Fluorescence intensity in the bleached strip drops and then recovers due to movement of new molecules into the area. (C) Quantitation of recovery of GFP-galactose transferase fluorescence in the photobleached strip in (B) allows its diffusion coefficient to be calculated. (D) Repeated photobleaching within the dashed box leads to depletion of GFP-galactose transferase fluorescence from a mitotic cell (MIT) but not from an interphase cell (INT). (E) Quantitation of the FLIP experiment in (D). Loss of GFP-galactose transferace fluorescence from surrounding regions of the mitotic cell is rapid (half-time = 5 minutes) suggesting that the protein is highly mobile. In contrast, 85% of this protein in interphase cells appears to be immobile.

Surrounding areas that do not drop in fluorescence contain proteins that are relatively immobile because they are either anchored or they are in another compartment. Use of FLIP is illustrated in Figure 13.18D in which two cells, both containing GalTase-GFP are repeatedly photobleached in the region demarcated by a dashed line. Imaging over the next few minutes reveals that the mitotic cell (MIT) is depleated of fluorescent GalTase-GFP while the interphase cell (INT) is not. Quantitation of these results (Figure 13.18E) demonstrates the nearly complete loss and high mobility of detectable GalTase-GFP in mitotic cells while in interphase cells at least 85% of this protein is immobile and not photobleached.

Fluorescence Resonance Energy Transfer (FRET)

If two chromophores are extremely close to one another (less than 10 nm) and the emission spectrum of one (the "donor") overlaps with the absorption spectrum of the other (the "acceptor"), then a transfer of energy can occur without a photon intermediate. Such a "radiationless" transfer of energy might be thought of as a temporary alignment of electron transitions at close range. The probability of this happening is inversely proportional to the distance between the chromophores to the fourth power; therefore, the monitoring of this event is an extremely sensitive measure of distance between chromophores.

Fortunately, FRET is fairly easily measured if it is carried out on a population scale, as shown in **Figure 13.19A** (see Plate 8 in the Color Addendum). Our donor molecule, A, absorbs light at 433 nm, and a proper excitation filter and dichroic mirror are used to isolate this wavelength just as in normal fluorescence microscopy. If there is no FRET occurring, then A will emit light at 475 nm as it normally would if there were no other chromophore present; however, if the acceptor molecule, B, is present and is located within 10 nm or less of A, FRET will occur. FRET in this case will result in light being emitted from B at 527 nm, while emission from A at 475 nm is reduced because of FRET siphoning off its energy and delivering it to B. Typical changes in the donor and acceptor emission spectra are shown in Figure 13.19B. If A and B do not interact at all, there is no FRET, and no emission is seen from B at 527 nm; if some but not all A molecules interact with B, a small peak will be seen at 527 nm (partial FRET); if all A and B molecules interact, maximal emission will be seen at 527 nm, and concurrently, a reduction in emission from A will be seen at 475 nm (maximal FRET). The detection of FRET therefore indicates that the donor and acceptor are linked closely in some manner either because (1) they are bound to one another, (2) they are a part of fusion proteins that are bound to one another, or (3) they are a part of the same macromolecule that has undergone a change in conformation that brings them close together.

FRET can be carried out either in live cells, fixed cells, or a cell-free system. Furthermore, in live cells, both the spatial and temporal dynamics of FRET can be observed. Figure 13.19C illustrates detection of cyclic AMP-dependent protein kinase (PKA) activity by using a "FRET probe." The probe is a genetically engineered fusion protein that contains a CFP domain, a phosphorylated serine binding domain, a PKA-specific substrate domain, and a YFP domain in that order (Figure 13.19D). Transfection and expression of this probe in cells, followed by activation of PKA results in phosphorylation of the substrate domain of the probe, binding of the substrate domain to the adjacent serine-phosphate binding domain, and a conformational change that brings the CFP and YFP domains together to produce FRET. Use of this "A-kinase activity reporter" (AKAR) in HeLa cells results in a marked increase in FRET when forskolin (Fsk) is used to activate adenylate cyclase, raise cyclic AMP levels inside the cell and increase PKA activity (Figure 13.19C). Subsequent washout of the forskolin and addition of the PKA inhibitor H-89 reverses activation of this enzyme, leading to removal of phosphate from the substrate domain of AKAR by cellular phosphatases, and the loss of FRET.

FRET can also be detected by a change in the lifetime of the fluorescence emission. Because transfer of the excited state from donor to acceptor is actually faster than the decay of the excited state to an emission state in the donor, the time between excitation and emission is reduced. This can be measured, pixel by pixel, using fluorescence lifetime microscopy (FLIM), a technique that utilizes exquisitely short pulses of laser light for excitation (picoseconds in duration) and sensitive detection of emitted light decay over a sub-nanosecond time course. The lifetime of emission is a sensitive indicator not only of FRET but of a whole set of factors in the chromophore

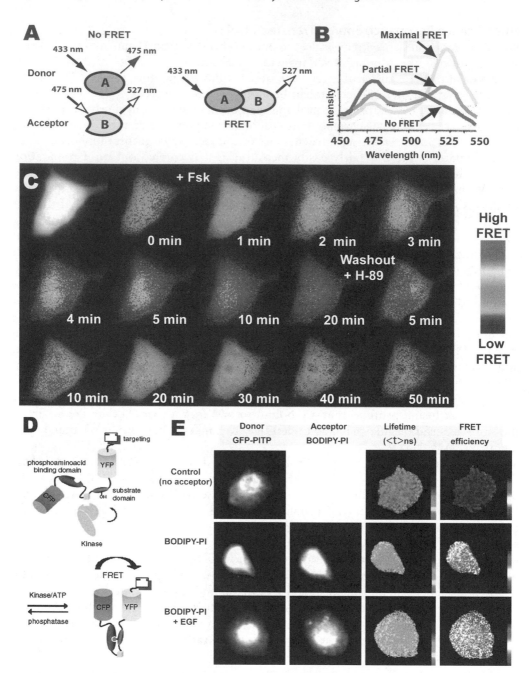

Figure 13.19 (See Plate 8 for the color version.)
Fluorescence resonance energy transfer (FRET). (A) Molecule A (cyan fluorescent protein; CFP) emits fluorescence at 475 nm while Molecule B (yellow fluorescent protein; YFP) emits fluorescence at 527 nm. If the molecules do not interact, excitation with 433-nm light will elicit fluorescence only from A. If the molecules closely interact, transfer of the excited state from A to B by FRET will result in fluorescence from B and a decrease in fluorescence from A. (B) Typical spectra from a CFP-YFP pair indicating no FRET, partial FRET, and maximal FRET. (C) Application of forskolin (Fsk) to HeLa cells results in protein kinase A activation as measured by FRET from an A-kinase activity reporter (AKAR). Washout of forskolin and addition of the protein kinase A inhibitor H-89 reduces FRET back to control levels. Levels of FRET are indicated by pseudocolor as calibrated on the right side. (D) Construction and function of the fusion protein (AKAR) for detecting FRET-based activity of protein kinase A. (E) Detecton of FRET between EGFP-phosphotidyl inositol transfer protein (PITP) and Bodipy-conjugated phosphotidyl inositol (Bodipy-PI) during epidermal growth factor (EGF) stimulaltion of COS-7 cells. Increased FRET, results in a decrease in fluorescence lifetime which can be measured by fluorescence lifetime microscopy (FLIM). Lifetime decrease and FRET efficiency increase, displayed in pseudocolor, is seen at the plasma membrane and in the Golgi region.

environment including solvation, quenching, charge interactions, and molecular motions placing this technique at the cutting edge of light microscopy advances; microscopes designed for such lifetime measurements have recently become available commercially.

Use of fluorescence lifetime to measure FRET between a phospholipid transport protein (PITP) and the phosphotidyl inositol (PI) transported is illustrated in Figure 13.19E. COS-7 cells were transfected with EGFP-PITP, Bodipy-PI incorporated by incubation, and FRET occurring between these two molecules measured before and after epidermal growth factor (EGF) stimulation. A pseudocolored image map of lifetime and FRET efficiency showed that in the presence of EGF, efficiency increased and (as expected) emission lifetime decreased near the plasma membrane.

FRET can also be carried out at the single molecule level. As a hypothetical example, if one wanted to image DNA polymerase activity, one could carry out synthesis of the new DNA strand using a DNA polymerase molecule to which an acceptor chromophore is attached and NTPs to which donor chromophores are attached. Each time the polymerase binds an NTP, FRET will be observed as the DNA chain is synthesized. Alternatively, one could label just one of the four nucleotide bases used and the dynamics of the FRET observed used to sequence the new chain being formed. Similar approaches can be used to look at interactions of any two types of molecules, providing that the proper chromophores can be attached without affecting their function.

Speckle Microscopy

The dynamics and interactions of proteins in large complexes can be monitored by labeling and tracking of a small number of the total population available. The labeled molecules act as "sentinel" proteins whose dynamics provide information on the dynamics of the entire population. Because only a small portion of proteins are labeled, their individual dynamics can easily be distinguished as "speckles" within a sea of unlabeled but otherwise identical proteins. An excellent application of this technique is to the study of microfilament and microtubule turnover in live cells. **Figure 13.20A** illustrates the appearance of a motile epithelial cell that has been injected with a modest amount of fluorescein-conjugated actin. The labeled actin is taken up into microfilaments by polymerization and appears as speckles of emitted light at random intervals along the actin polymer (arrows). By cataloging each speckle and following its path by fluorescence imaging at a frame rate of 30 per second, the bulk movement of each microfilament bundle as well as the intrabundle dynamics caused by polymerization and depolymerization of actin monomers can be followed. A vector diagram plotting such movement (Figure 13.20B) shows that the entire microfilament network is coordinated in its properties, resulting in zones such as the lamellipodium (Lp) and the lamellum (L) in which rapid retrograde flow of actin is observed, and interior regions (A) where anterograde actin flow occurs. These flows meet at a convergence zone (C) about 15 μm from the leading edge.

The specialized equipment necessary for speckle microscopy is similar to that required for standard single-molecule fluorescence studies. High numerical aperture objective lenses and a sensitive silicon-intensified camera are absolute requirements. Illumination, however, is usually by vapor arc lamp using standard optics; total internal reflection optics is not usually required. Rapid acquisition of digital images and computerized image analysis is a must. Specialized software using object-tracking routines that can be fine tuned to identifying speckles that emit photons intermittently is required. This software is custom written for the purpose and is not available commercially at this time.

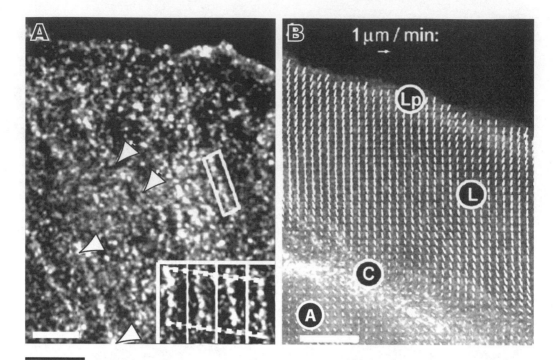

Figure 13.20
Dynamics of actin microfilaments in living cells using speckle microscopy. (A) Speckled fluorescence emission from dye-conjugated actin monomers incorporated into microfilaments. Strings of speckles (arrows and box) identify actin monomers in the same microfilament. (Inset) Movement of labeled actin monomers in a single microfilament. Bar = 5 μm. (B) Array of velocity vectors for speckles tracked by fluorescence video microscopy at the plasmalemma of a motile cell. At the leading edge of the cell, the lamellipodium (Lp) exhibits a high retrograde flow of actin, and the lamellum (L), a few micrometers inward, exhibits a slower retrograde flow, while the interior of the cell (A) exhibits anterograde flow. These opposing flows of actin meet in a zone of convergence (C) where contraction may be occurring. Bar = 5 μm.

■ Single Particle Tracking Using Video Microscopy

Because dyes are susceptible to photo destruction, it is seldom that a dye-conjugated macromolecule can be tracked continuously for long periods of time. In addition, fluorescence microscopy does not lend itself to continuous visualization of the cellular landmarks. When such tracking is desired, the usual method is to attach a visible marker to the molecule such as a latex bead or a colloidal gold particle. As an example, this strategy can be used to track movements of neurotransmitter receptors in the plasma membrane of neurons (**Figure 13.21**). For this purpose, cells were transfected with recombinant DNA in order to express a genetically engineered glycine receptor that contained a myc sequence (for antibody binding) in its extracellular domain and gephyrin binding sequence in its cytoplasmic domain (Figure 13.21A). Gephyrin is a protein in neurons that links plasma membrane receptors to microtubules. A 0.5-μm-diameter latex bead coated with an anti-myc antibody was then brought in contact with the cell surface so that it could bind to a recombinant glycine receptor. The movements of the bead were then tracked by video microscopy over a period of minutes as the receptor diffused laterally within the plane of the plasma membrane. The trajectory of the receptor and its velocity at each point in the path were analyzed and plotted as illustrated in Figure 13.21B. The path of the receptor could be divided into regions of free diffusion and regions of confinement. During periods of confinement, the velocity and distances moved by the

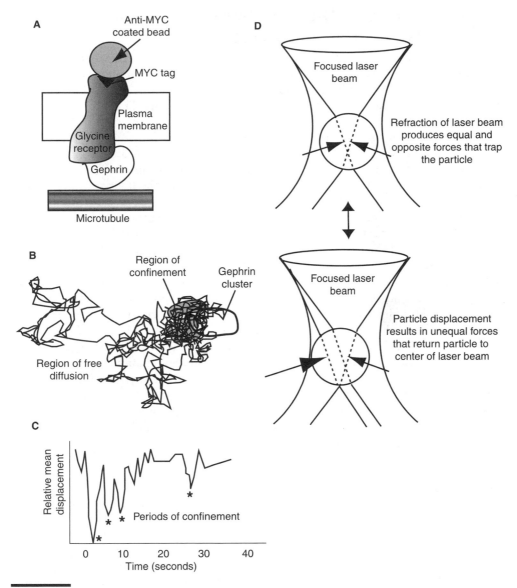

Figure 13.21

Single particle tracking. (A) Diagram of the sphere/receptor complex bound to gephyrin and microtubules during periods of confinement. (B) Trajectory of a latex sphere attached to glycine receptors as tracked by video microscopy. The sphere/receptor stalls and exhibits periods of confinement in regions containing the anchoring protein gephyrin. (C) Relative displacement of the sphere/receptor drops precipitously during periods of confinement. (D) Diagram of a focused laser beam used as "optical tweezers" to manipulate latex spheres into place on the cell surface. Forces created because of the refractive scattering of light off the sphere can trap and move the sphere (or any other organelle of sufficient index of refraction) as the laser beam is moved.

receptor were highly restricted (asterisks, Figure 13.21C). Periods of confinement took place whenever the receptor was in close proximity to clusters of gephyrin, which were localized by immunocytochemistry. This elegant example of particle tracking clearly illustrates that movements of single macromolecules can be followed for extended periods under the right conditions.

■ Optical Tweezers—The Laser Trapping of Small Organelles, Particles, and Motile Cells

In the single particle tracking experiment described, "optical tweezers" were used to place a latex bead on the cell surface so that it could interact with a membrane protein. This technique is used not only to manipulate organelles, latex beads, and other particles but can also be used to measure forces on particles and cells. The only requirement is that the object to be moved or measured should have an index of refraction sufficiently different from that of the surrounding medium. As shown in Figure 13.21D (upper panel), a narrowly focused laser beam of sufficient power, directed right at the particle, will be diffracted by the change in refractive index at the surface of the particle. Diffraction of light results in a change in momentum of the light ray and according to Newtonian mechanics the light will exert an equal and opposite pushing force on the particle as indicated by the arrows, thus, trapping the particle in the laser beam. If the laser beam moves, it will tend to drag the particle along with it. The reason is that any movement of the particle out of the beam will create an asymmetry in the light diffraction by the particle and therefore an asymmetry in the forces exerted on the particle by light. Specifically, light diffraction on the side of the particle **opposite** the direction of laser beam movement will become comparatively greater, and therefore, the increased force of light pushing on that side of the particle will become greater. As a result, the net force of light on the particle will push the particle back into the center of the laser beam (Figure 13.21D, lower panel). This effect is seen regardless of whether it is the laser beam or the particle that is moving. For example, sperm cells can be caught in a laser trap, and as the laser power is reduced, the sperm cell will, at some point, escape under its own power. The power of the laser at that point can be used to calculate the force of light exerted on the cell and thereby the motile force generated by the cell to break loose from the beam. In this manner, minute cellular forces in the piconewton range can be measured.

■ Fluorescence-Activated Flow Cytometry

Although fluorescence microscopy is an excellent technique for locating specific macromolecules within cells, it is not well suited to quantifying the level of macromolecules in a heterogeneous population of cells that may number in the tens of thousands. To quantitate rapidly the emission from multiple fluorescent probes in such a population, the technique of fluorescence-activated flow cytometry is without parallel. From an optical design standpoint, this instrument is somewhat like a laser scanning confocal microscope, except that instead of the laser beam moving, the cells are moving! As illustrated in **Figure 13.22A,** a stream of cells, previously labeled with fluorophores, is passed through a tubular optical chamber in single file. To maintain a steady stream, the cellular suspension is fed into a sheath of buffer that keeps the cells from hitting the walls of the tube and creating turbulence. As each cell passes through the tube, it is illuminated by a focused beam of laser light that scatters off the cell and excites the fluorescent dyes within the cell. Both the scattered light and the fluorescence emission is sampled at 90 degrees from the exciting beam by a series of dichroic mirrors and photomultiplier tubes almost identical to those found in the scanning head of a confocal microscope (Figure 13.22A). First, the short-wavelength exciting light that was scattered from the cell is stripped off. This signal is referred to as the "side-scattered" signal (SSC) and is related to the size, shape, and surface topology of the cell. Next, fluorescence emitted in the far-red is stripped off, followed by fluorescence in the orange and green regions of the spectrum. Typically, these fluorescence signals come from immunocytochemical

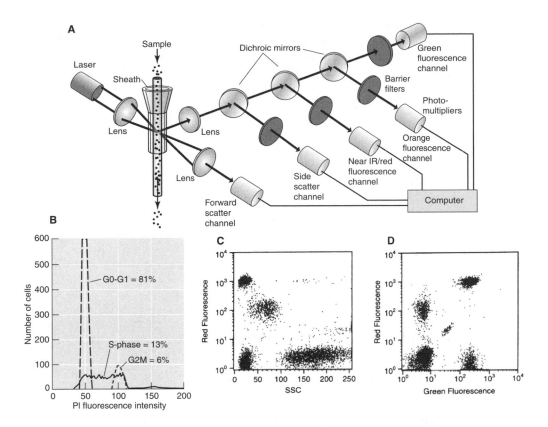

Figure 13.22

Flow cytometry. (A) A sample of cells is lined up and guided through an optical tube using a sheath of flowing buffer that separates the cells from the tube wall. As each individual cell passes through the laser beam, the light scattered off the cell and the fluorescence emissions from dyes associated with cellular structures are analyzed by a bank of dichroic mirrors and photomultiplier tubes much like those seen in a laser-scanning confocal microscope. (B) The cell population was stained with propidium iodide to quantitate DNA and analyzed by flow cytometry. Histogram data relating numbers of cells to intensity of emitted light exhibit a peak equivalent to 2N chromosomes showing that most cells are in the G1 phase and a peak equivalent to 4N chromosomes showing that some cells have already replicated their DNA and are in G2 or M phase. (C) A plot of two independent variables (red immunofluorescence and side scattered incident light) reveals four separate types of cells in this mixture. (D) Plot of immunofluorescence signals for two different cellular structures/proteins (red and green) indicates the presence of five different cell populations each having a characteristic level of the two antigens.

probes for specific proteins or from dyes staining for specific classes of macromolecules or organelles. Each dye signal is quantitated by a separate photomultiplier tube and the signals for each cell tallied by a computer. Also frequently measured is a "forward-scattering" signal (FSC) that measures the reduction of laser light because of the cell blocking the beam, a signal that is also related to cell size and shape. The data from thousands of cells can then be plotted out as histograms or two- or three-dimensional scatter plots showing the relationships between signals in various subpoplations of the cell suspension.

A typical use of such technology is to determine what proportions of cells in the population are in each phase of the cell cycle. Such a cell population can be stained with propidium iodide, a quantitative stain for DNA, and the staining intensity for each cell in the population measured by fluorescence-activated flow cytometry. If expressed as a histogram (Figure 13.22B), a large peak is seen for 81% of the cells in the G0 and G1 phases of the cell cycle that have the standard diploid number of chromosomes; however, those cells that have already replicated their DNA in preparation for mitosis (and are in the G2 and M phases of the cell cycle) will have twice as much DNA and twice as much propidium iodide staining, as seen in the smaller peak representing 6% of the cells. Cells that are in the S phase and have replicated part of their DNA have propidium iodide staining that is between these two peaks in intensity. In the scatter plots of Figures 13.22C and 13.22D, cells that have been immunocytochemically stained for two different proteins (resulting in a red signal and a green signal) and also measured for their SSC can be shown to consist of four major subpopulations of cells. These results can be observed in both the red signal/SSC plot in Figure 13.22C and the red signal/green signal plot in Figure 13.22D. The red signal/green signal plot, however, also reveals a small fifth subpopulation of cells that is not detected in Figure 13.22C.

Cells showing different scattering or fluorescence signals can actually be sorted into different tubes for further study. Just after passing through the laser beam and revealing its level of signal, each cell in a single drop will pass between two charged plates that can impart an electric charge to each drop. The drops are then separated by electromagnets into different tubes based on their charge. Surprisingly, electromagnets can bend not only electrons but entire drops that are orders of magnitude greater in mass!

References and Suggested Reading

General

Cox GC, ed. *Optical Imaging Techniques in Cell Biology.* Boca Raton, FL: CRC Press, 2006:288.

Tsien RY. Imaging imaging's future. *Nat Rev Mol Cell Biol.* 2003;(Suppl S):S16–S21.

Enzyme Cytochemistry and Histochemistry

Bancroft JD, Hand NM. *Enzyme Histochemistry (RMS Handbook).* New York: Springer Verlag, 1995.

Borgers M, Verheyen A. Enzyme cytochemistry. *Int Rev Cytol.* 1985;95:163–227.

Brooks SA, Leathem AJC, Schumacher U. *Lectin Histochemistry: A Concise Practical Handbook (RMS Handbook).* Oxfordshire, England: BIOS Scientific, 1997.

Hayat MA. *Stains and Cytochemical Methods.* New York: Springer, 1993, p. 474.

High OB. *Lipid Histochemistry (RMS Handbook).* Oxfordshire, England: BIOS Scientific, 1984.

Pearse AGE. *Histochemistry: Theoretical and Applied.* Vols. 1–3. Edinburgh: Churchill Livingston, 1985.

Van Noorden CF, Frederiks WM. *Enzyme Histochemistry: A Laboratory Manual of Current Methods (Microscopy Handbooks).* Oxford: Oxford University Press, 1993, p. 128.

Immunocytochemistry

Allan V, ed. *Protein Localization by Fluorescence Microscopy: A Practical Approach.* Oxford: Oxford University Press, 1999.

Beesley JE. *Colloidal Gold: A New Perspective for Cytochemical Marking. RMS Handbook No. 17.* Oxford: Oxford Science Publications, 1989.

Beesley JE, ed. *Immunocytochemistry and In Situ Hybridization in the Biomedical Sciences.* New York: Birkhauser, 2000, p. 267.

Bullock GR, Petrusz P, eds. *Techniques in Immunocytochemistry.* London: Academic Press, 1982.

Harlow E, Lane D, eds. *Antibodies: A Laboratory Manual.* New York: Cold Spring Harbor Laboratory, 1988.

Hayat MA. *Microscopy, Immunohistochemistry, and Antigen Retrieval Methods: For Light and Electron Microscopy.* New York: Springer, 2002, p. 360.

Polak JM, Van Noorden S, eds. *Introduction to Immunocytochemistry,* 3rd ed. London: Garland Science, 2003, p. 192.

Skepper JN. Immunocytochemical strategies for electron microscopy: choice or compromise. *J Microsc.* 2000;199:1–36.

Spehner D, Drillien D, Proamer F, Hanau D, Edelmann E. Embedding in Spurr's resin is a good choice for immunolabeling after freeze drying as shown with chemically unfixed dendritic cells. *J Microsc.* 2002;207:1–14.

Tokuyasu KT. Application of cryoultramicrotomy to immunocytochemistry. *J Microsc.* 1986;143:139–149.

Tokuyasu KT, Dutton AH, Singer SJ. Immunoelectron microscopic studies of desmin (skeletin) localization and intermediate filament organization in chicken skeletal muscle. *J Cell Biol.* 1983;96:1727–1735.

Tokuyasu KT, Dutton AH, Singer SJ. Immunoelectron microscopic studies of desmin (skeletin) localization and intermediate filament organization in chicken cardiac muscle. *J Cell Biol.* 1983;96:1736–1742.

Van der Loos CM. *Immunoenzyme Multiple Staining Methods (RMS Handbook).* Oxfordshire, England: BIOS Scientific, 1999.

Autoradiography

Baker JRJ. *Autoradiography: A Comprehensive Overview (RMS Handbook).* Oxford: Oxford University Press, 1989.

Castle JD, Jamieson JD, Palade, GE. Radioautographic analysis of the secretory process in the parotid acinar cell of the rabbit. *J Cell Biol.* 1972;53:290–311.

In Situ Hybridization

Abraham TW. Preparation of nonradioactive probes for in situ hybridization. *Methods* 2001;23:297–301.

Andreeff M, Pinkel D, eds. *Introduction to Fluorescence In-Situ Hybridization: Principles and Clinical Applications.* New York: John Wiley, 1999.

Butler K, Zorn AM, Gurdon JB. Non-radioactive in situ hybridization to Xenopus tissue sections. *Methods* 2001;23:303–312.

Maul G. *Hybridization Techniques for Electron Microscopy.* Boca Raton, FL: CRC Press, 1993.

Morel G, Cavalier A, Williams L, eds. *In Situ Hybridization in Electron Microscopy (Methods in Visualization).* Boca Raton, FL: CRC Press, 2001, p. 472.

Polak JM, McGee J, eds. *In Situ Hybridization: Principles and Practice.* Oxford, Oxford University Press, 1990, p. 258.

Single Molecule Fluorescence Microscopy

Hammes GG. *Spectroscopy for the Biological Sciences.* New York: Wiley–InterScience, 2005:184.

Hof M, Hutterer R, Fidler V. Fluorescence Spectroscopy in Biology: Advanced Methods and Their Applications to Membranes, Proteins, DNA, and Cells (Springer Series on Fluorescence). New York: Springer, 2005, p. 305.

Sako Y, Yanagida T. Single molecule visualization in cell biology. *Nat Rev Mol Cell Biol.* 2003;(Suppl S):S1–S5.

Smith A, Gell C, Brockwell D. *Handbook of Single Molecule Fluorescence Spectroscopy.* Oxford: Oxford University Press, 2006, p. 278.

Weiss S. Fluorescence spectroscopy of single biomolecules. *Science* 1999;283:1676–1683.

Yildiz A, Forkey JN, McKinney SA, Ha T, Goldman YE, Selvin PR. Myosin V walks hand over hand: single fluorophore imaging with 1.5 nm localization. *Science* 2003;33:2061–2065.

Fluorescence Return after Photobleaching

Lippincott-Schwarz J, Altan-Bonnet N, Patterson GH. Photobleaching and photoactivation: following protein dynamics in living cells. *Nature Rev. Mol Cell Biol.* 2003;(Suppl S):S7–S14.

Fluorescence Resonance Energy Transfer

Ha T. Single molecule fluorescence resonance energy transfer. *Methods* 2001;25:78–86.

Kenworthy A. Imaging protein-protein interactions using fluorescence resonance energy transfer microscopy. *Methods* 2001;24:289–296.

Periasamy A, Day R, eds. *Molecular Imaging: FRET Microscopy and Spectroscopy (Methods in Physiology Series).* New York: Academic Press, 2005:336.

Speckle Microscopy

Salmon W, Adams MC, Waterman-Storer CM. Dual-wavelength fluorescent speckle microscopy reveals coupling of microtubule and actin movements in migrating cells. *J Cell Biol.* 2002;158:31–37.

Vallotton P, Gupton SL, Waterman-Storer CM, Danuser G. Simultaneous mapping of filamentous actin flow and turnover in migrating cells by quantitative fluorescent speckle microscopy. *Proc Natl Acad Sci USA.* 2004;101:9660–9665.

Watanabe N, Mitchison TJ. Single-molecule speckle analysis of actin filament turnover in lamellipodia. *Science* 2002;295:1083–1086.

Fluorescent Proteins

Betzig E, Patterson GH, Rougrat R, et al. Imaging intracellular fluorescent proteins at nanometer resolution. *Science* 2006;313:1642–1645.

Chalfie M, Kain SR, eds. *Green Fluorescent Protein: Properties, Applications and Protocols (Methods of Biochemical Analysis).* Vol. 47, 2nd ed. New York: Wiley-Liss, 2005:464.

Goldman RD, Spector DL, eds. *Live Cell Imaging.* Woodbury, New York: Cold Spring Laboratory Press, 2004, p. 631.

Lippencott-Schwartz J, Patterson GH. Development and use of fluorescent protein markers in living cells. *Science* 2003;300:87–91.

Patterson GH, Lippincott-Schwartz J. Selective photolabeling of cells using photoactivatable GFP. *Methods* 2004;32:445–450.

Rothnagel JA, ed. *Fluorescent Proteins (Methods in Molecular Biology),* Philadelphia: Humana Press, 2008, p. 300.

Sullivan KF, ed. *Fluorescent Proteins,* 2nd ed. (Vol. 85, *Methods in Cell Biology*). New York: Academic Press, 2007, p. 660.

Single Particle Tracking

Kuo SC. Using optics to measure biological forces and mechanics. *Traffic* 2001;2:757–763.

Meier J, Vannier C, Serge A, Triller A, Choquet D. Fast and reversible trapping of surface glycine receptors by gephyrin. *Nat Neurosci.* 2001;4:253–260.

Flow Cytometry

Darzynkiewicz Z, Robinson JP, Crissman HA. *Cytometry, Part A (Methods in Cell Biology).* Vol. 63, 3rd ed. New York: Academic Press, 2001, p. 650.

Omerod MG. *Flow Cytometry: A Practical Approach,* 3rd ed. Oxford: Oxford University Press, 2000.

Shapiro HM. *Practical Flow Cytometry,* 4th ed. New York: Wiley-Liss, 2003.

Fluorescence Correlation Spectroscopy

Korlach J, Schwille P, Webb WW, Feigenson GW. Characterization of lipid bilayer phases by confocal microscopy and fluorescence correlation spectroscopy. *Proc Natl Acad Sci USA* 1999;96:8461–8466.

Medina MA, Schwille P. Fluorescence correlation spectroscopy for the detection and study of single molecules in biology. *Bioessays* 2002;24:758–764.

Spectral Imaging

Garini Y, Macville M, du Manoir S, et al. Spectral karyotyping. *Bioimaging* 1996;4:65–72.

Hiraoka Y, Shimi T, Haraguchi T. Multispectral imaging fluorescence microscopy for living cells. *Cell Struct Funct.* 2002;27:367–374.

Schrock E, du Manoir S, Veldman T, et al. Multicolor spectral karyotyping of human chromosomes. *Science* 1996;273:494–497.

Fluorescent Probes

Haugland RP. *A Guide to Fluorescent Probes and Labeling Techniques.* 10th ed. Eugene, OR: Molecular Probes, 2005, p. 1126. Available online at www.probes.invitrogen.com.

Hermanson GT. *Bioconjugate Techniques,* 2nd ed. New York: Academic Press, 2008, p. 1323.

Wolfbeis OS, ed. *Fluorescence Methods and Applications: Spectroscopy, Imaging, and Probes.* Malden, MA: Wiley-Blackwell, 2008, p. 452.

Optical Trapping

Conn PM. *Laser Capture in Microscopy and Microdissection. Methods in Enzymology.* Vol. 356. New York: Academic Press, 2002, p. 775.

Coppin CM, Finer JT, Spudich JA, Vale RD. Detection of sub-8-nm movements of kinesin by high-resolution optical-trap microscopy. *Proc Natl Acad Sci USA* 1996;93:1913–1917.

Coppin CM, Pierce DW, Hsu L, Vale RD. The load dependence of kinesin's mechanical cycle. *Proc Natl Acad Sci USA* 1997;94:8539–8544.

Color Addendum

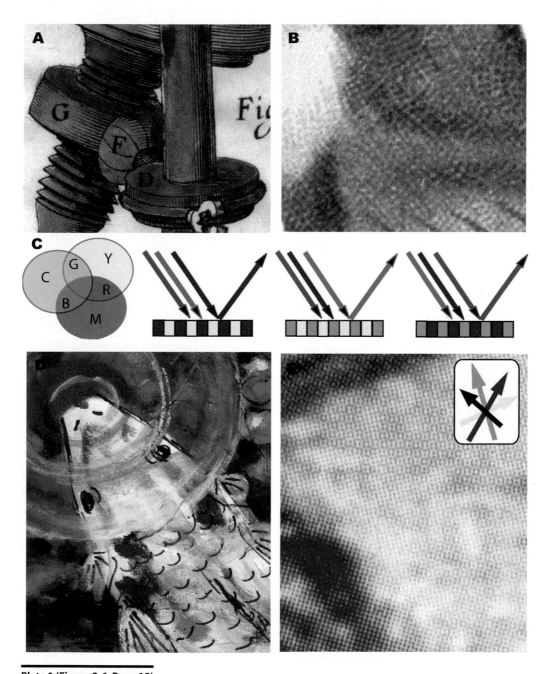

Plate 1 (Figure 2-6; Page 18)
Techniques for color illustration. (A) Hand coloring of an engraved illustration. (B) A chromolithograph at high magnification shows evidence of colored islands of ink. (C) A diagram showing which colors are reflected and seen in images printed with cyan, magenta, yellow, and black inks. (D) A four-color reproduction of a carp by a high-speed offset lithographic press. (E) At high magnification, the four-color print in (D) exhibits multicolored hexagonal patterns of dots composed of a screen pattern for each primary color. Inset: the orientation of each screen in E.

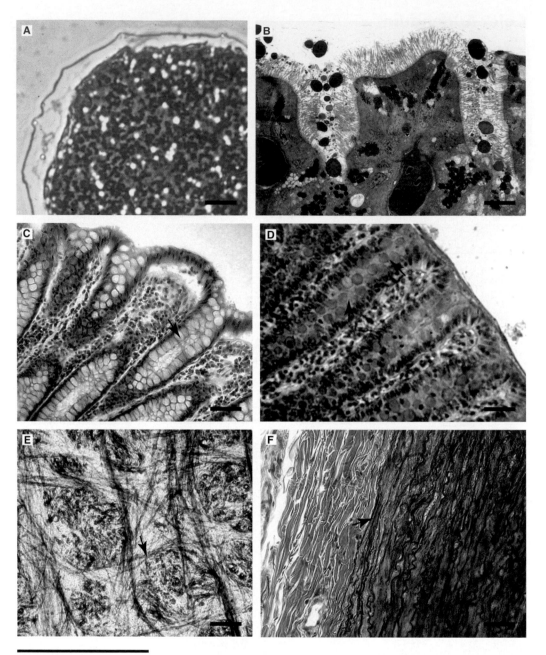

Plate 2 (Figure 3-6; Page 44)
Examples of stains for light microscopy. (A) Epoxy section of a fertilized sea urchin egg stained with toluidine blue. The fertilization envelope seen in cross-section protects the early embryo. (B) Epoxy section of the ciliated frog oviduct epithelium stained with a one-step staining mixture of eosin and toluidine blue. (C) Paraffin section of the large intestine stained with hematoxylin and eosin. The nuclei are stained dark purple with hematoxylin, while other proteinaceous structures are stained pink with eosin. The mucous-containing goblet cells in the epithelium (arrow) are clear because of lack of staining and extraction of the mucous. (D) Paraffin section of the large intestine stained by the periodic acid-Schiff (PAS) method. This technique stains the polysaccharides in mucous a brilliant pink to magenta (arrows). (E) Section of the hind brain stained with silver to reveal the numerous nerve axons (arrow) coursing through the tissue. (F) Transverse section through the wall of the aorta stained by van Gieson's method for elastin. The endothelium (left) contains a thick bed of collagen fibers that stain bright pink with eosin. The muscular layers (right) exhibit numerous concentric bands of elastin (arrow) that are stained dark violet to brown by this method. (A, B) Bars = 10 μm. (C–F) Bar = 20 μm.

A-2

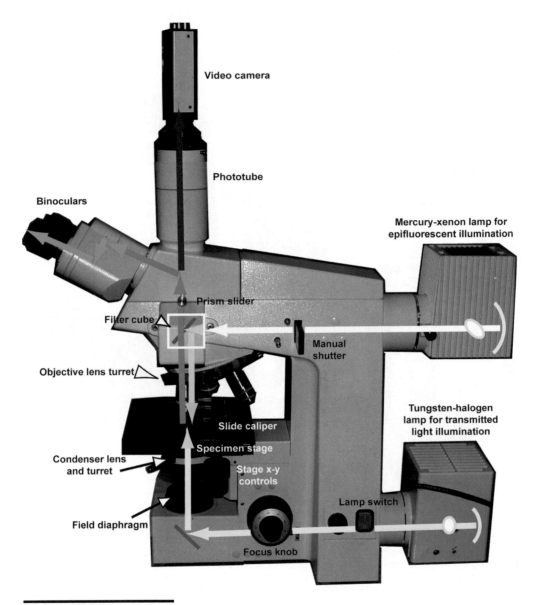

Plate 3 (Figure 6-14A; Page 131)
The design of a typical upright microscope equipped for both fluorescence and brightfield microscopy. Optical paths in the microscope are traced by arrows superimposed on the microscope image. Light from the tungsten-halogen lamp (yellow arrows) is used for transmitted light microscopy. After passage through the specimen, image-forming light is gathered by the objective lens and transmitted through the binocular eyepieces to our retinas to form a final image (green arrows). Alternatively, images can be sent to a camera for recording (red arrow). For fluorescence microscopy, light is provided by a mercury-xenon lamp (white arrows) and the wavelengths desired for chromophore excitation (blue arrows) selected by the filter cube. As a consequence, emitted light from the chromophores in the specimen is transmitted either to the binocular eyepieces (green arrows) or alternatively to a camera (red arrow).

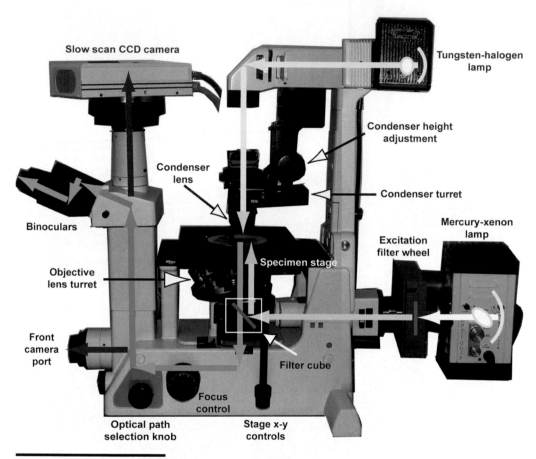

Slow scan CCD camera

Tungsten-halogen lamp

Condenser height adjustment

Condenser lens

Condenser turret

Binoculars

Mercury-xenon lamp

Excitation filter wheel

Specimen stage

Objective lens turret

Front camera port

Filter cube

Focus control

Optical path selection knob

Stage x-y controls

Plate 4 (Figure 6-14B; Page 132)

The design of a typical inverted microscope equipped for both fluorescence and brightfield microscopy. Optical paths in the microscope are traced by arrows superimposed on the microscope image. Light from the tungsten-halogen lamp (yellow arrows) is used for transmitted light microscopy. After passage through the specimen, image-forming light is gathered by the objective lens and transmitted to our retinas to form a final image (green arrows). Alternatively, images can be sent to a camera for recording (red arrows). For fluorescence microscopy, light is provided by a mercury-xenon lamp (white arrows) and the wavelengths desired for chromophore excitation (blue arrows) selected by the filter wheel or filter cube. As a consequence, emitted light from the chromophores in the specimen is transmitted either to the binocular eyepieces (green arrows) or alternatively to a camera (red arrows).

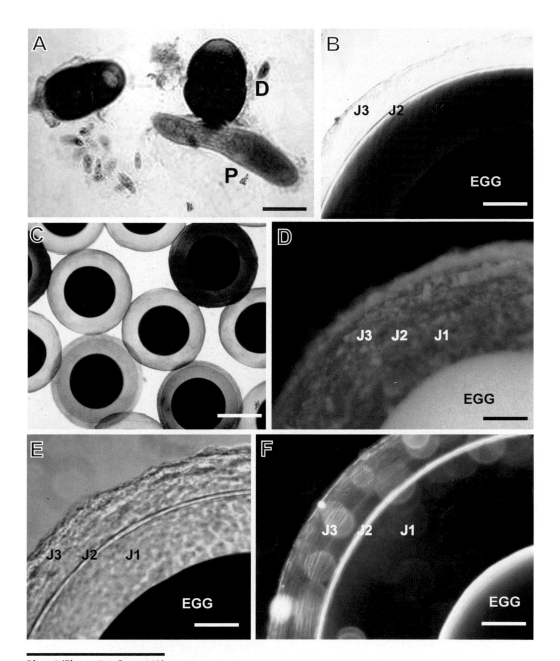

Plate 5 (Figure 7-1; Page 143)
Comparison of cells contrasted with vital dyes and optical contrast methods. (A) Three protozoa (a *Didinium* [D] is attacking a *Paramecium* [P] that it will soon eat while another *Didinium* stands by) are stained with neutral red. (B) The three jelly layers of a frog egg (J1, J2, and J3) are shown, as imaged by brightfield microscopy. The innermost layer, J1, has been selectively stained with methylene blue because of its high content of negatively charged sulfate groups. J3 and J2 are translucent, as are most biological structures in brightfield microscopy if they are unstained. Because of its size and pigment, the egg completely blocks transmitted light. (C) Alcian blue and methyl red staining of the frog egg jelly layers. (D) Fluorescence microscopy reveals staining by a dye-conjugated lectin that binds to sugars in all three jelly layers. The pigment of the egg fluoresces even brighter, an example of autofluorescence. (E) Phase-contrast microscopy reveals the boundaries and substructure of the three jelly layers without the use of dyes. (F) Darkfield microscopy (also without dyes) delineates the three layers because of the strong light scattering ability of J2. (A and B) Bars = 10 μm. (C–F) Bars = 100 μm.

A. Typical Protocol for Immunocytochemistry at the LM Level

1. Fix tissue or cells with 1% formaldehyde, then embed, section, and remove paraffin.
2. Expose to tris-buffered saline-tween 20 (TBS-T) with blocking protein at 20°C for 30 minutes.
3. Wash three times with TBS-T, 15 minutes each wash.
4. Incubate with TBS-T and blocker and primary antibody at 4°C overnight.
5. Wash three times with TBS-T, 15 minutes each wash.
6. Incubate with TBS-T and blocker and secondary antibody at 20°C for 2 hours.
7. Wash three times with TBS-T, 15 minutes each wash.
8. Observe by fluorescence/transmission light microscopy.

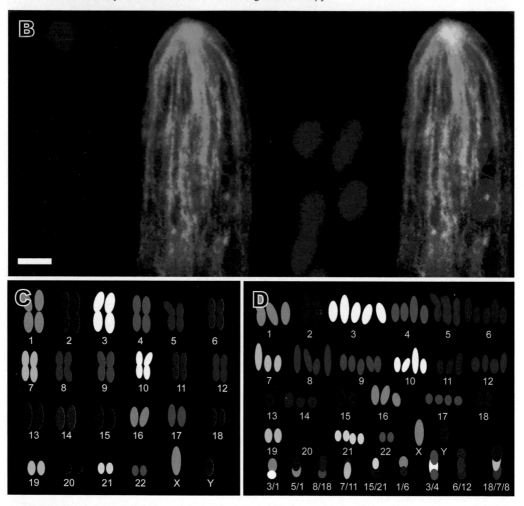

Plate 6 (Figure 13-6; Page 312)
Immunocytochemistry at the light microscopy level. (A) A typical protocol. (B) Laser scanning confocal microscopy illustrating a single hypha of the fungus *Allomyces macrogynus* labeled with antibodies for γ-tubulin (left, red) and β-tubulin (green) and DAPI for DNA (blue). γ-Tubulin identifies microtubule organizing centers. β-Tubulin identifies cytoplasmic microtubules, and DAPI identifies nuclei. An overlay of all three images is presented at the right. Bar = 2 μm. (C) Simulation of human chromosomes identified by "painting" with five oligonucleotide probes to repetitive nucleotide sequences. Chromosome numbering follows a standard convention from largest to smallest. The use of interferometric microscopy for this purpose is explained in the text and in Figure 13.11. (D) Simulation of human chromosomes from a malignant tumor. Chromosome "painting" indicates the presence of many chromosomal abnormalities including deletions, fusions, and multicopy regions. Chromosomes are numbered according to the origin of their chromatin as indicated by the painting process.

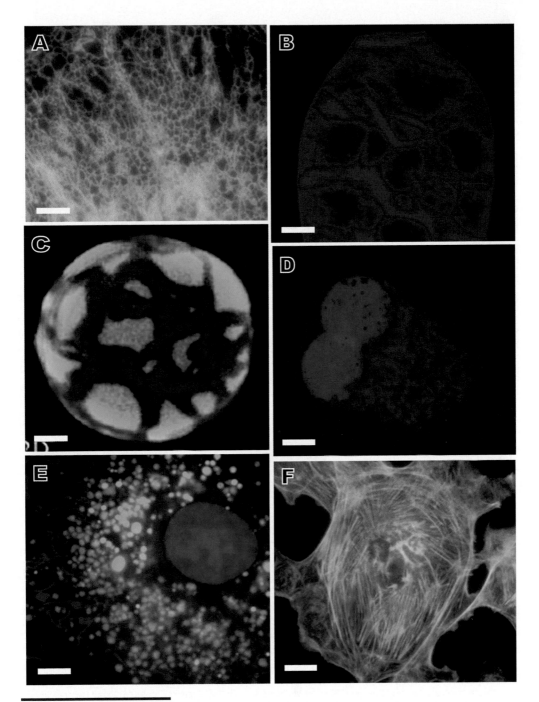

Plate 7 (Figure 13-12; Page 325)
Fluorescent probes for cell membranes and organelles. (A) Staining of mitochondria and the endoplasmic reticulum network in a bovine pulmonary artery endothelial cell using Mito-Tracker red and ER-Tracker Blue-White DPX (green). Bar = 1 μm. (B) FM-64 staining of the plasma membrane delineates newly formed fungal zoospores after septation of the fruiting body. Bar = 5 μm. (C) A giant unilamellar liposome stained with DilC$_{20}$ (red) and BODIPY FL C$_5$-HPC (green) to localize gel and fluid regions of the membrane respectively. Bar = 5 μm. (D) A BODIPY TR C$_5$-ceramide probe (red) is used in a bovine pulmonary artery endothelial cell to selectively stain the Golgi apparatus. The nucleus is counterstained blue with Hoescht 33342. Bar = 1 μm. (E) Selective staining of mitochondria and lysosomes in a bovine pulmonary endothelial cell using Mito-Fluor Far Red 680 and Lyso-Tracker Green DND-26. The nucleus is counterstained blue with Hoescht 33342. Bar = 2 μm. (F) Alexa 488-phalloidin staining of microfilaments (green) and antitubulin/Fab-Alexa 594 staining of microtubules (orange) in bovine pulmonary endothelial cells. Bar = 2 μm.

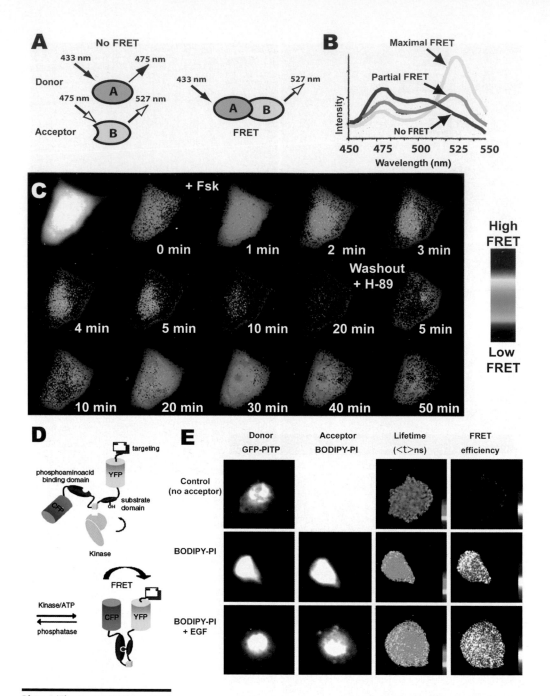

Plate 8 (Figure 13-19; Page 336)

Fluorescence resonance energy transfer (FRET). (A) Molecule A (cyan fluorescent protein; CFP) emits fluorescence at 475 nm while Molecule B (yellow fluorescent protein; YFP) emits fluorescence at 527 nm. If the molecules do not interact, excitation with 433-nm light will elicit fluorescence only from A. If the molecules closely interact, transfer of the excited state from A to B by FRET will result in fluorescence from B and a decrease in fluorescence from A. (B) Typical spectra from a CFP-YFP pair indicating no FRET, partial FRET, and maximal FRET. (C) Application of forskolin (Fsk) to HeLa cells results in protein kinase A activation as measured by FRET from an A-kinase activity reporter (AKAR). Washout of forskolin and addition of the protein kinase A inhibitor H-89 reduces FRET back to control levels. Levels of FRET are indicated by pseudocolor as calibrated on the right side. (D) Construction and function of the fusion protein (AKAR) for detecting FRET-based activity of protein kinase A. (E) Detecton of FRET between EGFP-phosphotidyl inositol transfer protein (PITP) and Bodipy-conjugated phosphotidyl inositol (Bodipy-PI) during epidermal growth factor (EGF) stimualtion of COS-7 cells. Increased FRET, results in a decrease in fluorescence lifetime which can be measured by fluorescence lifetime microscopy (FLIM). Lifetime decrease and FRET efficiency increase, displayed in pseudocolor, is seen at the plasma membrane and in the Golgi region.

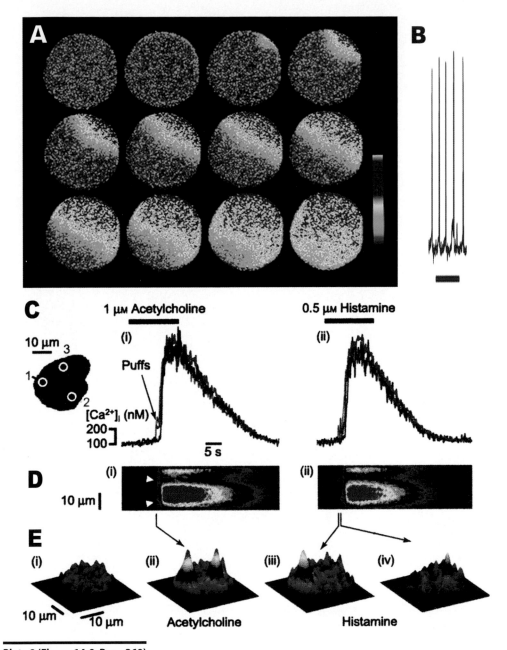

Plate 9 (Figure 14-9; Page 360)

Calcium waves, spikes and gradients detected by fluorescent probes. (A) Visualization of a wave of calcium traversing the cortex of a starfish oocyte at fertilization. The oocyte was microinjected with calcium green-dextran and images captured every 5 seconds after insemination. Bar = 50 μm. (B) Calcium spikes in a mouse egg after fertilization. The egg was injected with Oregon green BAPTA dextran for detection of calcium oscillations. The bar represents 1 hour. (C) Calcium signals measured in fura-2 loaded HeLa cells in response to acetylcholine and histamine. Three regions of interest (ROIs) have been chosen and intensity versus time measurements color-coded and plotted. Two small calcium puffs are seen in the red and green traces (arrow). (D) Line scans between ROIs 1 and 2 in the same cell were pseudocolored based on estimated calcium concentrations at each pixel and then plotted sequentially versus time. The time scale is identical to the traces in (C). (E) Three-dimensional plots of calculated calcium concentrations at each pixel. The left plot represents the unstimulated cell. Other plots each represent a specific time point indicated by the arrows leading from (D). Note the two small, transient, and highly localized calcium "puffs" that are thought to arise from regions having high sensitivity to inositol tris-phosphate. These "trigger" regions are thought to help initiate the more global calcium release seen moments later. Bars = 10 μm.

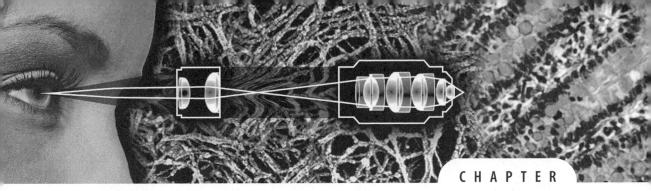

Imaging Ions and Intracellular Messengers

14

Ions such as sodium, potassium, and calcium and intracellular messengers such as cyclic AMP, cyclic GMP, and inositol phosphates are vitally important in cell regulation. It is not surprising that a considerable amount of work has been invested in imaging these molecules despite the fact that they are small and their distribution is easily disrupted during preparation of tissue for microscopy. Of almost universal importance are calcium ions and their localization and measurement will be used as the primary example of how small molecules have been hunted by microscopists.

Soon after calcium and cyclic AMP were discovered to be key intracellular messengers, the search began for ways to visualize these small molecules. It was known from physiological experiments that certain organelles in the cell, sarcoplasmic reticulum in muscle, and mitochondria in many cells contained large concentrations of calcium and that enzymes existed for the synthesis and breakdown of cyclic AMP. Furthermore, in the case of calcium, it was known that its rise and subsequent removal from the cytoplasm controlled muscle contraction. Thus, it made sense that one should be able to detect stores of intracellular calcium and physiological changes in their calcium content.

■ Precipitation Techniques

Because calcium is a soluble cytoplasmic component, it was immediately recognized that normal preparative procedures that fixed and permeabilized cells would lead to artifactual redistribution of these molecules; therefore, methods to identify calcium storing organelles centered on creating insoluble precipitates of calcium or using similar divalent cations such as strontium or barium to mark potential sites of calcium accumulation. These studies were carried out at the electron microscopy level and relied on formation of an electron dense precipitate as a marker. Uptake of strontium and barium into sarcoplasmic reticulum and mitochondria was visualized by dark precipitates confirming that these organelles do contain powerful divalent cation uptake systems (**Figure 14.1**). Similar results were obtained when calcium was precipitated by anions such oxalate and pyroantimonate to form insoluble salts.

These methods proved to be inadequate for three reasons. First, precipitation techniques were not quantitative, nor were they capable of resolving the dynamics of calcium movements. Second, clear changes in precipitation patterns that correlated with expected physiological events were not easily obtained. Third, it was recognized that the precipitation process itself could lead to artifactual redistribution of calcium.

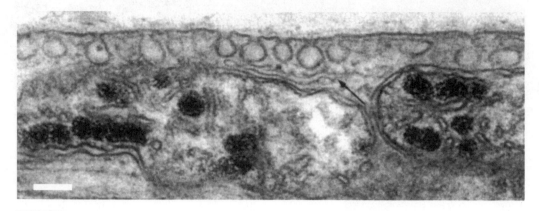

Figure 14.1
Localization of divalent cations by electron microscopy. Precipitates of barium phosphate in mitochondria of a smooth muscle cell after incubation in medium containing $BaCl_2$ are shown. Bar = 100 nm.

■ Rapid-Freezing Techniques

Rapid freezing of tissues to halt physiological events offered many advantages over precipitation. Freezing occurs within milliseconds with little redistribution of small molecules such as ions and intracellular messengers; therefore, this method offered the temporal and spatial resolution needed for physiological studies, as well as the ability to avoid artifactual redistribution of small molecules. Rapid freezing, however, provided a number of technical challenges. The tissue must first be frozen quickly to minimize ice crystal damage using cold metal block freezing (see Chapter 10). The tissue must then be processed in a manner that continues to limit redistribution of small ions. Finally, ion distribution must be detected at the electron microscopic (EM) level, usually by energy-dispersive x-ray microanalysis (EDAX) (see Chapter 9).

The technique considered best involved cutting frozen sections and then using a transmission electron microscope equipped with a cryostage to view the sections while still frozen. X-ray microanalysis of small regions within the specimen was then used to determine the relative calcium, sodium, and potassium contents of various cell organelles, as well as the cytoplasm. In this manner, estimates of ion distribution were obtained in a tissue that had been frozen during life and maintained in a frozen state before analysis. These studies demonstrated shifts in calcium content of sarcoplasmic reticulum during contraction and provided reliable estimates of intracellular ion concentrations. This technique was extremely labor intensive, however, and was eventually replaced by fluorescence microscopy as the method of choice.

■ Calcium Detection in Live Cells Using Fluorescent and Luminescent Probes

Fluorescence microscopy was seen as an easy alternative to electron microscopy because cells could be incubated with a calcium-sensitive dye and placed under the microscope within minutes. In addition, the observations made on single cells by microscopy could be correlated with population measurements using a fluorospectrometer. It only remained to find the right calcium-sensitive dye. Although a number of dyes, both absorption and fluorescence based, were tried with modest success, most did not provide the sensitivity and selectivity that was desired. Chlortetracycline, one of the best of the early dyes, serves as a good example.

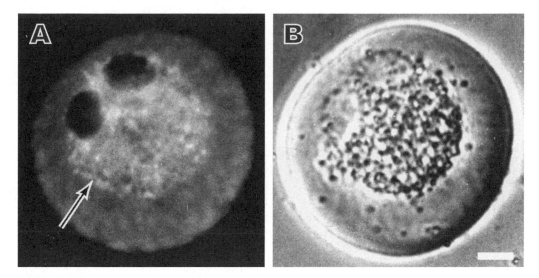

Figure 14.2
Chlortetracycline detection of calcium in organelles. An isolated pancreatic acinar cell after incubation in chlortetracycline. (A) The endoplasmic reticulum compartments throughout the cell fluoresce brightly, suggesting that they contain high concentrations of sequestered calcium (arrow). (B) The corresponding phase contrast image shows that this smooth endoplasmic reticulum is found in the secretory granule-containing region of the cell. Bar = 5 μm.

Chlortetracycline had been observed to undergo large increases in fluorescence in the presence of calcium, as first seen by its staining of bone matrix laid down in vivo during administration of this antibiotic. Subsequent studies with isolated cells showed that the dye was extremely fluorescent when taken up by calcium-accumulating organelles such as mitochondria and endoplasmic reticulum. Particularly revealing were rod cells from the retina that displayed intense fluorescence from the outer segment filled with endoplasmic reticulum and the middle segment filled with mitochondria. Furthermore, this dye was capable of following dynamics of calcium release from these organelles in real time in both contractile and secretory cells. For example, isolated pancreatic acinar cells exhibit a punctate pattern of chlortetracycline fluorescence in the smooth endoplasmic reticulum and mitochondria that surround its secretory granules (arrow, **Figure 14.2A**). In comparison, the secretory granules themselves are not fluorescent. Acetylcholine-induced secretion is accompanied by a decrease in chlortetracycline fluorescence that can be interpreted as resulting from calcium efflux from either endoplasmic reticulum or mitochondria.

Another approach to detecting calcium movements in live cells was by using the luminescent protein aequorin, isolated from jellyfish. This protein contains a cofactor coelenterazine that is held tightly within its three-dimensional structure and a specialized calcium-binding domain that is formed by an amino acid sequence related to the evolutionarily conserved EF-hand domains found in many calcium-binding proteins. After binding of calcium, aequorin changes its tertiary structure and as a result releases the cofactor and a photon of light via the reaction shown in **Figure 14.3A.**

The photons released can be counted by a photomultiplier tube or by a sensitive intensified slow-scan camera. Because aequorin must be microinjected into the cell of interest, it has usually been used in large cells such as muscle fibers or oocytes. A beautiful and early example of such an experiment was carried out by John Gilkey and Lionel Jaffe on eggs from the tropical fish Medaka. Eggs of many animal species contain a single layer of secretory granules—the cortical granules—that undergo exocytosis at fertilization. This wave of secretion starts at the point of sperm entry and propagates over the entire

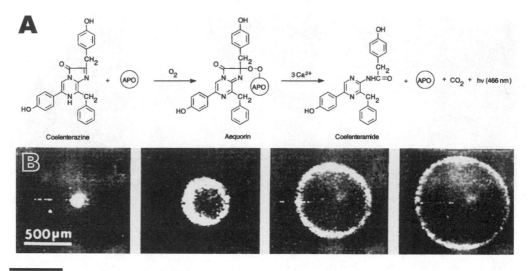

Figure 14.3
Detection of a calcium wave using the light-emitting protein aequorin. (A) Linkage of the cofactor coelenterazine to aequorin and its production of photons in the presence of calcium ions is the basis for this protein's use as an intracellular calcium reporter. (B) A wave of photoemission from aequorin-filled Medaka eggs just after fertilization. Aequorin in the egg is reporting a wave-like calcium signal that starts at the point of sperm entry and sweeps across the egg cortex triggering exocytosis of secretory granules. Bar = 500 μm.

egg surface before reaching the opposite pole. Just preceding this secretory wave is a wave of calcium release that was first visualized in the Medaka experiment (Figure 14.3B). The use of aequorin in a number of cells not only revealed the wide variety of calcium signals possible but also showed that live cells have extremely low levels of calcium in their cytoplasm at rest. Indeed, aequorin will generate photons at levels of calcium as low as 1 μM; cytoplasmic levels of calcium are lower than this because aequorin is completely "dark" in resting cells.

■ Calcium Homeostasis

These and more recent findings obtained by the fluorescent probes discussed later all point to a consistent model for calcium homeostasis in many cell types. As shown in **Figure 14.4,** the cell cytoplasm is held at a very low calcium concentration—about 10^{-7} M. This value is far below that found outside the cell. Microorganisms, plant cells, animal cells, and marine organisms all experience extracellular calcium levels that range from 10^{-4} to 10^{-2} M, that is, 1000 to 100,000 times higher than in the cytoplasm. This concentration gradient is maintained by an adenosine triphosphate (ATP)-driven pump in the plasma membrane and similar pumps in calcium-accumulating organelles within the cell. One of these organelles, the endoplasmic reticulum, accumulates calcium ions in an ATP-dependent manner to reach internal calcium concentrations of over 1 mM. Like the well-known example of sarcoplasmic reticulum in muscle, this organelle is called on to release calcium ions into the cell cytoplasm to trigger a response and later to sequester these ions to shut off the response. Another organelle with the potential to sequester calcium is the mitochondrion. The membrane potential of this organelle, formed by electron transport and proton pumping, is directly coupled to calcium uptake without the involvement of ATP. In contrast to the endoplasmic reticulum, mitochondria do not commonly store large amounts of calcium in healthy cells, nor do they usually release calcium to trigger physiological events. Rather, they seem to function as a reserve calcium sequestration unit in times of severe calcium load and stress.

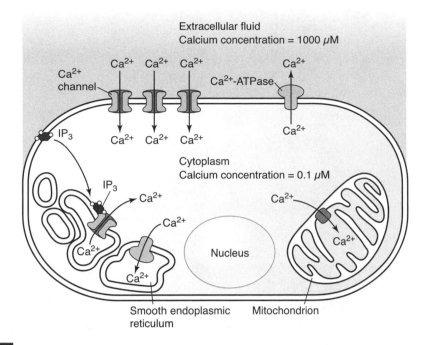

Figure 14.4
Calcium homeostasis in cells. Calcium fluxes in typical cells that are important in producing calcium signals.

The calcium homeostasis system is responsible for the production of calcium signals having a wide range of spatial and temporal variation from cell type to cell type. A temporary rise in cytoplasmic calcium levels to form these signals is due either to release of calcium from the endoplasmic reticulum or to entry of calcium from the extracellular space through stimulant-controlled calcium channels in the plasma membrane. The use of both of these sources is found in many cells, and it was not until a new generation of calcium probes was developed that this diversity in calcium signaling was appreciated.

■ Fluorescent Calcium Chelators: A New Generation of Probes

The calcium probes discussed above had a number of technical disadvantages. Chlortetracycline, for example, was found to increase its fluorescence about 50-fold in the presence of calcium—a very respectable figure; however, this increase occurred only in the presence of membranes into which the probe could intercalate, and therefore, this probe was sensitive to calcium within membrane-bound organelles and not to calcium ions in the cytoplasm that represent the calcium signal. In addition, magnesium, as well as calcium, could elicit a fluorescence increase. Aequorin, on the other hand, was specific for calcium but was capable of being used only in large cells and having the further disadvantage that it had to be used in the dark because it is light sensitive! Both probes suffered from the fact that the calcium signals detected were not easily quantitated.

For these reasons, Dr. Roger Tsien, in the late 1970s, pioneered the synthesis of fluorescent calcium chelators that could be used in live cells to monitor calcium signals. These probes consisted of a fluorophore linked to the well-known calcium chelator ethylene glycol-bis(beta-aminoethyl ether)-N,N,N′,N′-tetraacetic acid (EGTA), which coordinates calcium very tightly and selectively (**Figure 14.5A**). The first widely used probe of this type was Quin-2, a probe that could be injected into the cell, providing the cell was large enough, or that could be incubated with a cell suspension in its esterified form.

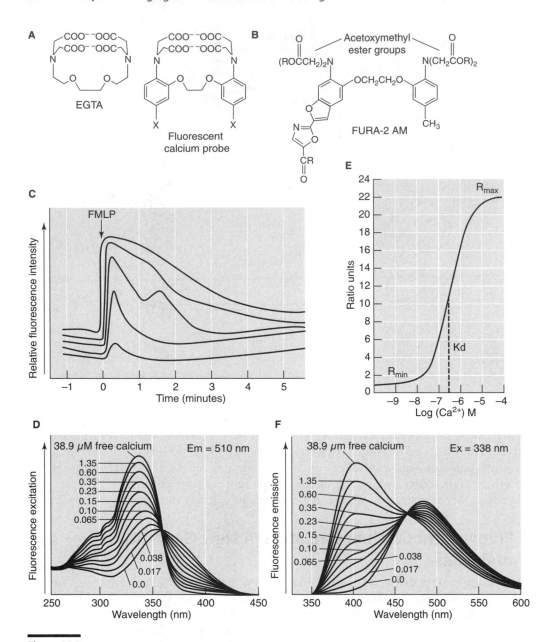

Figure 14.5
Spectral properties of fura-2 and indo-1. (A) Structure of a typical calcium-sensitive fluorescent dye (right) compared with the structure of the calcium chelator EGTA (left). (B) Structure of cell-permeant fura-2 ester. (C) Neutrophils loaded with indo-1 acetoxymethylester were stirred in a fluorometer cuvette at 37°C and indo-1 emission monitored 400 nm during application of the chemotactic peptide formyl-leucyl-phenylalanine (FMLP). Concentrations of FMLP used (top to bottom trace) are 10^{-8} M, 3×10^{-9} M, 10^{-9} M, 3×10^{-10} M, and 3×10^{-11} M. The initial peak of calcium at low FMLP concentrations was due to intracellular release from the smooth endoplasmic reticulum, while the shoulder and long plateau of elevated calcium levels at higher FMLP concentrations were due to calcium influx from the extracellular space. (D) Calibration of the calcium-dependent excitation spectrum of fura-2 with EGTA calcium buffers. (E) The emission ratio of fura-2 is increased 22-fold in the presence of a saturating amount of calcium. Accurate calibration of the calcium dependence of this ratio can be used to estimate cytoplasmic calcium concentrations in vivo. (F) Calcium-dependent emission spectrum of indo-1, a dye excited by an ultraviolet argon laser and frequently visualized using confocal microscopy.

Because all probes based on the structure of EGTA chelate calcium ions using four carboxyl groups strategically arranged like two claws, these probes are highly negatively charged. After esterification of the carboxyl groups, however, these probes become uncharged and capable of diffusing through the plasma membrane of virtually all cells (note ester groups in the structure of fura-2 in Figure 14.5B). The alcohol used for esterification is usually acetoxymethanol; after getting inside the cell, the ester is spontaneously hydrolyzed to the charged form of the probe. In many cells, hydrolysis is aided by cytoplasmic esterases, making this conversion particularly efficient.

Quin-2 (and second-generation probes such as fura-2 and indo-1) was frequently used in cell population studies carried out in fluorospectrometers. An example of such a study is shown in Figure 14.5C, wherein a population of rabbit neutrophils exhibits time-dependent and dose-dependent changes in probe fluorescence during response to the tripeptide formylmethionine-leucyl-phenylalanine, a chemotactic agent. The complex calcium signal detected exhibited at least three phases that differed in their dependence on extracellular calcium.

■ Fura-2 and Indo-1: Second-Generation Calcium Probes

Quin-2 was an important advance in intracellular calcium measurement but suffered from a technical problem. Its quantum efficiency was so low that high concentrations of the probe were needed to obtain an adequate signal. As a result, the intracellular probe concentration was sufficiently great to buffer the calcium signal and damp its fluctuations.

These problems were overcome by the second-generation dyes fura-2 and indo-1, which exhibited high quantum efficiency and, in addition, had dissociation constants for calcium binding in a range appropriate to the study of submicromolar calcium signals (discussed later) (**Table 14.1**). First, the quantum efficiencies of fura-2 and indo-1 allow lower intracellular dye concentrations to be used. Because cells themselves have significant calcium buffering capacity, the dye represents only a minor increase in buffering capacity that has little effect on the dynamics of the calcium signal being measured.

Another important advance in second-generation calcium probes is the ability to monitor calcium concentration changes at two independent wavelengths. The classic example of this is found in fura-2, whose excitation spectra are provided in Figure 14.5D. As calcium levels are raised, a strong intensity increase in the excitation spectrum is seen at 340 nm, while a marked decrease is seen at 380 nm. There are several advantages to monitoring the changes at both wavelengths and expressing the result as a ratio. First, the ratio F340/F380 is more sensitive to changes in calcium concentration than either single wavelength measurement alone. As shown in Figure 14.5E, this ratio changes by a factor of 1:22 as the dye is titrated with calcium, while F340 and F380 change by 1:10 and 1:2.5, respectively. Second, expression of the data as a ratio automatically normalizes the data for any differences in path length or dye concentration. These problems are of minor importance in fluorometric measurements on cell populations but can be very important in microscopic studies where differences in cell thickness and compartmentalization of the dye in subcellular organelles may complicate the fluorescence intensity pattern being observed.

To obtain data at two different excitation wavelengths, one usually obtains an image first at one wavelength and then at the other. The intensity at each pixel of the first image is then divided by the intensity of the corresponding pixel of the second image. The first requirement for "ratio imaging" of fura-2 fluorescence is for a continuous-spectrum ultraviolet light source (usually a vapor arc lamp) and the ability to change the excitation wavelength quickly using a computer-controlled filter wheel placed between the lamp house and the dichroic mirror used to reflect exciting light onto the specimen. Changing the excitation filter between each frame usually limits the speed of data

Table 14.1	Fluorescent Calcium Probes		
Probe	**EM/EX Wavelength**	**K_d (nM)**	**Comments**
Quin-2	Ex 333/353 Em 495	60	Older dye; single wavelength; low quantum yield and low K_d results in calcium buffering
Indo-1	Ex 346/330 Em 475/401	230	Emission ratio at 400/485 used; ultraviolet excitation, often by laser
Fura-2	Ex 363/335 Em 512/505	145	Excitation ratio at 340/380 used; ultraviolet excitation; most commonly used
Fluo-3/Fluo-4	Ex 506 Em 526	390/345	Single wavelength; visible excitation
Fluo-5F	Ex 491 Em 518	2300	Single wavelength; used in the low micromolar range
Fluo-4FF	Ex 491 Em 516	9700	Single wavelength; used in the high micromolar range
Fluo-5N	Ex 491 Em 518	90,000	Single wavelength; used in the sub millimolar range
Calcium Green-1	Ex 506 Em 531	190	Single wavelength; used in the sub micromolar range
Calcium Green-2	Ex 506 Em 536	550	Single wavelength; used in the low micromolar range
Calcium orange	Ex 549 Em 575	185	Single wavelength; used when the green emission channel is being used
Calcium crimson	Ex 587 Em 615	185	Long-wavelength emission; used when green and yellow channels are taken

Values are taken from the *Handbook of Fluorescent Probes* at probes.invitrogen.com/handbook. Excitation and emission maxima are in aqueous solution.

collection to about three to four raw images or one or two ratioed images per second. This speed can be increased to about 10 ratioed frames per second by using a separate light source and fixed excitation filter for each wavelength in combination with electronically controlled shutters.

Ratio imaging can be carried out by laser scanning confocal microscopy in which case indo-1 or similar dyes are used rather than fura-2. These dyes exhibit a shift in their emission spectrum after calcium binding rather than in their excitation spectrum. This allows continuous scanning with one laser line while images at two different emission wavelengths are captured simultaneously using appropriate dichroic mirrors and barrier filters. Image capture is faster and can be further increased in speed by using line scanning rather than point scanning (see Chapter 12) with speeds up to 50 ratioed images per second being achievable. Ratio imaging can also be facilitated either by use of an acousto-optic tunable filter (AOTF) to switch excitation wavelengths rapidly (see Chapter 12) if fura-2 is being used or by use of an interferometric microscope capable of doing multispectral imaging as described in Chapter 13.

■ Requirements for Live Cell Imaging

The use of such probes successfully entails important technical requirements regardless of whether they are monitored by fluorospectrometry or fluorescence microscopy. Critical for live cell experiments is provision for a physiological environment, including the

Table 14.2 Hydrogen Ion Buffers for Live Cell Imaging		
Buffer	**pK at 20°C**	**Comments**
Bicarbonate/CO_2	6.35	Requires continual O_2/CO_2 gassing
ADA	6.6	Gassing with O_2 optional
PIPES	6.8	Gassing with O_2 optional
ACES	6.9	Gassing with O_2 optional
BES	7.15	Gassing with O_2 optional
MOPS	7.2	Gassing with O_2 optional
TES	7.5	Gassing with O_2 optional
HEPES	7.55	Gassing with O_2 optional
Values for nonbicarbonate buffers are from MaxChelator at www.stanford.edu/~cpatton/maxc.html.		

control of ion concentrations, pH, temperature, oxygenation, and buffer flow. Ion concentrations and osmotic strength are usually set to mimic the normal extracellular fluid composition that the tissue experiences. Hydrogen ion concentration is maintained at physiological levels by adequate buffering commonly by using a bicarbonate, an organic amine, or a phosphate buffer. **Table 14.2** lists the common buffering agents used for this purpose. Organic amine buffers and phosphate buffers offer the advantage that pH can be held stable without gas equilibration. These are the buffers of choice when using an open system whether it is a flow-through chamber or a simple slide and cover glass arrangement. In particular, the organic amine buffers, chemical cousins of amino acids, are tolerated by tissues very well. The 4-(2-hydroxyethyl)-1-piperazine-ethanesulfonic acid (HEPES), piperazine-1,4-bis(2-ethanesulfonic acid) (PIPES), and 3-(N-morpholino)propanesulfonic acid (MOPS) versions are arguably the most common buffers used in biologic research (Table 14.2). Nevertheless, for mammalian cells, bicarbonate buffers are often chosen because they mimic the normal buffering system found in mammalian tissue fluids. Bicarbonate buffering systems require continuous equilibration with CO_2, which is usually achieved by maintaining a reservoir of buffer through which is bubbled a CO_2/O_2 gas mixture. The gassed buffer is then supplied to a flow-through chamber containing the tissue or cells undergoing microscopic observation. When bubbling CO_2/O_2 through media for live-cell imaging the media must first be equilibrated in an incubator before starting the bubbler. Following equilibration, a flow rate of one bubble every 5 to 10 seconds is all that is necessary to maintain a steady buffer pH.

Flow-through chambers can be open or closed as illustrated in **Figures 14.6A** and **B** respectively. Open chambers allow access to the tissue, e.g., for purposes of electrophysiology or microinjection. Open chambers also offer the advantage that any bubbles in the buffer are quickly eliminated at the chamber before they can disrupt microscopy. On the other hand, closed chambers, although they prevent access to the tissue, do have the advantage that buffers equilibrated with gas are not exposed to the room atmosphere until after they exit from the chamber; therefore, there is no opportunity for the buffer to become deoxygenated or carbon dioxide poor. Closed chambers also prevent contact of the objective lens with buffer or its vapor. Objective lenses can fog up when used with warm buffers in an open chamber if used on an upright microscope, a problem that can be reduced by using an objective heater to keep the objective at temperature sufficient to prevent condensation on the front lens. Heating the objective also prevents heat loss

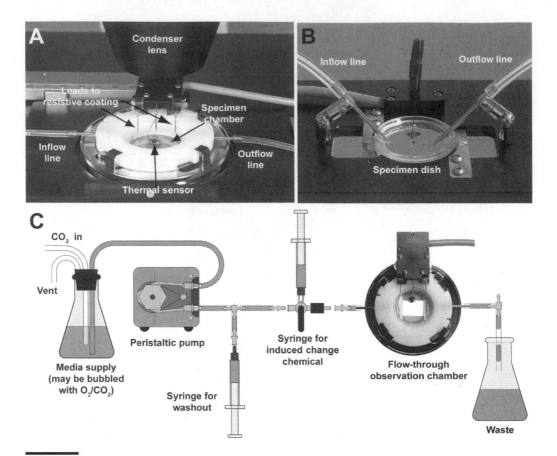

Figure 14.6
Flow-through chambers for live cell imaging. (A) A Bioptechs FCS2 closed flow-through chamber designed for accurate and uniform temperature control by a combination of peripheral heat and the use of an optically clear, electrically conductive coating on the chamber's upper surface. The FCS2 is also available in an upright version called the FCS3. It has the same capabilities and is ideal if only an upright imaging system is available. Both chambers allow user definable flow characteristics. (B) A Bioptechs Delta T4 open flow-through chamber allowing accurate temperature control and access to cells for microinjection or routine time-lapse imaging of a wide variety of specimen types such as adherent cells, suspensions and tissue. Both open and closed chambers are typically used on an inverted microscope. (C) Design of a complete flow-through system for the FCS2 closed chamber unit. The medium reservoir can be gassed with a O_2/CO_2 mixture if appropriate and biochemicals or drugs can be added transiently by in-line syringes.

from the specimen (see below). Provision for buffer perfusion of either type of chamber requires a medium reservoir, a peristaltic pump, inflow tubing and port, the chamber itself, and outflow port and tubing leading to a waster container (Figure 14.6C). Also useful are hypodermic syringes and valves in the intake line used for transient application of biochemicals or drugs to cells during microscopic observation. In the case of a closed-system chamber one must make sure the outflow is open to the atmosphere at the same level as the specimen to prevent a water column from forming that can create either pressure or partial vacuum in the chamber. The presence of any variation in pressure will cause the coverslip to flex.

Of particular importance is the maintenance of proper temperature; this can be technically challenging during long-term viewing and in the face of differing flow rates. During short-term observations in which temperature control does not have to be precise, specimen temperature can be controlled by flow-through buffer temperature or alternatively by warming or cooling of the stage. If the former is used, buffer in the reservoir is maintained at the desired temperature by warming or cooling before entry into the chamber. This technique only works when the cells are tolerant of continual perfusion. If the latter, stage warming is usually achieved by rheostat-controlled flow of current through resistance wires embedded in the stage. Proper adjustment necessitates continuous measurement of specimen temperature by a thermal sensor and either manual or automatic feedback control of the current flow. Be aware that heating of the stage, instead of heating the specimen only, can cause z-axis drift. To avoid this, a specimen warmer, thermally insulated from the stage and designed to minimize thermal expansion in the z-axis, has been introduced recently by Bioptechs for warm-specimen work in plastic dishes. Although cooling of microscopic stages could be, in theory, achieved by Peltier cooling (also referred to as thermoelectric cooling; a bimetallic junction abstracts heat from its surroundings as current is passed), this is seldom used because heat abstraction by this process is too low in capacity. Rather, more efficient cooling of the stage requires plumbing for a flow-through refrigerant precooled to an appropriate temperature.

Accurate and uniform temperature control, however, especially over long periods of time, requires a more sophisticated approach used in the Bioptechs stages pictured in Figures 14.6A and B. In these units, a specially designed glass plate of high optical quality is coated uniformly with a thin but transparent and conductive layer of indium tin oxide. When current is passed through this layer, uniform resistive heating is created directly at the specimen-glass interface. This transmits heat by means of conductivity which is far more efficient than radiation. A thermal sensor in contact with the plate sends a feedback signal to the heater controller that allows heating to be continuously adjusted to maintain a set temperature even in the face of external thermal influences such as perfusion and surface evaporation. In applications where cooling below ambient is necessary an inexpensive yet elegant means of cooling can be achieved by a ring shaped tube in fluid communication with a chiller bath to withdraw heat from the specimen.

In the case that high numeric objectives are used, an objective heater is necessary to prevent heat from being carried away from the specimen by fluid contact between the objective and coverslip. If the specimen is to be observed at temperatures below ambient an objective cooler and thermal isolator is also necessary so that heat from the nosepiece is not drawn up through the objective. The thermal isolator also provides a dry air trap so that condensation does not form on the lower element of the objective.

■ Quantitation of Fluorescence Signals

A second technical consideration is calibration of the calcium-binding dye in order to convert fluorescence intensity into estimates of calcium concentrations. To do this, one must be able to measure dye fluorescence at very low but known calcium concentrations. One difficulty encountered is that even distilled water contains trace amounts of calcium in the micromolar concentration range. Because probes useful for detection of calcium signals must bind calcium at concentrations ranging from 0.05 μm to 10 μm, this degree of calcium contamination is problematic. This problem is solved by using the chelator EGTA as a calcium buffer. As shown in **Table 14.3**, the presence of 10 mM EGTA and calcium chloride in the range of 1 to 8 mM results in free calcium ion concentrations in the micromolar range because of the fact that most of the calcium is bound to the EGTA and, therefore, not accessible to the fluorescent dye. In addition to

Table 14.3 EGTA Buffers for Low Calcium Levels			
[EGTA] (mM)	**[Ca^{2+}] (Total) (mM)**	**[Ca^{2+}] (Free) (μM)**	**Comment**
10	1	0.04	
10	2	0.11	
10	4	0.29	
10	6	0.65	
10	8	1.7	
10	9	3.8	Out of good buffering range
10	9.5	8.0	Out of good buffering range
10	9.7	13.3	Out of good buffering range

Free calcium concentrations assume an ionic strength of 0.15 N, a pH of 7.0, and a temperature of 20°C and are calculated from MaxChelator at www.stanford.edu/~cpatton/maxc.html.

appropriate EGTA and calcium levels, a calibration buffer must also mimic intracellular conditions in regard to ions, pH, and ionic strength. Thus, for a mammalian cell, the buffer would most likely consist of 150 mM KCl, 20 mM NaCl, 10 mM HEPES, pH 7.0, 10 mM EGTA, and an appropriate concentration of $CaCl_2$. The pH is of particular importance. Because protons and calcium ions compete for the same binding sites on EGTA, alterations in pH will change the EGTA dissociation constants for calcium binding. Furthermore, if pH is not buffered, the pH will drop because of the release of protons from EGTA upon addition of calcium. For this reason, pH must not only be buffered but must also be adjusted by addition of base after the proper amount of calcium is added to each buffer.

Using an EGTA/calcium buffer, one can obtain a series of spectra for any calcium dye that demonstrates calcium dependence in its fluorescence (see Figure 14.5D for fura-2 calibration curves). In the case of indo-1, increased calcium levels result not only in an increase in fluorescence intensity, but also a shift in its emission spectrum to longer wavelengths (Figure 14.5F).

The dissociation constant for a calcium binding dye determines the calcium concentration range over which the probe is useful for monitoring a signal. This constant is equal to the concentration of calcium that results in half of the dye molecules being bound with calcium. It is measured by titrating the dye with calcium and determining what calcium concentration results in a half-maximal increase in fluorescence of the probe. If one assumes that the fluorescence observed is linearly proportional to the calcium-bound dye species, then the largest fluorescence changes will be observed at calcium concentrations close to the dissociation constant for the dye. Fluorescence changes large enough to be quantitated will be found over calcium concentrations that range from $K_d/10$ to $10K_d$—a hundred-fold range. **Figures** 14.5 and **14.7** illustrate such calibrations for a number of dyes. One interesting comparison can be made for the two versions of calcium green. The low-K_d version (Figure 14.7A) displays usable fluorescence at calcium concentrations from 0.05 μM to 1 μM, while the high-K_d version (Figure 14.7B) exhibits useful signals at calcium concentrations from 0.5 to 10 μM; therefore, the low-K_d version might be used for neutrophils in which calcium levels rarely exceed 1 μM, while the high-K_d version might be suitable for visualizing the much stronger calcium wave in Medaka eggs.

Although single wavelength fluorescence determinations do not allow correction for variations in path length and dye concentration, estimates can be made that relate calcium concentration to fluorescence intensity. To derive a mathematical relationship,

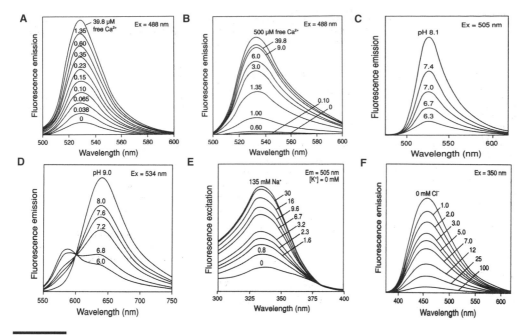

Figure 14.7
Spectral properties of other ion-sensitive fluorescent dyes. (A) Calcium-dependent emission spectrum of calcium green 1. (B) Emission spectrum of calcium green 2. It binds calcium more weakly than calcium green 1 and therefore can be used to monitor calcium signals in the 1- to 10-μM range rather than in the 0.1- to 1-μM range. (C) The pH-dependent emission of BCECF. (D) Emission of SNARF-1, a second-generation pH probe. (E) Calibrated emission spectra of the sodium ion probe SBFI. (F) Calibrated spectra of the chloride ion probe N-(ethoxycarbonylmethyl)-6-methoxyquinolinium bromide (MQAE).

one assumes that calcium binding by the dye remains in equilibrium throughout the reporting period and that there is a conservation of dye molecules. From the assumptions, these equations follow:

$$K_d = [Ca^{2+}]_{in}\, [Dye^{2-}]/[Ca{:}Dye] \tag{14.1}$$

$$[Dye]_{TOTAL} = [Dye^{2-}] + [Ca{:}Dye] \tag{14.2}$$

where K_d is the dissociation constant for the binding of calcium to the dye; $[Ca^{2+}]_{in}$, $[Dye^{2-}]$, and $[Ca{:}Dye]$ are the concentrations of free cytoplasmic calcium ions, unbound dye, and calcium-bound dye, respectively; combining these equations, we arrive at this relationship:

$$[Ca^{2+}]_{in} = K_d[F_{ex} - F_{min}]/[F_{max} - F_{ex}] \tag{14.3}$$

where F_{min} is the fluorescence intensity of the unbound dye, F_{max} the fluorescence intensity of the calcium bound dye, and F_{ex} the experimental fluorescence observed at each pixel as a function of time providing the dye remains in equilibrium with changes in $[Ca^{2+}]_{in}$ at that location.

F_{min} can be estimated by treating cells with a divalent cation ionophore plus extracellular Mn^{2+}. Flooding of the cell cytoplasm with high concentrations of Mn^{2+} forms a nonfluorescent manganese complex with the probe. F_{max} can be estimated by treating cells with a divalent cation ionophore plus extracellular calcium. Flooding of the cell

Processing of Ratiometric Data for Ion Detection Using Fura-2

1. Barrier filter (525 nm longpass) constant throughout the experiment

2. Load cells with Fura-2.

3. Capture control image at 340 nm excitation using a bandpass filter.

4. Capture control image at 380 nm using a bandpass filter.

5. Start capturing images alternating between excitation at 340 nm and 380 nm.

6. Stimulate cells to respond with calcium signals.

7. Determine R(max) and R(min) using standard EGTA buffers or with cells using calcium ionophore with calcium and manganese.

8. Postcapture processing for each pixel:

   For each pixel at (x, y, t):

   $$\frac{\text{Intensity (340 nm)} - \text{Intensity (340 nm Control)}}{\text{Intensity (380 nm)} - \text{Intensity (380 nm Control)}} = R(x, y, t)$$

   $$[Ca^{2+}(x, y, t)] = K_d[F_{\text{free } 380}/F_{\text{bound } 380}][R(x, y, t) - R_{min}]/[(R_{max} - R(x, y, t)]$$

   Assign pseudocolor to each pixel based on $[Ca^{2+}]$ using look up table

Figure 14.8
Flow chart for ratiometric calcium determination from image pixel intensities.

cytoplasm with high concentrations of calcium forms fluorescent complexes from all probe molecules. K_d can be either determined experimentally by using EGTA-Ca^{2+} buffers in vitro or by using values found in the literature or referenced by the manufacturer. One must be careful, however, to use a value that has been determined under conditions similar to that found in the cell cytoplasm because dissociation constants are sensitive to ionic strength, temperature, and particularly pH because hydrogen ions compete with calcium ions for probe binding.

The equations previously derived can be modified in a straight forward manner to use ratioed values of fluorescence for those probes that lend themselves to quantitation at two wavelengths, designated w1 and w2. In this case, the following equation is used:

$$[Ca^{2+}]_{in} = K_d \, [S_{f2}/S_{b2}][R_{ex} - R_{min}]/[R_{max} - R_{ex}] \qquad (14.4)$$

where R_{min}, R_{max}, and R_{ex} are equal to $F_{min}(w1)/F_{min}(w2)$, $F_{max}(w1)/F_{max}(w2)$, and $F_{ex}(w1)/F_{ex}(w2)$, respectively. The ratio S_{f2}/S_{b2} is an experimentally determined constant equal to the fold change in fluorescence of the dye at wavelength 2 on binding saturating amounts of calcium. Based on these equations, dynamic computations can provide estimates of free calcium ion concentrations at each pixel as a function of time. A flowchart for such calculations with Fura-2 is provided in **Figure 14.8.** Calculations are usually made during postcapture processing because of the need to acquire control frames for F_{min} and F_{max} determinations either before or after the experimental sequence.

The usual way in which such massive quantities of data are presented is by pseudo color. Using a lookup table to represent calcium concentrations in color, one can assemble a video sequence of calcium signal dynamics in both space and time. Such data may be three dimensional (x, y, t) or four dimensional (x, y, z, t) if stacks of optical sections are being acquired on a confocal microscope. Hot colors (white, red, and yellow) are generally used for high calcium concentrations and cool colors (green and blue) for low calcium concentrations.

Estimates of calcium concentrations obtained by these means are subject to several well-known artifacts. The most important are those associated with obtaining accurate control frames to estimate F_{min} and F_{max}. Because these frames are usually taken either before or after the experiment, movement of the cell or its organelles could result in pixels being mismatched when calcium calculations are carried out. These problems are of less consequence if F_{min} is low in all cases or if ratios are used rather than single wavelength fluorescence values. R_{max} and R_{min} should, in theory, be fairly consistent from pixel to pixel because differences in path length and dye concentration have already been accounted for.

A second problem concerns regions of low signal. In this case, calculations of differences between F_{ex} and F_{max} and F_{ex} and F_{min} are subject to great fluctuation and may even produce negative (therefore impossible) concentrations. For this reason, values below a certain threshold regardless of sign are set to zero. A third problem is due to superimposition of compartments in a wide-field microscope. For example, superimposition of cytoplasm and extracellular space at the edge of the cell will lead to an artifactual reduction in the perceived calcium level in the regions just inside of the plasma membrane.

A fourth problem is probe distribution. Because cytoplasmic calcium signals are usually of interest, the cleanest experimental situation would be to have all of the probe molecules in that one compartment, with none being present inside organelles or in the extracellular space. In practice, this goal is hard to achieve, especially if cells have been loaded with probe using an acetoxymethyl ester. The nonpolar ester can diffuse through organelle membranes as well as the plasma membrane, and multiple compartments may end up being loaded. In some cases, the compartments receiving the highest concentration of probe are those that have the highest concentrations of esterases to cleave off the ester group and generate an active calcium-sensing molecule. The most common way of limiting a probe to the cytoplasmic compartment is to microinject the cell with the probe covalently linked to a dextran polymer. The size of the polymer prevents passage of the dye through organelle membranes.

■ Calcium Signals Visualized by Fluorescent Probes

The ability of fluorescent probes for calcium to visualize calcium movements in both space and time has led to an astounding array of calcium signals being documented. The four most common types are waves, oscillations, puffs, and global rises. Just after fertilization, echinoderm eggs exhibit a wave of calcium release that sweeps through the egg cortex from the site of sperm entry to the opposite pole. The spatial and temporal characteristics of this wave are efficiently recorded in a *Pisaster ochraceus* starfish oocyte that was microinjected with calcium green-1 dextran, as illustrated in **Figure 14.9A** (see Plate 9 in the color addendum). Spike-like oscillations of calcium are seen in both isolated and epithelial cells. In Figure 14.9B, a mouse egg previously injected with Oregon green BAPTA dextran and fertilized with sperm exhibits regular spike-like increases in cytosolic calcium levels. Spikes occur at frequencies between 4 and 10 per hour and each spike lasts 2 to 3 minutes. A global rise in calcium can be seen in Hela cells simulated with either acetylcholine or histamine (Figure 14.9C). Signals from three different regions of the cell

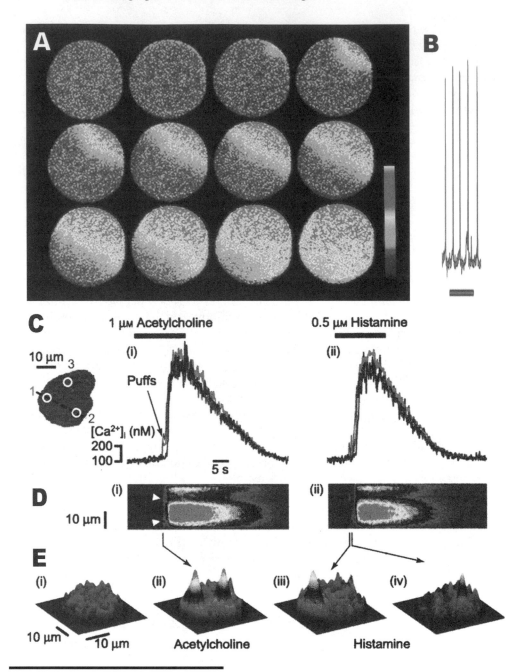

Figure 14.9 (See Plate 9 for the color version.)
Calcium waves, spikes and gradients detected by fluorescent probes. (A) Visualization of a wave of calcium traversing the cortex of a starfish oocyte at fertilization. The oocyte was microinjected with calcium green-dextran and images captured every 5 seconds after insemination. Bar = 50 μm. (B) Calcium spikes in a mouse egg after fertilization. The egg was injected with Oregon green BAPTA dextran for detection of calcium oscillations. The bar represents 1 hour. (C) Calcium signals measured in fura-2 loaded HeLa cells in response to acetylcholine and histamine. Three regions of interest (ROIs) have been chosen and intensity versus time measurements color-coded and plotted. Two small calcium puffs are seen in the red and green traces (arrow). (D) Line scans between ROIs 1 and 2 in the same cell were pseudocolored based on estimated calcium concentrations at each pixel and then plotted sequentially versus time. The time scale is identical to the traces in (C). (E) Three-dimensional plots of calculated calcium concentrations at each pixel. The left plot represents the unstimulated cell. Other plots each represent a specific time point indicated by the arrows leading from (D). Note the two small, transient, and highly localized calcium "puffs" that are thought to arise from regions having high sensitivity to inositol tris-phosphate. These "trigger" regions are thought to help initiate the more global calcium release seen moments later. Bars = 10 μm.

(red, blue, and green) show a similar rapid rise and slower tapering off of the calcium signal; however, the red and green traces each show a small transient "puff" of calcium just before the global rise (arrow). If calcium levels at that moment are plotted as a three-dimensional map of the cell (intensity, x and y; Figure 14.9E), several puffs can be seen that are very localized, short-lived increases in calcium released from the smooth endoplasmic reticulum (SER). One surprise was that calcium release was initiated at a site distant from the plasma membrane containing the cholinergic receptors responsible for initiating the signaling pathway. This suggested that inositol triphosphate, the intracellular messenger initiating calcium release, diffused through the cell cytoplasm until particularly sensitive inositol 1,4,5-trisphosphate (IP3) receptors were encountered. Finally, Figure 14.9D shows that the dynamics of the calcium signal can also be expressed as a time series of line scans across the diameter of the cell. Although providing some essential features of the data, line scans are not as immediately interpretable as the three-dimensional presentation in Figure 14.9E.

■ Fluorescent Probes for Other Ions

The development of fluorescent probes for calcium has led to development of probes for other ions using similar principles. **Tables** 14.1, **14.4,** and **14.5** summarize probes available for hydrogen, sodium, potassium, magnesium, zinc, and chloride ions, as well additional probes for calcium not previously discussed. Most of these probes are available in both their salt and acetoxymethyl (AM) ester forms and their dissociation constants for ion binding are usually consistent with the intracellular concentrations encountered for each ion.

Particularly noteworthy are probes for hydrogen ions because responses of many cells are accompanied by cytoplasmic pH changes. These probes contain an acidic group linked to a fluorochrome that undergoes large changes in emission upon dissociation of a hydrogen ion. Titration of 2'-7'-bis(carboxyethyl)-5(6)-carboxyfluorescein (BCECF) shows that it reaches half maximal fluorescence change at pH 7.0, its K_d, a value similar to that of the cytoplasm in many cells (Figure 14.7C). The use of BCECF in sea urchin eggs, for example, shows that the cytoplasmic pH changes at fertilization from 6.8 to 7.2

Table 14.4 Hydrogen Ion (pH) probes			
Probe	Em/Ex Wavelengths	pK (pH Range)	Comments
SNARF-1	Ex 548/576 Em 587/635	7.5 (6.5–8.0)	Ratiometric using emission at 580/640
BCECF	Ex 482/503 Em 520/528	7.0 (6.5–7.5)	Ratiometric using excitation at 490/440
DCF	Ex 475/492 Em 517	6.4 (6.0–7.2)	Ratiometric using excitation at 490/450
LysoSensor green	Ex 442 Em 505	5.2 (4.5–6.0)	Single wavelength
Oregon green	Ex 489/506 Em 526	4.7 (4.2–5.7)	Ratiometric using excitation at 490/440
LysoSensor yellow	Ex 384/329 Em 540/440	4.2 (3.5–5.0)	Ratiometric using emission at 450/510
LysoSensor blue	Ex 373 Em 425	5.1 (4.3–6.0)	Single wavelength
Values are taken from the *Handbook of Fluorescent Probes* at probes.invitrogen.com/handbook. Excitation and emission maxima are in aqueous solution.			

Probe	Em/Ex Wavelength	K_d (mM)	Comments
Magnesium green	Ex 506 Em 531	1.0	Mg^{2+} probe; single wavelength
Magnesium orange	Ex 550 Em 575	3.9	Mg^{2+} probe; single wavelength
Mag-fluo-4	Ex 493 Em 517	4.7	Mg^{2+} probe, single wavelength; fluorescence increases more than 100 fold when Mg^{2+} is bound
FluoZin-3	Ex 494 Em 516	15 nM	Zn^{2+} probe; single wavelength; fluorescence increases more than 100 fold when Zn^{2+} is bound
RhodZin-3	Ex 552 Em 576	65 nM	Zn^{2+} probe; single wavelength; used when green channel is taken; fluorescence increases more than 100 fold when Zn^{2+} is bound
SBFI	Ex 333 Em 539	3.8	Na^+ probe; single wavelength; ultraviolet excitation
Sodium green	Ex 507 Em 532	6.0	Na^+ probe; single wavelength; visible excitation
PBFI	Ex 338 Em 507	5.1	K^+ probe; single wavelength; ultraviolet excitation
SPQ	Ex 344 Em 443	85*	Cl^- probe; single wavelength; quenching of excited state by chloride

Table 14.5 Probes for Other Ions

*Reciprocal of the Stern-Volmer quenching constant.

Values are taken from the *Handbook of Fluorescent Probes* at probes.invitrogen.com/handbook. Excitation and emission maxima are in aqueous solution.

because of activation of a sodium-hydrogen cotransporter. The cytoplasmic alkalinization detected by BCECF is in agreement with measurements by ion-specific microelectrodes and from a physiological perspective is known to be critical for activation of protein and DNA synthesis in the fertilized egg. SNARF-2, a second-generation pH probe, has the advantage that its emission spectrum (Figure 14.7D) exhibits an isobestic (inversion) point and, therefore, can be evaluated by ratioed imaging to correct for differences in path length. Probes for magnesium, sodium, and chloride are also available, having dissociation constants for binding of their ions at 4, 5, and 6 mM, respectively, appropriate to intracellular concentrations found for these ion species (Figures 14.7E and 14.7F).

■ Potential-Sensitive Probes

Neurobiologists for many years have sought optical techniques to monitor membrane potentials. Two types of probes have been developed that exhibit potential-dependent fluorescence, fast-acting membrane intercalating probes, and slow-acting lipophilic ions that redistribute across membranes in a manner consistent with transmembrane potential (Table 14.6). Fast-acting probes such as di-4-ANEPPS intercalate into the outer monolayer of the plasma membrane much like a phospholipid would and do not cross the

Table 14.6	Major Classes of Potentiometric Probes		
Probe	Em/Ex Wavelength	Uses	Comments
di-4-ANEPPS	Ex 497 Em 705	Mapping of membrane potentials in neurons and muscle fibers	Fast response; styryl family of dyes; excitation ratio 440/505 decreases in response to membrane hyperpolarization
di-18:2-ANEPPS	Ex 501 Em 705	Mapping of membrane potentials in neurons and muscle fibers	Fast response; tethered to membrane by fatty acid anchors
RH 414	Ex 532 Em 716	Synaptic and ion channel activity	Fast response; styryl family of dyes; fluorescence decreases in response to depolarization
rhodamine 123	Ex 507 Em 529	Mitochondrial membrane potential	Slow response; ratiometric
DiSBAC2	Ex 535 Em 560	Combined optical potential and Ca^{2+} measurements; confocal imaging of membrane potential	Slow response; oxanol family of dyes; fluorescence decreases in response to hyperpolarization
DiOC$_6$	Ex 484 Em 501	Mitochondrial activity; labeling of endoplasmic reticulum	Slow response; carbocyanine family of dyes; fluorescence response to depolarization depends on concentration
Merocyanine 540	Ex 555 Em 578	Membrane potential in muscle fibers and mitochondria	Biphasic response (fast and slow)

Values are taken from the *Handbook of Fluorescent Probes* at probes.invitrogen.com/handbook. Excitation and emission maxima are in aqueous solution.

membrane because they are charged (**Figure 14.10A**). In this environment, voltage changes across the membrane can result in reconfiguration of electronic orbitals permissive to fluorescence within milliseconds. Depolarization of the membrane results a radical shift in the emission spectrum; green fluorescence at 540 nm is increased, while red fluorescence at 610 nm is decreased (Figure 14.10B). Changes in fluorescence at both wavelengths can be ratioed to produce a parameter that follows the membrane potential very closely. For example, action potentials in guinea pig cardiac muscle cells can be followed either by di-4-ANEPPS or by microelectrode recordings with excellent correlation in the measurements (compare Figures 14.10C and 14.10D).

Slow-acting probes actually pass through the membrane when responding to changes in membrane potential, thereby accounting for why their response takes seconds rather than milliseconds. Because these dyes fluoresce considerably stronger in the extracellular space than in the cytoplasm, shifts of these dyes caused by depolarization of the plasma membrane results in an increase in fluorescence. Probes in this class include carbocyanines and rhodamines. As demonstrated in Figure 14.10E, Mito Tracker Red selectively stains mitochondria that are actively respiring and exhibit a membrane potential of approximately -150 mV. Dissipation of the membrane potential by addition of mitochondrial uncoupling agents results in exit of the dye from mitochondrial membranes and

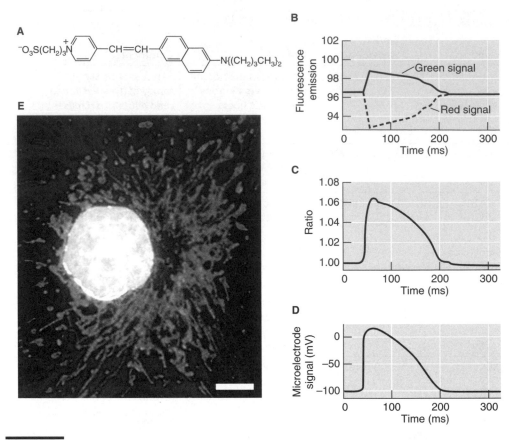

Figure 14.10
Structure and use of membrane potentiometric probes. (A) Structure of Di-4-ANEPPS a fast response potentiometric probe. (B) Fluorescence emission from a Di-4-ANEPPS–loaded rabbit heart during an action potential. The green and red signals are obtained at 540 nm and 610 nm and above, respectively. (C and D) The red/green signal ratio versus time is nearly identical to an electrical recording of the potential during the same period. (E) Bovine pulmonary artery endothelial cell stained with MitoTracker Red CMX Ros and SYTOX Green, a nucleic acid stain. The MitoTracker dye is sensitive to the membrane potential of the mitochondria (arrows). Bar = 2 μm.

redistribution to other regions of the cell. These probes are used either for detecting stable membrane potentials in nonexcitable cells or for localizing subcellular organelles such as mitochondria that maintain a substantial membrane potential. Because of the dependence of membrane potentials on metabolic activity and membrane transport, these probes are also used to distinguish live cells from dead cells.

■ Cyclic Nucleotide and Protein Kinase Probes

Small molecule cyclic AMP probes, analogous to the fluorescent chelators used for Ca^{2+} and other ions, have not yet been developed. Instead, artificial protein constructs that produce fluorescence resonance energy transfer (FRET) on binding cAMP have been developed. These probes contain a protein domain that binds cyclic AMP sandwiched between a cyan fluorescent protein domain and a yellow fluorescent protein domain. After cyclic AMP binding, the probe undergoes a conformational change that brings the CFP and YFP domains in together to produce FRET that can be measured in both time and space. Thus, FRET becomes a measure of the distribution of increasing cyclic AMP concentrations within the cell.

A

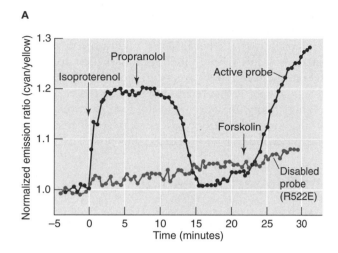

B

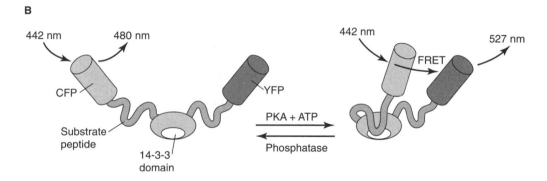

Figure 14.11

Design and use of FRET probes in living cells. (A) Cells in which a FRET probe is used to detect the rise in cAMP levels during isoproterenol stimulation (black circles). The subsequent addition of propranolol blocks isoproterenol stimulation and the addition of forskolin directly stimulates cAMP production by adenyl cyclase. As a control, the FRET probe, disabled by a site-directed mutation, shows no fluorescence change (gray circles). (B) Design of a FRET probe for sensing cAMP-dependent protein kinase phosphorylation activity. Phosphorylation of the substrate peptide results in its binding to the 14-3-3 domain thus bringing the YFP and CFP domains into close contact to produce FRET.

As shown in **Figure 14.11A,** cells expressing such a reporter protein, when stimulated by isoproterenol to produce cyclic AMP, exhibit a rapid increase of FRET. Subsequent addition of propranolol, a blocker of isoproterenol, results in FRET decreasing back to baseline levels. Further addition of forskolin, a direct simulator of adenylate cyclase, again results in cyclic AMP production and the observation of FRET. In contrast, these changes in FRET are not seen in cells expressing a reporter protein that does not bind cyclic AMP due to a change in one amino acid (R to E at residue 522).

FRET probes of similar design can be used to detect activity of cyclic AMP-dependent kinase (PKA). In this case, the hinge region of the reporter construct contains a substrate domain and a "14-3-3" domain that is known to bind tightly to serine residues that have been phosphorylated (Figure 14.11B). After a serine residue in the substrate domain is phosphorylated by PKA, it becomes bound by the 14-3-3 domain initiating a conformation change that brings CFP and YFP domains together to produce FRET.

The fact that these reporter proteins can be expressed intracellularly without need for loading can have great advantages. First, by using protein localization signals, the reporters can be targeted to virtually any cell compartment, including the cytoplasm, nucleus, endoplasmic reticulum, Golgi apparatus, mitochondria, and lysosomes. Likewise, this technology is extremely versatile because the linker protein can be designed to monitor virtually any small molecule whose binding induces a conformational change. Currently, probes based on this technology are available for both Ca^{2+} and cyclic GMP as well as cyclic AMP. Extension of this technology to use the variable regions of specific antibodies as linkers may allow its widespread use for detection of almost any antigenic molecule. Likewise, this approach can be extended to localization of virtually any protein kinase activity and beyond that to other enzyme classes.

References and Suggested Reading

General

Tsien RY. Building and breeding molecules to spy on cells and tumors. *FEBS Lett.* 2005;579:927–932.

Weijer CJ. Visualizing signals moving in cells. *Science* 2003;300:96–100.

Zhang J, Campbell RE, Ting AY, Tsien RY. Creating new fluorescent probes for cell biology. *Nat Rev Mol Cell Biol.* 2002;3:906–918.

Live Cell Imaging

Gerlich D, Ellenberg J. 4D imaging to assay complex dynamics in live specimens. *Nat Rev Mol Cell Biol.* 2003;(Suppl S):S14–S19.

Goldman RD, Spector DL, eds. *Live Cell Imaging.* Woodbury, New York: Cold Spring Laboratory Press, 2004, p. 631.

Kam Z, Zamir E, Geiger B. Probing molecular processes in live cells by quantitative multidimensional microscopy. *Trends Cell Biol.* 2001;11:329–334.

Stephens DJ, Allan VJ. Light microscopy techniques for live cell imaging. *Science* 2004; 300:82–86.

Calcium Imaging

Berridge MJ, Bootman MD, Roderick HL. Calcium signaling: dynamics, homeostasis and remodeling. *Nature* 2003;4:517–529.

Chandler DE, Williams JA. Intracellular divalent cation release in pancreatic acinar cells during stimulus-secretion coupling: I use of chlortetracycline as fluorescent probe. *J Cell Biol.* 1978;76:371–385.

Grynkiewicz G, Poenie M, Tsien RY. A new generation of Ca^{2+} indicators with greatly improved fluorescence properties. *J Biol Chem.* 1985;260:3440–3450.

Happel RD, Simpson JAV. Distribution of mitochondrial calcium: pyroantimonates precipitation and atomic absorption spectroscopy. *J Histochem Cytochem.* 1982;30:305–311.

Kazilek CJ, Merkle CJ, Chandler DE. Hyperosmotic inhibition of calcium signals and exocytosis in rabbit neutrophils. *Am J Physiol.* 1988;254(5 Pt 1):C709–C718.

Minta A, Kao JP, Tsien RY. Fluorescent indicators for cytosolic calcium based on rhodamine and fluorescein chromophores. *J Biol Chem.* 1989;264:8171–8178.

Putney JW, ed. *Calcium Signaling,* 2nd ed. Boca Raton, FL: CRC Press, 2005, p. 536.

Rudolf R, Mongillo M, Rizzuto R, Pozzan T. Looking forward to seeing calcium. *Nat Rev.* 2003;4:579–585.

Other Second Messenger Probes

DiPilato LM, Cheng X, Zhang J. Fluorescent indicators of cAMP and Epac activation reveal differential dynamics of cAMP signaling within discrete subcellular compartments. *Proc Natl Acad Sci USA* 2004;101:16513–16518.

Zhang J, Ma Y, Taylor SS, Tsien RY. Genetically encoded reporters of protein kinase A activity reveal impact of substrate tethering. *Proc Natl Acad Sci USA* 2001;98:14997–15002.

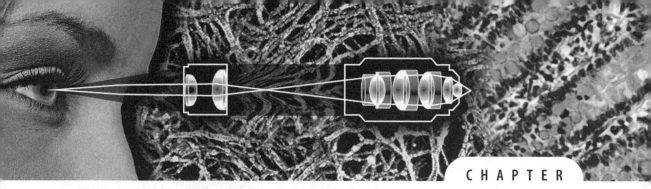

Imaging Macromolecules and Supermolecular Complexes

Although molecules can be detected and localized at the light microscopy level, they cannot be imaged because of the diffraction-based limit of resolution being on the order of 0.25 μm. Techniques capable of imaging molecules must have a resolution of 100 to 1000 times better than this, and the usual choices are cryoelectron microscopy, electron tomography, electron and x-ray diffraction, atomic force microscopy, and transmission electron microscopy combined with either rotary platinum shadowing or negative staining. We discuss each of these techniques in this chapter.

■ Rotary Platinum Shadowing

In Chapter 10, we learned that quick freezing, deep etching, and rotary shadowing could produce three-dimensional specimens that when viewed by transmission electron microscopy provided elaborate subcellular detail. This same method has been modified and applied to macromolecules with excellent results. Preparation of the specimen requires four steps that are outlined in **Figure 15.1**: (1) macromolecules or supermolecular structures are purified, experimentally manipulated in vitro, and adhered to mica; (2) the specimen is quick frozen, usually by a metal mirror block; (3) the specimen is fractured, etched, and shadowed; and (4) the replica is cleaned and viewed by transmission electron microscopy.

For macromolecules, mica is the substratum of choice because it provides a smooth surface at the molecular level. To prepare the mica, sheets are freshly cleaved, cut into pieces about 3 mm square, and then homogenized with a Deunce homogenizer to produce a slurry of fine chips. The homogenization buffer is aqueous, about pH 7.0, and typically contains about 0.15 M KCl. Adsorption of potassium and chloride ions to the mica surface during preparation results in a surface with less negative charge than bare mica and one that adheres proteins and protein complexes well. If negatively charged macromolecules such as DNA or RNA are being imaged, then the mica would be exposed to positively charged polymers such a polylysine, Alcian blue, or cytochrome C to provide a positively charged surface. Another option is to derivatize covalently

Procedure for Visualizing Molecules by Quick-Freezing, Rotary-Shadowing, and Deep-Etching

1. Freshly cleave mica and homogenize in blender. Soak mica chips in KCl.

2. Wash mica chips in water and resuspend as a slurry.

3. Mix macromolecules to be visualized wiith mica slurry to allow adsorption.

4. Place droplet on cushion and ultrarapidly freeze.

5. Place frozen specimen in freeze-fracture unit, cleave, and etch at −95°C.

6. Shadow with platinum and carbon while specimen is rotating.

7. Clean replica with hydrofluoric acid, wash, and view by transmission EM.

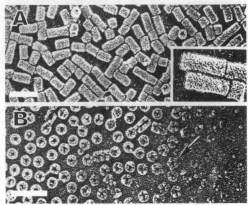

Figure 15.1
Procedure and use of quick-freezing, deep-etching and rotary-shadowing (QF-DE-RS) to visualize macromolecules. (A) Hemocyanin, a hollow, cylindrically shaped protein from the blood of the horseshoe crab, is visualized after QF-DE-RS. Bar = 100 nm. (B) Cross-fracture of these proteins standing on end (arrow) shows that the cylinder is hollow. Bar = 100 nm.

the mica surface to produce either fixed negative or fixed positive charges. An excellent example of derivatization is the use of aminopropylsilatrane-linked mica for imaging of nucleic acid polymers, a technique that has been widely used in atomic force microscopy studies.

Proteins, immediately after in vitro treatment to provide the desired physiological state, may be either incubated with mica chips directly or first fixed with formaldehyde or glutaraldehyde before being mixed with mica. The protein–mica slurry is then centrifuged to produce a pellet; the pellet is washed with 15% methanol to remove salts and is immediately quick frozen using a "slammer" type of cryofixation device (see Chapter 10). The rapidly frozen specimen can then be stored for the long term (weeks or months) in liquid nitrogen if desired. The specimen is then mounted in a freeze-fracture unit equipped with a microtome blade and fractured at −95°C, and the specimen is deep etched by ice sublimation onto the liquid nitrogen-cooled knife parked above the specimen. The etching period, usually 3 to 10 minutes, is followed by application of platinum–carbon from an electron beam gun, followed by carbon evaporation to provide a backing. During platinum–carbon evaporation, the specimen is rotated at a speed of 20 rpm to coat all sides of each molecular complex. Platinum–carbon evaporation is typically done from a low angle, 10 to 25 degrees, to produce a decorating effect whereby more platinum adheres to the molecules to be imaged than to the surrounding mica.

The replica is usually stripped from the mica by gentle immersion into weak hydrofluoric acid, which actually dissolves some of the mica. Without use of hydrofluoric acid, the replica generally sticks too tightly to the mica and can seldom be removed intact; however, because hydrofluoric acid is volatile and can cause tissue damage, it must be used with care. After cleaning, the replica is mounted on a grid and observed by transmission electron microscopy, usually at an accelerating voltage of 80 kV. Figure 15.1 shows hemocyanin, a blood protein isolated from horseshoe crabs, prepared with this technique. The unique, cylinder-shaped structure of this protein is easily visualized from both side and end-on views. The end-on view (Figure 15.1, lower panel)

Procedure for Visualizing Macromolecules by Glycerol Spraying

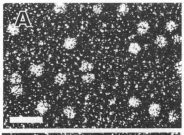

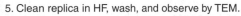
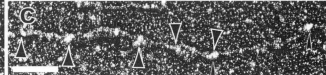

1. Mix macromolecule suspension with 50% glycerol v/v.

2. Load artist's airbrush with 25 to 50 μl of glycerol suspension.

3. Using compressed air, spray fine mist of droplets onto freshly cleaved mica target positioned 25 cm from airbrush.

4. Place in freeze fracture unit at room temperature, and shadow with platinum and carbon while rotating.

5. Clean replica in HF, wash, and observe by TEM.

Figure 15.2
Procedure for visualizing macromolecular complexes by glycerol spraying. (A) Ovoperoxidase, a 70-kD globular protein isolated from sea urchin egg cortical granules. Bar = 50 nm. (B) Proteoliasin–ovoperoxidase complex. Rotary shadowing demonstrates that a 1:1 complex forms with ovoperoxidase (single arrowhead) binding at one end of the fibrous proteoliasin protein (double arrowhead). In vivo, proteoliasin serves to tether ovoperoxidase to the egg extracellular matrix. Bar = 50 nm. (C) In the presence of calcium, these two proteins form long chain-like heteromultimers with bound ovoperoxidase being seen at regular intervals (arrowheads). Bar = 200 nm.

depicts a fracture plane that starts at the mica surface and then rises to cleave a number of cylinders mid drift, revealing that they are hollow. This technique is particularly well suited to imaging protein–protein interactions as well as individual macromolecules.

Platinum-rotary shadowing of macromolecules does not necessarily require them to be rapidly frozen. An alternative that was actually developed before the freezing and etching method described above uses an artist's airbrush to spray macromolecules onto a mica surface at room temperature. First, macromolecules are suspended in a buffer containing 50% glycerol. Second, this solution is sprayed downward as a fine mist at a mica target about 25 cm away, being careful to spray only enough droplets to produce a widely spaced array. As each droplet hits the mica surface, the force of impact causes it to spread and then retract, leaving the macromolecular solute scattered randomly on the mica surface surrounding the droplet. The mica is then evaporated under vacuum at room temperature, and its surface is replicated with platinum–carbon. In this case, shadowing is done at room temperature, usually from an angle of 6 to 10 degrees, and can be either unidirectional, bidirectional, or rotary.

This technique produces a decorated molecular specimen like that shown in **Figure 15.2**. Imaged are two extracellular matrix molecules from sea urchin eggs, ovoperoxidase, a globular protein (Figure 15.2A), and proteoliasin, a fibrous protein (Figure 15.2B). When incubated together before spraying and replication, these two proteins form a 1:1 complex with ovoperoxidase (single arrowhead) binding to one terminus of proteoliasin (double arrowheads) (Figure 15.2B). In the presence of calcium, these proteins form end-to-end polymers, with each proteoliasin monomer binding one globular ovoperoxidase (arrowheads) (Figure 15.2C). These images are consistent with biochemical data indicating that proteoliasin is an ovoperoxidase-binding protein that serves to link this enzyme to the newly forming extracellular matrix of the sea urchin embryo just after fertilization.

Thus, protein–protein interactions are frequently maintained during incubation in glycerol and spraying onto mica.

For larger protein complexes, organelles, and cell fragments, critical point drying can be used to prepare the specimen for rotary shadowing with platinum. Cells and cell fragments are usually adhered to a glass surface using poly-L-lysine. The glass surface is cleaned by sonication in methanol. A 0.1% solution of poly-L-lysine is pipetted onto its surface, allowed to stand a minute or two, and then rinsed away with distilled water. The excess water is then wicked away. The glass is allowed to dry and then is stored in a Petri dish for use within a few days. During use, a suspension of cells or cell fragments is pipetted onto the coated glass surface and allowed to adhere for several minutes, and then the unbound material is washed away with a gentle stream of buffer. Alternatively, adherent cells can be sheared open with a stream of buffer from a syringe leaving cell cortex fragments adhering to the glass (**Figure 15.3A**). Typically, a plastic squirt bottle and an appropriate "intracellular" buffer are used for the purpose. Intracellular buffers usually have KCl as their major salt, a pH of about 7.0, and a calcium EGTA buffer to maintain

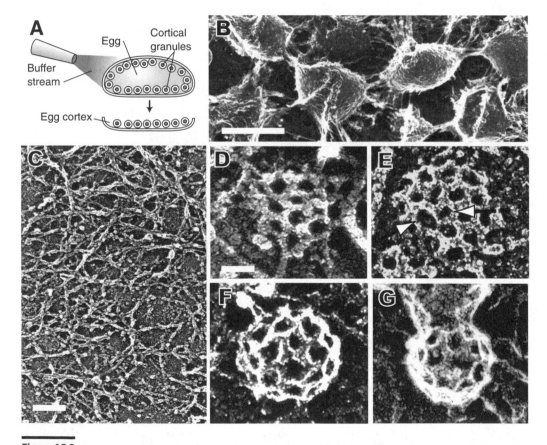

Figure 15.3
Visualization of protein complexes in tissues prepared by critical point drying and by freeze drying and rotary shadowing. (A) Preparation of cell cortex fragments by buffer shearing of glass-adherent cells. (B) Secretory granules adhering to an isolated sea urchin egg cortex. The specimen was prepared by critical point drying and rotary shadowing. Bar = 0.5 μm. (C) Neutrophil cortex prepared by buffer shearing. In this aerial view of the inside of the plasma membrane, numerous microfilaments can be seen criss-crossing the membrane surface. Bar = 100 nm. (D–G) Stages in the formation of clathrin-coated endocytic vesicles on the inner side of the neutrophil plasma membrane. The hexagonal array of clathrin trimers rearranges to form a heptagon and pentagon (between arrows, E) as the basket work produces the curvature needed to pinch off membrane. Bar = 50 nm.

low calcium levels. Mg-ATP and an ATP-generating system are included in the buffer if the cell cortices are expected to carry out any physiologic processes in vitro. The adherent structures (cells or cortices) are then fixed by immersion in 1% glutaraldehyde for 5 to 10 minutes and are washed and then either dehydrated through a series of ethanols and critical point dried using CO_2 (see Chapter 9) or rapidly frozen and freeze dried (see Chapter 10).

Sea urchin egg cortices prepared in this manner, critical point dried, and rotary shadowed with platinum exhibit secretory granules linked by cytoskeletal filaments on the inner side of the plasma membrane (Figure 15.3B). The addition of micromolar concentrations of calcium triggers these granules to fuse with the plasma membrane, representing an in vitro model for exocytosis that can readily be analyzed by light and electron microscopy.

Cell cortices visualized by rapid freezing, freeze drying, and rotary platinum shadowing tend to result in cleaner specimens than those prepared by critical point drying. For example, visualization of clathrin-mediated endocytosis in cortices of neutrophils provides a good example of using this method to identify structural dynamics in supermolecular complexes. In Figure 15.3C, the cytoplasmic surface of a neutrophil cortex reveals dense patches of microfilaments interspersed with endocytotic foci covered with a basketlike network of clathrin. At higher magnification, one can find a series of stages in the process of pinching off of an endocytic vesicle (Figures 15.3D–15.3G). The clathrin coat, initially a flat network having a repeating hexagon pattern (Figure 15.3E), gradually bends to form a sphere (Figure 15.3F). This transformation in shape powers the invagination of the plasma membrane to form a round clathrin-coated endocytic vesicle (Figures 15.3G). As the coat bends, the array of hexagons rearranges to form heptagon–pentagon pairs at key locations to induce curvature (arrows, Figure 15.3E). Recently, rotary shadowing has revealed the interaction of the clathrin coat with adaptor and receptor proteins that are thought to initiate these changes in clathrin structure during receptor-mediated endocytosis.

Macromolecules, individual and in complexes, can be visualized in tissues that have been prepared by quick-freezing, deep-etching, and rotary-shadowing. In this case, the macromolecule is visualized in situ without purification. This technique, providing that the molecules of interest are not obscured by other structures, has two advantages: (1) The macromolecule is seen in context, that is, while interacting with its functional partners, and (2) molecules in the cell interior can be readily visualized because the frozen cells can be cleaved open before etching and shadowing. **Figure 15.4A** provides a low-magnification aerial view of the surface of a sea urchin egg a few minutes after fertilization. The egg surface is covered with newly formed microvilli, and above the egg surface hovers an extracellular envelope that will protect the early embryo. This "fertilization envelope" is formed through elevation of a lace-like "vitelline envelope" from the egg surface, which then undergoes a macromolecular self-assembly process elegantly revealed in quick-frozen, deep-etched, and rotary-shadowed specimens. At first, the envelope retains its initial lacy appearance (Figure 15.4B), but within minutes, it is converted to a coat that has Matterhorn-like peaks and is coated with rows of "paracrystalline" protein forming a sheet-like armor (Figure 15.4C). The assembly of the sheets occurs first at the peaks wherein individual protein monomers form chain-like multimers of four to six units (circles, Figure 15.4D). These "S"-shaped multimers, after assembly on an underlying bed of filaments, act as nucleation sites for the formation of a two-dimensional quasicrystalline array seen in Figure 15.4E at high magnification. Diffraction patterns produced by these lattices, when analyzed by Fourier transform mathematics, provide an averaged lattice having a parallelogram-shaped unit cell that measures 13 by 19 nm (Figure15.4F).

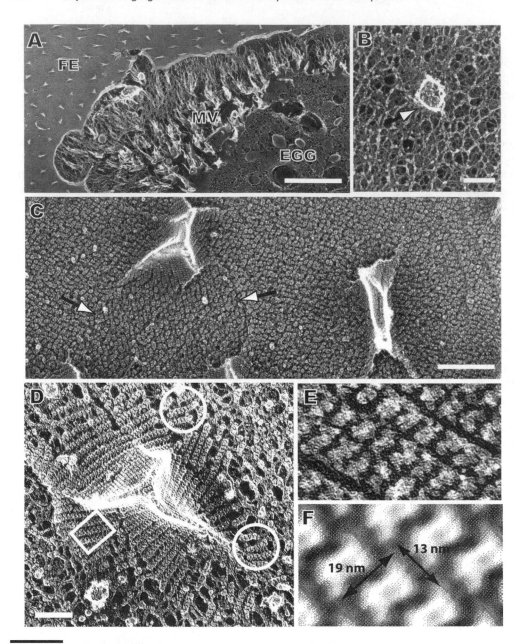

Figure 15.4

Visualization of protein complexes in tissues prepared by quick-freezing deep-etching and rotary-shadowing. (A) Aerial view of sea urchin egg just after fertilization. The specimen, prepared by QF-DE-RS, illustrates the fertilization envelope (FE) that has elevated off of the egg surface that is now studded with microvilli (MV). The FE exhibits a regular array of protruding "casts" that once covered the microvilli on the unfertilized egg surface. Bar = 5 μm. (B) The FE 15 seconds after elevation consists of a lace-like network of filaments. This network serves as a matrix on which secreted proteins will be deposited. The central feature (arrow) is a microvillar "cast." Bar = 0.5 μm. (C) One minute after elevation, the envelope has been coated with "paracrystalline" protein to form a protective plating. Microvillar casts has now changed from a rounded to a peak-shaped protrusion. The paracrystalline coating forms rows in which the orientation changes at borders between domains (arrows). Bar = 0.3 μm. (D) The FE 2 minutes after elevation. At high magnification, rows of "S"-shaped paracrystalline proteins form sheets that coat the "Matterhorn"-like peaks. Circles call attention to small multimers that appear to nucleate the self-assembly process. The box indicates location of detail in (E). Bar = 100 nm. (E) High-magnification view of the "S"-shaped pattern of units in sheets of paracrystalline protein. (F) Averaged unit cell dimensions in the paracrystalline protein array. The unit cell reconstruction was carried out by obtaining diffraction patterns from electron microscopy negatives, masking weak diffraction spots and then using a Fourier mathematical transformation on the masked diffraction pattern to compute the average unit cell intensities.

■ Negative Staining

One of the oldest techniques for visualizing macromolecular structure, negative staining, is simple and effective. Purified proteins, viruses, bacteria, and subcellular organelles are typical specimens because the structures to be visualized must be no thicker than about 0.1 μm in their final form. The specimen is chemically fixed and pelleted by centrifugation, and a suspension is prepared by mixing the pellet with either 1% uranyl acetate, 16% ammonium molybdate, or 1% phosphotungstic acid. The suspension is then pipetted onto a formvar-coated grid and allowed to settle and adhere. Alternatively, the macromolecular suspension can be treated with uranyl acetate (1%) and/or tannic acid (0.25%) directly on the grid without prior chemical fixation. Finally, the specimen and stain are allowed to dry on the grid, producing a thin layer of macromolecules embedded in a layer of electron dense stain. As shown in **Figure 15.5A**, a single actin microfilament, 7 nm in diameter, is outlined by stain. The electron translucent filament appears white against a black background of stain, hence the term "negative staining." Providing that the layer of stain is thin enough (i.e., no thicker that the specimen itself) and that the stain has been able to penetrate the crevices in the molecules, details of the macromolecular structure are highlighted. For example, addition of myosin S1 (a proteolytic fragment of the myosin head containing an actin binding site) to such microfilaments produces a delicate arrowhead pattern that reveals the directionality of the microfilament and the

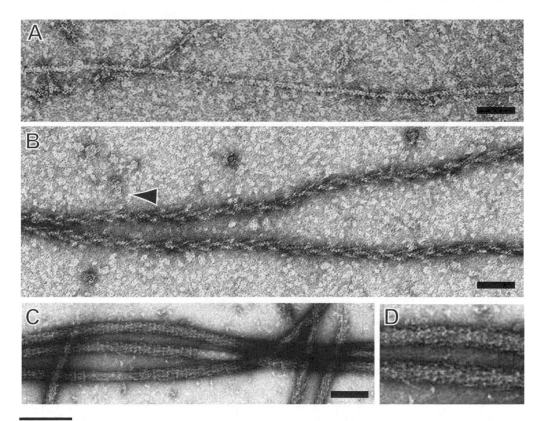

Figure 15.5
Negative staining of molecular structures. (A) A single branched microfilament visualized by negative staining with uranyl acetate. Bar = 100 nm. (B) Microfilaments decorated with S1 myosin heads to produce a classic "arrowhead" pattern pointing in the direction indicated. Visualized by negative staining with uranyl acetate and tannic acid. Bar = 50 nm. (C) Thick filaments composed of scallop muscle myosin stained with uranyl acetate and tannic acid. Bar = 100 nm. (D) Higher magnification view of (C) shows an orderly pattern of myosin heads extending from the thick filament surface.

underlying double helical structure of the filament (Figure 15.5B). Muscle myosin itself forms highly ordered "thick" filaments composed of hundreds of individual myosin molecules (Figure 15.5C). Within these filaments, the myosin tails interact to form an orderly ensemble that ensures that the myosin head domains form a precise pattern on the surface of the filament perfectly positioned to interact with surrounding microfilaments in the intact muscle (Figure 15.5D).

Negative staining is quick and inexpensive and requires little preparative know-how. For these reasons, it has long been a favorite among structural biologists seeking to image a new macromolecular preparation for the first time or to use as a quality-control measure for monitoring the overall structure of their molecular preparations. There are several drawbacks of the technique when used for elucidating structural changes. First, the specimen has been air dried, a procedure that can distort molecular structures and alter the spatial relationships in protein complexes. Second, fine structure in the specimen may be lost if the stain cannot sufficiently penetrate into the specimen topology. Finally, most electron microscopists consider that the technique is rather "dirty" and can easily contaminate the electron microscope column with heavy metal stain.

■ Single Particle Analysis of Frozen Macromolecules by Cryoelectron Microscopy

For those macromolecules that cannot be crystallized for diffraction studies or for relatively large and complex structures such as viruses or ribosomes that are not easily ordered in any manner, cryoelectron microscopy is often the method used to provide a structural model. The basic steps in this technique are to (1) rapidly freeze the structure in a thin layer of vitreous ice, (2) image the frozen specimen using a transmission electron microscope equipped with a cryostage, (3) computer analyze images and group them into major classes whose members can be averaged to reduce noise, and (4) use class images to reconstruct a three-dimensional model.

Rapid freezing of macromolecules in vitreous ice—that is, ice that is amorphous and contains no crystal structure—requires a very thin specimen. A fine mesh grid is dipped into an aqueous suspension of the structures to be imaged; excess fluid is wicked off, and the grid is immediately plunged into a cryogen such as liquid ethane. Wicking off the right amount of fluid from the grid is critical to success. Leaving too much liquid results in a layer of water too thick to be frozen to the vitreous state; removing too much fluid leaves a layer so thin that it breaks before it can be frozen. The grid is fastened into a freezing machine at the tip of spring-loaded tweezers (**Figure 15.6A**). The tweezers are then accelerated by gravity into an insulated well of cryogen; the well is covered with a shutter to prevent premature cooling of the specimen during acceleration, and the shutter opens just before arrival of the specimen. This device ensures that the specimen continually comes in contact with fresh cryogen as it freezes, thus providing freezing rates of greater than 10,000°C per second.

The grid is then mounted into a cryostage and inserted into the column of a transmission electron microscope. Throughout the process, the role of the cryostage is to maintain the specimen at liquid-nitrogen temperatures to avoid crystallization. In order to cause as little radiation and heat damage to the specimen as possible, images are taken with as low a dose of electrons as possible using a relatively high electron accelerating voltage (100 or 120 kV). Focusing, beam intensity, and exposure are all computer controlled so as to achieve the best possible result with the least radiation. Generally, a large objective aperture is used to increase resolution, although this comes at the cost of image contrast, as one can see in Figure 15.6B, an electron micrograph of ribosomes in ice. Images are captured by a digital slow-scan camera and then analyzed by specialized

Procedure for Structural Determination of Macromolecular Complexes by Single-Particle Cyro–Electron Microscopy

1. Dip microscope grid in aqueous macromolecule suspension; wick off excess sample.

2. Ultrarapidly freeze to vitreous ice by high-velocity immersion into liquid ethane. Transfer to the liquid nitrogen–cooled stage of a cryo-TEM unit.

3. Image specimen using computer-controlled, low-dose illumination.

4. Align single particle images. Sort images into classes, and average to reduce noise.

5. Orient class images to produce a low-noise, 3D consensus image.

6. Working 3D model is then used for error analysis to evaluate and reformulate better class images.

7. Steps 5 and 6 are repeated a number of times to arrive at the minimum error 3D model.

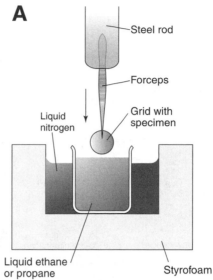

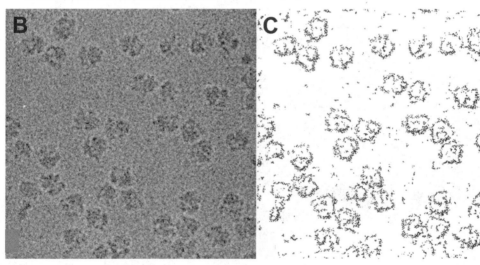

Figure 15.6
Procedure for cryoelectron microscopic determination of macromolecular structure. (A) Machine for rapidly freezing thin specimens on grids. (B) Low-contrast image of ribosomal subunits frozen in vitreous ice. (C) Single-particle recognition by edge detection software.

computer software. An initial task for such software is particle recognition often requiring edge detection followed by automated comparison of candidate particles with an idealized model that can be rotated in one or two axis to test for superimposition (Figure 15.6C). Once particles are chosen their images serve as data evaluated in a multi-step process to obtain the "best fit" three dimensional model that can account for the two dimensional data with least error. **Figure 15.7A** illustrates diagrammatically the steps involved in image analysis and three-dimensional reconstruction by using a familiar macroscopic object— a house—as the specimen. Even after correct particle recognition, one must choose the best images. This may sound simple, but because thousands of images are required, this is often done by computer software using density, contrast, and a lack of defects or

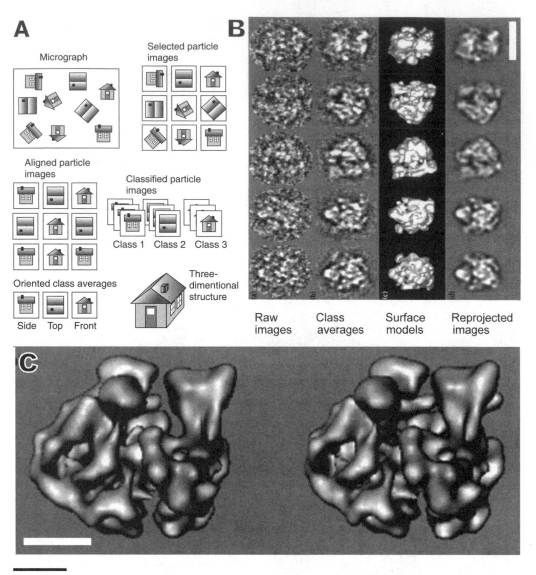

Figure 15.7
Protocol for three-dimensional modeling based on single-particle cryoelectron microscopy images of ribosomes. (A) Illustration of image processing procedures using a house as a familiar macroscopic specimen. (B) Comparison of oriented raw images (column 1), classified and averaged images (column 2), surface model equivalents of the classified images (column 3), and error-minimized class images (column 4). The surface models and reprojected images were obtained by working backward from the final model and are shown for comparison. Bar = 20 nm. (C) Stereo pair of final three-dimensional model of the *E. coli* 70S ribosome. Bar = 10 nm.

noise as criteria. A wide variety of object orientations increase data complexity but are absolutely necessary to providing adequate structural information for three-dimensional reconstruction. (Could you accurately reconstruct an elephant from only one photograph of it?) Second, the structures imaged are at completely random orientations relative to the optical axis. This necessitates aligning the images by computer-aided rotation and translation. Next, images showing similar views of the specimen, that is, from the same side or angle, are grouped into classes. This is carried out by pattern-recognition software. The images within each class are averaged to reduce noise and some inherent biological variability in the structural details. The class average images are then "classified" by assigning an orientation to each class image and are used to create a trial three-dimensional model. An error analysis is carried out by comparing the original aligned images and class averaged images to the provisional three-dimensional model and determining whether any changes can be made in the classes to lower the error. This process is reiterated a number of times, as indicated by the protocol presented in Figure 15.6.

Now let us look at the process in more detail using real data from cryoelectron microscopy studies of ribosomes. In Figure 15.7B, the raw images of ribosomes show that these images are considerably more complex than the house diagrams. This is because macromolecules are translucent; each image contains both internal and external information, not just a surface view. Computer analysis chooses similar images, presumably having a similar object orientation, to be averaged for the production of class images. As shown in Figure 15.7B (second column), this results in a significant noise reduction. For purposes of comparison, the third column shows how these class average images correlate to surface views of the final ribosome model.

In arriving at class-average images, the software had to make decisions on how much intraclass variation can be allowed versus how many classes can be tolerated. Certain images may have had to be eliminated from analysis because they do not fit well into any well developed class. This process is akin to a much more complicated version of drawing a "best-fit" curve for a group of data points. Groups of data points that are similar in pattern will have a much stronger influence on what the curve looks like than any individual outlying points that have little influence on the curve.

Finding class images is complicated enough that a reiterative process is required. Each round of error calculations strengthens some image classes while weakening others; in addition, these calculations are used to modify each class image to increase its statistical validity. This new information is then used in the next round of calculations to provide even better class images. This process results in class images that have minimized error, and a set of these "reprojected images" is shown in the last column of Figure 15.7B.

Then the process of three-dimensional reconstruction begins. Reconstruction, although computationally intense, is relatively straightforward if the spatial relationship between "class images" is known. An example of using orientations predetermined by experimental procedure is that of three-dimensional models readily constructed from images taken at set angles during electron microscopic tomography (see next section) or at set heights during laser scanning confocal microscopy (see Chapter 12). Here, our goal is an order of magnitude more difficult—finding the correct but unknown spatial relationships between class-averaged images. Again, a reiterative process is used that tests out orientations of each class relative to the others to produce a reconstructed trial model that incorporates the least amount of error. In each iteration, a subset of orientations is optimized, and the new orientations used as the next subset is optimized.

The resulting three-dimensional reconstruction can provide information for a surface model, a cutaway model, or a stereo model, as shown in Figure 15.7C. The resolution of the resulting model depends on the resolution of the original component images and on the class-averaging process—that is, how much error is present in each class. For example, if the macromolecule showed true randomness in its orientation within the specimen, all images would be different and lead to larger errors in each class; designation of classes would be arbitrary. On the other hand, if the macromolecule had several preferred orientations and these orientations were at set angles, say orthogonal or triangulated, the class averages would have much lower errors and lead to a model with a somewhat higher resolution. Under the best of circumstances, resolution is in the 2- to 4-nm range—similar to that of rotary shadowing and electron tomography, but worse than that of X-ray or electron diffraction methods.

■ Electron Microscope Tomography

Although "single-particle" analysis using cryoelectron microscopy to construct three-dimensional structures is a powerful technique, there is an inherent limitation to its resolution of fine structure. This limitation is due to the fact that thousands of molecular objects must be queried, each slightly different from each other and each having an orientation that is initially unknown. In contrast, electron microscope tomography is based on projection images of a single object oriented at numerous predetermined angles to the optical axis. Typically, a specimen is placed on a goniometer stage, and the specimen is tilted around an axis perpendicular to the electron beam. The thickness of the specimen is limited by the penetration depth of the electron beam. Thus, electron microscopes having an intermediate range of accelerating voltage (200 to 400 kV) are used allowing for a specimen thickness of 0.5 to 1 μm—10 times thicker than the typical specimen for a 120-kV microscope. In addition, the microscope is best equipped with an energy filter that can discard electrons that have been inelastically scattered by the specimen. These lower energy, non–image-forming electrons can form a substantial background in specimens as thick as those needed for tomography.

The specimen itself can be an epoxy resin section, a thin frozen specimen, or more rarely a frozen thick section. Resin-embedded specimens are usually rapidly frozen and freeze substituted before embedding and sectioning to minimize fixation artifacts. Specimens are imaged at multiple tilt angles both positive and negative from horizontal. The highest resolution in three-dimensional reconstruction would, in theory, be obtained by tilting the specimen a total of 180 degrees and photographing the specimen at an infinite number of different angles within that range. In practice, the specimen is tilted through a range of about 120 degrees, and an image is obtained at every 2.5 degrees to provide a resolution of approximately 2 to 5 nm in the x, y, and z directions. The specimen takes on a large accumulated dose of electrons during the process that in resin sections leads to thinning as much as 30% to 50% and in frozen specimens can lead to both distortion and destruction of the specimen. Recent automation of the tomography process by computer control of specimen tilt, focus, exposure, and imaging sequence has provided the ability to carry out such work using relatively low doses of electrons with less than 3% of the total dose being used for nonimage purposes, such as focusing.

Computer analysis of data, as well as computer control of data acquisition, has revolutionized electron tomography, and its use has seen an exponential growth in recent years. Starting with image capture by a large-field digital still camera (e.g., 4000 × 4000 pixels), construction of the three-dimensional model, its storage, and output of its features can now be done entirely in silico.

Providing that section thickness and angles of tilt are known precisely, the model can be correctly scaled in all dimensions and volume information derived from the model. Just as the term *pixel* can be used to describe a unit of defined location and size within a planar image, the term *voxel* (for "volume element") can be used to identify a unit of defined location and size within a three-dimensional model. For example, a substructure within such a model can be said to have a volume of N number of voxels. Ultimately, calibration of the input data must be done to convert pixel or voxel units to absolute values for linear distance, area, and volume. For these quantities to be useful in three dimensions, however, the software must be able to identify objects, compartments, and surfaces in three dimensions.

Common methods for doing this are segmentation routines. Using a two-dimensional example, one can describe this process as edge-finding algorithms that identify changes in pixel intensity in a given search area. After evidence of an edge is found, a search is made for continuation of the edge in nearby pixel arrays. This search continues until an area is entirely enclosed by edge predictions and designated as a separate object. The definition of objects can be very strict, in which case the edge must be continuous or very near to continuous. Alternatively, edge prediction can be much more liberal depending on the size of the search area allowed for finding a continuation of the edge. In a liberal case, the edge could be interrupted by noise or discontinuities without rejecting the area as an object. After an edge or object area is recognized, the position of the edge can be determined by joining pixels that have the same gray level within a certain range. Such edge finding and object defining processes are useful in analysis of any images that have adequate contrast whether they are fluorescence, phase contrast, or electron micrographs.

Applying the same concepts to three dimensions and electron tomography, the software must search a three-dimensional data set for changes in voxel density to detect surfaces and boundaries. Starting with a specimen volume that contains such changes in density (designated by either the user or the software), a search for similar changes in density is initiated in the surrounding voxels. After these are found a surface has begun to be determined. Subsequently, the surface is extended in all directions by further searches. The surface detected may, if extended sufficiently, enclose a compartment, for example, a membrane-bound vesicle. The positioning of the surface rendered can be determined by connecting voxels that have similar densities—the so-called isodensity method.

Currently, three software programs are in fairly widespread use for 3-D model construction: IMOD, shareware developed by McIntosh and colleagues at the University of Colorado that uses isodensity and user-defined manual methods for segmentation; AMIRA, marketed by the Indeed-Visual Concepts GmbH, Berlin; and AVS, offered by Advance Visual Systems in Waltham, MA. The commercial programs use both isodensity and other proprietary approaches for visualizing surfaces. More recently, Ress and colleagues at Stanford University have developed a fourth approach that provides not only for visualization but also for estimations of error that are referred to as "spatial uncertainty measurements." Model reconstruction by all programs can be aided by randomly placed "fiducial" markers such as colloidal gold beads to act as reference voxels in all data sets used.

An example of such visualization is the three-dimensional, computer-reconstructed model of a microtubule array with transport vesicles at a fungal hyphal tip (**Figure 15.8A**). This model was created from a set of intermediate-voltage transmission electron micrographs like that in Figure 15.8B using a stack of 200-nm-thick sections. Images of each thick section, taken at a defined set of angles, were then entered into the software program IMOD to generate the model. Structures were segmented manually, and the output represents objects such as vesicles and microtubules as geometric shapes that only approximate the actual object surfaces.

The presynaptic terminal of the frog neuromuscular junction also offers a fascinating three-dimensional puzzle of physiologic importance. The plasma membrane of the terminal exhibits "active zones," each zone consisting of a long strip of docking sites for synaptic vesicles. Understanding the geometry of how vesicles dock and then fuse with the membrane to release neurotransmitter is an important goal in neurobiology. A thick section through an active zone was analyzed by electron tomography, and a three-dimensional virtual model of electron density was calculated. As illustrated in Figure 15.8C, virtual sections of this model in any orientation are easily calculated and displayed. The central panel reveals an electron density map in a plane perpendicular to the electron beam. The dashed lines show where this plane intersects with two other planes of view shown in the left and bottom panels. A surface-rendered illustration of the model (Figure 15.8D) was created using the software developed at Stanford and is at high enough resolution to show the surface irregularities of synaptic vesicle membranes and the plasma membrane at which they are docked. Also visualized are the protein complexes that retain the vesicle at the plasma membrane and that mediate membrane fusion when exocytosis is triggered by a nerve impulse. These docking complexes can be isolated by segmentation and viewed separately, as shown in Figure 15.8E. Each point on the surface of this structure can be assigned an

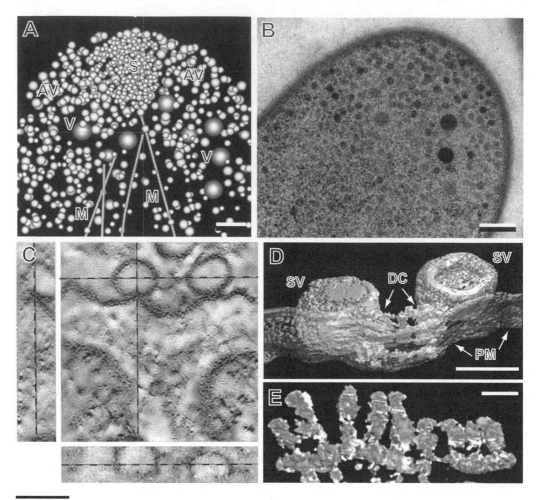

Figure 15.8
Electron microscopy tomography. (A) Model of the microtubule and transport vesicle array at the tip of a fungal hypha constructed by computer-aided electron tomography. Included in the model are microtubules (M), Spitzenkörper-associated vesicles (S), apical vesicles (AV), and Woronin bodies (V). (B) Intermediate-voltage electron micrograph of a 200-nm-thick section of a fungal hypha taken at an accelerating voltage of 200 kV. This specimen is similar to those used to create the tomographic model in (A). (C) Virtual orthogonal slices of a three-dimensional tomographic model of synaptic vesicles docked at the presynaptic membrane of a frog neuromuscular junction. The dashed lines in the central, face-on view represent the planes of section for the two other orthogonal views. (D) Surface-rendered model of synaptic vesicles (SVs) docked at the active zone generated from the same tomographic data set. Active zones of the presynaptic membrane (PM) exhibit docking complexes (DCs) that tether the vesicle to specialized sites for exocytosis. (E) En face view of a docking complex on the cytoplasmic face of the presynaptic membrane. The gray levels represent the magnitude of "spatial uncertainty" values, with darker gray representing low values (1 nm), and white areas representing high values (5 nm). Bars = 0.5 μm in (A), 1 μm in (B), 50 nm in (D), and 20 nm in (E).

uncertainty value that is a statistical measure of how well the position of the surface is defined at that point. Regions with little uncertainty (<2 nm) are shown in dark gray, while areas with greater uncertainty (3 to 5 nm) are shown in lighter gray or white.

Computer-aided tomography, now teamed up with the superior preservation of rapid freezing and automated data acquisition from fragile, frozen, hydrated specimens, has become an extremely important technique in contemporary bioimaging studies. The 2- to 5-nm resolution of this technique in three dimensions makes it a superior choice for modeling the features of large protein complexes and organelles as well as small volumes within whole cells. The strength of this approach is that the full complexity of cellular architecture can be visualized in situ without resorting to cell disruption or biochemical purification.

■ X-Ray and Electron Diffraction Methods

These methods use the inherent order of a three-dimensional or two-dimensional crystal to amplify submicroscopic structural features into macroscopic diffraction patterns. Because a crystal consists of thousands if not millions of macromolecules oriented in almost exactly the same way, even subtle structural features can give rise to diffracted radiation of measurable intensity. The major steps in the process, illustrated in **Figure 15.9,**

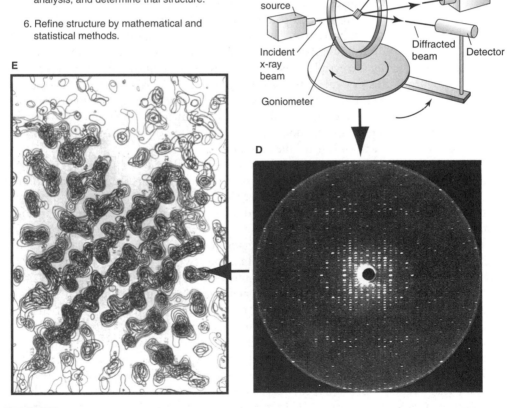

A. X-Ray Crystallography Procedure

1. Grow crystals.

2. Perform preliminary diffraction to determine crystal quality and unit space group.

3. Collect diffraction data on film or diffractometer.

4. Determine phase information using heavy atom derivatives or directly.

5. Calculate electron density map using Fourier analysis, and determine trial structure.

6. Refine structure by mathematical and statistical methods.

Figure 15.9
Procedure for x-ray diffraction determination of macromolecular structure. (A) Procedure for x-ray diffraction crystallography. (B) Crystals of aspartate 2-oxoglutarate aminotransferase. (C) X-ray diffractometer and detector. A single crystal is attached to a goniometer head that can position and rotate the crystal in all three dimensions. The detector can be rotated through any angle relative to the x-ray source. The entire three dimensional diffraction pattern can be captured by systematic rotation of the crystal relative to the x-ray source. (D) X-ray diffreacton pattern from a crystal of aspartate 2-oxoglutarate transferase having a resolution of 2.8 angstroms. (E) Electron density pattern in three dimensions calculated by Fourier transform analysis of diffraction data.

are (1) formation of high-quality crystals of the macromolecule, (2) positioning of the crystal in a collimated and in-phase beam of x-rays or electrons, (3) recording the diffraction pattern produced, and (4) analysis of the diffraction pattern by Fourier transform mathematics to produce an electron density map in a three-dimensional space.

Production of macromolecular crystals is an art. Crystallization is favored by presence of a nucleating structure, a high (supersaturating) concentration of the macromolecule, a near-zero balance of charge on the macromolecule, and conditions that favor a very slow growth of the crystal. The optimal parameters are sufficiently complex and incapable of being predicted that crystallization of a new macromolecule is usually carried out by systematic variation of the four major parameters—concentration, pH, ionic strength, and time. The high concentrations of solute required and the need for a whole matrix of experimental conditions usually mean that several milligrams of the purified macromolecule must be available. Because a balance of positive and negative charge on the macromolecules is favorable for molecule–molecule interactions, pH is usually varied near the isoelectric point of the macromolecule. In order to coax the crystallization process along as slowly as possible, concentration of the macromolecule is usually near saturation to begin and subsequently solvent is gradually removed by controlled evaporation, thus slowly concentrating the macromolecule over a period of hours or days.

Crystals that can be used for diffraction usually range from 0.2 to 1 mm on each face as shown in Figure 15.9B. The formation of any crystal this size is good news, but before success is in hand, the crystal must be tested for production of a good diffraction pattern under standard conditions. A good diffraction pattern consists of strong and well-defined spots arraigned in a number of consecutive tiers. Diffraction spots at the largest angles from the optical axis determine the resolution that the data will provide; thus, it is important to have a distinct pattern as far into the periphery as possible. An example of such a diffraction pattern is shown in Figure 15.9D. Crystals that provide such patterns are referred to as "diffraction-grade" crystals and are extremely well ordered with very few flaws caused by misoriented macromolecules or empty unit cells. Crystals that have many flaws lead to an indistinct pattern of spots at the periphery and are not suitable for structure determination. Further work must be done to improve the fidelity of the crystallization process.

For electron diffraction, two-dimensional crystals can be used if three-dimensional crystals are not available. A few proteins naturally form two-dimensional arrays in vivo (e.g., bacterial rhodopsin), and these are easier to obtain in diffraction quality arrays. In most cases, however, other tricks must be used to coax crystallization. Two-dimensional crystals can sometimes be formed on an artificial support, having molecular regularity that encourages an orderly deposition of the macromolecule. In other cases, the presence of ions or ligands can induce crystallization in either two or three dimensions. A good example is tubulin. In the presence of zinc ions, this protein forms two-dimensional sheets that are further stabilized by the addition of taxol, a drug that specifically binds to β-tubulin, thereby locking it into a crystal-forming conformation. Two-dimensional crystals can be further stabilized by placing them in an "embedding" medium that contains compounds such as tanins or glucose that hydrogen bond with the protein, thereby taking the place of water that would normally be present.

Radiation that is monochromatic, collimated, and in phase must be directed at the crystal. In the case of x-rays, this is not easy because solid lenses that will refract x-rays do not exist. Instead, x-rays can be focused by Fresnel lenses, which use diffraction by an array of concentric circular slits to do the job. The x-ray beam is focused onto the crystal that is held in a goniometer, a device that can position and rotate the crystal in three dimensions (Figure 15.9C). Thus, instead of collecting data from one angle onto a flat sheet of film, as was the case in early crystallography, the entire diffraction pattern is collected with an electronic x-ray detector and x-ray beam at fixed positions while the crystal is rotated in a precise pattern of movements. This arraignment allows higher order diffractions to be observed, with the detector remaining at a constant distance from the crystal regardless of angle. The electronic data are then digitized and analyzed by computer software.

In the case of electron diffraction, the beam provided by the tungsten filament in an electron microscope is naturally monochromatic and is focused onto the crystal by the condenser lenses.

Electrons that are scattered elastically by the atomic nuclei within the crystal constitute the signal that is projected onto the camera screen by the objective and projector lenses of the microscope. In the imaging mode of an electron microscope (**Figure 15.10A**),

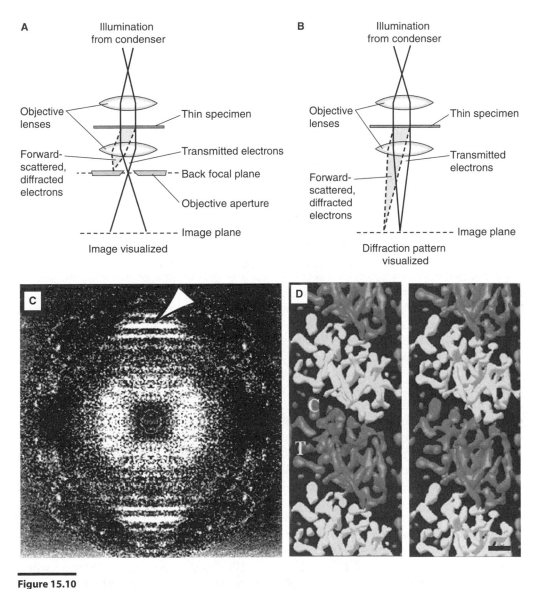

Figure 15.10

Diffraction in an electron microscope. (A) Ray path in an electron microscope set up for image acquisition. Most of the electrons diffracted by the specimen are blocked by the objective aperture. The projector lenses are used at full strength to form an image at the viewing screen and camera. (B) Ray path in an electron microscope set up for diffraction pattern acquisition. The objective aperture has been withdrawn and the projector lenses set at low power to focus the diffraction pattern onto the viewing screen and camera. (C) An electron diffraction pattern produced by sheets of the microtubule protein tubulin. The arrow points to a peripheral reflection indicative of 2.5-angstrom resolution. (D) The 6-angstrom resolution three-dimensional model of the tubulin lattice calculated from the Fourier transformation of diffraction data like that shown in (C). Stereo pair. Bar = 100 nm.

many of the electrons diffracted by the specimen are blocked by the objective aperture so as to gain contrast and cut down on spherical aberration. Because these electrons are absolutely essential to formation of a high-resolution diffraction pattern, the objective aperture must be removed (Figure 15.10B). In addition, the power of the objective and projector lenses have to be reduced considerably so that the back focal plane of the projector lens rather than the image plane coincides with the photographic plate or charge-coupled device (CCD) chip (Figure 15.10B). The image recorded is a diffraction pattern such as that shown for a two-dimensional array of tubulin in Figure 15.10C.

One major consideration in data collection is the so-called phase problem. This refers to the fact that diffraction spot data, regardless of its resolution, are intensity data only. The phases of diffracted rays contain essential information that cannot be obtained from the recorded intensities alone. This is much like the situation in light microscopy in which the structure of biological specimens results in phase changes that cannot be detected by our eyes or cameras. There are several approaches to solving this problem. One is to grow isomorphic crystals that have a heavy metal atom such as uranium lodged at a specific point in the tertiary structure of the macromolecule. The heavy metal atom is a much stronger site of diffraction than any other atom in the macromolecule (C, H, O, and N are too light to compete), and the perturbations caused by its presence provide phase data. A related method is to use two or more wavelengths of x-rays, and by comparing the diffraction patterns obtained, one can determine phases.

The availability of both intensity and phase data now allows mathematic reconstruction of an electron density space using a Fourier mathematical transform. This mathematical transform is similar to that used to convert phase-related interference data into spectral data for chromosome painting in Chapter 13. Diffraction data and image data (in this case electron densities in three dimensions) are inversely related, and for that reason, diffraction data are said to occur in "reciprocal" space. Appropriate software allows this data conversion to be carried out on a high-end computer. The electron density space calculated is then used to identify and predict the positions of atoms, making up the protein backbone, which includes all peptide bonds. Next, amino acids at each position are identified by the size and shape of their side chains wherever the resolution achieved allows. Finally, in maps with the highest resolution (1.5 to 3 angstroms; Figure 13.9E), individual atomic coordinates are determined, including those of tightly bound solvent molecules. This process of identifying and locating individual atoms is much facilitated if the amino acid sequence of the protein is known.

Atomic coordinates are then deposited into the Protein Data Bank, an international database that provides access to structural data by scientists throughout the world. Currently, over 10,000 protein structures are available in this database, with over 80% of the structures having been determined by x-ray crystallography and less than 2% having been determined by electron diffraction. The remainder of the structures has been determined by either nuclear magnetic resonance in solution (16%) or theoretical calculations (2%). These atomic coordinates can also be entered into molecular modeling programs that are able to present the three-dimensional model on a computer screen or as an image file. Such software can usually present a color-coded model in any one of the many forms used to present stereo or depth-perceived images or movies discussed in Chapter 12. An example of such a model is shown in Figure 15.10D as a stereo pair. The protein structure shown is that of alternating α- and β-tubulin, one of the few proteins whose structure has been determined by electron diffraction. Although x-ray crystallography usually produces higher resolution models in less time than electron diffraction, crystallographers were unable to produce diffraction-grade crystals of this protein after years of effort. Thus, the determination of its structure using a two-dimensional crystalline array and electron diffraction represents a particularly noteworthy achievement.

■ Scanning Probe Microscopy

Scanning probe microscopy is not an imaging technique in the narrow sense because no radiation is used as a probe of structure. Instead, a minute tip of atomic dimensions is used to scan the surface of a specimen by near or actual contact. Depending on the type of tip, the data collected may be topology, charge distribution, magnetic fields, chemical reactivity, or location of specific molecules. Although these microscopies were first used for materials sciences, the last 10 years has seen increasing use and adaptation of the technology to biological specimens. And rightfully so! The fact that molecules can be imaged at atomic resolution at atmospheric pressure and even immersed in physiological salt solutions is a molecular biologists dream! A further advantage is that the "microscope" itself is small and simple in design, is relatively inexpensive (one fifth the price of an electron microscope), and does not require a costly service contract because most problems can be dealt with by a reasonably handy scientist. The scanning probe microscopy that is the most compatible with biological systems is atomic force microscopy, and we limit our discussion to this technique.

Atomic force microscopy uses a molded probe of silicon nitrate shaped like a pyramid that comes to an apex that under optimal conditions has only a few atoms at its point. This tip is molded into a triangular shaped cantilever that attaches it firmly to the instrument (**Figures 15.11A** and **15.11B**). The tip is held stationary in the x- and y-axis but is rapidly vibrated in the z-axis so that it can repeatedly contact the specimen. As the probe approaches the surface of the specimen, the atoms at the apex interact with individual atoms of the specimen. The forces experienced as two atoms approach one another include both attraction due to London–Van der Waals forces (dipole-dipole interactions of electron orbitals) and repulsion due to nuclear charge. These forces are a function of interatomic distance, as plotted in Figure 15.11C.

Thus, as the tip approaches the specimen surface, the interatomic forces between the two will temporarily bend the cantilever, which acts like a miniature spring. Bending of the cantilever can be measured by a very sensitive (and clever) method. A laser beam directed at the upper surface of the cantilever is reflected to a photoelectric detector. The detector compares the laser light falling on two adjacent regions and thereby provides a sensitive measure of beam deflection. Using the spring constant of the cantilever, one can convert deflection to force and a readout of force versus time as the tip interacts with the specimen surface can provide topologic information. As shown in Figure 15.11D (top panel), a tip interacting with a solid surface results in an attractive force (brief downward deflection), followed by an increasing repulsive force (upward deflection), as predicted by the London–Van der Waals equation (Figure 15.11C) and as illustrated by the bending of the cantilever in Figure 5.11D. If the specimen topology changes between "taps" of the probe, this force curve will be displaced either up or down, thus providing a readout of this parameter.

The topology should not change unless a new x, y location on the specimen is probed. In order to obtain a topology map, the specimen is rastered past the tip position in a typical series of horizontal scans, data at each x, y position collected, and computer analysis used to produce a pseudocolored two- or three-dimensional representation of the surface. Precise movement of the specimen is carried out by a piezoelectric stage consisting of a ceramic column that can be flexed in any direction by application of precise voltage differences across the x- and y-axes, as demonstrated in Figure 15.11E. Movement of the stage is monitored by sensors in all three axes, x, y, and z, to provide feedback control to circuits driving the stage, thereby ensuring subnanometer precision in the data map (Figure 15.11F). The fields rastered can be as large 100 μm square (as in probing whole cells) to as small as 100 nm square (as in probing small molecules), depending on the design of the piezoelectric stage. Typically, an atomic force microscope will have a number of stages available for use in different magnification ranges.

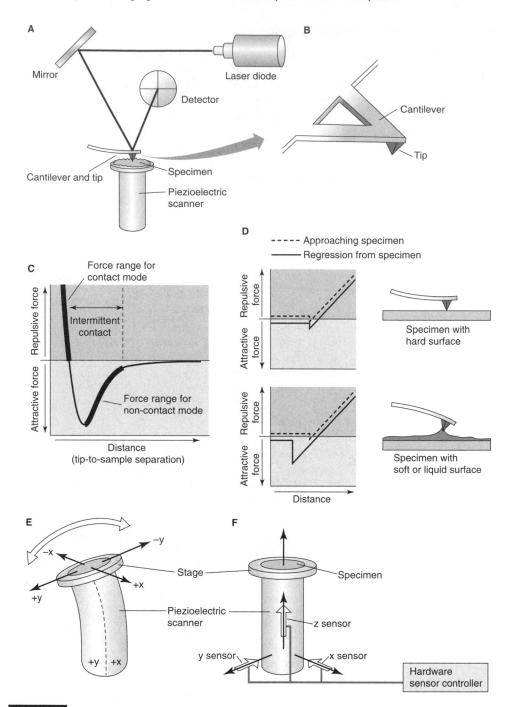

Figure 15.11
Design of the atomic force microscope (AFM). (A) Laser beam readout of the tip–specimen interaction using a double photodiode detector. (B) Diagram of the cantilever tip that contacts the specimen. (C) Attractive and repulsive forces experienced as atoms approach one another on the atomic scale. Tip–specimen interactions can be regulated to key off of either repulsive forces in the contact mode or attractive forces in the noncontact mode. (D) Force versus distance diagrams for tip contact with a specimen. For a hard surface, the tip experiences the same forces on approach and regression from the specimen surface (upper panel). For soft surfaces, liquid surfaces, or surfaces decorated with adhesive molecules, the approach and regression forces may differ due to adhesive interactions of the tip with the specimen (lower panel). Such traces can be used to calculate the magnitude of adhesive forces. (E) Cartoon of the piezoelectric stage used to move the specimen during scanning. Precise voltages applied to the ceramic column regulate its bending in the x and y directions on a microsecond time scale. (F) Sensors of movement in all three dimensions provide feedback control of the voltages applied to the specimen stage.

The tip itself is driven in repeatedly in the z-axis each time it probes the specimen using a high-frequency oscillating magnetic field. Clearly, tip excursion in the z-axis has to be coordinated with x, y movement of the specimen, and both of these are under computer control. Because tip interaction with the specimen involves physical forces, a delicate biological specimen can actually be distorted or moved if the tip employs too much force or is dragged across the surface. Because this can lead to artifacts, one can employ a "noncontact" mode in which tip–specimen interaction is halted by feedback control of the z-axis excursion as soon as the attractive force is registered and before large repulsive forces are generated. In other cases, dragging of the tip across the specimen surface can actually provide additional data about the topology or consistency of the specimen. Such data are referred to as "lateral force" imaging, with the data being generated by quantitation of the twisting motion of the tip in the x or y direction.

The resolution of the atomic force microscope is remarkable. Individual gold atoms of a monolayer can be visualized, producing a hexagonal packing pattern characteristic of crystalline gold. The resolution achieved in favorable specimens is less than 1 nm, and gold, graphite, and other surfaces having regular structure are often used to test resolution of the microscope. Graphite, composed of low atomic number carbon atoms, would be difficult to image in an electron microscope. Although the resolution obtained in biological samples is not as great, AFM is routinely used to visualize macromolecules and macromolecular complexes either *in vitro* or *in situ*. An array of gap junctions on the surface of a cell offers a good example (**Figure 15.12A**). As shown at higher magnification in the inset, each junction is composed of six "connexions" that form the wall of a cylindrical tunnel whose opening is clearly visualized. The structural details rendered by AFM have been shown to be comparable to those obtained by both cryo-electron microscopy and negative staining.

A tour-de-force in AFM imaging of biological macromolecules is provided by using DNA as an example. In Figures 15.12B and 15.12C, circular plasmid DNA is imaged in an aqueous buffer at a set concentration of magnesium chloride and pH. Supercoiling of the plasmid DNA can be monitored as the magnesium concentration is increased, the pH is decreased, and or the AT/GC base ratio is changed by genetic engineering. In addition, a specialized, four-strand "Holiday" junction can be seen (arrows, Figures 15.12B and 15.12C) whose structure is important in DNA recombination. If the plasmid DNA is tethered to the mica surface in one region, the rest of the molecule is free to change its architecture in a dynamic manner. As a result, the Holiday junction can be observed to undergo conformational changes from scan to scan (arrows, Figure 15.12D). Scanning is carried out in the noncontact mode to avoid artifactual movement induced by tip contact. Certain regions of the DNA strands are seen to move frequently, whereas other regions remain still. These dynamics appear to be nucleotide sequence dependent with regions rich in AT exhibiting greater flexibility and increased movement. Currently, visualization of dynamics is limited by a scan rate of about 1 frame per 30 seconds in typical AFM units, but recent advances show that this rate can be increased 10- to 100-fold in microscopes of specialized design.

Atomic force microscopy produces useful medium-range resolution images of protein assembly processes as well. In Figure 15.12E, amyloid-like fibrils composed of prion protein are imaged during aggregation, a process that is thought to be critical to the fatal brain pathologies associated with these proteins. At higher magnification, these fibrils are seen to have a variety of segmented, branched and twisted morphologies (Figure 15.12F).

The atomic force microscope is also able to measure forces between specific individual macromolecules. Given two interacting molecules, if one is adsorbed to the mica surface and the other adsorbed to the tip, the force generated by their interaction can be measured at the resolution of single-bond energies in the pico-newton range. A practical comparison

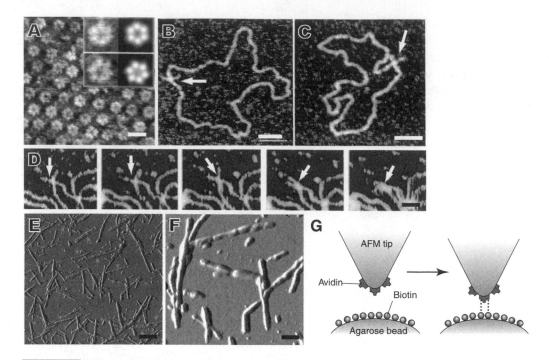

Figure 15.12
Atomic force microscopy (AFM) visualization of atomic and macromolecular structure. (A) AFM image of the extracellular aspect of gap junctions expressed and assembled in HeLa cells. The inset shows two gap junctions at higher magnification (left) and their images rotationally averaged based on six-fold symmetry. Bar = 10 nm. (B and C) Circular plasmids of DNA containing "Holiday" junctions (arrows) thought to be important in homologous recombination. Bars = 50 nm. (D) AFM scan series demonstrating movements of the Holiday junction. The plasmid was imaged in a physiological buffer. Bar = 30 nm. (E) Aggregation of the encephalopathy prion protein into amyloid-like fibrils as visualized by AFM. Bar = 1000 nm. (F) Prion protein amyloid fibrils as viewed by AFM at higher magnification. Bar = 250 nm. (G) Diagram illustrating force measurements between proteins. Avidin bound to the AFM tip will interact with biotin on the surface of the specimen beads. Force measurements between tip and specimen are sensitive enough to detect single avidin-biotin bond interactions.

by which to gauge this sensitivity is that a pico-newton is roughly equivalent to the force of a single noncovalent bond. Figure 15.12G demonstrates the concept of measuring forces between interacting molecules. One member of the interacting pair, in this case the protein avidin, is attached to the atomic force microscope tip. The other member of the interacting pair, in this case biotinylated agarose, is attached to the specimen surface. As the tip contacts the surface, one or more noncovalent interactions are formed between the interacting molecules. In the example illustrated, biotin, a small organic molecule, will bind strongly to the avidin on the tip, each avidin having four binding sites for biotin. As the tip is pulled away, the forces incurred are recorded by the bending of the cantilever. Sudden decreases in the attractive forces holding the tip at the specimen can be attributed to breaking of single, noncovalent bonds as biotin is wrenched from its binding site.

Application of this technique to hybridization of complementary single-stranded DNAs can, for example, detect the unzipping of the hydrogen bonds holding two oligonucleotides together. As the oligonucleotide attached to the tip is pulled away from the oligonucleotide attached to the mica substratum, force on the two increases until one or more hydrogen bonds break. At the rupture of each hydrogen bond, the two nucleotides slide relative to one another, thereby relieving the force on the tip. Binding strength between proteins can also be measured by atomic force techniques similar to those described above for measuring the attractive force of hydrogen bonding between oligonucleotides. For example, binding forces between ligands and protein receptors and between antibodies and antigens can be examined. Increased adhesion of the tip to the surface, caused by the interaction of these two proteins, is measured by the increased force that is required to pull the tip away. A map of this parameter would detect "hot spots" that represent the location of antigens on the surface.

■ Near-Field Microscopy

Microscopy with light through the use of lenses is limited in resolution by diffraction, as described in Chapter 6, and cannot be used to obtain an image of individual macromolecules. The use of light by these familiar techniques is sometimes referred to as "far-field" microscopy. In contrast, if light is used as a small probe for scanning microscopy akin to the approach used for atomic force microscopy described above, the diffraction limitations to resolution do not apply. The necessary features of the system are as follows: (1) the probe must be maintained at a distance from the specimen that is less than the wavelength of the light, (2) the light must be directed through a probe that is no larger in diameter than the resolution desired—usually tens of nanometers, and (3) the small amount of light used in the probe must be capable of being detected. Imaging under these conditions is referred to as "near-field" microscopy, and features of the instrumental design are diagrammed in **Figures 15.13A** and **15.13B.**

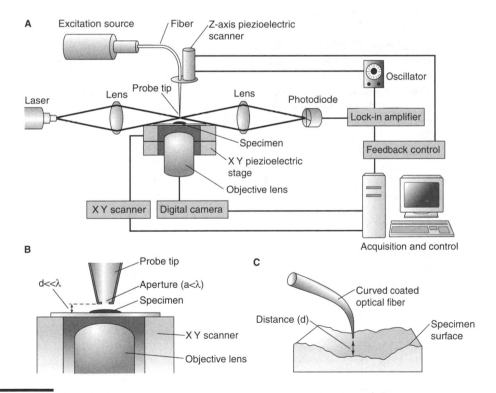

Figure 15.13
Design of a scanning near-field optical microscope. (A) Major components of the instrument include a laser light source coupled to a microscopic delivery tip, a piezoelectric device for rastering the probe tip, a laser–photodiode system for detecting tip position and an objective and detector for sensing specimen–light interactions. (B) The probe tip delivering the light must be maintained at a distance d << λ, with d typically being about 20 nm, and the probe tip diameter a < λ, typically about 50 nm. (C) The distance d is sensed by piezoelectric oscillation of the probe, with its amplitude and phase of oscillation (as detected by a laser beam reflection) being used to determine the sheer forces exerted on the probe by the specimen surface. These forces are very sensitive to d and guide probe movement in the z-axis so as to maintain d constant and small.

A high-intensity beam of light from a laser is brought in through a bent glass light pipe or quartz crystal that ends in a narrow tip. The probe is coated with aluminum to prevent escape of light except at the very tip. The distance between the tip and the specimen is maintained at a critical distance by oscillation of the tip by a piezoelectric device to detect shear forces near the specimen surface and the oscillation detected by a laser beam reflecting off the tip onto a quadrant photo diode detector much as cantilever position in an AFM is sensed (Figure 15.13C). The amplitude and phase of the tip are used to direct feedback control of probe position in the z-axis. The evanescent field set up due to interaction of the light with the specimen is gathered at a close range by an objective lens and focused onto a CCD chip or the front window of a photomultiplier tube. The scanned image produced by computer synchronization of the amplitude data with the scanning commands can have a resolution of 20 nm or less if the probe-specimen distance is carefully controlled. Recently, even better resolution has been achieved by guidance of the light beam using solid, finely tapered probes rather than the tubular guidance system described above.

References and Suggested Reading

General

McIntosh JR, ed. *Cellular Electron Microscopy*. New York: Academic Press, 2007, p. 880.

Sali A, Glaeser R, Earnest T, Baumeister W. From words to literature in structural proteomics. *Nature* 2003;422:216–225.

Quick Freezing, Deep Etching, and Rotary Shadowing

Chandler DE, Heuser, J. The vitelline layer of the sea urchin egg and its modification during fertilization: a freeze-fracture study using quick-freezing and deep-etching. *J Cell Biol*. 1980;84:618–632.

Heuser, JE. Development of the quick-freeze, deep-etch, rotary-replication technique of sample preparation for 3-D electron microscopy. *Prog Clin Biol Res*. 1989;295:71–83.

Heuser JE. Procedure for freeze-drying molecules adsorbed to mica flakes. *J Mol Biol*. 1983;169:155–195.

Heuser JE. Protocol for 3-D visualization of molecules on mica via the quick-freeze, deep-etch technique. *J Elect Microsc Tech*. 1989;13:244–263.

Hirokawa N, Heuser, JE Quick-freeze, deep-etch visualization of the cytoskeleton beneath surface differentiations of intestinal epithelial cells. *J Cell Biol*. 1981;91(2 Pt 1):399–409.

Mozingo NM, Somers CE, Chandler DE. Ultrastructure of the proteoliasin–ovoperoxidase complex and its spatial organization within the *Strongylocentrotus purpuratus* fertilization envelope. *J Cell Sci*. 1994;107(Pt 10):2769–2777.

Negative Staining

Bremer A, Henn C, Engel A, Baumeister W, Aebi U. Has negative staining still a place in biomacromolecular electron microscopy? *Ultramicroscopy* 1992;46:85–111.

Harris JR. *Negative staining and cryoelectron microscopy (Royal Society Microscopy Handbooks)*. Oxfordshire, England: BIOS Scientific, 1997.

Harris JR. Negative staining of thinly spread biological particulates. *Methods Mol Biol*. 1999;117:13–30.

Harris JR, Scheffler D. Routine preparation of air-dried negatively stained and unstained specimens on holey carbon support films: a review of applications. *Micron*. 2002;33:461–480.

Kiselev NA, Sherman MB, Tsuprun VL. Negative staining of proteins. *Electron Microsc Rev*. 1990;3:43–72.

Ohi M, Li Y, Cheng Y, Walz T. Negative staining and image classification—powerful tools in modern electron microscopy. *Biol Proc Online* 2004;6:23–34.

Zhao FQ, Craig R. Capturing time-resolved changes in molecular structure by negative staining. *J Struct Biol*. 2003;141:43–52.

Cryoelectron Microscopy and Single Particle Analysis

Chiu W, Baker ML, Jiang W, Dougherty M, Schmid MF. Electron cryomicroscopy of biological machines at subnanometer resolution. *Structure* 2005;13:363–372.

Frank J. Single-particle imaging of macromolecules by cryo-electron microscopy. *Annu Rev Biophys Biomol Struct.* 2002;31:303–319.

Frank J. *Three-Dimensional Electron Microscopy of Macromolecular Assemblies: Visualization of Biological Molecules in Their Native State.* Oxford: Oxford University Press, 2006, p. 432.

Frank J, Zhu J, Penczek P, Li Y, Srivastava S, Verschoor A, Radermacher M, Grassucci R, Lata RK, Agrawal RK. A model of protein synthesis based on cryo-electron microscopy of the E. coli ribosome. *Nature* 1995;376:441–444.

Frank J, Penczek P, Grassucci R, Srivastava S. Three-dimensional reconstruction of the 70S *Escherichia coli* ribosome in ice: distribution of ribosomal RNA. *J Cell Biol.* 1991;115:597–605.

Ren G, Reddy VS, Cheng A, Melnyk P, Mitra AK. Visualization of a water-selective pore by electron crystallography in vitreous ice. *Proc Natl Acad Sci USA* 2001;98:1398–1403.

Roos N, Morgan AJ. *Cryopreparation of Thin Biological Specimens for Electron Microscopy: Methods and Applications (Royal Society Microscopy Handbook).* Oxford: Oxford University Press, 1990.

Stark H, Mueller F, Orlova EV, Schatz M, Dube P, Erdemir T, Zemlin F, Brimacombe R, van Heel M. The 70S *Escherichia coli* ribosome at 23 angstrom resolution: fitting the ribosomal RNA. *Structure* 1995;3:815–821.

Thuman-Commike PA. Single particle macromolecular structure determination via electron microscopy. *FEBS Lett.* 2001;505:199–205.

Unger VM. Electron cryomicroscopy methods. *Curr Opin Struct Biol.* 2001;11:548–554.

Yonekura K, Maki-Yonekura S, Namba K. Building the atomic model for the bacterial flagellar filament by electron cryomicroscopy and image analysis. *Structure* 2005;13:407–412.

Electron Tomography

Baumeister W. Electron tomography: towards visualizing the molecular organization of the cytoplasm. *Curr Opin Struct Biol.* 2002;12:679–684.

Frangakis AS, Forster F. Computational exploration of structural information from cryo-electron tomograms. *Curr Opin Struct Biol.* 2004;14:325–331.

Frank J. *Electron Tomography: Methods for Three-Dimensional Visualization of Structures in the Cell,* 2nd ed. New York: Springer, 2006, p. 466.

Frank J, Wagenknecht T, McEwen BF, Marko M, Hsieh C-E, Mannella CA. Three-dimensional imaging of biological complexity. *J Struct Biol.* 2002;138:85–91.

Medalia O, Weber I, Frangakis AS, Nicastro D, Gerisch G, Baumeister W. Macromolecular architecture in eukaryotic cells visualized by cryoelectron tomography. *Science* 2002;298:1209–1213.

Ress DB, Harlow ML, Marshall RM, McMahan UJ. Methods for generating high-resolution structural models from electron microscope tomography data. *Structure* 2004;12:1763–1774.

Stevens AC, Aebi U. The next ice age: cryo-electron tomography of intact cells. *Trends Cell Biol.* 2003;13:107–110.

X-Ray and Electron Diffraction

Champness PE. Electron Diffraction in the Transmission Electron Microscope (Royal Society Microscopy Handbook). Oxfordshire, England: BIOS Scientific, 2001.

Glaeser R, Chiu W, Frank J, DeRosier D. *Electron Crystallography of Biological Macromolecules.* Oxford: Oxford University Press, 2007, p. 448.

Grayor-Wolf S, Nogales E, Kikkawa M, Gratzinger D, Hirokawa N, Downing KH. Interpreting a medium-resolution model of tubulin: comparison of zinc-sheet and microtubule structure. *J Mol Biol.* 1996;262:485–501.

Hammond C. *Introduction to Crystallography (Royal Society Microscopy Handbooks).* Oxford: Oxford University Press, 1992.

Lowe J, Li H, Downing KH, Nogales E. Refined structure of $\alpha\beta$-tubulin at 3.5 angstroms resolution. *J Mol Biol.* 2001;313:1045–1057.

Nogales E, Grayer-Wolf S, Khan IA, Luduena RF, Downing KH. Structure of tubulin at 6.5 angstroms and location of the taxol-binding site. *Nature* 1995;375:424–427.

Pickworth-Glusker J, Trueblood KN. *Crystal Structure Analysis: A Primer.* Oxford: Oxford University Press, 1985.

Protein Data Bank website. The nature of 3-D structural data. www.rcsb.org/pdb/experimental_methods.

Pusey ML, Liu ZJ, Tempel W. Life in the fast lane for protein crystallization and X-ray crystallography. *Prog Biophys Mol Biol.* 2005;88:359–386.

Rhodes G. *Crystallography Made Crystal Clear,* 2nd ed. New York: Academic Press (Elsevier), 1999.

Rossman MG, Morais MC, Leiman PG, Zhang W. Combining X-ray crystallography and electron microscopy. *Structure* 2005;13:355–362.

Vainshtein BK. *Modern Crystallography 1: Fundamentals of Crystals. Symmetry, and Methods of Structural Crystallography (Modern Crystallography),* 2nd ed. New York: Springer, 2003, p. 482.

Watt IM. *The Principles and Practice of Electron Microscopy,* 2nd ed. Cambridge: Cambridge University Press, 1997, pp. 263–274.

Atomic Force Microscopy

Bonnell D, ed. *Scanning Probe Microscopy and Spectroscopy, Theory, Techniques and Applications,* 2nd ed. New York: Wiley-VCH, 2001.

Braga PC, Ricci D, eds. *Atomic Force Microscopy: Biomedical Methods and Applications (Methods in Molecular Biology).* Philadelphia: Humana Press, 2003.

El Kirat K, Burton I, Dupres V, Dufrene YF. Sample preparation procedures for biological atomic force microscopy. *J Microsc.* 2005;218(Pt 3):199–207.

Hansma HG, Kasuya K, Oroudjev E. Atomic force microscopy imaging and pulling of nucleic acids. *Curr Opin Struct Biol.* 2004;14:380–385.

Jena BP, Hörber JKH, eds. *Force Microscopy: Applications in Biology and Medicine.* New York: Wiley-Liss, 2006, p. 300.

Jena BP, Horber JKH, eds. *Atomic Force Microscopy in Cell Biology (Methods in Cell Biology, Volume 68).* New York: Academic Press, 2002:300.

Lushnikov AY, Bogdanov A, Lyubchenko YL. DNA recombination: holliday junctions dynamics and branch migration. *J Biol Chem.* 2003;278:43130–43134.

Silva LP. Imaging proteins with atomic force microscopy: an overview. *Curr Protein Pept Sci.* 2005;6:387–395.

Scanning Near Field Microscopy

Courjon D. *Near Field Microscopy and Near Field Optics.* London: Imperial College Press, 2003, p. 340.

Edidin M. Near-field scanning optical microscopy, a siren call to biology. *Traffic* 2001;2:797–803.

Garcia-Parajo M, Veeman JA, Ruiter A, Van Hulst N. Near-field optical and shear-force microscopy of single fluorophores and DNA molecules. *Ultramicroscopy* 1998;73:331–319.

Lewis A, Radko A, Ben-Ami N, Palanker D, Lieberman K. Near-field scanning optical microscopy in cell biology. *Trends Cell Biol.* 1999;9:70–72.

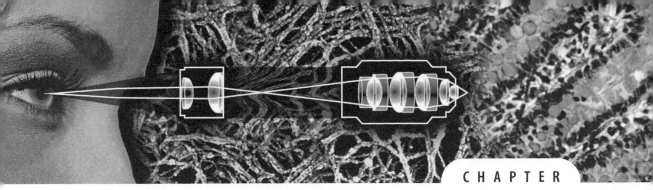

Image Processing and Presentation

Although image content is of primary scientific importance, its processing and presentation to the reader/viewer are key steps in providing information that is both accurate and compelling. It is probably best said that the *right* image is worth a thousand words. Presenting the right image begins with choosing the right specimen and microscopy technique, executing the microscopy in an optimal manner, editing/processing the image in an appropriate manner, and finally presenting high-quality images in either photographic or electronic format. Common forms of presentation include electronic journal articles, projected slides or electronic files, large format posters, and single copy wall illustrations. This chapter summarizes the most important considerations in producing high-quality images for a scientific audience.

■ Start with the Right Specimen and Microscopy Technique

Do the processes or structures to be studied require the high resolution of electron microscopy, or can they be best seen at lower resolution but with a wider field of view in light microscopy? The use of both light microscopy (LM) and electron microscopy (EM) is desirable in many biological problems so that findings at each level of resolution can be correlated with one another. It is usually best to start with light microscopy, which provides an overview of the specimen structure and aids in evaluating whether the biological specimen is being prepared in an appropriate manner. An important consideration is whether the right tissues or cells are present, their location in the specimen, and whether they are at a high enough density to make their study feasible. For example, if the cell type to be visualized occurs only once every 100 cells and is deep within the tissue specimen, one can predict that many tissue specimens will be required to produce a robust study at the light level and that at the EM level one may find the desired cell very rarely, if at all.

Does the process to be studied require visualization of live cells? If so, one is likely to be restricted to using various forms of light microscopy in conjunction with specimen stages that provide the proper temperature, buffer, and oxygenation (see Chapter 14). Clearly, the light microscopy technique used must be best suited to the purpose, for example, video microscopy to study cell motion (intracellular or extracellular), fluorescence microscopy to locate individual organelles or macromolecules, or laser scanning confocal microscopy to study the three-dimensional structure of cells or tissues.

What are the best contrasting optics to use for live cells? Phase contrast, differential interference contrast, Hoffman interference contrast, and staining with vital dyes can each be the best choice, depending on the specimen used and the objectives of the study. Phase contrast, for example, has a greater depth of focus and might best be used for larger or moving cells so as to have as much of the specimen in focus as possible. Differential interference contrast has a narrow depth of focus and might best be used on cell monolayers or for visualization of intracellular organelles. Vital staining may allow the dynamics of a specific organelle to be followed in real time.

If working with stabilized cells, what is the best method of fixation and tissue preparation? If excellent structural preservation is required, one should use a cross-linking chemical fixative such as glutaraldehyde; if antigenicity or enzyme activity is required for immunocytochemistry or histochemistry, then a non–cross-linking fixative such as formaldehyde is needed. Are the advantages of rapidly frozen specimens, such as stopping biological actions fast or avoidance of fixation artifacts, of paramount importance? In that case, the considerable labor of doing rapid freezing work must be considered.

Are specialized techniques in either light or electron microscopy to be used? The use of specialized techniques is best carried out in conjunction with or after study of the specimen with standard fixation or live cell-imaging methods. The principle here is to use the least time-consuming preparative techniques first so one can become familiar with the normal structure of the specimen and to allow initial decisions and mistakes to be made in as an efficient manner as possible. For example, one would be ill-advised to carry out immunocytochemistry on a specimen whose structure was not known through standard techniques designed to optimize structural preservation. Likewise, one would have a very difficult time to interpret correctly freeze-fracture replicas from a tissue that had not been studied by standard light and electron microscopic techniques.

In summary, the most scientifically informative images come from studies in which the tissue preparation and microscopic methods have been optimized by a considerable attention to preliminary experiments. As in most scientific experimentation, bioimaging usually involves at least as much or more time spent in developing the specimens and methods as in obtaining and processing presentable data. Thus, any advantage that can be gained from prior literature describing tissue preparation techniques and the normal structure of the specimen is to be sought.

■ Getting the Most Out of Your Microscopy

The most agonizing experience for a scientist is to collect large amounts of data only to find that it was collected under the wrong conditions or is of low quality because the instrumentation was not set up properly. Bioimaging is no exception. In this section, we give advice on how to work efficiently and how to avoid a disaster like that just mentioned.

1. Obtain the best possible training on how to use the instrumentation from an experienced professional. Reading books can be helpful, but this is no substitute for seeking the tutelage of an experienced person. Would you expect to drive an automobile by reading a book? Although training can be informal, as from a fellow student, formal training with a bioimaging laboratory manager either through ad hoc sessions or hands-on course work is highly recommended.

2. The microscope and quality of your work are best served by using an instrument startup and shutdown manual. If one is not available, then we suggest that you write a brief list of steps for this purpose and make this list available to yourself

whenever using the microscope. The startup procedure *must* include all procedures required to optimize the image, such as adjusting for Köhler illumination or phase ring alignment on a light microscope or adjusting the aperture alignment and astigmatism on an electron microscope.

3. Image capture is best standardized. First, all operating parameters of the camera must be optimized for *your* standard work and clearly written in *your* laboratory notebook. Some studies such as immunocytochemistry or morphometrics demand such standardization as a matter of procedure. For example, to be comparable, fluorescent micrographs must have been taken with the same optical components, light intensity, exposure time, and detector sensitivity. For morphometrics, micrographs must be taken at the same magnification, and fields must be chosen either randomly or in a set pattern to avoid bias.

4. Make every image count! Out-of-focus images can never be salvaged. Practice your ability to focus until you are very competent. Always check your focus on all of the images you produce to give you feedback on your ability. Never check focus by looking at a monitor screen or at film by eye—both methods are too low resolution to detect focus problems. Instead, magnify digital images electronically or view film negatives with a loupe to assess focal clarity.

5. Never hesitate to capture as many high-quality images as possible, whether they are on film or are digital! The time and cost you spend in preparing the tissue for microscopy far outweigh any cost in film or electronic media. After you have left a particular feature on the grid or slide you are viewing, you may not be able to locate it again. Thus, it is important to record excellent or unusual examples at first sight.

6. Be sure to document your images thoroughly in your laboratory notebook and in your electronic files (if applicable). The specimen preparation warrants recording in detail, and every image captured must be adequately documented for optical parameters, exposure, subject, and image identification number. Any image captured with nonstandard parameters needs these to be documented. Titles and storage locations for both electronic files and film negatives should be standardized for easy access. It is particularly important to record magnifications of each image and to calibrate the magnification scale on the same instrument producing your images.

Calibration on a light microscope can be carried out using a diffraction grating having precisely ruled lines whose interline distance can be determined spectroscopically. Calibration of an electron microscope can be carried out using replicas of diffraction gratings or monolayers of heavy-metal crystals of known dimensions. These calibrating specimens are easily purchased from microscopy supply houses.

■ Postproduction Analysis and Optimization of Images

Typically, a microscopist will capture hundreds or thousands of images for each image presented or published to the scientific public. Each image presented must then be chosen carefully to represent the richness and breadth of the microscopist's experience with the specimen. The first step (unless a systematic approach is being used for morphometrics) is to review "working" images to cull out those that do not clearly show the point to be illustrated or have obvious preparation artifacts. Second, the best images are cropped to remove extraneous information that distracts the viewer without adding information. Finally, the resulting cropped images are adjusted for optimal performance from both an informational and aesthetic point of view.

Adjustments are carried out in either a program such as Photoshop for electronic images or the darkroom during printing of photographs. Common adjustments are for brightness (density), contrast, and gamma. Digital images can be further processed in specialized manners not possible in analog photographic printing.

Digital images (and analog photographic images that have been scanned) can be expressed as numerical intensity values for each picture element (pixel). The numerical data for the entire collection of pixels are stored in a spreadsheet referred to as a "look-up table" (LUT). Information for each pixel is in the form of 8 bits (1 byte) (consumer grade video and still cameras), 10 or 12 bits (high end still cameras), or 16 bits (2 bytes) for scanning devices and some imaging software (e.g., MetaMorph and PhotoShop). The common format of 8 bits allows information for location and for 256 gray intensity levels, while 12 bits allow up to 4096 gray levels. Color information is coded simply as three 8- or 16-bit channels, that is, one channel and one LUT for each of the three primary colors (see Chapters 2 and 5). These formats are referred to as 24- and 48-bit color, respectively.

All image-processing operations perform arithmetic operations on LUT data. Some of the common operations are listed in **Table 16.1**. Although some programs routinely retain a raw, unchanged version of all LUTs (e.g., Scanlytics), some programs such as Adobe Photoshop do not. After a LUT is changed, the original data are gone forever. It is extremely important, therefore, that the original image file be kept unaltered in a non-degradable file format such as TIFF and that only copies of the original file be used for image processing.

To demonstrate the effects of common processing operations, we use as a test image a fluorescence micrograph of an isolated hepatocyte stained with the antibiotic chlortetracycline to visualize the numerous mitochondria found in these cells (inset, **Figure 16.1A**). The LUT data for this image are most easily expressed as a population of values referred to as an *image histogram*. As shown in Figure 16.1A, this histogram plots the number of pixels displaying a certain gray-level intensity (y-axis) for all the gray levels in order (x-axis). For an 8-bit channel, the plot starts with 0 (black) and ends at 255 (white). The histogram data for the hepatocyte image show that the dark background is represented by a peak at gray levels 0 to 25 and the fluorescence signal by a broad range of gray levels from 30 to about 200, and the brightest highlights (mitochondria in this case) represented by gray levels between 200 and 255. This image has been adjusted for optimal brightness and contrast, as shown by the fact that it contains pixel intensities all of the way from black to white, with a broadly distributed range of gray levels in between.

LUT values can be changed by any operation whether it be the initial capture of the image (by film, a solid state chip, or our eyes), the processing of the image by analog or digital techniques, or the storage or transmission of the image. The changes made in the LUT can be expressed by a plot in which the gray levels in the input image are plotted on the x-axis and the corresponding gray levels in the output image after the operation are plotted on the y-axis. Thus, if the operation made no changes in gray levels, the output image would be identical to the input image, and this process would be represented by a 45-degree line (Figure 16.1B). Changes in image brightness would be represented by a shift in this line toward either lower gray levels resulting in a darker image or higher gray levels resulting in a lighter image. There would be no change in slope. On the other hand, changes in image contrast would result in changes in slope; a gentler slope would indicate a reduced contrast, while a steeper slope would indicate higher contrast in the output image compared with the input image. Clearly, changes in brightness (y intercept) can be made independently of changes in contrast (slope). For example, the identity line might represent the capture of light by a solid-state photodiode chip because such chips provide an electron signal that is linearly proportional to the

			Frequency
Operation	**Effect or Use**	**LUT Changes**	**of Use**
Frame averaging	Removes noise and other random fluctuations	Pixel intensity is an average of pixel intensities at the same location in multiple input images	Common
Frame subtraction	Removes undesirable background features; exhibits changes between frames	Intensity is the difference in pixel intensity between two images at the same location	Common
Brightness change	Makes image lighter or darker	All pixel intensities are increased or decreased by a constant amount	Common
Contrast change	Increases or decreases the range of gray levels in the image; image is "snappier" with more contrast and more uniform with lower contrast	All pixel intensities are changed according to a linear equation; the slope of the equation determines the level of contrast	Common
Gamma change	Gamma increase emphasizes details in bright areas; gamma decrease emphasizes details in darker areas	All pixel intensities are changed according to an exponential equation; the exponent (gamma) determines which part of the gray level scale is expanded	Common
Color hue and range adjustment	Colors shifted in spectrum to compensate for instrument, user or printing bias	Pixel intensities in each primary color channel are varied independently to emphasize or de-emphasize certain hues	Uncommon
Pseudo-coloring	Ranges in gray level are expressed in colors; emphasizes differences in pixel intensity that correspond to physiological or geometric parameters	Each pixel is assigned a color based on its intensity and on a separate pseudocoloring LUT	Common
Edge sharpening	Increases contrast at edges for emphasis	Each pixel intensity is compared with those of adjacent or neighboring pixels and changed to emphasize differences	Uncommon
Blurring	Decreases contrast at edges to "soften" them	Each pixel intensity is compared with those of adjacent or neighboring pixels and changed to de-emphasize differences	Rare

Table 16.1 Common Image Processing Operations

number of photons captured. Lines representing changes in brightness or contrast might represent either digital image processing operations or photographic darkroom techniques using the right combination of emulsion, developer, exposure time, and developing time (see Chapter 2).

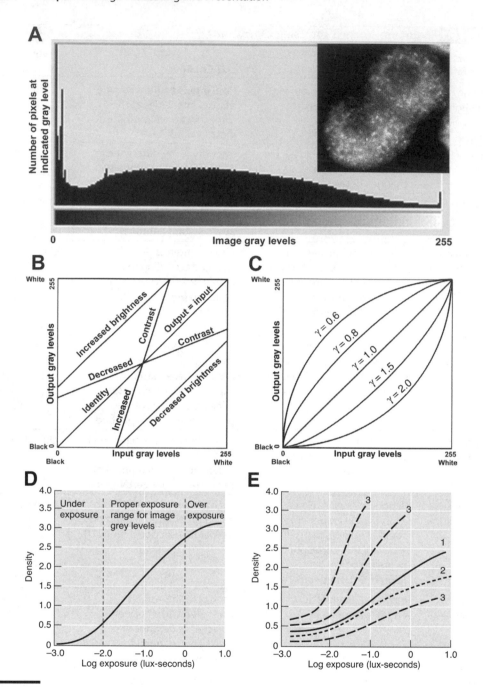

Figure 16.1

The image histogram and the effect of image-processing operations on the lookup table. (A) Histogram of the gray-level frequencies in the inset image of chlortetracycline-loaded hepatocytes. The bright fluorescent spots in the image correspond to the substantial number of mitochondria in these cells. (B) The relationship between pixel intensities in an input image to those in an output image during changes in brightness and contrast. Each axis represents gray levels that vary from black (0) to white (255) in an 8-bit monochrome image. Changes in brightness shift the identity line right or left, while changes in contrast alter the slope of the line. (C) Exponential relationships in the input–output pixel intensities, as determined by Equation 16.2. If the exponent gamma is below 1, dark areas of the image become richer in the range of intensities; if gamma is above 1, bright areas of the image become richer in the range of intensities. (D) Characteristic curve for the exposure–density relationship of a photographic film. The curve is a logarithmic relationship and the slope of the curve in the linear range is equal to the gamma of the film. (E) A comparison of characteristic curves for selected black and white films. 1 = Kodak T-Max 100, 2 = Ilford XP-2 Super, 3 = Kodak Tech Pan processed with three different developers.

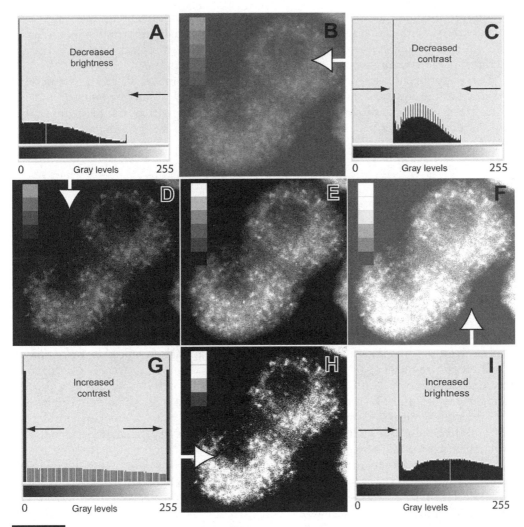

Figure 16.2
Effect of changes in brightness and contrast on the image, standard gray level tiles, and the image histogram.
(A and D) Decrease in brightness is accompanied by a shift of the histogram toward low numbered gray levels
(arrow) and a darker image. (B and C) The decrease in contrast is accompanied by a compression of the histogram
toward midrange gray levels (arrows) and a "graying" of the image. (E) Starting image of two chlortetracycline
loaded hepatocytes well adjusted for brightness and contrast. The histogram of this image is shown in Figure 16.1
(F and I). Increased brightness is accompanied by a shift of the histogram toward higher numbered gray levels
(arrow) and a bright image. (G and H) The increased contrast is accompanied by expansion of the histogram toward
low and high number gray levels (arrows) and an image that emphasizes bright objects on a dark background.

Effects of changes in brightness and contrast on both the image and the image histogram are shown in **Figure 16.2**. As brightness is decreased, the image histogram is compressed toward low-numbered gray levels (arrow, Figure 16.2A), and there is a buildup in the number of black pixels. Because this process has converted a range of gray levels (the darkest ones) to pure black, there is a permanent loss of information in the darker areas of the image (Figure 16.2D). As brightness is increased, the image histogram is compressed toward high numbered gray levels (arrow, Figure 16.2I), and there is a buildup in the number of pure white pixels and a permanent loss of information in the brighter areas of the image (Figure 16.2F). Decreased contrast compresses the image histogram from both extremes (arrows, Figure 16.2C) into the middle gray levels, resulting in a "grayer" image (Figure 15.2B). Increased contrast expands the image histogram over a larger range of gray levels (arrows, Figure 16.2G), potentially resulting in a buildup of both white and black pixels in the image (Figure 16.2H). Again, both of these processes can result in permanent loss of information if carried too far.

These routine adjustments are linear (i.e., the changes in the LUT can be described by a straight line equation); however, other changes in the LUT can be nonlinear and can be described only by curved relationships that change one part of the histogram more dramatically than other parts. The typical equation used to describe the relationship of output values to input values is as follows:

$$\text{output value} = [\text{input value} - \text{max/min} - \text{max}]^{\gamma} \tag{16.1}$$

where min and max are minimum and maximum gray levels (e.g., 0 and 255) and γ is an exponent. This equation can be also expressed as follows:

$$\text{normalized output values} = [\text{normalized input values}]^{\gamma} \tag{16.2}$$

If $\gamma = 1.0$, the equation is the linear relationship that we discussed previously (Figure 16.1C). On the other hand, if $\gamma > 1.0$, the equation is nonlinear with high numbered gray levels (near white values) being selectively spread out over a much greater number of gray levels in the output image, while the darker areas of the image get short changed. Conversely, if $\gamma < 1.0$, details in darker areas of the image are enhanced by being spread over a greater range of gray values. Changes in γ are nominally reversible, as demonstrated in Equation 16.2. An image process having a $\gamma = 3$ can be reversed by a process having a $\gamma = 1/3 = 0.33$. In practice many image processes (e.g., changes in brightness, contrast, and gamma) are reversible only if there is an infinite number of gray levels; short of that there is usually some information loss and image degradation due to repeated expansion and compression of the histogram.

The effect of γ on the image and image histogram is shown in **Figure 16.3.** If one starts with an image that has a histogram compressed in the darker pixels (Figures 16.3A and 16.3B), low gamma processes can expand the histogram in this region over a large range of gray levels, thus providing much more detail in these regions (Figures 16.3C and 16.3D). For example, as one progresses from B to D to F, one sees increasing detail over the nuclear regions of these two cells. If one starts with an image that has a histogram compressed in the brighter pixels (Figures 16.3E and 16.3F), a series of high gamma processes can expand the histogram in this region, providing much more gray-level detail in the bright regions.

The term γ is also used to describe input/output relationships in light detecting devices. CCD chips in solid-state cameras have a γ close to 1.0 (i.e., the output signal compared with light input is remarkably linear). On the other hand, photographic films can have a wide range of γ values that are measured by the slope of a "characteristic curve" resulting from plotting the film density versus exposure (Figure 16.1D). This slope is dependent on both the film characteristics and on the developer and developing time used, as illustrated in Figure 16.1E.

Our eyes have a $\gamma = 0.7$, demonstrating that our visual system has the ability to bring out faint objects in low light areas of the visual field while keeping the bright areas from glaring at us. This helps us detect details over a wide range of light intensities in the same field. Indeed, a $\gamma < 1$ could be very useful in the detection or illustration of both faint and bright emitting objects in the same fluorescence micrograph; however, too large of a gamma can result in a loss of information.

The image-processing routines discussed to this point have been global (i.e., they are likely to affect every pixel within the image). We now discuss processes that are not

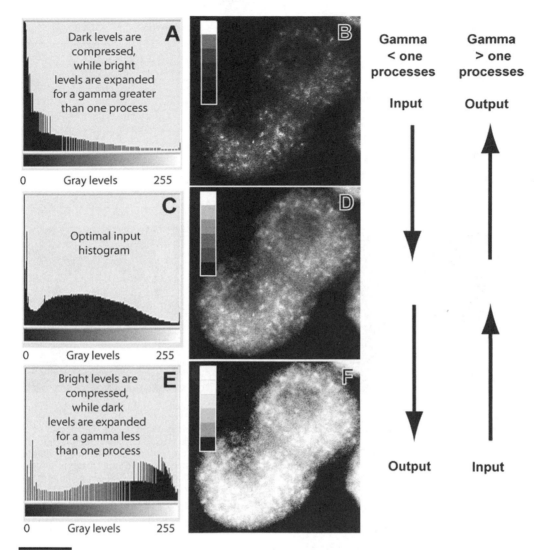

Figure 16.3
The effect of gamma on the image, standard gray level tiles, and image histogram. (A) An image histogram that has been shifted by a high gamma operation to darker gray levels while expanding the range of gray levels in bright regions of the image. (B) The image corresponding to the histogram in (A). The standard gray-level tiles have shifted to darker values. (C and D) Image histogram and image before application of gamma operations. (E) Image histogram that has been shifted by a low gamma operation to brighter gray levels while expanding the range of gray levels in darker regions of the image. (F) Image corresponding to the histogram in (E). The standard gray level tiles have shifted to lighter values.

global because they affect only those pixels at certain image features or pixels that show a large variation from one image to the next. Examples of a few common nonglobal processes are included in Table 16.1.

Most nonglobal processes are aimed at better detection of subtle details or in elimination of instrumentation or process artifact. Examples of the latter include blurring to eliminate noise or screening or segmentation effects. Blurring can be accomplished as illustrated in **Figure 16.4A,** in which each pixel intensity is converted to an average of itself with the surrounding eight pixels, thereby making each pixel intensity more similar to

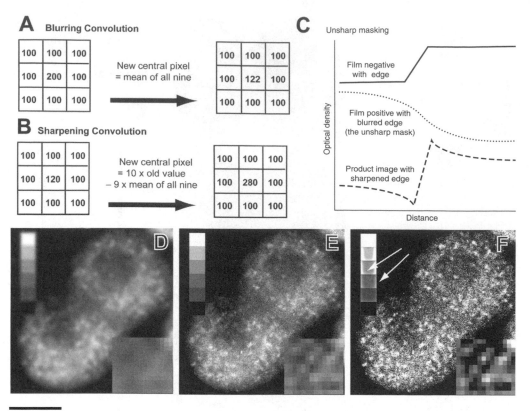

Figure 16.4

Blurring and sharpening operations. (A) Effect of a blurring operation using intensity averaging. The intensity of the central pixel becomes more like that of surrounding pixels in the output image. (B) Effect of a sharpening operation using intensity averaging. The difference between the intensity of the central pixel and its surrounding pixels is increased. (C) Line scans of an image showing the increase in pixel intensity as the scan crosses an edge. Unsharp masking enhances the change in pixel intensities at the edge (fine dashed line) when compared with the original image (solid line). (D) The image in (E) has been blurred by a Gaussian averaging operation (A). (E) Input image of hepatocytes with fluorescent mitochondria. (F) Image in (E) that has been sharpened by the unsharp mask process illustrated in (C). (D–F) Insets at the lower right show a small area of each image magnified to show pixels. The pattern of pixel intensity is enhanced in (F) but is obscured in (D).

surrounding intensities. Alternatively, the intensity of each pixel can be determined by averaging with pixel intensities any number of pixels away weighted according to a Gaussian distribution as a function of distance. Gaussian blurring is illustrated in Figure 16.4D; the gray level tiles in the inset have edges that are smoothly blurred with the surroundings over a number of pixels. Both are examples of a "convolution" operation in which each pixel intensity is altered based on other pixels that may lie along either the x-, y-, or z-axis (or a combination thereof) relative to the pixel being operated on.

Other processes that are effective for eliminating noise (generally white or black pixels) include median blurring (each pixel intensity converted to the median of the nine pixel group) and the averaging of frames. Median blurring is particularly effective for removing pixel defects at specific locations, while frame averaging is best used for the elimination of random noise. Also important is the tool of background subtraction in which a frame without specimen is subtracted from a frame with specimen, thereby eliminating uneven background illumination, dust or scratches in the optics, and other unwanted image features that are not a part of the specimen.

Operations aimed at detection of image details include edge sharpening and edge masking in which neighboring pixels that show differences in intensity are selectively altered so as to emphasize or de-emphasize these differences. For example, Figure 16.4B illustrates an edge-sharpening routine in which the input intensity of each pixel is compared with the average intensity of all adjacent pixels and then modified to make its intensity difference from the surround even greater in the output than it was in the input. Another technique to sharpen edges, well known from analog image processing (photography), is unsharp masking. This process involves taking an in-focus image or negative and exactly superimposing it on the same image that has been blurred. If the two images are averaged, pixel by pixel (or printed from superimposed negatives), the resulting image will have enhanced edge features. As illustrated in the line scans of Figure 16.4C, an edge denoted by a change in pixel intensity (solid line), when averaged with the same edge blurred (dotted line), results in sharper edges (dashed line). The effects of unsharp masking can be seen in Figure 16.4F, wherein the fluorescence pattern of the hepatocytes is not only sharpened but also the edges of the gray-level tiles are flared with white over a number of pixels to enhance their edges (arrows).

Finally, the ability to process digital images by user-defined operations in programs such as Adobe Photoshop leads to almost limitless possibilities in how to manipulate image information. Examples of user-defined operations on an image LUT (illustrated in **Figure 16.5**)

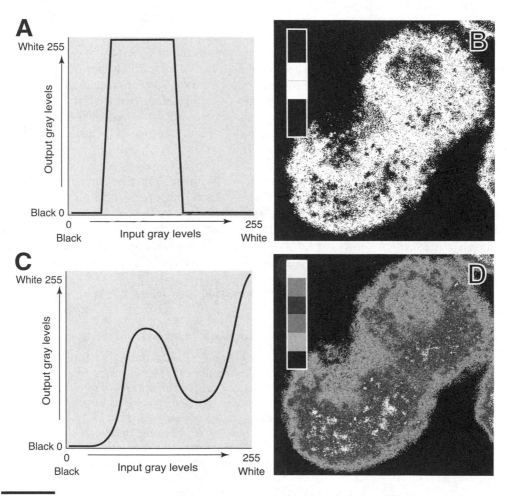

Figure 16.5
Examples of user-defined operations on the LUT. (A) Histogram of an operation that highlights pixels with gray levels between 51 and 150 in white against a black background. (B) Result of the operation in (A). (C) Histogram of a user-defined input–output relationship created in Adobe Photoshop. This operation is useful only for aesthetic purposes because multiple gray levels in the input image are mapped to the same gray level in the output image. (D) Result of the operation in (C).

include highlighting in white a specific pixel intensity range using a binary operation referred to as "thresholding" (Figure 16.5A and 16.5B) and using a user-determined curvilinear input–output relationship that assigns multiple input levels to the same output gray level (Figure 16.5C) to produce special image effects (Figure 16.5D).

■ When Is Image Processing Reasonable?

Some scientists say one should never process images and liken it to the falsification of data. This view is clearly due to a lack of information and understanding of what science can accomplish. First, virtually all scientific observations are made indirectly through the use of instruments that modify the data that our own senses alone would detect. Thus, image processing is just the last of many steps designed to detect and image structures that cannot otherwise be studied. Furthermore, even our own senses are not without their data-processing capabilities. If "seeing is believing," our own visual system has been demonstrated to use amplification, γ reduction, edge sharpening, motion detection, background subtraction, and frame averaging to enhance our visual perception!

Rather, it is a matter of when to use image processing and how to document the processing you have used. Because the goal of scientific publication is to archive data that stand the test of time (as opposed to archiving interpretations of data that do not always hold up), it is important that images presented in professional journals represent faithfully both the breadth of findings that the microscopist has made and the structural details that the methods used can achieve. Image processing should aim at revealing structural features that are reproducible, even if they cannot be easily observed in the absence of processing. Image processing is also legitimately used to remove systematic artifacts such as noise, to enhance pattern recognition (as in pseudocoloring), and even to achieve a more aesthetically pleasing image if the processing does not affect the actual perception or understanding of the structures visualized. On the other hand, image processing that obscures structural features or alters the nature of these features should not be used regardless of purpose. If additional information of value seems to be possible through image processing but its affect on the data is not entirely understood, it is important for the investigator to (1) show data before and after processing and (2) to use control and model specimens to understand better the effect of the image processing methods used.

A useful statement on ethics of image processing is provided by the *Journal of Cell Biology* in their instructions to authors:

> The following information must be provided about the acquisition and processing of images:
>
> 1. Make and model of microscope 2. Type, magnification, and numerical aperture of the objective lenses 3. Temperature 4. Imaging medium 5. Fluorochromes 6. Camera make and model 7. Acquisition software 8. Any subsequent software used for image processing, with details about types of operations involved (e.g., type of deconvolution, 3D reconstructions, surface or volume rendering, gamma adjustments, etc.).
>
> No specific feature within an image may be enhanced, obscured, moved, removed, or introduced. The grouping of images from different parts of the same gel, or from different gels, fields, or exposures must be made explicit by the arrangement of the figure (i.e., using dividing lines) and in the text of the figure legend. . . . Adjustments of brightness, contrast, or color balance are acceptable if they are applied to the whole image and as long as they do not obscure, eliminate, or misrepresent any information present in the original, including backgrounds. . . . Non-linear adjustments (e.g., changes to gamma settings) must be disclosed in the figure legend.

More liberal use of image processing would seem to be warranted in those instances in which the presentation is not intended to be for archival purposes, for example in seminar slides, poster presentations, or journal covers and wall decorations that have a largely aesthetic role to play.

■ Obtaining Quantitative Data from Images

Images have direct impact on our senses, allowing us to use the substantial pattern-recognition abilities of the human brain that are unlike those of any technology. By nature, our pattern-recognition routines are holistic and nonnumeric; however, additional quantitative information can be obtained from images, and increasing emphasis is being placed on this type of information in research publications. The most common types of quantitative information for structural features include relative counts, counts per unit distance or area, relative area occupied, size and shape distribution, relative volume occupied, and relative optical density. A more comprehensive list of such parameters is given in **Table 16.2**. Production and analysis of such numerical data are termed *stereology* or *morphometrics*.

Meaningful measurements require that input images be gathered in a systematic, unbiased manner at a consistent magnification and under standardized conditions. Before computer analysis of such images, grids of points or lines were superimposed on each image, and data were gathered systematically as to the organelle type present at each location (**Figure 16.6A**). Patterns frequently used include the square lattice (left), the parallel line test system (right), and the Mertz curvilinear test system (center). Recording of organelle identity at points on either a square lattice or parallel line test system allows calculation of the relative volume of each organelle. Recording of how many times a membrane crosses a line in the parallel line system allows calculation of the surface area and relative surface area of specialized cellular membranes. This same determination can be done with the Mertz curvilinear test system, which is required if the membranes are not randomly distributed but rather occur within set structures at nonrandom intervals—a good example being Golgi membranes. An alternative method for measuring image area for calculation of organellar volumes was to cut out these features and weigh the paper.

Today, morphometrics is all done by computer software with the help of a mouse. Programs such NIH Image and MetaMorph are capable of importing electronic images, automatically identifying features by shape, pixel intensity, and color, and then analyzing these features for size, perimeter length, numerical density, area, relative area, and implied volumes. Of course, one can also manually identify features to be analyzed by point-and-click routines. An example of such measurements is shown in Figures 16.6B and 16.6C using MetaMorph. The hepatocyte depicted contains numerous mitochondria. One can first "threshold" the image, thereby identifying all pixels that have an intensity that is above a certain gray level or within a certain gray level range. Because this process does a good but not perfect job of picking out the mitochondria, one next analyzes the selected objects by pixel count (i.e., area), resulting in the histogram shown in Figure 16.6D. The selection of the proper size range within the histogram via a set of "calipers" results in elimination of small pixel groups and highlighting a fairly "pure" mitochondrial "fraction" (white, Figure 16.6C). This fraction can now be analyzed to produce parameters such as area, relative area, perimeter, numerical density, and as extensions, relative, and absolute volumes.

Table 16.2 Common Parameters Calculated in Morphometric and Motion Analysis Studies		
Parameter	**Symbol**	**Units**
Still Image Analysis (Stereology)		
Volume of a structure	V	μm^3
Fractional volume	V_V	$\mu m^3/\mu m^3$
Mean volume of an element	v	μm^3
Area	A	μm^2
Fractional area	A_A	$\mu m^2/\mu m^2$
Mean area	A_A/N_A	μm^2
Length of a test line	L	μm
Linear fraction (length of a line on a feature per unit test line length)	L_L	$\mu m/\mu m$
Length per unit test area or per unit test volume	L_A, L_V	$\mu m/\mu m^2, \mu m/\mu m^3$
Surface area	S	μm^2
Surface area per unit test volume	S_V	$\mu m^2/\mu m^3$
Surface area to volume ratio	S_V	μm^{-1}
Number of test points	P	integer
Number of points on structure/total points	P_P	integer
Number of features	N	integer
Number of features per unit test area or per unit test volume	N_A, N_V	$\mu m^{-2}, \mu m^{-3}$
Profile diameter	d	μm
Mean diameter	D	μm
Section thickness	T	μm
Video Analysis (Motion Analysis, Two-Dimensional)		
Time between tracking points	t	sec
Change in distance along x- or y-axis between Nth-1 and Nth tracking points	X_N, Y_N	μm
Distance traveled between tracking points	D_N	μm
Mean distance between tracking points for a single entity	D	μm
Total linear distance traveled for a single entity	D_T	μm
Actual distance traveled	D_A	μm
Velocity along the x- or y-axis between tracking points	V_{XN}, V_{YN}	$\mu m/sec$
Forward velocity between tracking points	V_N	$\mu m/sec$
Average velocity for a single entity	V	$\mu m/sec$
Direction of travel relative to x-axis between tracking points	Θ_N	degrees
Change in direction of travel	Θ_N	degrees
Overall direction of travel	Θ_T	degrees

This table is constructed partially from data in Table 13.2, Bozolla and Russell.

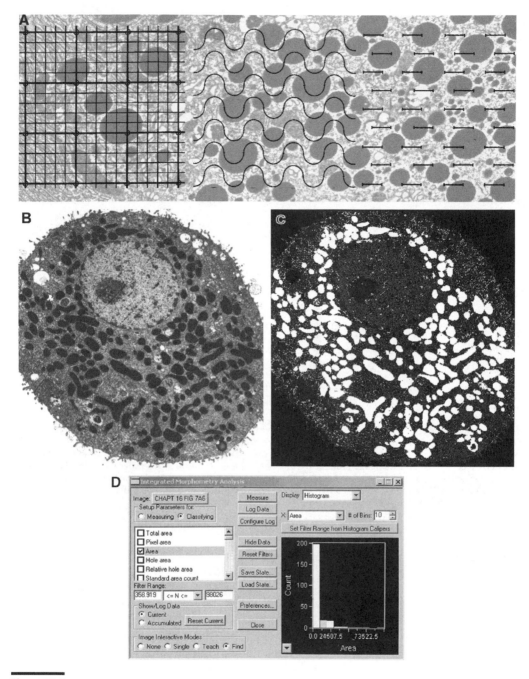

Figure 16.6

Examples of morphometric analysis. (A) Three different patterns used for morphometric analysis of micrographs: the square lattice (left), the parallel line test system (right), and the Mertz curvilinear test system (center). (B) Electron micrograph of an isolated hepatocyte showing a large number of darkly stained mitochondria. (C) Selection of the mitochondrial cross-sections in (B) using density and area classification algorithms in MetaMorph. (D) Histogram of the size distribution of mitochondria cross-sections in (C).

■ Motion Analysis

Video or time-lapse images can be analyzed for time-dependent changes in structural features such as movement of cells or organelles, growth or depletion of object classes, or the "morphing" of one compartment shape into another. This information can be

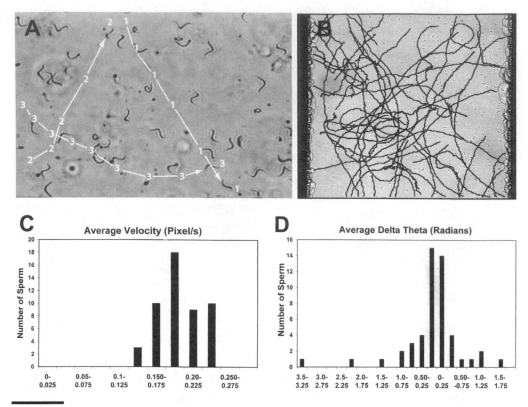

Figure 16.7
Examples of motion analysis. (A) Trajectories of three sperm cells superimposed on the last frame of a video sequence simulating output of a particle tracking feature of Metamorph software. Each data point represents an x, y, t data set. Based on an object recognition algorithm, the software calculates the geometrical center of user-defined objects in selected frames to produce the data sets. (B) Trajectories of 50 frog sperm displayed by the SpermTrak program using point-and-click data collection. (C) Histogram of average forward velocities (V) for the above 50 sperm. (D) Average change in turning angle (2) for 50 sperm. (C and D) were plotted in Microsoft Excel using data loaded from the SpermTrak program.

obtained manually using transparent acetate sheets that when properly overlaid using registration marks provide a clear indication of movement or shape changes between successive frames of a video or time-lapse sequence. This method gets the job done but is extremely labor intensive and not recommended. Rather, computer-aided motion analysis is the only efficient way to approach these data.

Software to automate cell or organelle identification and tracking is available commercially as "particle tracking" programs and, although somewhat expensive, provides a straightforward way to cut the labor involved at least an order of magnitude. An example of such a tracking program is an add-on feature of Metamorph, simulated output of which is used to illustrate the process in **Figure 16.7A.** Screen capture of three sperm cells combined with their trajectories recorded in previous frames is shown. The program first had to recognize the cells or a feature in the cell, usually of high contrast, by a characteristic pixel intensity pattern or "signature," a process referred to as "object recognition." In each frame, the software located this signature within a user-defined search range surrounding the position of the object in the previous frame. In this manner, objects to be tracked are recognized, their trajectories plotted, and x, y, t data sets automatically output to a spreadsheet all without operator input other than initiating the run.

Alternatively, a number of user-developed programs for motion analysis have been written that carry out motion analysis using point-and-click methods for data input. An example is the program "SpermTrak," written by Lindsey Burnett at Arizona State University. Figure 16.7B illustrates trajectories of frog sperm in a chemoattractant gradient obtained with this program. From x, y, t data sets, the program calculates and graphs a number of motion analysis parameters such as those listed in Table 16.2. The program can also plot histograms of the data obtained at a number of time points for a single cell or data obtained for a population of tracked cells, as shown in Figures 16.7C and 16.7D. The first plot shows the range of average velocities in a population of 50 sperm; the second shows the distribution of changes in turning angle theta for the same 50 sperm.

■ Three-Dimensional Reconstructions

Three-dimensional information requires imaging a specimen at multiple planes or angles. To achieve this, the specimen itself must be three dimensional; a single thin section or a dried, negatively stained specimen will not provide such information. On the other hand, three-dimensional information can be obtained from many different types of specimens, as illustrated in **Table 16.3**. A live specimen, a series of sections cut successively from a tissue

Table 16.3	Methods Used for Three-Dimensional Computer-Aided Reconstruction and Volume Measurements		
Type of Specimen	**Method of Preparation**	**Imaging Technique**	**Postcapture Data Analysis**
Isolated organelles, viruses, bacteria	Oriented and resin embedded	Electron microscopic tomography	Images at different tilt angles processed to form model
Isolated organelles, viruses, bacteria	Rapidly frozen in random orientation	Cryoelectron microscopic tomography	Similar images grouped in classes; class images tested for orientations that construct model with least error
Cell monolayers and isolated cells	Live or fixed; no embedding	Differential interference contrast	Z-axis control of stage to produce optical serial sections with which to form model
Tissues	Live or fixed stained with fluorophores	Confocal microscopy	Z-axis control of stage to produce optical serial sections with which to form model
Tissues	Embedded sections physically cut in a serial manner	Transmission electron or light microscopy	Images of sequential sections placed in registration to form model
Tissues	Thick sections	High-voltage transmission electron microscopy	Images at different tilt angles processed to form model
Tissues	Freeze fracture and replication	Transmission electron microscopy	Replica tilted at two angles to provide stereo-paired views of replica surface

block, a single thick section encompassing a significant volume of the specimen, and a freeze-fracture replica are all examples of specimens from which three-dimensional information can be harvested and volumes calculated. Whole specimens (live or fixed) up to 300 µm thick are studied by laser-scanning confocal microscopy (see Chapter 12); images from serial sections can be manipulated by computer software to produce three-dimensional models and volume information. Freeze-fracture replicas and thick sections can be tilted at a variety of angles using a goniometer stage in a transmission electron microscope, and the sets of images captured analyzed by computer software and then by the investigator. Broadly speaking, the use of computer analysis to generate three-dimensional models from either physical or optical sections is termed "computed axial tomography," or CAT. Conceptually, this is the same type of analysis that is applied to sets of x-ray images obtained for medical diagnostic purposes and referred to as "CAT scans." Computed axial tomography has become an extremely important technique in contemporary bioimaging studies.

Presentation of three-dimensional models in either journals or a seminar carries with it its own set of requirements. Methods of presentation in hard copy are discussed in Chapter 12, while methods for presentation in a seminar are discussed in the next section.

■ Methods Used for Presentation of Images

Developing Figure Layouts

In our opinion, the figure layouts for a manuscript are best sketched first in cartoon form even before the text is written. Because little time is expended in this process, these cartoons are a good basis for discussion between authors concerning the subject matter, number, order and magnification of figures, the placement of figures on the page (logic, aesthetics, and economy of space are all considerations here), and any use of color or line drawings. At this stage, the arrangement is very fluid, offering no resistance to optimization. Cartooning has the additional advantage that one often identifies what controls need to be illustrated and what key experiments might be missing whose inclusion would produce a stronger manuscript. Cartooning can also form a basis for writing the text because it is easier to develop a story that flows if the figures being used for illustration also flow smoothly. The second step is to use a program such as Microsoft Power Point or Adobe Illustrator to make electronic formats that can be altered and printed out at any stage of manuscript production.

Journal Figures

Traditionally, glossy photographs produced in the darkroom were submitted for publication. The photographs were of good contrast and density and the same physical size as they would appear in the journal. The "publication prints" submitted would be labeled with "press-on" transfer symbols, would be "camera ready," and would be labeled on the back in pencil with identifying information. Reviewer's prints could be of reasonable but slightly lower quality, such as those produced from Polaroid, Type 55 negative film.

Today, virtually all journals require images to be submitted electronically whether they were originally obtained by analog or digital techniques. One should identify an appropriate journal and study the instructions for authors before starting preparation of any figures for publication. In most cases, electronic images in a tagged image file format (TIFF) and having at least 300 to 500 pixels per inch are required for halftone figures. This requirement is based on the fact that production of halftone figures requires the use of a screen having 80 to 150 dots per inch; any electronic file that has less than 300 pixels per inch will produce disturbing moiré patterns that will be seen superimposed on the actual illustration. Figures to be reproduced as black and white line drawings (no gray levels) require a resolution of at least 1000 pixels per inch to obtain sharp edges.

Images originally obtained on film or printed on photographic paper can be scanned to produce an electronic file. The common reflective scanning devices can be used for prints, but only specialized transmitted light scanners can be used to scan negatives. If a print is to be scanned to produce the electronic file, be sure that it has not been printed by a laser jet printer, a photocopier, or any other method that uses screening because this will ultimately result in the moiré patterns described above regardless of the resolution of the electronic file. Of special importance is the fact that any figure or text prepared for the web or taken off the web, for example as a part of a PDF file, is at very low resolution and cannot be republished in print under most circumstances.

Labeling of electronic image files is carried out by software programs such as Photo-Shop, Illustrator, or PowerPoint. Labels should be used to number the figures, point out important features of the image, and present a scale bar that documents the magnification of the image. The scale bar is essential in allowing the viewer to determine immediately the approximate size of any structure in the image and serves as a "ruler" that will remain correct regardless of whether the image is enlarged or reduced in size during the publication process. Be cautious, however, as publishers usually do not accept PowerPoint or PDF files because these formats do not allows conversion to high resolution TIFF or encapsulated PostScript (ESP) files that are industry-wide standards for image file format. If necessary, one can submit individual image or analog layouts of figures that have been scanned to produce TIFF files. Because TIFF files can be quite large, some publishers allow you to send these files on CD by regular or express mail. Alternatively, many publishers support file transfer protocol (FTP) sites for uploading large files via the Internet. In contrast, figure layouts to be used solely for review can be produced in PowerPoint or Word, converted to PDF files, and submitted. The journal editors will then typically send the PDF file(s) to reviewers either as an e-mail attachment or to be downloaded from a journal website that is password protected.

■ Poster Presentations

The first step is to consult the instructions to authors to determine the size and any layout restrictions that have been imposed. Next, decide whether the poster will be printed item by item on a regular printer or printed as one large sheet on a large-format printer. If the item by item approach is chosen, text can be printed from Microsoft Word on 36- to 60-pound paper and then applied with adhesive to colored mat board. Images printed photographically or by dye sublimation, ink jet, or high-quality laser jet printer are also applied to mat board. Either spray or stick adhesive can be used. Mat board color should be chosen for aesthetic appeal and ease of viewing (e.g., orange is not recommended). Do not forget your pushpins!

If the large format approach is chosen, one must first determine the file formats that the large printer accepts for input. Common formats are PowerPoint or EPS files. Next determine whether there are size restrictions based on the software or printer capability, and taking this into account, determine the size of your electronic layout. Next, import or create your title and text, and import images to be used. Usually, we use PowerPoint and first create a draft set of figures as a "slide show." In this way, each figure can be composed and checked for accuracy and completeness individually without being concerned with the overall poster layout. Images are usually imported as Joint Photographic Experts Group (JPEG) files because PowerPoint automatically increases pixel number and uses interpolation methods to avoid pixilation as the image size is increased to produce figures that are easily seen at a distance of 2 to 3 feet. Individual figures precomposed in this manner are then imported to main poster-sized layout as a "picture" so that all of the features (text, line drawing, and image) can be scaled proportionally when size changes are needed. An efficient but rough layout can then be made by dragging figures around with the mouse cursor. Final positioning of all elements, however, should be done by typing in exact position parameters rather than by cursor.

Text is best created using a sans serif font such as Arial and lettering that is at least 18 point so that it can be easily read from a distance of 2 to 3 feet. Nothing is more aggravating that trying to read a poster with small font when you are in the second row of viewers. Attracting customers from a distance with reasonably large text is always good policy.

Before the large-format poster print is made at considerable cost, it is wise to have a small draft version printed and given a final review for errors, text size, line thicknesses, and proper color rendition. The color pallet that you have used on your computer monitor in composing the poster can be altered significantly in density, contrast, or color balance and hue by any printer. The paper used for printing should be glossy or semiglossy, and the inks used should be relatively waterproof and light stable if you expect your poster to be used for more than a "one night stand."

■ Seminars/Slide Shows

A slide show should be designed to tell a smooth, flowing story in a logical order one bit at a time so that the audience can actually understand and remember some of what you said. Composition of slides should be done electronically using software such as Power-Point. Guidelines on making an effective slide presentation include the following:

1. Each slide should present a limited amount of information that can be conveyed in three to four bullets or one to four images. Images should be sufficiently large and labeled properly to allow important features to be easily seen.

2. The use of color in a tasteful manner is a key to keeping the audience alert, and color will never be cheaper to use. If you are using black and white images, place them on a colored background or label them with colored fonts.

3. Series of slides that build off of one another by adding new information to an already existing backdrop during each slide change can be particularly effective.

4. Images imported into a slide show do not have to be at resolutions as high as those submitted for publication. A typical computer screen will display 72 pixels per inch in the VGA mode, and digital projectors use similar formats. This is considerably lower than the 300 pixels per inch minimum required for publication.

5. For studies of organelle or cell movement, insertion of video clips into your presentation can be essential. Linkage of a slide to the video clip in PowerPoint has two requirements. First, the video file must be in an uncompressed format, such as Audio Video Interleave (AVI) (for PCs) or QuickTime (for Macs). Video footage captured in compressed formats such as Moving Pictures Expert Group (MPEG) or digital video (DV) must be converted to AVIs or QuickTime using video-editing software such as Adobe Premier. These programs can also be used to cut or splice video segments to form a clip. Second, the AVI or QuickTime file *must* be stored in the same folder as the PowerPoint presentation, and the entire folder must be moved as a unit to the desktop of the presentation computer. Remember that video files are large, and one should not attempt to access these files from a CD during a presentation because the rate of data transfer may not be sufficient to allow a smooth delivery. Should you need lengthy video clips (more than 1 minute), access from a DVD may work best.

6. For presentation of three-dimensional images, planning is required ahead of time. There are two common methods, both of which rely on delivering a slightly different

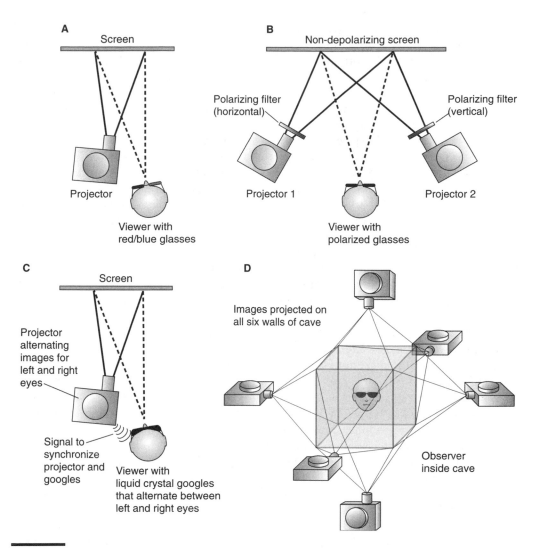

Figure 16.8
Setups used for projection of three dimensional still images and video/film. (A) Stereo pairs projected by red–blue anagrams. (B) Stereo pairs projected by polarized light. (C) Video/film projected using synchronized liquid–crystal filters. (D) Video/film projection using a visual "cave."

image to each eye (see Chapter 12). In the anagram method, each image of the stereo pair is projected in a different color, and the two images superimposed on the screen. This method uses one projector and one "slide" but requires that red/blue colored glasses be handed out to each audience member ahead of time (**Figure 16.8A**). Only black and white images can be projected using this method. The other method, using polarizing filters, can project three-dimensional color images but requires more equipment. Two slides, each having one stereo pair image, are projected onto a smooth (nondepolarizing) screen and viewed by an audience through polarizing glasses. Two projectors are required, with each projector having a polarizing filter that is oriented 90 degrees from that of the other projector. The two projectors must be carefully positioned to exactly superimpose the two images on the screen (Figure 16.8B). This second method is more complex but has the advantage that color images can be projected. Nevertheless, it is usually used only by the most dedicated three-dimensional bioimagers.

After your presentation is ready, the folder can be stored either on the hard drive of your laptop computer, on a compact disk, or on a portable universal serial bus (USB) drive. Some people consistently bring their laptop to the seminar site because they know that the presentation will run smoothly on their own system. The only risk involved is whether the computer will interface successfully with the digital projection equipment at a "foreign" site. This risk is modest for those using PCs but in the past has been much higher for those presenters using Macintosh computers. The alternative, loading your presentation onto the "host" computer from CD or portable USB drive, can be much easier, especially when air travel is required to reach the presentation site. This method has its own glitches, the most common being (1) the host computer is running outdated software or has too little memory or processor speed, (2) the presenter has not burned the CD or loaded the USB drive properly, or (3) the presentation has not been loaded into the host computer properly (i.e., in the same folder with no required files omitted). These problems can usually be avoided by checking that host site's computer has adequate hardware and software and by downloading your presentation to computers other than your own ahead of time to test for proper file transfer.

Beyond the reach of most scientific presenters are several elaborate immersion techniques for visual communication. A very sophisticated and highly technical projection system for three-dimensional images is that used by IMAX, flight simulators, and high-end action simulators (Figure 16.8C). This system uses precisely timed, rapid projection of stereo pairs as images alternating in their axis of polarization. The audience wears goggles that have liquid crystal filters that expose images to the left and right eyes in an alternating pattern. The image train on the screen is precisely synchronized with the filter changes in the goggles via infrared signaling. The frequency of the image train is faster than the flicker frequency of our visual system, allowing our brain to fuse the stereo pairs into a three-dimensional image train.

Another exciting development in image presentation is the use of "visual caves" to surround the viewer with imagery, giving the impression of being transported to any space, microscopic or macroscopic (Figure 16.8D). This technique requires four to six streams of video (at a minimum) precisely synchronized and projected onto the walls, ceiling, and floor of the "cave." This provides a viewing experience that vastly expands the older cinema scope and 360-degree surround techniques that only modified the horizon. Scientific "caves" can now give virtual video tours of any cellular or tissue compartment using data derived from three-dimensional imaging techniques such as confocal microscopy.

What will be the next frontier in bioimaging data capture, processing, or presentation? We hope that you will be there and will make it happen!

References and Suggested Reading

Gonzalez RE, Woods RC. *Digital Image Processing,* 4th ed. Upper Saddle River, New Jersey: Prentice Hall, 2007, p. 976.

Howard V, Reed MG. *Unbiased Stereology: Three-Dimensional Measurement in Microscopy.* BIOS Scientific, 1998, p. 246.

Jain AK. *Fundamentals of Digital Image Processing.* Upper Saddle River, New Jersey: Prentice Hall, 2004.

Pluta M. *Advanced Light Microscopy: Measuring Techniques.* Amsterdam, Netherlands: Elsevier, 1993, p. 718.

Russ JC, Dehoff RT. *Practical Stereology,* 2nd ed. New York: Springer, 2000, p. 382.

Tarantino C. *Digital Photo Processing,* Boston: Thompson Course Technology, 2003, p. 224.

Umbaugh SE. *Computer Imaging: Digital Image Analysis and Processing.* Boca Raton: CRC Press, 2005:659.

Wu Q, Merchant F, Castleman K. *Microscope Image Processing.* New York: Academic Press, 2008, p. 576.

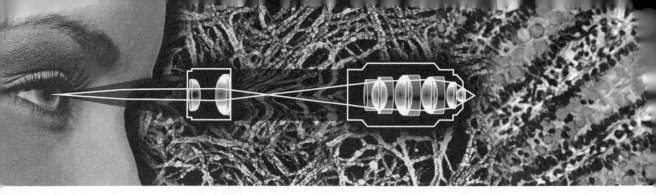

Index

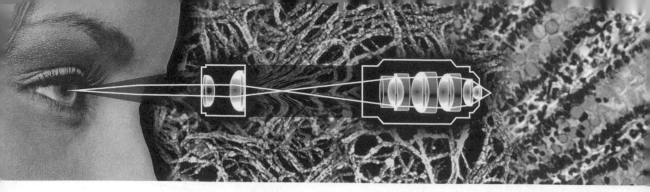

Figure Credits

Note: Unless otherwise indicated, photographs or illustrations are under copyright of Jones and Bartlett Publishers, or have been provided by the authors.

Chapter opener art design by Jacob Sahertian.

Chapter 1
1.1 © Library of Congress [LC-USZ62-95187]; **1.2a–b** Courtesy of Ladd Research Industries; **1.2c** Courtesy of William Sharp.

Chapter 2
2.1 Reproduced from Hartsoeke, Nicolaas. *Essai de dioptrique.* 1694; **2.2, 2.4a–b** © Library of Congress [LC-USZ62-95187]; **2.5a–b** Courtesy of Linda Chandler; **2.6d** Courtesy of Douglas Chandler; **2.7b, 2.7f** Reproduced from *The Journal of Cell Biology*, 1980, 86: 666–674. Copyright 1980, Rockefeller University Press; **2.8a–b** Adapted from Astrua M. *Manual of Colour Reproduction.* Fountain Press, Kings Langley, 1973.

Chapter 3
3.1b, 3.3e From *Electron Microscopy, Principles and Techniques for Biologists, Second edition*, courtesy of John J. Bozzola; **3.5g** Courtesy of Denton Vacuum, Inc; **3.7d** Courtesy of Ventana Medical Systems/RMC, Inc.; **3.8c–d** From *Electron Microscopy, Principles and Techniques for Biologists, Second edition*, courtesy of John J. Bozzola; **3.10b–c** Adapted from Hammersen, F. *Histology: Color Atlas of Microscopic Anatomy, Third edition.* Urban and Schwartzenberg, 1985.

Chapter 4
4.2b–c Reproduced from *Journal of Ultrastructure Research*, Vol. 82, Chandler DE et al., "Freeze-fracture studies of chemotactic…," pp. 221–232, Copyright 1983, with permission from Elsevier; **4.2e** Courtesy of Dr. Richard Trelease; **4.2f** Reproduced from *Developmental Biology*, Vol. 131, Larabell C., Chandler DE, "The coelomic envelope of *Xenopus laevis* eggs…," pp. 126–135, Copyright 1989, with permission from Elsevier; **4.3a–b** From *Electron Microscopy, Principles and Techniques for Biologists, Second edition*, used with permission from the Trust of Lonnie Russell; **4.3c** Courtesy of R. Dallai; **4.3d** Adapted from Alberts et al. *Essentials Cell Biology, Second edition.* Garland Science, 2004; **4.4d** Reproduced from Roberson, R.W., *Mycologica* 84 (1992): 41–51. Reprinted with permission from *Mycologica*. © The Mycological Society of America; **4.4e** Reproduced

from *The Journal of Cell Biology*, 1980, 86: 666–674. Copyright 1980 Rockefeller University Press; **4.4f1** Reproduced from *Journal of Molecular Biology*, Vol. 169, Heuser, J.E. and R. Cooke. "Actin-myosin interactions visualized by the quick-freeze...," pp. 26, Copyright 1983, with permission from Elsevier; **4.4f2–3** Reproduced from *The Journal of Cell Biology*, 1982, 94: 129–142. Copyright 1980 Rockefeller University Press; **4.5a** Reproduced from *Development Biology*, Vol. 82, Chandler, D.E. and J. Heuser, "Postfertilization growth of microvilli in the sea urchin egg...," pp. 08, Copyright 1981, with permission from Elsevier; **4.5b** Reproduced from *Developmental Biology*, Vol. 131, Larabell, C.A. and Chandler, D.E., "The coelomic envelope of *Xenopus laevis* eggs..." pp. 10, Copyright 1989, with permission from Elsevier; **4.5c–d** Reproduced from *Development Biology*, Vol. 82, Chandler, D.E. and J. Heuser, "Postfertilization growth of microvilli in the sea urchin egg...," pp. 08, Copyright 1981, with permission from Elsevier; **4.6d:** Courtesy of W. Dougherty; **4.7a** From *Electron Microscopy, Principles and Techniques for Biologists, Second edition*, used with permission from the Trust of Lonnie Russell; **4.7c–d** Reproduced from Roberson, R.W. and M.S. Fuller, *Protoplasm* 146 (1988): 143–149, with kind permission of Springer Science and Business Media; **4.7e** Courtesy of Dr. Richard Trelease; **4.7f** From *Electron Microscopy, Principles and Techniques for Biologists, Second edition*, used with permission from the Trust of Lonnie Russell; **4.9b,4.10a** Courtesy of Dr. Richard Trelease; **4.10b** From *Electron Microscopy, Principles and Techniques for Biologists, Second edition*, used with permission from the Trust of Lonnie Russell; **4.10c** Courtesy of Dr. Richard Trelease; **4.11d** From *Electron Microscopy, Principles and Techniques for Biologists, Second edition*, used with permission from the Trust of Lonnie Russell; **4.12b-c** Reproduced from D.E. Chandler, *The Journal of Cell Biology*, 1980, 86: 666–674. Copyright 1980 Rockefeller University Press; **4.13e** From *Electron Microscopy, Principles and Techniques for Biologists, Second edition*, used with permission from the Trust of Lonnie Russell; **4.13f** Reproduced from *Journal of Ultrastructure Research*, Vol. 82, Chandler DE et al., "Freeze-fracture studies of chemotactic...," pp. 221–232, Copyright 1983, with permission from Elsevier; **4.13g** From *Electron Microscopy, Principles and Techniques for Biologists, Second edition*, used with permission from the Trust of Lonnie Russell; **4.14c, 4.14e–f, 4.15a–b, 4.15c–d, 4.16a–d,** Courtesy of Dr. Richard Trelease; **4.17** Courtesy of Robert W. Roberson; **4.18a–b** Courtesy of Dr. Gregory J. Brewer; **4.18c** Adapted from Jotlik, W.K., et al. *Zinsser Microbiology, Twentieth edition*. Appleton & Lange, 1995; **4.19a–b** Courtesy of Dr. Daniel Friend.

Chapter 5
5.3 Adapted from Inoue S and K. Spring. *Video Microscopy, Second edition*. Plenum Press, 1997; **5.17** Adapted from Douglas B. Murphy. *Fundamentals of Light Microscopy and Electronic Imaging*. Wiley-Liss, 2001.

Chapter 6
6.2, 6.5a–b, 6.8 Adapted from Douglas B. Murphy. *Fundamentals of Light Microscopy and Electronic Imaging*. Wiley-Liss, 2001; **6.9** Adapted from the Molecular Expressions, Florida State University; **6.11a–c, 6.13, 6.15a–b** Adapted from Douglas B. Murphy. *Fundamentals of Light Microscopy and Electronic Imaging*. Wiley-Liss, 2001; **6.18** Adapted from the Molecular Expressions, Florida State University.

Chapter 7
7.1a Courtesy of Mike Dingley and MicScape, www.microscopy-uk.org.uk/mag/art97/ dingley3.html; **7.2b–c** Modified from Spencer, M. *Fundamentals Light Microscopy*. Cambridge University Press, 1982. Reprinted with permission of Cambridge University Press; **7.3d** Reproduced from D.E. Chandler and J.A. Williams, *The Journal of Cell Biology*,

1978, 76: 386–399, Copyright 1978 Rockefeller University Press; **7.3e–g** Courtesy of Drs. R. Lopez-Franco and C Bracker; **7.4a** Modified from Spencer, M. *Fundamentals Light Microscopy*. Cambridge University Press, 1982. Reprinted with permission of Cambridge University Press; **7.4b** Reproduced from D.E. Chandler and J.A. Williams *The Journal of Cell Biology*, 1978, 76: 386–399, Copyright 1978 Rockefeller University Press; **7.5c–g** Adapted from Douglas B. Murphy. *Fundamentals of Light Microscopy and Electronic Imaging*. Wiley-Liss, 2001; **7.6** Reproduced from H. Sato et al., *The Journal of Cell Biology*, 1975, 67: 501–517. Copyright 1975 Rockefeller University Press; **7.7a–b** Adapted from Douglas B. Murphy. *Fundamentals of Light Microscopy and Electronic Imaging*. Wiley-Liss, 2001; **7.8a** Reproduced from *Fungal Genetics and Biology*, Vol. 37, Riquelme M, et al., "The effect of ropy-1 mutation on cytoplasmic...," pp. 9, Copyright 2002 with permission from Elsevier; **7.9a** Adapted from Douglas B. Murphy. *Fundamentals of Light Microscopy and Electronic Imaging*. Wiley-Liss, 2001.

Chapter 8
8.12a–b From *Electron Microscopy, Principles and Techniques for Biologists, Second edition*, courtesy of John J. Bozzola.

Chapter 9
9.2 Adapted from Hitachi Scientific Instruments; **9.5** Adapted from Postek et al. *Scanning Electron Microscopy: A Students Handbook*. Ladd Research Industries, Inc., 1980; **9.8** Adapted from Dykstra MJ. *Biological Electron Microscopy: Theory, Techniques and Troubleshooting*. Plenum Press, 1999; **9.9** Adapted from Wischnitzer, S. *Introduction to Electron Microscopy, Third edition*. Pergamon Press, 1988; **9.11a–j** From *Electron Microscopy, Principles and Techniques for Biologists, Second edition*, courtesy of Steve Schmitt, Randy Tindall, and John Bozzola; **9.12, 9.13** Adapted from Postek et al. *Scanning Electron Microscopy: A Students Handbook*. Ladd Research Industries, Inc., 1980.

Chapter 10
10.2a Reproduced from *Methods in Enzymology*, Vol. 221, Merkle CJ, Chandler DE, "Visualization of exocytosis by quick freezing...," pp. 12, Copyright 1993, with permission from Elsevier; **10.2b** Courtesy of Ventana Medical Systems/RMC, Inc.; **10.2d** Reproduced from *Methods in Enzymology*, Vol. 221, Merkle CJ, Chandler DE, "Visualization of exocytosis by quick freezing...," pp. 12, Copyright 1993, with permission from Elsevier; **10.2e** Courtesy of Ventana Medical Systems/RMC, Inc.; **10.3a–d** Reproduced from *Methods in Enzymology*, Vol. 221, Merkle CJ, Chandler DE, "Visualization of exocytosis by quick freezing...," pp. 12, Copyright 1993, with permission from Elsevier; **10.4b** Reproduced from R. Dahl and L.A. Jtaehelin, *The Journal of Electron Microscopy Technique*, Vol. 13, No. 3, 1989, pp. 165–174. Copyright 1989, Reprinted with permission of Wiley-Liss, Inc., a subsidiary of John Wiley and Sons, Inc.; **10.4c–d** Courtesy of William Sharp; **10.5b** Reproduced from *Fungal Genetics and Biology*, Vol. 37, Riquelme M, et al., "The effect of ropy-1 mutation on cytoplasmic...," pp. 9, Copyright 2002 with permission from Elsevier; **10.5c** Reproduced from *Methods in Enzymology*, Vol. 221, Merkle CJ, Chandler DE, "Visualization of exocytosis by quick freezing...," pp. 12, Copyright 1993, with permission from Elsevier; **10.5e–g** Reproduced from D.E. Chandler and J.E. Heuser, *The Journal of Cell Biology*, 1980, 84: 618–632. Copyright 1980 Rockefeller University Press; **10.6c** Reproduced from Fisher, K.E. et al. "Cytoplasmic cleavage in zoosporangia of *Allomyces macrogynus*." *Journal of Microscopy* 2000, 199: 260–270 by copyright permission from Blackwell Publishing; **10.6d** Courtesy of David Lowry; **10.8a** Reproduced from *The Journal of Cell Biology*, 1980, 84: 618–1632. Copyright 1980 Rockefeller University Press; **10.9** Reproduced from Van de Meene et al., *Archives of Microbiology* 184 (2006): 259–270, with kind permission of Springer Science

and Business Media; **10.12a–d** From *Electron Microscopy, Principles and Techniques for Biologists, Second edition*, used with permission from the Trust of Lonnie Russell; **10.13a–d** Reproduced from D.E. Chandler and J.E. Heuser, *The Journal of Cell Biology*, 1980, 84: 618–632. Copyright 1980 Rockefeller University Press; **10.14a** Reproduced from *The Journal of Cell Biology*, 1980, 86: 666–674. Copyright 1980 Rockefeller University Press; **10.15c** Reproduced from *The Journal of Cell Biology*, 1980, 84: 618–632. Copyright 1980 permission from the Rockefeller University Press.

Chapter 11

11.2a–e Reproduced from Reinhart D et al, *Zygote* 6:173–182 (1998) by copyright permission from Cambridge University Press; **11.3a–b, 11.4a–b, 11.5a–b, 11.6** Adapted from Inoue S. and K. Spring. *Video Microscopy, Second edition*. Plenum Press, 1997; **11.7a–b** Courtesy of Microimaging Applications Group/Photometric; **11.8a–c** Adapted from Inoue S. and K. Spring. *Video Microscopy, Second edition*. Plenum Press, 1997; **11.9** Adapted from Inoue S and K. *Spring. Video Microscopy, Second edition*. Plenum Press, 1997; **11.10, 11.12a–b** Adapted from Inglis, A.F. and A.C. Luther. *Video Engineering, Second edition*. McGraw-Hills, 1996; **11.13a–b, 11.14a–b** Adapted from Inoue S and K. Spring. *Video Microscopy, Second edition*. Plenum Press, 1997; **11.15a–c, 11.16a–b** Modified with permission from Noll AM, *Television Technology: Fundamentals and Future Prospects*, Norwood, MH: Artech House, Inc. (1988), © 2003–2007 by Artech House, Inc; **11.17a–b** From *Digital Video Primer*, © 2006 Adobe Systems Incorporated. All rights reserved. Adobe is either (a) registered trademark or a trademark of Adobe Systems Incorporated in the United States and/or other countries.

Chapter 12

12.2a–b Modified from Haugland, R.P., *Handbook of Fluorescent Probes and Research Chemicals, Sixth edition*, Eugene, OR (1996) by copyright permission from Molecular Probes, Inc. Available on the website probes.invitrogen.com; **12.3b** Adapted from Douglas B. Murphy. *Fundamentals of Light Microscopy and Electronic Imaging*. Wiley-Liss, 2001; **12.5, 12.6a–b** Adapted from Abramowitz M. *Fluorescence Microscopy: The Essentials* (1993) by copyright permission from; Olympus-America, New York; **12.8a–c, 12.9a** Courtesy of Carl Zeiss Microimaging, Inc.; **12.9b** Adapted from Douglas B. Murphy. *Fundamentals of Light Microscopy and Electronic Imaging*. Wiley-Liss, 2001; **12.10c** Courtesy of Carl Zeiss Microimaging, Inc.; **12.11a, 12.11c** Adapted from the Rockwell Laser Industries website, **12.12b** Adapted from the Molecular Expressions website, Florida State University; **12.12d–e** Courtesy of Leica Microsystems, Inc.; **12.13b** Adapted from Douglas B. Murphy. *Fundamentals of Light Microscopy and Electronic Imaging*. Wiley-Liss, 2001; **12.14a** Courtesy of Carl Zeiss Microimaging, Inc.; **12.14b** Reproduced from Chandler, D.E. et al., *Cell Tissue Research* 246 (1986): 153–161, with kind permission of Springer Science and Business Media; **12.14c** Reproduced from X. Xiang, et al., *Molecular Development and Reproduction*, Vol. 70, No. 3, 2005, pp. 344–360. Copyright 2005, Reprinted with permission of Wiley-Liss, Inc., a subsidiary of John Wiley and Sons, Inc.

Chapter 13

13.1a Reproduced from the image gallery of the website probes.invitrogen.com by copyright permission of Molecular Probes, Inc.; **13.1b** Courtesy of Dr. Daniel Friend; **13.1c** Courtesy of Richard Trelease; **13.3b–d** Reproduced from J.D. Castle, et al., *The Journal of Cell Biology*, 1972, 53:290–311. Copyright 1972 Rockefeller University Press; **13.4** Modified with from Haugland, R.P., *Handbook of Fluorescent Probes and Research Chemicals, Sixth edition*, Eugene, OR (1996) by copyright permission from Molecular Probes, Inc. Available on the website probes.invitrogen.com; **13.5d1–3** Reproduced from X. Xiang, et al., *Molecular Development and Reproduction*, Vol. 70, No. 3, 2005, pp. 344–360. Copyright 2005, Reprinted with permission of Wiley-Liss, Inc., a subsidiary of John Wiley and Sons, Inc.; **13.6b** Reproduced from McDaniel, D.P. and R.W. Roberson, *Protoplasma* 203 (1998): 118–123, with kind permission of Springer Science and Business Media; **13.6c–d** Adapted from Garni Y et al. *Bioimaging* 4:65–72 (1996). IOP Publishing, Ltd.; **13.7, 13.7 insert, 13.8** Reproduced from K.T. Tokuyasu, et al., *The Journal of Cell Biology*, 1983, 96: 1727–1735. Copyright 1983 Rockefeller University Press; **13.9a** Adapted from Douglas B. Murphy. *Fundamentals of Light Microscopy and Electronic Imaging*. Wiley-Liss, 2001; **13.9b** Reproduced from Lisenbee CS et al., 2003, "Overexpression and mislocation of a tail-anchored GFP…" *Traffic* 4:491–501 by copyright permission from Wiley-Blackwell Publishing; **13.9c** Adapted from Douglas B. Murphy. *Fundamentals of Light Microscopy and Electronic Imaging*. Wiley-Liss, 2001; **13.9d** Reproduced from Lisenbee CS et al., 2003, "Overexpression and mislocation of a tail-anchored GFP…" *Traffic* 4:491–501 by copyright permission from Wiley-Blackwell Publishing; **13.10c–e** Reproduced from the image gallery of the website probes.invitrogen.com by copyright permission of Molecular Probes, Inc.; **13.11a–c** Adapted from Garni Y et al. *Bioimaging* 4:65–72 (1996). IOP Publishing, Ltd.; **13.12a** Reproduced from the image gallery of the website probes.invitrogen.com by copyright permission of Molecular Probes, Inc.; **13.12b** Reproduced from Fisher, K.E. et al. 2000. "Cytoplasmic cleavage in zoosporangia of *Allomyces macrogynus*." *Journal of Microscopy* 199:260–270 by copyright permission from Blackwell Publishing; **13.12c** Reproduced from Korlach K et al. 1999. "Characterization of lipid bilayer phases by confocal microscopy and fluorescence correlation spectroscopy." *Proceedings of the National Academy of Sciences*, 96: 8461–8466. Copyright (1999) National Academy of Sciences, U.S.A; **13.12d–f** Reproduced from the image gallery of the website probes.invitrogen.com by copyright permission of Molecular Probes, Inc.; **13.13a–c** Reproduced from B.S. Bonnell and Chandler, D.E., *Molecular Development and Reproduction*, Vol. 44, No.2, 1996, pp. 212–220. Copyright 2005, Reprinted with permission of Wiley-Liss, Inc., a subsidiary of John Wiley and Sons, Inc.; **13.15a–c** Reproduced from Fix M et al. 2004. "Imaging single membrane fusion events mediated by SNARE proteins." *Proceedings of the National Academy of Sciences*, 101: 7311–7316. Copyright (2004) National Academy of Sciences, U.S.A; **13.17a–b, 13.17d** Adapted from Medina MA and P. Schwille, *BioEssays* 24:758–764 (2002); **13.18a** Adapted from Alberts B et al. *Molecular Biology of the Cell, Fourth edition*. Garland Science, 2002; **13.18b–e** Reprinted from *Cell*, Vol. 99, Zaal Kristien JM et al., "Golgi membranes are absorbed into and reemerge from the ER during mitosis," pp. 589–601, Copyright 1999, with permission from Elsevier; **13.19c–d** Reprinted from *Methods*, Vol. 40, Ni Q et al., "Analyzing proten kinase dynamics in living cells with FRET reporters," pp. 279–286, Copyright 2006, with permission from Elsevier; **13.19e** Reprinted from *Current Biology*, Vol. 13, Larijani Banafshe et al., "EGF Regulation of PITP…" pp. 7, Copyright 2003, with permission from Elsevier; **13.20a, 13.20a insert** Reproduced from W.C. Salmon, et al., *The Journal of Cell Biology*, 2002, 158: 31–37. Copyright 2002 The Rockefeller University Press; **13.20b** Reproduced from Vallotton P et al. 2004. "Simultaneous mapping of filamentous actin flow and turnover in migrating cells by quantitative fluorescent speckle microscopy." *Proceedings of the National Academy of Sciences*, 101: 9660–9665. Copyright (2004) National Academy of Sciences, U.S.A; **13.21b–c** Adapted from Meier J et al., *Nature Neuroscience* 4:253–460 (2001).

Chapter 14
14.1 Reproduced from A.P. Jomlyo, et al., *The Journal of Cell Biology*, 1974, 61: 723–742. Copyright 1974 Rockefeller University Press; **14.2a–b** Reproduced from D.E. Chandler and J.A. Williams, *The Journal of Cell Biology*, 1978, 76: 386–399. Copyright 1978 Rockefeller

University Press; **14.3a** Modified from Haugland, R.P., *Handbook of Fluorescent Probes and Research Chemicals, Sixth edition*, Eugene, OR (1996) by copyright permission from Molecular Probes, Inc. Available on the website probes.invitrogen.com; **14.3b** Reproduced from J.C. Gilkey, et al., *The Journal of Cell Biology*, 1978, 76: 448–466. Copyright 1978 Rockefeller University Press; **14.5b** Modified from Haugland, R.P., *Handbook of Fluorescent Probes and Research Chemicals, Sixth edition*, Eugene, OR (1996) by copyright permission from Molecular Probes, Inc. Available on the website probes.invitrogen.com; **14.5c** Modified from Kazilek CJ et al. 1988. "Hyperosmotic inhibition of calcium signals and exocytosis in rabbit neutrophils." *American Journal of Physiology* 254: C709–C718. Used with permission of The American Physiological Society; **14.5d, 14.5f** Modified from Haugland, R.P., Handbook *of Fluorescent Probes and Research Chemicals, Sixth edition*, Eugene, OR (1996) by copyright permission from Molecular Probes, Inc. Available on the website probes.invitrogen.com; **14.6** Courtesy of BiopTechs, Inc.; **14.7a–f** Modified from Haugland, R.P., Handbook *of Fluorescent Probes and Research Chemicals, Sixth edition*, Eugene, OR (1996) by copyright permission from Molecular Probes, Inc. Available on the website probes.invitrogen.com; **14.9a** Reproduced from the image gallery of the website probes.invitrogen.com, courtesy of Steven Stricker: **14.9b** Reprinted from *Developmental Biology*, Vol. 298, Campbell K, Swann K, "Ca^{2+} oscillations stimulate an ATP...," pp. 225–233, Copyright 2006, with permission from Elsevier; **14.9c–e** Reprinted from *Current Biology*, Vol. 10, David Thomas et al., "Microscopic properties of elementary..." pp. 225–233, Copyright 2008, with permission from Elsevier; **14.10a** Modified from Haugland, R.P., Handbook *of Fluorescent Probes and Research Chemicals, Sixth edition*, Eugene, OR (1996) by copyright permission from Molecular Probes, Inc. Available on the website probes.invitrogen.com; **14.10b–d** Modified from Knisley SB et al. 2000. "Ratiometry of transmembrane voltage-sensitive fluorescent dye emission in hearts." *American Journal of Physiology* 279: H1421–H14332. Used with permission of The American Physiological Society; **14.10e** Reproduced from the image gallery of the website probes.invitrogen.com by copyright permission of Molecular Probes, Inc.; **14.11a** Reproduced from Zhang J et al. 2001. "Genetically encoded reporters of protein kinase a activity reveal impact of substrate tethering." *Proceedings of the National Academy of Sciences*, 98: 14997–15002. Copyright (2001) National Academy of Sciences, U.S.A.; **14.11b** Reproduced from DiPilato LM et al. 2004. "Fluorescent indicators of cAMP and Epac activation reveal differential dynamics of cAMP signaling within discrete subcellular compartments." *Proceedings of the National Academy of Sciences*, 101:16513–16518. Copyright (2004) National Academy of Sciences U.S.A.

Chapter 15
15.1a–b Reprinted from *The Journal of Molecular Biology*, Vol. 169, Heuser JE, "Procedure for freeze-drying molecules absorbed to mica flakes," pp. 155–195. Copyright 1983, with permission from Elsevier; **15.2a–c** Reproduced from Mozingo NM et al., *Journal of Cell Science*, 107:2769–2777 (1994) by copyright permission of The Company of Biologists, Ltd.; **15.3b** Reprinted from *The Journal of Ultrastructure Research*, Vol. 89, Chandler, D.E., "Exocytosis in vitro: Ultrastructure of the isolated sea urchin...," pp. 198–211, Copyright 1984, with permission from Elsevier; **15.4a–f** Reproduced from D.E. Chandler and J.E. Heuser, *The Journal of Cell Biology*, 1980, 84: 618–632. Copyright 1980 Rockefeller University Press; **15.5a–d** Reprinted from *The Journal of Structural Biology*, Vol. 141, Zhao F, Roger C, Capturing time-resolved changes in molecular structure by negative staining, pp. 43–52, Copyright 2003, with permission from Elsevier; **15.6a** Adapted from Stewart, M. *Electron Microscopy—A Practical Approach*, Harris JR, ed. IRL, Oxford, 1991; **15.6b–c** Reprinted from *The Journal of Structural Biology*, Vol. 133, Nicholson WV, Glaeser RM, Automatic particle detection in electron microscopy,

pp. 90–101, Copyright 2001, with permission from Elsevier; **15.7a** Reprinted from *FEBS Letters*, Vol. 505, "Thuman-Commike PA, Single particle macromlecular structure determination via electron microscopy," pp. 199–205, Copyright 2001, with permission from Elsevier; **15.7b–c** Reprinted from *Structure*, Vol. 3, Stark H et al., "The 70S *Escherichia coli* ribosome at 23 angstrom resolution: fitting the ribosomal RNA," pp. 815–821, Copyright 1995, with permission from Elsevier; **15.8a** Reproduced from Harris SD et al, *Eukaryotic Cell* 4 (2004): 225–229, by copyright permission of the American Society of Microbiology; **15.8c–e** Reprinted from *Structure*, Vol. 12, Ress DB et al, "Methods for generating high-resolution structural models…," pp. 1763–1774, Copyright 2004, with permission from Elsevier; **15.9b** Reprinted from *Journal of Molecular Biology*, Vol. 112, Arnone A et al, "Preliminary crystallographic study of aspartate…," pp. 509–513, Copyright 1977, with permission from Elsevier; **15.9c** Adapted from Glusker, J.P. and K.N. Trueblood. *Crystal Structure Analysis—A Primer, Second edition.* Oxford University Press, 1985; **15.9d** Reprinted from *Journal of Molecular Biology*, Vol. 112, Arnone A et al, "Preliminary crystallographic study of aspartate…," pp. 509–513, Copyright 1977, with permission from Elsevier; **15.9e** Reprinted from *Journal of Molecular Biology*, Vol. 112, Blake, C. F. et al, "Structure of prealbumin…," pp. 339–359, Copyright 1978, with permission from Elsevier; **15.10c–d** Reproduced from *Journal Molecular Biology*, Vol. 262, Wolf SG et al, "Interpreting a Medium-resolution Model of Tubulin…," pp. 485–501, Copyright 1996, with permission from Elsevier Limited; **15.12a** Reprinted from *Journal of Molecular Biology*, Vol. 315, Hand GM et al, "Isolation and characterization of gap junctions from tissue culture cells," pp. 587–600, Copyright 2002, with permission of Elsevier; **15.12b–d** Reproduced from Lyubchenko YL, Shlyakhtenko LS. 1997. "Visualization of supercoiled DNA with atomic force microscopy in situ." *Proceedings of the National Academy of Sciences,* 94:496–501. Copyright (1997) National Academy of Sciences, U.S.A; **15.12e–f** Reprinted from M. Anderson, et al., *Journal of Molecular Biology*, Vol. 358, "Polymorphism and ultrastructural organization of prion protein amyloid fibrils…," pp. 580–596, Copyright 2006, with permission from Elsevier; **15.12g** Adapted from Digital Instruments; **15.13a–c** Adapted from the Molecular Expressions website, Florida State University.

Chapter 16
16.1a insert Reproduced from D.E. Chandler and J.A. Williams, *The Journal of Cell Biology* 76: 386–399 (1978). Copyright 1978 Rockefeller University Press; **16.1d–e** Adapted from the Molecular Expressions website, Florida State University; **16.2b, 16.2d, 16.2e–f, 16.2h, 16.3b, 16.3d, 16.3f** Reproduced from D.E. Chandler and J.A. Williams, *The Journal of Cell Biology* 76: 386–399 (1978). Copyright 1978 Rockefeller University Press; **16.4c** Adapted from Douglas B. Murphy. *Fundamentals of Light Microscopy and Electronic Imaging.* Wiley-Liss, 2001; **16.4d–f inserts, 16.5b, 16.5d** Reproduced from *The Journal of Cell Biology* 76: 386–399 (1978). Copyright 1978 Rockefeller University Press; **16.6a** Adapted from Bozzola, J.J and L.D. Russell. *Electron Microscopy. Principles and Techniques for Biologists, Second edition.*